Basin Analysis

Academic Press Geology Series

Mineral Deposits and Global Tectonic Settings
A. H. G. Mitchell and M. S. Garson
1981

Applied Environmental Geochemistry
I. Thornton (ed.)
1983

Geology and Radwaste
A. G. Milnes
1985

Mantle Metasomatism
M. A. Menzies and C. J. Hawkesworth (eds.)
1987

The Structure of the Planets
J. W. Elder
1987

Fracture Mechanics of Rock
B. K. Atkinson (ed.)
1987

Isotope Chronostratigraphy: Theory and Methods
Douglas F. Williams, Ian Lerche, and W. E. Full
1988

Basin Analysis: Quantitative Methods, Volume 1
Ian Lerche
1990

Basin Analysis: Quantitative Methods, Volume 2 (in preparation)
Ian Lerche
1990

Basin Analysis

Quantitative Methods

Volume 1

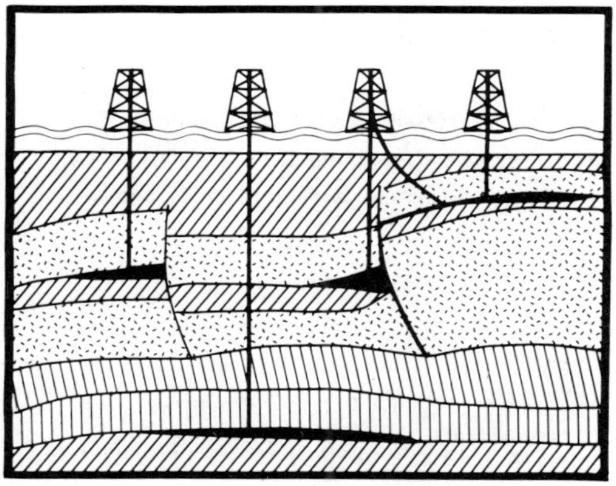

Ian Lerche

Department of Geological Sciences

University of South Carolina

Columbia, South Carolina

Academic Press, Inc.

Harcourt Brace Jovanovich, Publishers

San Diego New York Berkeley Boston

London Sydney Tokyo Toronto

This book is printed on acid-free paper. ∞

Academic Press, Inc.
San Diego, California 92101

United Kingdom Edition published by
Academic Press Limited
24–28 Oval Road, London NW1 7DX

Library of Congress Cataloging-in-Publication Data

Lerche, I. (Ian)
Basin analysis : quantitative methods / I. Lerche.
p. cm. — (Academic Press geology series)
Bibliography: p.
Includes index.
ISBN 0-12-444172-6 (alk. paper)
1. Petroleum—Geology. 2. Basins (Geology)—Analysis.
3. Sediments (Geology)—Analysis. I. Title. II. Series.
TN870.5.L37 1989
622'.1828—dc20 89-31669
CIP

Printed in the United States of America
89 90 91 92 9 8 7 6 5 4 3 2 1

Contents

Contents of Volume 2

(in preparation)

Preface

THE EXPLORATION FOR OIL is a high-risk game. Worldwide drilling success is around 5–10%, the average cost of drilling is around $1000/ft, and the average well is now around 15,000 ft deep. Typical costs per well are thus about $15 million, and with about 10–20 wells per economic reservoir found, the cost approaches $150–300 million for break even. It is then no wonder that over the years the oil industry has actively encouraged research and development of methods that cut the risk factor. Such endeavors have also revealed a vast amount of basic knowledge about the earth that might otherwise have been considerably delayed.

Two fundamental avenues of attack have been developed over the years: (1) methods designed to locate oil *in situ* from direct measurement ahead of the drill (including seismic techniques; electrical and magnetic, gravity, and heat-flux surveys; and shallow probing for "tracer" signals such as in soil gas geochemistry) and (2) methods focusing on the dynamical evolution of a sedimentary basin in relation to the timing of hydrocarbon generation, migration, and accumulation to provide an assessment of which areas in a basin might be the most prospective for oil accumulations today.

The direct detection methods suffer from the usual drawback that their resolution capability rapidly declines with increasing depth. The basin analysis methods suffer from the drawbacks that they start with limited, imprecise data, and the temporal reconstruction of sedimentary basinal evolution is by no means a clear-cut jigsaw puzzle, as we shall see. Yet, these are the only weapons that we have in our meager arsenal that we can use to decrease the risk factors. We are therefore faced with the problem of determining (a) what we can resolve in a basin without equivocation, (b) what we can resolve with a determined degree of error, and (c) what we cannot resolve without further data—and we must also suggest the type of data needed to clarify the unknown.

This book addresses the problem of quantitative basin analysis in relation to oil accumulations. Three major factors contribute to oil occurrences. First is the burial history of sedimentary units driven by sediment supply, mechanical and chemical compaction, tectonic forces, erosional events, intrusive events, sea-level changes, and so on. A knowledge of this dynamical evolution of sediments is crucial to an understanding of basin analysis because allied phenomena, such as paleostructures, porosity, sedimentary thermal conductivity, fluid flow, solubility, and fracturing and faulting, are all dependent to a greater or lesser extent on the sedimentary patterns of behavior.

If the burial history of the sediments can indeed be decoded, we must then squarely face determination of the second factor: how to provide knowledge of the thermal history of the sediments. Two options are available: Either we assume a priori a model or models for heat-flux evolution, in which case the problem is solved by fiat, or we use present-day data that contain some cumulative measure of thermal history and attempt to "invert" such data to reconstruct the sedimentary thermal history. In this case we are limited by the degree to which the thermal indicator data store the impressed signature of thermal history.

If the thermal history of the sediments can be unraveled, we must then add the third major factor: hydrocarbon generation, migration, and accumulation. Here we face two quandries: How do we provide thermokinetic models of hydrocarbon production from organic material, and how do we assess their accuracy?

These three major factors taken together are the linchpins in all attempts to date to come to grips with the discipline of quantitative basin analysis. In addition, two subordinate classes of problems arise in discussions of the three major factors. In order to provide quantitative models of evolution based on data, we require constitutive relations for each lithology type connecting several quantities (e.g., permeability to porosity, thermal conductivity to porosity, solubility to pressure and temperature). These constitutive equations depend on measurements of rock properties that are often difficult to obtain, confounded by several factors, or change from basinal setting to basinal setting. So, problems arise concerning the sensitivity of basin analysis results to the functional form and parameter dependence of such constitutive relations.

In addition, even if the underlying relationships were known precisely, the systematic and statistical errors in present-day data, against which we must compare model predictions, guarantee a degree of "spread" in assessing models of basinal evolution. We must allow for these vagaries of real data in attempts to reconstruct the geohistory, thermal history, or hydrocarbon history of sedimentary basins.

We address these issues here in several major sections. The first section of the book is devoted to examining the quantitative methods of reconstructing the burial history of sedimentary material in basins, allowing for ever increasing degrees of rigor and complexity as more processes are added to the description of the system.

The second section of the book addresses the problem of quantitative reconstruction of the thermal history of sedimentary basins based on both model-derived heat fluxes and on inversion of present-day thermal indicator data.

The third section of the book integrates the hydrocarbon generation, migration, and accumulation modeling aspects with the burial history and thermal history as underpinning buttresses. We examine some of the repercussions of different hydrocarbon models for present-day oil accumulation sites. In addition, we show how both burial history and thermal information can be combined to determine timing, emplacement, and erosion of geological events.

The fourth major section of the book applies sensitivity analysis to the first three sections and, in addition, addresses special problems initiated or enhanced by the behavior of eustatic sea level and clastic sediment loading behaviors.

Appendixes are included at the end of several chapters, where some of the basic physical properties, some mathematical problems, and some methods for manipulating present-day data are examined in more detail.

Each section of the book contains applications and illustrations to demonstrate the sort of behavior that can be derived from the modeling procedures. We also provide case histories illuminating comprehensive pictures of evolution for a few basins, so that the application of combined facets of basin analysis can be seen interacting.

While several good books are available on the *qualitative* application of basin analysis methods (e.g., Hunt, *Petroleum Geochemistry and Geology*), there does not appear to be available a book that examines the quantitative aspects of basin analysis viewed as an integrated picture. The purpose of this book is to attempt to provide such a picture, allowing for the uncertainties and inaccuracies of our understanding and of the data to which we are beholden.

The perceived audience is twofold: (1) professional geologists, geophysicists, and petroleum exploration scientists who are actively involved in the business of assessing the hydrocarbon potential of basins and

(2) graduate students in petroleum exploration departments who are committed to the quantitative methods of analysis as integrative procedures for analyzing geological, geochemical, and geophysical data.

The book is set at two levels. Sufficient details of the mathematical, physical, chemical, and geological modeling developments are included in each chapter so that, in combination with the appendixes, detailed knowledge can be obtained by a professional interested in further developing such ideas. At the same time, those who wish to obtain more of an overview of the subject can do so by skipping most of the mathematical development and concentrating on the case history examples presented.

The difficulty of addressing every aspect of every subdiscipline of every detail of basin analysis is one we clearly recognize. Such a development would call for a very long book indeed. Aware of our own limitations and of book length, we have, perforce, been somewhat cavalier in our treatment of some subjects that, in our opinion, play a secondary role with respect to the major theme of presenting an integrated picture of quantitative basin analysis techniques.

We had intended to include major sections dealing with the impact of tectonic forces on basinal evolution (e.g., viscoelastic plate models), with carbonate evolution, salt dynamics, shale diapiric effects, igneous intrusives, overthrust domains, and so on. But the book is overly long already. Perhaps, then, the excuse that the book is devoted to quantitative *methods* in basin analysis will suffice. The logic development—quantitatively model, constrain by data, invert the procedure, and determine resolution—applies well beyond the limited foray we provide here, and would seem to be the appropriate protocol for assessing the applicability of any particular arena in basin analysis quantification.

Whether or not we have been successful in our aims is for the reader to decide. We are painfully aware that this book is nowhere near a complete treatise on the subject matter, which is a rapidly developing field. But perhaps it will have served its purpose if others can develop the subject matter to its ultimate resolution in time. We hope the concepts, ideas, and examples in this book will help in such endeavors.

This book would not have been possible without considerable amounts of help, aid, and succor from a truly enormous number of people and organizations.

Original development of many of the basic ideas was initiated during my days at Gulf Oil before its demise. I thank all at Gulf who contributed so much to this area of exploration research, particularly Bill Glezen and also C. Kendall, who was involved in the genesis of the Basin Analysis Group at the University of South Carolina. Over the years, the Industrial Associates of the Basin Analysis Group at U.S.C. have provided financial and technical support as well as some outstanding ideas for further development and quantification. These associates include Sun Oil, Marathon, Conoco, Texaco, Chevron, Saga Petroleum, Statoil, Unocal, Norsk Hydro, Western Geophysical, Arco, Japex, Petrobras, Broken Hill Proprietary, and Phillips Petroleum. Dominant among the professionals who supplied preprints, reprints, and discussions of some of the problems are A. Yukler, D. Welte, C. Bethke, P. Chenet, Li Tai-Ming, M. Bennett, M. Zeitlin, W. G. Lyon, J. Cunliffe, P. Green, I. Duddy, M. Harrison, S. Larter, R. Clark, B. Russell, and T. Cross. To my students over the years I extend my grateful thanks. Truly, without their help, nothing would have developed. They are C. Armagnac, J. Bucci, S. Cao, C. Kuo, Z. P. Wei, Z. He, J. Cogan, J. Pantano, J. Iliffe, K. Nakayama, C. Hermanrud, T. McKenna, X. Wang, and R. Thomsen. My secretary, Donna Black, did a remarkable job in coverting my scrawl to readable prose and remained unfailingly cheerful during many rewrites. Thank you, Donna. This undertaking would not have been possible without the encouragement of Dean J. Durig, D. F. Williams, and other colleagues at U.S.C., as well as support from Academic Press, particularly in moments when panic and despondency were rampant. Elaine Hadaddin drafted many of the figures, while photographic services were ably performed by Sue Gager. Thank you both.

Friends and family suffered, but were remarkably forbearing, during the long, long months of writing. To all of you, I offer my appreciation and thanks.

IAN LERCHE

1

Introduction

THERMAL DEGRADATION through geologic time of organic material in fine-grained sediments is generally accepted as the dominant hydrocarbon generation mechanism (Hunt, 1979; Waples, 1984). That single statement triggers a plethora of problems that must be addressed if we are to improve our chances of finding hydrocarbon-bearing reservoirs. The problems can be split conveniently into three major groups: burial history; thermal history; and hydrocarbon generation, migration and accumulation history. As we shall see, these groups are not isolated but are very much interdependent.

I. Burial History

Put to one side for the moment the question of hydrocarbon production and concentrate on dynamical evolution. At deposition, sediments typically have porosities in the 50–75% range. When sediments are drilled at depth, say 15,000 ft, porosities are typically of order 10% or less. Thus an enormous amount of water has escaped from the sediments during their deposition and later evolution. (For the example above, a 15,000-ft section all at 10% porosity would have been 27,000 ft thick if rehydrated to 50% porosity, so 12,000 columnar feet of water has been "lost.") Because of the fluid escape, the grain-to-grain contact pressure must increase to support the overlying sediment weight. Clearly, a problem of fundamental concern, underlying all other aspects of basin analysis, is to understand the evolution of compaction in a sedimentary basin and to track the dynamical evolution of fluid escape. In addition, the sediments must be supported on an underlying basement rock. As sediment is added to a basin, the basement will also respond to the applied load (rough values are 0.5 psi/ft for water loading and 1 psi/ft for fully compacted rock loading). If the basement material behaves as an incompressible fluid, then full Airy (1856) isostatic compensation occurs, creating a surficial volume accommodation for further sediments. However, if the basement behaves as a rigid unmoving plate, no response occurs. Reality must lie between these two extremes, leading to basinal evolution models which encompass tectonic effects due to flexural plate behavior, tectonic rifting, wrenching, compression, and so on. Dynamical fluid escape depends intricately on the permeability behavior of the deposited, and evolving, lithologic units. [Rough magnitudes of permeability are about a darcy for sand and millidarcies to microdarcies for shales (1 darcy $\approx 10^{-8}$ cm^2).] As fluid loss proceeds porosity decreases, so permeabilities become smaller, leading to an ever-increasing delay in extracting the residual fluids. The addition of more sedimentary overburden is then compensated for by an increase in the retained fluid pressure over the hydrostatic value, leading to an overpressured situation. Rough estimates for fluid escape rates are about 50–200 ft/Myr from a shale of millidarcy permeability at 0.2 psi/ft overpressure. If a sedimentary deposition rate of greater than, roughly, 100 ft/Myr occurs, then there is a very strong possibility of significant overpressuring with geologic time—again dependent on lithology. Since, on average, about 75% of all deposition is shale, overpressure is not a minor, rarely occurring phenomenon, but rather has a major, dominant role to play and must be allowed for in burial history calculations. If overpressure occurs at any time in a basin's history, several consequences are immediate. First, the compaction of overpressured formations is less than it would be without the overpressure, so a formation retains a thickness closer to its depositional value for a longer period of time. In turn, this effect changes the structural and stratigraphic shaping of sedimentary units and provides a wider (laterally) thoroughfare for fluid migration. Second, the higher porosity relative to that of a nonoverpressured region lowers the thermal conductivity of an overpressured region. In turn, this increases the vertical temperature gradient (often by as much as about 1°C/100 m from the "normal" value of around 2–2.5°C/100 m) in order to preserve the heat flux. Thus overpressured regions tend to be hotter than nonoverpressured regions at the same subsurface depth. Third, as a consequence of the increased temperatures, the solubilities of carbonates and sulfates (dominantly calcium, magnesium, barium) are lowered, leading to calcite, gypsum, marl, and anhydrite precipitation, which tends to strengthen formations by diagenetic cementation. Fourth, isolated sand bodies in a surrounding shale matrix will also be overpressured so that a parametric geometry dependence on overpressuring exists. Fifth, the thermal expansion of hot water in the overpressured formations also provides an excess pressure that acts to divert fluid flow and to pressure the encompassing rocks. Since rocks have a finite mechanical strength, if the total fluid pressure exceeds a limiting value the rock must fracture, thereby releasing fluid. Empirically, the total fluid pressure limit appears to be roughly around 0.8–0.85 psi/ft (excess fluid pressure of 0.3–0.35 psi/ft). Thus the dynamical evolution of sediments is firmly tied to the dynamical fluid flow. As deposition continues fluid pressure builds, rocks fracture and then reseal, and pressure rebuilds, leading to ongoing episodic pulses of overpressure with time. The ability to account for this behavior is a crucial ingredient of any model.

In addition, an increased temperature speeds hydrocarbon generation from kerogen, providing increased conversion from solid material to liquid (oil) and gaseous components. These nonsolid components provide a generated pressure, can flow either in solution with the prevailing waters or as separate phases, and so affect the dynamical burial history of the sediments. A further, occasionally dominant, interaction occurs from clay diagenesis as smectites dewater to illites and kaolinites, releasing their "bound" water into the fluid system. The illitization is temperature and pressure dependent and is triggered by potassium catalysis but can be catalytically "poisoned" by sodium and calcium ions. Hence the dissolved minerals and ions in the subsurface waters also have a role to play in influencing sedimentary evolution.

Apart from fluid effects, major components in burial history evolution are also provided by depositional variation laterally in a basin, erosional events, overthrusting, rotation of fault blocks, intrinsic faulting of the sediments as syndepositional growth faults, tectonically induced faulting, or shear failure caused by differential stress on consolidated, lithified formations as the basin evolves. The presence of faults can lead to fluid escape up the faults, if the faults stay open for any length of time, and thereby influence the sediment evolution.

In addition to sand and/or shale, two other major lithological types of rock fabric—salt and carbonates—can be dominant controls on basinal evolution. Salt normally occurs in an evaporitic basin which is later open to clastic input. The evaporation can leave behind great thicknesses of salt (in the Gulf Coast of the United States, estimates of the original Louann (Jurassic) salt thickness are in the several thousand feet range), overcoated with evaporites. Later sediment deposition on the underlying mobile, constant-density salt causes salt movement laterally and vertically, giving rise to sediment distortion, faulting, and changes in the pattern of later sediment deposition through topographic expression from the buoyant salt and salt dissolution (Lerche and O'Brien, 1987). Indeed, in some places (e.g., the Red Sea) evaporation is so rapid that salt is precipitating today on top of sediments.

Carbonates play a different role (Milliman, 1974). Deposited as the skeletal remains of marine organisms, carbonates basically "grow" by secretion of calcium. Hence carbonate reefs, platforms, pinnacle reefs, and so forth represent shallow marine deposition in warm water where nutrient supply is abundant and the optical depth to sunlight penetration is high. Large, laterally extensive, carbonate deposition is possible because the weight of the deposited carbonates causes basinal sag, leaving a volume accommodation for further carbonate deposition. Indeed in the Dampier Basin of Western Australia upward of 3000 ft of carbonate material, several tens of miles wide and about 100 miles in length, was deposited in the Tertiary. Apart from the continuing impact such a weight has on the evolution of basinal structure, the stress on neighboring formations can often be sufficient to produce major faulting. In addition, the porous carbonates can act as good reservoir sites, can alter further depositional patterns of sedimentation, and can act as aquifers for ground water transport into the deep subsurface, and for deep basinal waters to reach the surface.

In any attempt to determine the dynamical burial of sediments in a basin we have four major controls at each well, or pseudowell, location. These controls are (1) the observed formation thicknesses, (2) the observed porosity with depth, (3) the observed fluid pressure with depth, and (4) the observed formation permeability with depth. To be acceptable, a model must obtain agreement and consistency with at least these four controls at each location. As we shall show in later chapters, it is not always easy to constrain models to honor these control criteria.

II. Thermal History

Put to one side for the moment the problems of dynamical burial history and hydrocarbon production and concentrate on the thermal history of the sediments. As we shall see in the next subsection, knowledge of the thermal history of the sediments is a crucial ingredient in assessing the hydrocarbon generation and migration history. On the assumption that the dynamical burial history has been determined precisely, we need to determine the temperature history of each and every sedimentary unit with time and burial depth. In turn, determination of the temperature history rests on two fundamental factors: a determination of the sources of heat flux through, and within, the sediments and how those heat fluxes change with time and spatial location, and a determination of the transfer of heat (conduction, convection) through the sediments. Significant sources of heat internal to the sedimentary sequence can be highly radioactive layers (e.g., the Kimmeridge clay in the North Sea basin), igneous intrusives, frictional heating by overthrusting, and so on. By and large it would appear that such internal heat sources, while often dominant locally, are secondary in magnitude and regional character to heat flux variations through the basement rocks.

Present-day estimates of sedimentary rock thermal conductivities vary in the rough range 7 ± 3 mcal cm^{-1} s^{-1} °C^{-1} with the low values being for high-porosity unconsolidated rocks and the high values for low-porosity, nearly fully compacted sediments (Hunt, 1979). Ignoring lithology effects, a rough quasi-empirical estimate connecting fractional porosity ϕ and bulk thermal conductivity can be obtained from (Lewis and Rose, 1970)

$$\kappa = K_R^{1-\phi} K_w^{\phi}$$

where K_w is the thermal conductivity of water (~1 mcal cm^{-1} sec^{-1} °C^{-1}) and K_R the thermal conductivity of fully consolidated sediments (~10 mcal cm^{-1} sec^{-1} °C^{-1}). Present-day temperature gradients are around 2°C/100 m, so that, if conduction dominates, estimates of basement heat flux today are around $Q = 1$ HFU (heat flux unit) (1 HFU $\cong$ 42 W m^{-2}). A dynamical range of about 0.5–3 HFU is not inappropriate for most sedimentary basins today. Thus, while we can estimate from measurements the spatial variation of heat flux in a basin today, we have to determine how it varied in the past. Two end-member approaches are possible:

1. We can be guided by our knowledge of the dynamical evolution of a basin, posit a model of basement motion *and* a model relation between basement motion and basement heat flux variations, determine any parameters in the proposed basement motion, and apply these directly to the heat flux. Such schemes give rise to a variety of popular models of the stretching-rift sort (McKenzie, 1978), phase change models (Falvey and Middleton, 1981), and the like, which we discuss in more detail in Chapter 2. Most such models end up predicting some sort of high heating in the past ($Q \approx$ 4–10 HFU) with a cool-down to the present day. The weakness here is that a

dynamical–thermal model connection must be obtained, as well as a dynamical model from which to extract salient parameters.

2. We can be guided by the fact that various thermal indicators exist in wells today. These thermal indicators usually vary in some systematic way with depth. The general sense of argument here is that if the temperature-dependent physics/chemistry of the indicators along the known burial paths is given, the evolution of temperature with time can be determined. For instance, Lopatin (1971) and later Waples (1980), have popularized the idea that vitrinite reflectance $R(z)$, measured with depth z, can be related to a time–temperature integral (TTI) of the form

$$\mathrm{TTI} = \int_0^t \exp[(T(t') - T_c) / T_D] \, dt'$$

where the integral is taken along the burial path of the sediments from deposition to the present day and T_c and T_D are two scaling constants. The basic idea is to make TTI a hydrocarbon measure. To foster such ends, it has been proposed (Waples, 1984) that T_c is best set at 105°C and T_D at $10/\ln_e 2$°C. But a connection still needs to be found between $R(z)$ and TTI. If the geothermal gradient is constant for all time, then TTI can, of course, be *calculated* for each sedimentary layer (but not measured) and plotted versus the *measured* $R(z)$ values. A quasi-empirical relation is then generated connecting TTI and R. Unfortunately, there is no way to know whether the geothermal gradient is constant with time, nor is there any way to assess the choice of parameter values for T_c and T_D. The point is that there is no fundamental requirement for the maturity of vitrinite reflectance to be slaved one-to-one to that of hydrocarbons. What is needed is a physical/chemical equation connecting vitrinite reflectance to its own TTI (and not the TTI purported to represent hydrocarbon maturity). Then we stand a chance of being able to determine the equivalents of T_c and T_D, as well as the heat flux variation in time. The same point has also been made by Waples (1988) in a strong plea for a kinetic description to replace the Lopatin (1971) TTI. As we shall see in later chapters, every thermal indicator (vitrinite, steranes, pollen translucency, hopanes, $^{39}Ar/^{40}Ar$, apatite fission tracks, . . .) has its own particular TTI, each related in divers ways to the special nature of the individual indicator. As a consequence, inverse schemes, based on multiple thermal indicators, are rapidly replacing the primitive thermal maturity procedure of "calibrating" the Lopatin–Waples TTI to vitrinite reflectance (for which values of TTI = 15 and TTI = 75 are often taken as marking the onset of entry to the "oil window" and "gas window," respectively—or nominally equivalently, $R = 0.6$ and $R = 1.1$).

Suppose that such devices have provided a basement heat flux variation with time at each location in a basin. We then have to be concerned with a good determination of paleo-temperature evolution at each location. Two major problems still confound the issue.

First is the problem of assessing the thermal conductivity variation in the basin, because lithology and porosity differences spatially will act as focusing and defocusing domains for heat transport by conduction. Second is the problem that high-permeability sedimentary beds can provide excellent pathways for convective transport of heat carried by flowing fluids, thereby preferentially heating and/or cooling regions in amounts considerably different from those that could be provided by conduction alone. Hydrocarbon generation is affected by such convective heat transfer, as are hydrocarbon transport and further generation of oil to gas after location in a reservoir site. In addition, there is a change in mineral solubility by high-temperature fluid flow leading to cementation and/or dissolution, which influence the migration pathways by diagenetic evolution. In turn, the transport of heat affects fluid expansion and rock failure by pressure, as well as affecting the porosity. But the porosity variations relate directly to the thermal conductivity, thereby altering the temperature gradient and thus changing the convective heat transfer, and also reflect directly on the burial history. Thus the circle is complete—thermal and burial histories are intimately intertwined and cannot be considered in isolation but must be addressed simultaneously.

The only direct constraints on the thermal history that we have at the present day are measurements of multiple thermal indicators with depth as well as present day-temperature gradient. To be acceptable, any thermal evolution model must, at the least, be consistent with all such data.

III. Hydrocarbon Generation, Migration, and Accumulation Histories

Put to one side the dynamical and thermal histories and concentrate on the hydrocarbons. General considerations (Hunt, 1979; Brooks and Welte, 1984) suggest that the dominant mode of hydrocarbon production is by thermal degradation of kerogen after sedimentary burial. A "rich" source rock might contain around 1–5% TOC (total organic carbon) capable of producing hydrocarbons. Three major problems need to be addressed: (1) how are hydrocarbons produced from kerogen? (2) how do hydrocarbons migrate out of the (dominantly shale) source beds, and then how do they move thereafter (primary and secondary migration)? (3) what sort of accumulation sites and sealing conditions are available for the hydrocarbons, and do those sites persist to the present day?

Hydrocarbon production is a peculiar phase change in which solid material (kerogen) is turned into mobile liquids (oils) and mobile gases (mainly methane but with some ethane, H_2S, CO_2, N_2). Furthermore, oils are virtually incompressible and normally have a specific gravity around 0.8–1.01—usually less than that of subsurface brines (~ 1.0–1.3). Gases are highly compressible and normally have a specific gravity of around 0.1–0.3 at production at depth. Thus oils and gas are buoyant relative to water (buoyancy pressure is ~ 0.1 psi/ft for oils and ~0.3–0.5 psi/ft for gas). Some oils (e.g., isopropyl alcohol) are completely miscible with water, others are not (Hunt, 1979); while gas tends to be completely miscible until the "bubble point" pressure and temperature are reached, when gas exsolves and is transported as a separate phase at lower temperatures and pressures. It is these large phase change properties that make the job of assessing potential hydrocarbon reservoir sites so difficult, as we shall see later.

The ability to provide a detailed kinetic scheme relating the thermocatalytic breakdown of kerogen to hydrocarbons is fraught with major difficulties, perhaps the least of which is to provide a quantitative scheme handling the breakdown of organic macromolecule chemistry to methane (molecular weight 16) in the subsurface when ionic, mineral, and surficial chemistry influence and modify the chemical kinetics that otherwise might be thought appropriate on the basis of laboratory measurements. From van Krevelen diagrams of H/C and O/C we recognize (Hunt, 1979) that there are at least three types of kerogen—presumably having their own kinetic breakdown schemes, and yielding different production efficiencies for oil and gas generation. We also know that as the subsurface temperature increases with depth, oil gradually converts to gas (Hunt, 1979).

Attempts to bypass the detailed kinetics then revolve around the idea of providing a kinetic breakdown scheme which encompasses the gross scale of the observations but not the details. Thus questions concerning the amount of oil generated are fair game, but not questions concerning the amounts of *n*-butane production versus paraffin wax; that is, the kinetic frameworks are "lumped impedance" types of devices attempting to bracket the overall trend of behavior. Such a lumped kinematic scheme was introduced by Tissot (1969), refined by Tissot and Espitalie (1975), and popularized in Tissot and Welte (1978). Since then, many other lumped kinetic frameworks have been suggested. A common theme among them is that they all require multichannel reactions, in series or parallel, with specification of many reaction rate constants, and activation energies which relate kerogen to oil and gas, and oil to gas. All are temperature dependent and most involve first-order chemical kinetics.

By whatever scheme, then, various sets of thermochemical kinetic breakdown equations are available for assessing hydrocarbon production from kerogen. Typical *total* production estimates range from 1 to 500 mg of hydrocarbons per gram of kerogen, while typical estimates of the *onset* of significant oil (gas) production *rates* are around 100°C (150°C).

Having converted kerogen to mobile hydrocarbons, the question arises of how the liquids escape the fine-grained clastic source beds of generally low permeability (primary migration). Several mechanisms have been proposed. Thermal expansion coefficients of fluid hydrocarbons are larger than rock values. Hence the conversion of kerogen to fluid puts an additional fluid pressure on the encompassing source rock, leading to the possibility of microfracturing and fluid loss through the high-permeability fractures. An alternative suggestion is that the volume occupied by produced hydrocarbons is larger than that of the original kerogen, leading directly to an increase in pressure, again driving fluid motion and microfracturing. A third suggestion is that capillary pressure at sand–shale interfaces has a dominant role to play. Typical pore radii in fine-grained materials are around 10^{-5}–10^{-6} cm, while in sands the size is around 10^{-4} cm. The surface tension difference T between oil and water is around 100 dyn/cm, so that the capillary pressure p at a sand/shale interface is $p \approx 2T(1/r_{\text{fine}} - 1/r_{\text{coarse}}) \approx$ 20–200 atmospheres ($\equiv$ 300–3000 psi). Thus capillary pressure is the largest force at the boundary but drops rapidly away from the boundary. Hence oils tend to displace water in coarse-grained beds abutting the fine-grained source bed, while water displaces oil in the source.

By whichever process, generated oil and gas now find themselves in a relatively high-permeability bed. They can then move (secondary migration). Several forces act on the hydrocarbons. Any excess fluid pressure in an overpressured formation causes a hydrodynamic drive acting not only on the ambient subsurface fluids but also on the hydrocarbons. The fractions of oil and gas which go into solution move with the prevailing fluid. But the nonsoluble gas/oil fractions have a vertical buoyancy pressure drive acting on them of around 0.1–0.5 psi/ft. Resistance from overlying less permeable beds then means that the nonsoluble fractions tend to move along the path of least resistance, which is not necessarily vertical. Thus the dynamical evolution of the structural aspects of bedding geometry now come into play, with up-dip migration to structural culminations being favored by oil and gas, and with gas motion attempting to outstrip oil motion (by about a factor of 5–10 in speed) because of its greater buoyancy drive.

On the other hand, it is not just the pressure gradient which controls fluid motion. Darcy's law notes that the speed of motion is proportional to the permeability as well as to the driving pressure gradient. As the hydrocarbons move from their source location to potential accumulation sites, any significant decrease in permeability will effectively bar their further migration. Such decreases can be caused by compaction as the basin develops, by diagenetic cementation, by spatial variations in lithology, and, of course, by tectonic forces altering the basin. Thus, as well as the desire of hydrocarbons to seek out structural highs as accumulation sites, there are also stratigraphic sites of potential accumulation which may, or may not, be associated with structural evolution and development. And the stratigraphic variations also develop with time.

Three further possibilities cause ongoing variations in hydrocarbon accumulation sites: hydrodynamic versus hydrostatic conditions, sealing conditions, and continued generation of oil to gas in a reservoir.

Suppose that oil has succeeded in finding a place to accumulate at some time in a basin's history. Continued evolution of the basin can lead to aquifer flow at later times attempting to flush the oil from its original accumulation site. The hydrodynamic forces can flush the oil down-dip or up-dip, depending on their direction. The prevailing milieu of physical dynamical transport plays a major role in some hydrocarbon regions, as has been amply demonstrated by Hubbert (1953) and Dahlberg (1982) both theoretically and with case histories.

In addition, the sealing conditions around a filled reservoir at some instant of time, such as an overlying impermeable shale, can change significantly as later deposition and basinal development affect the stress conditions of the seal—occasionally leading to fracturing and faulting of the seal and consequent loss of the hydrocarbons.

We also note that the models of kinetic degradation of kerogen to oil and gas contain components linking oil degradation to gas production. So an oil-filled reservoir, even if kept sealed, will in the course of time gradually convert from oil to gas. But this conversion increases the pressure on the seal (because the gas buoyancy pressure is, at a minimum, about five times that of oil), leading to the distinct possibility of later rupture and hydrocarbon loss.

IV. Summary

The simple recognition that hydrocarbons are produced by thermal degradation of organic material forces an examination of the interlocked evolution of sedimentary burial and dynamical interaction, of the evolving thermal regime in the subsurface, and of hydrocarbon generation, migration, and accumulation. These three dominant components are by no means capable of being examined in isolation—each has an impact on the others.

The major concerns of quantitative methods in basin analysis are to examine the interrelations among the various components, to quantify the dynamical behavior so that basinal processes can be modeled, to use present-day data to constrain generic processes and their applications to particular basins, and to provide an assessment of the most likely sites of hydrocarbon accumulations at the present day ahead of expensive drilling. Such assessments must also provide a measure of their probability of being right based on the best sensitivity studies possible with the data and models to hand.

In the long-range thrust to reach that goal we shall, perforce, be addressing a number of individual problems, with others being less well developed, in order to exhibit as sharply as possible the role of the individual facet under consideration. But we will also provide a gradually more integrated picture of interwoven development as we proceed.

2

Basin Evaluation and a Case History

I. Basin Evaluation Using Burial History Calculations

Burial history calculations available to the petroleum geologist include plots of burial history and geohistory, porosity, and thermal conductivity as functions of depth and lithology, as well as the results obtained from algorithms to handle unconformities and backstripped crustal subsidence; calculations that predict the breakdown of kerogen to hydrocarbons in terms of time and temperature; and methods for determining heat flow from geologic models and thermal indicators such as vitrinite reflectance. Data inputs required for these calculations include depths of formation tops, ages of formation tops, lithologies of formations, depths of water at deposition of the different formations, porosities of the formations as a function of depth, bottom-hole temperatures and/or formation temperatures, and any kerogen content and vitrinite reflectance values as functions of depth. Sources of many of these data may be well logs, well reports, geologic papers, and/or seismic sections.

This chapter considers how parameters can be extracted from burial history calculations to construct maps that can be compared to the current locations of oil and gas fields and so used to locate and rank prospective acreage. These parameters include the potential source rock maturity of a formation plotted as a function of time and depth, the thermal history of any hydrocarbons associated with the potential source rock, the rate of sediment accumulation of a formation as a function of time, and subsidence rate of a formation as a function of time. Subsidence rate includes total subsidence, compaction-induced subsidence, isostatic response to sediment load, and backstripped crustal subsidence (total subsidence minus the effects of compaction and isostatic response). The technical key to this burgeoning mode of exploration is the development of "user-friendly" computer software that can manipulate easily both well and seismic data. If burial history is plotted graphically, the resulting curves can be used to prepare paleostructure maps and cross sections of reservoir and source rock units, on which thermal maturity can be shown as a function of time. Tied to seismic stratigraphic studies, burial and thermal histories provide a basis for analyzing the timing of hydrocarbon generation and identifying possible paths of oil migration with respect to trap history.

A. Burial History, Geohistory, and Backstripped Subsidence

Geohistory (Fig. 2.1) and burial history plots (Fig. 2.2) can be used to predict the age and thickness of formations (van Hinte, 1978), estimate the amount of material removed at unconformities by erosion (Fig. 2.3), and estimate the rate of sediment deposition as a function of time (Fig. 2.2). In addition, thicknesses of geologic layers can be used to predict the

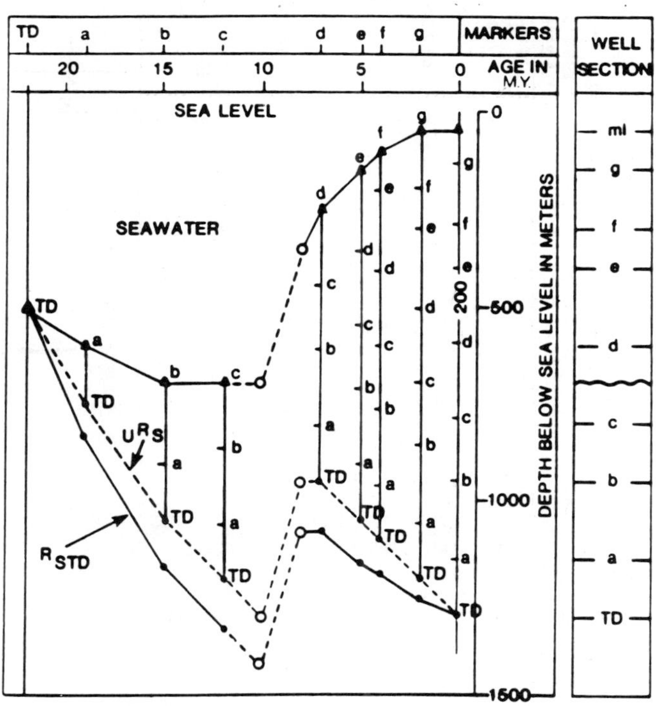

FIGURE 2.1 Geohistory diagram for a hypothetical well. Upper curve shows water depth history for location; lower curves show subsidence and uplift at that location. Times TD through c represent era of tectonic subsidence; c through g represent era of uplift. (After van Hinte, 1978.)

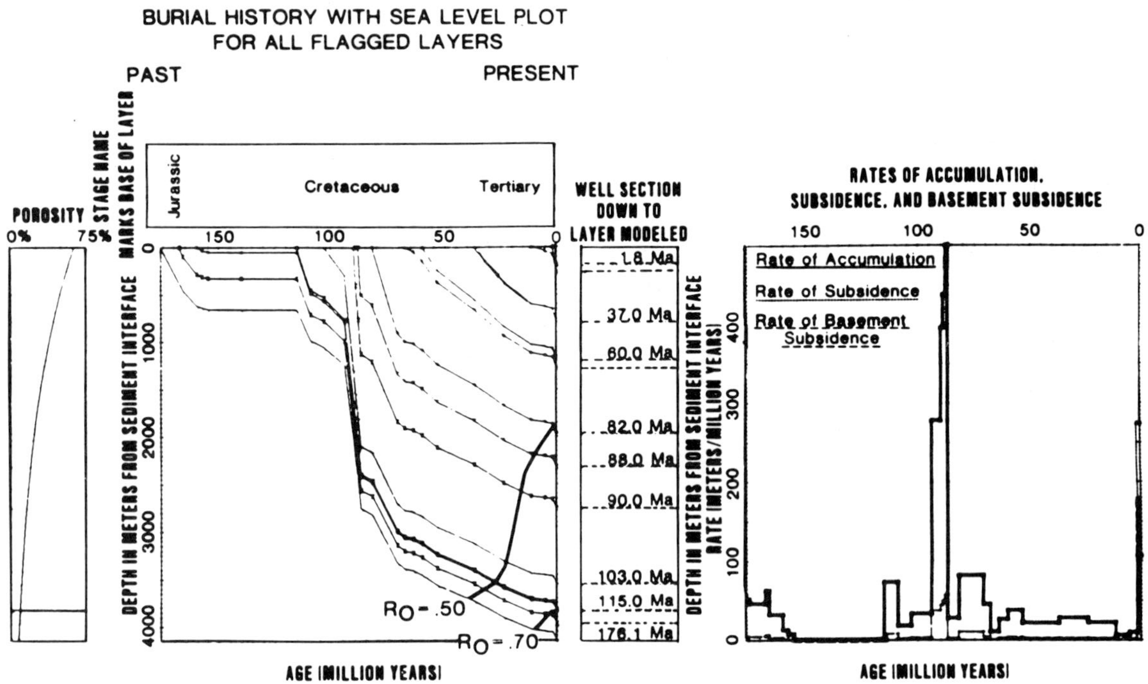

FIGURE 2.2 Machine-generated burial history plot of Norway well; $R_0 = 0.5$ and $R_0 = 0.7$ curves represent Waples thresholds for onset and maximum generation of oil, respectively. Rates of sediment accumulation, subsidence, and basement subsidence as a function of age are also shown.

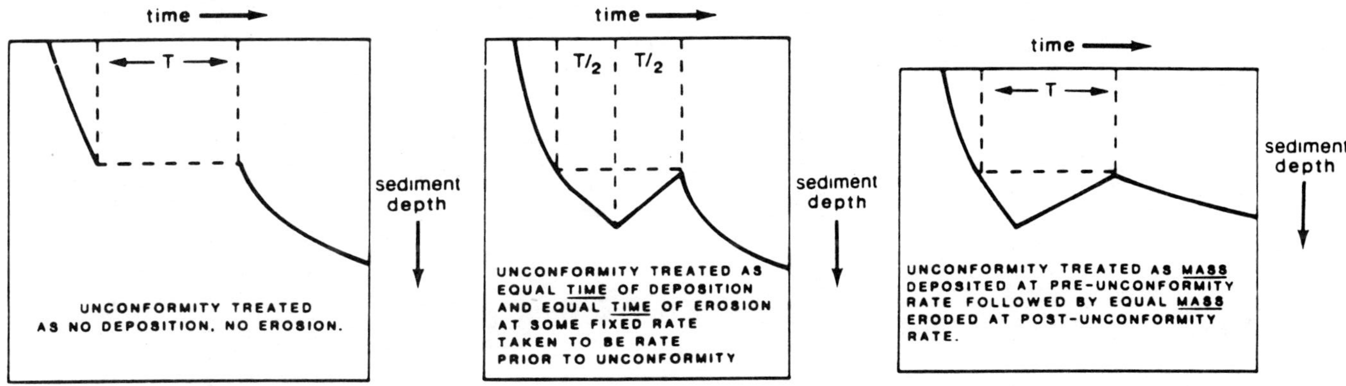

FIGURE 2.3 Geohistory diagram for hypothetical well with unconformities handled in different ways.

thermal history of geologic formations from either backstripped subsidence or thermal indicators.

The data needed to construct burial curves that do *not* model changes in thickness (as in curve ${}_{u}R_{s}$ in Fig. 2.1) are the numerical geologic ages of the layers, their depths, and the water depth of each layer at its time of deposition. The burial history plot (Fig. 2.2) is then simply obtained by linearly interpolating between the depth at the present time for the first layer and the time at which this layer was on the surface. Ignoring compaction effects, all lower layers move up at the same rate as the first layer. At the time when the first layer was on the surface, we know the depths of all lower layers. Then it is simply a matter of repeating the process until all layers have been moved to the surface. The geohistory plot (Fig. 2.1) (i.e., the plot of the depth below sea level) is then constructed by adding the depth of water at each time of deposition. This is the backstripping process.

However, in order to calculate burial history and geohistory and construct plots in which compaction *is* modeled, it is necessary to model the changes in rock porosity with depth. This modeling is illustrated by considering the rate of sediment accumulation (R) with respect to time given by van Hinte (1978):

$$R = T(1 - \phi)/A(1 - \phi_0) \tag{2.1}$$

where T is the thickness at present porosity (ϕ), A the time duration, and ϕ_0 the porosity at time of deposition.

Thus, to model layer thickness as a function of time, it is necessary to model the change in porosity as a function of time or, equivalently, as a function of depth. Van Hinte (1978) has illustrated how to do this for present porosities taken from well data and for arbitrarily chosen initial porosities. In the absence of porosity data from wells, phenomenological models can be used to relate porosity to depth or load for various lithologies. One such model, proposed by Sclater and Christie (1980), supposes an exponential decrease in porosity with depth. Even though actual values of porosity are variable and depend on compositional variations of lithology, effects of overpressure or undercompaction, and/or diagenesis, they generally decrease with depth. Sclater and Christie's (1980) model approximates this overall depth behavior in porosity. Based on the data of Athy (1930a) and Hedberg (1936) for normal pressures, the porosity (ϕ) can be represented by

$$\phi = \phi_0 e^{-cz} \tag{2.2}$$

where ϕ_0 is the surface ($z = 0$) porosity. The constant c must be given for each different lithology. A table of values for ϕ_0 and c is given by Sclater and Christie (1980, Table A1a). Values of ϕ_0 and c from actual borehole porosity measurements can be obtained by fitting an exponential function via least-squares techniques.

An alternative phenomenological model of porosity versus depth has been proposed by Falvey and Middleton (1981). In their model, the porosity decreases as a function of the load on the layer being considered. Falvey and Middleton (1981) argued that in most cases exponential porosity–depth relationships do not fit shallower-depth data particularly well. They formulated a relationship by assuming the incremental change in porosity ($d\phi/\phi$) is proportional to the change in load (dL) and the void ratio (e) as

$$d\phi/\phi = -ke\ dL \tag{2.3}$$

where

$$e = \phi/(1 - \phi) \quad \text{and} \quad dL = (1 - \phi)\ dz$$

with dz being the change in depth of sediment due to the change in load and k a lithology-dependent constant. Equation (2.3) has the solution

$$1/\phi = 1/\phi_0 + kz \tag{2.4}$$

where ϕ_0 is the initial uncompacted porosity and z is the depth.

Equations (2.2) and (2.3) make it relatively easy to correct well data for the effects of compaction. First, the total amount (h_s) of conserved skeletal or solid material in an interval (z_1 to z_2) is given in the Sclater–Christie model by

$$h_s = \int_{z_1}^{z_2} [1 - \phi(z)]\ dz \tag{2.5}$$

that is,

$$h_s = (z_2 - z_1) - (\phi_0 / c)(e^{-cz_1} - e^{-cz_2}) \tag{2.6}$$

Hence, knowing that a rock layer was on the surface at a previous time, where $z_1 = 0$, we can calculate its base by solving equation (2.6) self-consistently. Then the depths and thicknesses of all lower layers can be obtained sequentially and self-consistently. A similar equation is obtained in the Falvey–Middleton analysis:

$$z_4 - z_3 = h_s + (1/k)\ \ln[(1/\phi_0 + kz_4)/(1/\phi_0 + kz_3)] \qquad (2.7)$$

Given the compaction parameters and drill-hole values of depth, age, and paleowater depth, a computer program can be designed that will give the burial history and geohistory curves as shown in Figs. 2.1 and 2.2.

Another important factor to consider in conjunction with porosity is the modeling of unconformities. There are at least three possible ways to model an unconformity (Fig. 2.3):

1. As a depositional hiatus.
2. As a period of deposition and erosion in which the thickness of eroded section and the absolute age marking the beginning of erosion are determined from a knowledge of the local geology (see van Hinte, 1978); these parameters may be varied to see which model best fits other observed data (e.g., present-day vitrinite reflectance and/or behavior of velocity or density logs).
3. As a period of deposition and erosion in which the thickness of eroded section and the age at the beginning of erosion are calculated as follows:

$$\mathrm{AGE_E} = (R_1 \times \mathrm{AGE_1} + R_2 \times \mathrm{AGE_2})/(R_1 + R_2)$$
$$\text{Eroded thickness} = R_1 \times (\mathrm{AGE_E} - \mathrm{AGE_1})$$

where $\mathrm{AGE_E}$ is the age at the beginning of erosion, $\mathrm{AGE_1}$ and $\mathrm{AGE_2}$ are the observed ages of sediment immediately above and below the unconformity, and R_1 and R_2 are the calculated sedimentation rates in the immediately overlying and underlying sediment layers.

Knowing the depth and thickness of each layer as a function of time, and also knowing the lithologies of the layers and their densities, it is possible to determine the basement subsidence as a function of time by applying Airy (1856) isostasy calculations. The depth of the earth's crust into the medium is determined by the load on the crust, as has been discussed by Bomford (1971). We can then apply isostasy arguments, as discussed by Steckler and Watts (1978), to backstrip (remove) the layers of sediment and replace these layers by water. The basement subsidence (Y) in terms of the water depth at time of sediment deposition (W_d) and sediment thickness (S) is given by (see also Fig. 2.4):

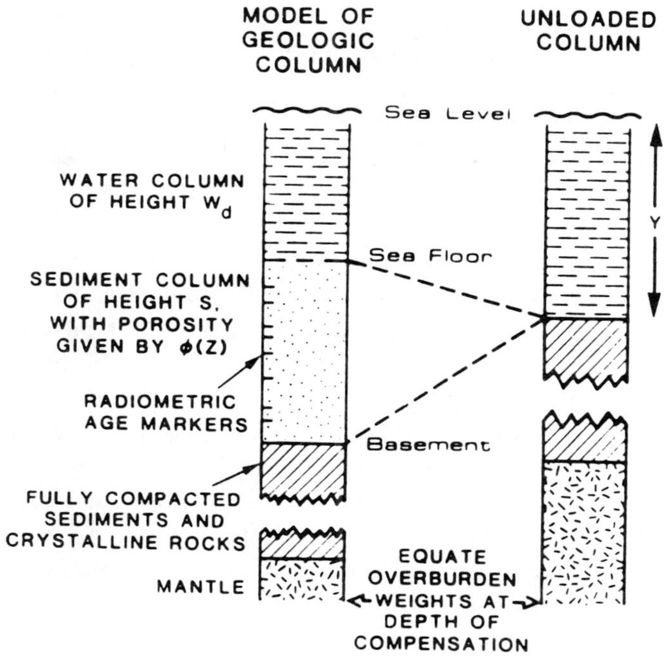

FIGURE 2.4 Schematic diagram of reconstructed (loaded) sedimentary section and backstripped (unloaded) sedimentary section. Parameters are defined in text. (From Steckler and Watts, 1978.)

$$Y = S[(\rho_m - \rho_s)/(\rho_m - \rho_w)] + W_d - \Delta_{SL}[\rho_m/(\rho_m - \rho_w)] \qquad (2.8)$$

where ρ_m is the average mantle density, ρ_w the average water density, ρ_s the average sediment density, and Δ_{SL} is the change in elevation of mean sea level; Y is the depth to basement without sediment and water loads and represents the subsidence caused by tectonic effects.

The terms W_d and Δ_{SL} must be determined from sediments by paleobathymetric analysis.

B. Hydrocarbon Generation and Thermal History

One of the principal goals of explorationists using burial history technology is the prediction of source rock maturity and the ranking of prospects.

Involved in the evaluation of a sedimentary basin is the determination of the amounts of oil and gas that have been generated and accumulated and the locations in which they occur. To do this, the following information is needed: (1) amount of oil and gas generated per square kilometer in each source rock and in every part of the basin; (2) timing of hydrocarbon formation, timing of the development of impermeable seals, and timing of folding and faulting, so traps can be identified; (3) amounts of oil and gas that have been expelled from the source rock into the porous reservoirs and how much is still retained in the source rock; and (4) evaluation of the ultimate oil and gas reserves of a sedimentary basin.

Only a quantitative approach, allowing a computation of the amount of oil and gas generated in any place in the basin as a function of time, can provide this information. However, because many of the intermediate steps in petroleum and gas generation from kerogen degradation are unknown, such a scheme has to be phenomenological; that is, parameters have to be determined by least-squares fitting to basin and laboratory data. Such a mathematical model of petroleum generation accounting explicitly for geologic time was introduced by Tissot (1969) and is discussed fully by Tissot and Espitalie (1975). The model is based on kinetics of kerogen degradation and uses the general scheme of first-order chemical kinetics. Kerogen is a macromolecule composed of polycondensed nuclei bearing alkyl chains and functional groups—the links between nuclei being heteroatomic bonds or carbon chains. As the burial depth and temperature increase, the bonds are successively broken, roughly in order of increasing rupture energy. The products generated are heavy heteroatomic compounds, carbon dioxide, and water; then progressively smaller molecules; and finally hydrocarbons. At the same time, the residual kerogen becomes progressively more aromatic and evolves toward a carbon residue.

Usefulness of a phenomenological model depends on two factors: (1) the validity of the physics and chemistry involved in the calculation, and (2) the values of the parameters chosen. Tissot and Welte (1978) compared the figures computed by their model with the corresponding amounts of petroleum generated in various basins. They claimed a quadratic deviation lower than 10^{-2} and a correlation coefficient better than 0.9. The model has been claimed to simulate experimental heating during various times from 1 hour to 1 year—again with satisfactory agreement with laboratory experiments. The same set of constants given by Tissot and Welte (1978, Table V.4.1) is said to account for all conditions of kerogen degradation including: (1) evolution at relatively low temperatures (50–150°C; 122–302°F) over a time of 10 to 400 Myr, (2) artificial evolution through laboratory experiments (180–250°C; 356–482°F), and (3) high-temperature (400–500°C; 752–932°F) retorting of oil shales.

The constants, however, were determined using constant geothermal gradients. Strictly speaking, the use of Q (heat flux) or bottom-hole temperature data generates changing thermal gradients with depth and time and requires a recalibration of values for greater applicability.

To apply the phenomenological model of Tissot and Welte (1978) to the prediction of change in hydrocarbon content of rocks with respect to time and temperature, first-order chemical reaction kinetics are assumed for the breakdown of kerogen to hydrocarbons as governed by

$$-dx_i/dt = k_i x_i \qquad (2.9a)$$

$$k_i = A_i e^{-E_i/k_B T(t)} \qquad (2.9b)$$

where x_i is the number density of component i, k_i the reaction rate for this reaction, k_B Boltzmann's constant, A_i the reaction rate in the limit of high temperature, E_i the activation energy, and $T(t)$ the temperature as a function of time acting on component x_i.

In the Tissot and Welte scheme, three types of initial kerogens are considered. The three types differ in their hydrogen-to-carbon ratio. Loosely speaking, these three types can be considered type 1 (algal) with a high petroleum potential, type 2 (amorphous) with an intermediate petroleum potential, and type 3 (woody) with a low petroleum potential. Each of these types of kerogen is assumed to have six different bond types, having strengths or activation energies in the range 10 to 80 kcal/mol. The amount of organic material (x_i) reacting in the ith reaction (breaking of the ith bond type) for i = 1 to 6 is specified for each of the three kerogen types. Calibrations for the formation of gas have been made from laboratory experiments on hydrocarbon cracking. Consideration of a single reaction with an activation energy of 80 kcal/mol was thought convenient to account for gas generation in the deep parts of sedimentary basins. Tables of the activation energies E_i, the reaction constants, A_i, and the initial values, x_i for each kerogen type have been given by Tissot and Welte (1978, Table V.4.1).

To calculate the quantity of hydrocarbons produced by potential source rocks, it is necessary to use the above calculations and measurements or estimates of the percentage of organic material in those source rocks. Nakayama and Van Siclen (1981) detailed the methodology of this system to work out a hydrocarbon budget for a basin. In order to make predictions about the hydrocarbon history of a basin, it is also necessary to determine the temperature as a function of time. Tissot and Welte (1978) used constant geothermal gradients in their models. However, by allowing the heat flux to vary with time, their model is improved. The temperature, as a function of time, for a basin with a variable basement heat source $Q(t)$, located at $z = Z_{\text{basement}}$ is given by the heat diffusion equation:

$$\partial T/\partial t = (\rho C)^{-1}\nabla \cdot (K\,\nabla T) + (\rho C)^{-1}Q(t)\delta\,(z - Z_{\text{basement}}) \qquad (2.10)$$

where C is the specific heat, ρ the density, K the thermal conductivity, and $\delta(z)$ the Dirac δ function.

If $Q(t)$ varies slowly as a function of time compared with the diffusion time scale through the top 3–5 km (9,800–16,400 ft) of the earth's crust and sediment (i.e., slowly on a time scale compared to about 4 or 5 Myr), then the solution of the heat diffusion equation is approximated by

$$T(t) = T_t(t) + Q(t)D(t)/K(t, z) \qquad (2.11)$$

where $T_t(t)$ is the temperature at the top of the layer as a function of time, $D(t)$ the layer thickness as a function of time, and $K(t, z)$ the average thermal conductivity of the layer as a function of time and depth.[1] Thus, to predict basin maturity as a function of time, it is necessary to know $Q(t)$, $D(t)$ and $K(t, z)$. Although $D(t)$ and $K(t, z)$ can be determined from the burial history and lithology, in order to make accurate predictions of hydrocarbon maturity, $Q(t)$ must be determined also over the range of the age of the basin to the present. The previous section showed how to determine layer thickness $D(t)$ using porosity–depth relations and a given decompaction relation.

To estimate paleotemperature, it is necessary to know the thermal conductivity as a function of time. For two reasons it is difficult to extrapolate the data in a particular basin back into the past. First, there are large differences in lithology; second, most extrapolation techniques involving empirical relations connecting thermal conductivity to porosity have been worked out only for sandstones. Sandstones often occur only

---

[1] The precise solution is $T = T_s + Q \int_0^z dz'/K(z', t)$ where T_s is the surface temperature. For a layer of thickness $D(t)$ at depth z with a temperature on the top of the layer of $T_t(z)$, we can equally write

$$T = T_t(z) + Q(t)D(t)/\langle K(t,z)\rangle$$

where the average K, $\langle K\rangle$, is given by

$$\langle K\rangle = \left[\int_z^{z+D} dz'/K(t,z')\right]^{-1} D(t)$$

in limited areas and may have been deposited during relatively brief time intervals in a basin. However, these empirical techniques provide the basis for tackling other lithologies, such as shales or limestones. Evans (1977) and Sclater and Christie (1980) have shown that such empirical relations have considerable validity for the pre-Tertiary shales, marls, and chalks of the North Sea. Thus, if a reasonable estimate of the porosity can be determined, it is then possible to relate thermal conductivity to porosity. The point here is that if a sediment contains water in its matrix, at 100% porosity the sediment will have the conductivity of water, whereas, when totally compacted, it will have the conductivity of the matrix. The conductivity will vary between the two extremes, and it is necessary only to obtain some values in between in order to obtain a relation between porosity and conductivity that is reasonably valid throughout the expected range of porosity. Such phenomenological relations have been obtained from the data of Sclater and Christie (1980) and Evans (1977) for shale, sands, chalks, shaly sand, and salt. The dependence of thermal conductivity on temperature is not included in those calculations, but it is of importance.

To estimate the paleotemperature, it is necessary to know thermal conductivity as a function of depth and time. Because this dependence cannot be measured, it is necessary to rely on a phenomenological relation between thermal conductivity and some other parameter that can be deduced at previous times. A particularly useful relation exists between thermal conductivity and porosity, because porosity as a function of time is calculated in the burial history computation. Sclater and Christie (1980, Appendix B) have given a relation between thermal conductivity and porosity for North Sea chalks and shales. Their porosity values are taken from deep-sea clays and shallow-water North Atlantic sediments (Ratcliffe, 1960). Observed thermal conductivities are from Evans (1977). The relation between these two quantities is summarized in Fig. 2.5 (Sclater and Christie, 1980, Appendix B). For porosities intermediate between the data points shown in Fig. 2.5, a linear interpolation scheme is used to obtain thermal conductivities. When few or no data are available, about the best that can be done is to write $K_s = K_r(K_w/K_r)^{\phi}$, where K_r, K_w, and K_s are the thermal conductivities of rock matrix, water, and sediment, respectively, and ϕ is the fractional porosity (Lewis and Rose, 1970).

These values of the thermal conductivity have been inferred from the relationship between porosity and conductivity; their reliability depends on porosity calculations. Thermal conductivity is also a function of

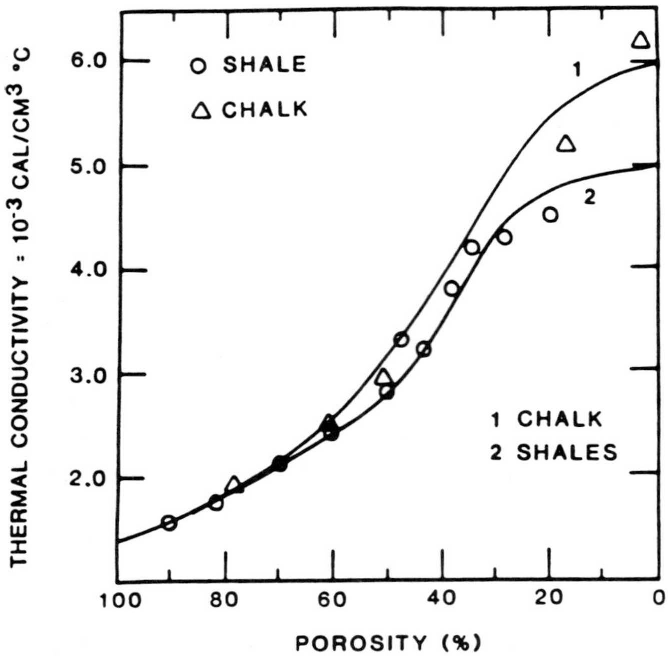

FIGURE 2.5 Relation between thermal conductivity and porosity for North Sea chalks and shales. For porosities greater than 40%, values are taken from deep-sea clays and shallow-water North Atlantic sediments of Ratcliffe (1960). Observed conductivities are from Evans (1977), with porosities estimated using relation of Sclater and Christie (1980). (From Sclater and Christie, 1980.)

temperature; however, to our knowledge this dependence has not been tested extensively for its effect on real data. Kappelmeyer and Haenel (1974) have summarized the available data on variation of thermal conductivity with temperature for several rock types. The thermal conductivities of most materials decrease with increasing temperature. The only material recorded whose thermal conductivity increases with increasing temperature is shale. Shale is the most common sedimentary rock occurring in most basins. Using a small thermal conductivity (K) has the effect of increasing the temperature calculated at a given depth. Thus, the effect would be to predict that oil is mature, whereas the oil actually may not yet be mature, or to predict that gas occurs, whereas the rock actually may be still in the oil window. Thus, to make accurate predictions of hydrocarbon maturity, it is necessary to include the temperature dependence of thermal conductivity in calculating temperatures via

$$K_L = (a + bT)^{-1} \tag{2.12}$$

where a is a positive constant and b is a positive constant for all materials except shale, where it is negative. Total thermal conductivity is the sum of a lattice conductivity K_L, which decreases with increasing temperature, and a radiative part K_R, which increases with increasing temperature. For most materials, it is possible to ignore K_R and consider only K_L. According to Kappelmeyer and Haenel (1974), thermal conductivities typically decrease by a factor of about two between 0 and 430°C (32 and 806°F). This large effect occurs over the temperature range that is the prime window for oil generation. It therefore appears necessary to include the effects of temperature variation of thermal conductivity in thermal maturation calculations.

Any model of paleoheat flow must satisfy the condition that it agrees with the present-day heat flow as determined from bottom-hole or formation temperatures, surface temperatures, and thermal conductivities determined from present-day lithologies. For example, we found that heat flows for the North Sea, which we determined using the measured surface temperature, the corrected bottom-hole temperatures, and thermal conductivities based on the empirical porosity–thermal conductivity relations, are in the range 1.0 to 1.8 heat flux units (HFU). These numbers are in fair agreement with the 1.19 to 1.48 HFU cited by Harper (1971). In contrast, Sclater and Christie (1980) cited a heat flow of 1.7 HFU with no quoted error.

C. Backstripped Subsidence and Heat Flow History

Backstripped subsidence (Y) can be predicted by geologic models. By comparing the backstripped subsidence with the predicted subsidence of a particular geologic model, we can (1) evaluate the validity of that geologic model by examining the root-mean-square (r.m.s.) difference between the predicted subsidence and the backstripped subsidence; (2) make predictions of heat flow history using the geologic model; (3) calibrate the geologic model against thermal indicators such as vitrinite reflectance or the annealing of fission track scars; and (4) predict the maturity of potential source rocks, as a function of time, away from wells using the crustal subsidence derived from seismic sections.

Thus, geologic models are an important source of information on the change in heat flow as a function of time. However, the several available models do not agree even in areas like the North Sea, where the heat flux could be increasing for one model and be decreasing for another. Also, for the Michigan basin, Cercone (1984) produced a model of heat flow history that suggests much higher heat flows in the past than those of Nunn (1981), Nunn *et al.* (1984), and others. Thus, it is important to be able to assess quantitatively geologic models for basin formation by comparing their predictions for subsidence and calibrating their heat flow to independent measurements.

In this chapter, we examine four basic models: the McKenzie lithosphere stretching model (McKenzie, 1978), the dike intrusion model (Royden *et al.,* 1980), the thermal expansion model (Sleep, 1971), and the deep crustal metamorphism model (Falvey, 1974). A model not reviewed here is the two-layer extensional model described by Royden and Keen (1980) and by Hellinger and Sclater (1983), which expands on the McKenzie (1978) model and has been used by them to explain problems they have with the models we discuss.

We begin with the thermal expansion model of Sleep (1971). This model was designed to give the thermal history of the Atlantic continental margin. This margin formed when spreading began between Africa, Europe, and the Americas. It is proposed that the thermal history of the margin resembles that of an oceanic crust as it spreads away from a mid-oceanic ridge, so the thermal structure of continental margins is deduced by studying midocean ridges. The most obvious feature of a mid-oceanic ridge is its topographic anomaly. This anomaly is interpreted as isostatically compensated thermal expansion of hot material that intrudes along the axis of the ridge. Due to thermal expansion, an initial uplift occurs. Then, as the continental edge is removed from the source of the thermal anomaly at the ridge crest, thermal contraction and, consequently, subsidence occur. The elevated surface gradually subsides to its original position as the lithosphere cools. However, to produce the configuration of many continental margins, material must be removed on a large scale by erosion when uplift occurs.

In the Sleep model, the temperature is given by

$$T(x) = g(x) \sum_{n \text{ odd}} \exp(-\kappa\pi^2 n^2 t c^{-2}) A_n \sin(n\pi z/c) \tag{2.13}$$

where

$$A_n = (2/c) \int_0^c T(z) \sin(n\pi z/c)\, dz \tag{2.14}$$

and κ is the thermal diffusivity, x the horizontal extent, z the depth, and c the maximum depth (i.e., thickness of the crust).

The function $g(x)$ is a measure of the width of the ridge. The amount of uplift $u(x)$ caused by isostatically compensated thermal expansion is then given by

$$u(x) = (2\alpha c/\pi) g(x) \sum_{n \text{ odd}} (A_n/n) \exp(-\kappa \pi^2 n^2 t c^{-2}) \tag{2.15}$$

where α is the volume coefficient of thermal expansion.

Because the first overtone attenuates nine times faster than the fundamental, it is an excellent approximation to keep only the first terms in equations (2.14) and (2.15). Then

$$u(x) = u_0(x) \exp(-\kappa \pi^2 c^{-2} t) \tag{2.16}$$

where

$$u_0(x) = (2\alpha c/\tau) A_1 g(x)$$

No sediments would accumulate if the lithosphere of the newly formed continental margin expanded and then contracted without any crustal thinning. One way to thin the crust, as suggested by Sleep, is to erode the crust while it is uplifted. The change in elevation with respect to base level is the sum of the changes due to thermal contraction and to erosion. In the limit that the erosion time constant and the heat flow time constant are equal, Sleep gives a solution for the elevation (E) with respect to the final level as a function of time:

$$E = E_0[1 - (\kappa \pi^2/c^2)t] \exp(-\kappa \pi^2 c^{-2} t) \tag{2.17}$$

where E_0 is the initial elevation.

From the elevation, it is possible to calculate the subsidence. Thus, by doing a least-squares comparison (discussed below in greater detail), it is possible to use equation (2.13) to obtain parameters determining the heat flow as a function of time, depth, and horizontal extent. The thermal cooling model has been further considered by Turcotte and Ahern (1977), as well as by Keen (1978). These authors found that the subsidence as a function of time is proportional to the square root of the time elapsed since cooling began.

Several authors, including Hellinger and Sclater (1983), have criticized the Sleep model. Within the Sleep model, material must be removed on a large scale by erosion when uplift occurs. For example, there seems to be no evidence that 2 km (6500 ft) or more of rock was removed from the North Sea after block faulting during the Jurassic and Cretaceous (McKenzie, 1978). Erosion that did occur is local and associated with the horsts produced by normal faulting.

Another model for the development of sedimentary basins, proposed by McKenzie (1978), avoids the necessity for material removal by erosion. In the McKenzie model, a rapid stretching of continental lithosphere produces thinning and passive upwelling of hot asthenosphere. This stage is associated with block faulting and subsidence. The lithosphere then thickens by cooling, and slow subsidence occurs not associated with faulting. The slow subsidence and heat flow are explained as depending only on the amount of stretching, which can be estimated numerically. In the McKenzie model, the initial subsidence (S_i) is given by

$$S_1 = a\{(\rho_0 - \rho_c)(t_c/a)[1 - \alpha T_1(t_c/2a)] - (\alpha T_1\rho_0/2)\} \times (1 - 1/\beta)/[\rho_0(1 - \alpha T_1) - \rho_w] \quad (2.18)$$

where a is the thickness of the lithosphere, t_c the initial thickness of the continental crust, ρ_0 the density of the mantle, ρ_c the density of the continent, ρ_w the density of the seawater, α the thermal expansion coefficient of the mantle, T_1 the temperature of the asthenosphere, and β the crustal stretching factor.

For large values of time and small values of β, the elevation is given by

$$E(t) = [4a\rho_0\alpha T_1/\pi^2(\rho_0 - \rho_w)]e^{-t/\tau} \quad (2.19)$$

where

$$\tau = a^2/\pi^2\kappa \quad (2.20)$$

and κ is the thermal diffusivity.

The heat flow $Q(t)$ is given by:

$$Q(t) = (KT_1/a)[1 + (2\beta/\pi)\sin(\pi/\beta)e^{-t/\tau}] \quad (2.21)$$

For small values of β and large values of time, a straight-line fit which best approximates equation (2.19) for the subsidence is given by

$$Y = E_0 r[(2t/e\tau)^{1/2} - 2e^{1/2} + 1] \qquad (2.22)$$

where

$$E_0 = 4a\rho_0\alpha T_1/\pi^2(\rho_0 - \rho_w) \qquad (2.23a)$$
$$r = (\beta/\pi)\sin(\pi/\beta) \qquad (2.23b)$$

and

$$\tau = 62.8 \text{ Myr} \qquad (2.23c)$$

The subsidence predicted by the McKenzie model is similar in form to that predicted by the Turcotte–Ahern model.

In the McKenzie model, prior to the time of stretching, the heat flow is given by the leading coefficient in equation (2.21) (i.e., KT_1/a). Sclater and Christie (1980) pointed out that the present-day heat flow (approximately 0.8 HFU) predicted by the McKenzie model (see Fig. 2.6) is too low to agree with the present measurements of heat flow measured in the North Sea (approximately 1.5 HFU). Sclater and Christie have proposed a modification of the McKenzie model in which the heat flow has an additional component of 0.8 HFU coming from the decay of radioactive elements within the crust. Another modification has been proposed by Keen *et al.* (1981), in which the effect of thermal blanketing by low-thermal-conductivity sediments is included, slowing the cooling of the lithosphere. This effect is claimed to be particularly important to studies of older rifted margins, as in the North Sea, where large thicknesses on the order of 15 km (49,000 ft) of sediments have been deposited. Royden and Keen (1980) and Hellinger and Sclater (1983) have proposed modified stretching models encompassing two layers, which handle the problems, not confronted by McKenzie, of uplift and erosion on the flanks of a rift basin.

Another model, proposed by Royden *et al.* (1980), that provides an exponential subsidence roughly compatible with observational data involves the fracture of the continental lithosphere and its intrusion by dikes from the mantle. Replacement of light crustal rocks by denser ultramafic material results in initial subsidence and avoids the general problem of uplift as in Sleep's theory. If γ_d is the fraction of lithosphere

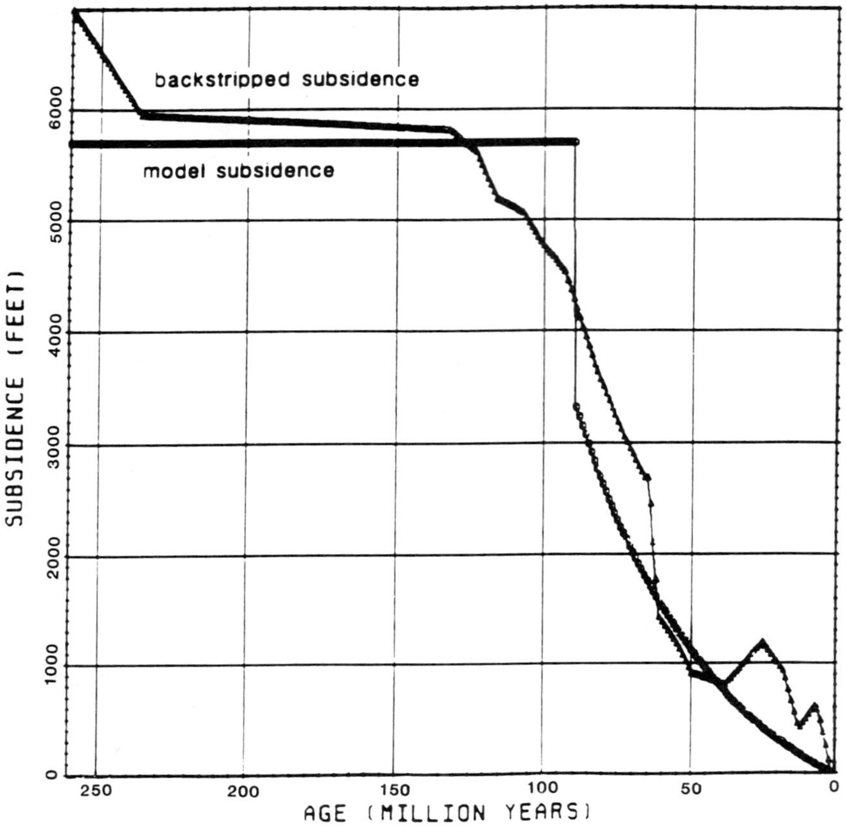

FIGURE 2.6 Comparison of backstripped and model subsidence for well A in the North Sea. Backstripped subsidence was obtained from a relation of Steckler and Watts (1978), assuming $\Delta_{SL} = 0$. Model subsidence is that predicted by least-squares optimized McKenzie crustal stretching model, with constraint that present heat flow $= 1.27 \times 10^{-2}$ cal m^{-2} sec^{-1}, as obtained from bottom-hole measurements.

composed of dike material intruded from the asthenosphere, then the heat flow $Q(t)$, and surface elevation $u(t)$, respectively, are given by

$$Q(t) = (T_m K/l)(1 + 2\gamma_d \sum_{n=1}^{\infty} e^{-n^2 \pi^2 \kappa t/l^2}) \tag{2.24}$$

$$u(t) = [\alpha l \rho_m T_m/(\rho_m - \rho_w)](4\gamma_d/\pi^2) \times \sum_{n=0}^{\infty} [1/(2n+1)^2] \exp[-(2n+1)^2 \pi^2 \kappa t/l^2] \tag{2.25}$$

where T_m is the temperature at the base of the mantle, l the thickness of the lithosphere, κ is the thermal diffusivity of the lithosphere, α the thermal expansion coefficient of the mantle, ρ_m the density of the mantle, ρ_w the density of water, and K the thermal conductivity.

The theories of basin formation thus far considered all rely on the thermal contraction of the crust to produce subsidence. An alternative explanation has been proposed by Falvey (1974) and elaborated by Middleton (1980) and Falvey and Middleton (1981). This explanation suggests that the initial stage of subsidence is caused by deep crustal metamorphism in basins of relatively young, or unaltered, crust. Falvey (1974) has demonstrated that elevation of the geotherm during the rifting of continental margins may produce subsidence of the margin due to deep crustal metamorphism and has also suggested that formation of intracratonic basins may entail such a metamorphism. The deep crustal metamorphism mechanism implies a subsidence of

$$[\rho_m(\rho_2 - \rho_1)/\rho_1(\rho_m - \rho_s)]L(t) \qquad (2.26)$$

where $L(t)$ is the thickness of greenschist facies rocks metamorphosed to amphibolite facies rocks at time t, ρ_1 the density of the greenschist facies, ρ_2 the density of the amphibolite facies, ρ_s the density of the sediments, and ρ_m the density of upper mantle rocks. Falvey (1974) has suggested that the metamorphism that drives the basin subsidence occurs predominantly at the greenschist–amphibolite facies boundary. The model entails a period of heating followed by a period of cooling. Subsidence due to deep crustal metamorphism occurs during the later part of the heating period and is followed immediately by subsidence due to thermal contraction during the cooling period. The time of transition between the heating phase and the cooling phase is t_1.

Subsidence is proposed to occur in two stages (see Fig. 2.7). Initially, subsidence is thought to be due to deep crustal metamorphism and will continue as long as crustal temperatures increase. This is followed by subsidence due to thermal contraction of the lithosphere, occurring over periods of time greater than t_1 (i.e., for the cooling phase). The downward movement, $l(t)$, of the earth's surface due to metamorphism of the heating stage of the thermal event is offset by the tendency for uplift, $h(t)$. Total displacement, taking these two factors into account and neglecting erosion, is given by

$$S(t) = l(t) - h(t) \qquad (2.27)$$

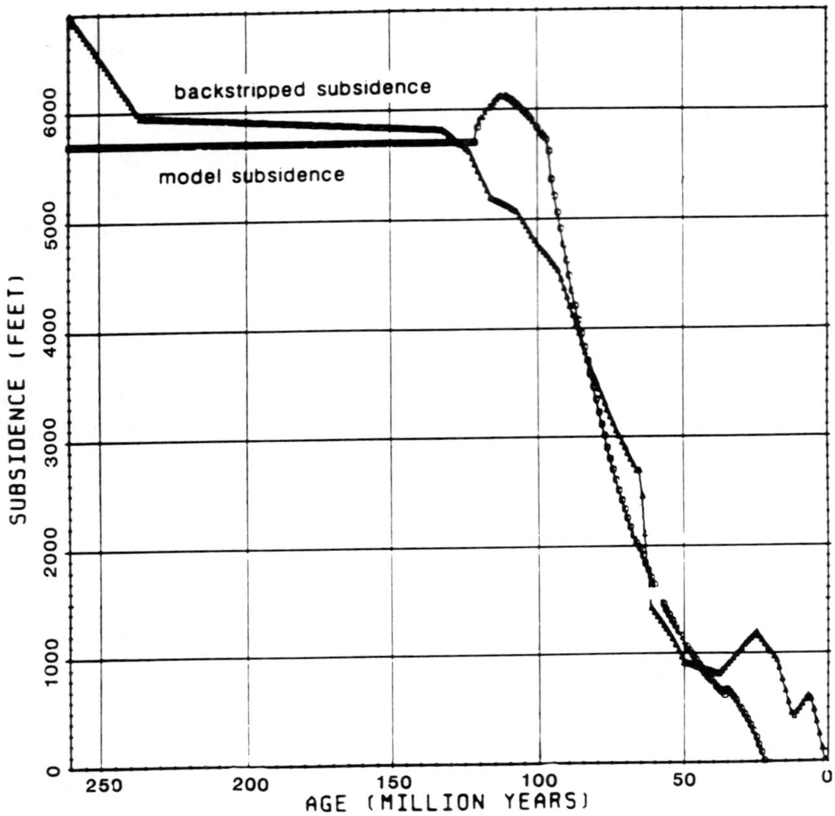

FIGURE 2.7 Comparison of backstripped and model subsidence for well A in the North Sea. Backstripped subsidence was obtained from relation of Steckler and Watts (1978), assuming $\Delta_{SL} = 0$. Model subsidence is that predicted by least-squares optimized Falvey deep crustal metamorphism model (Middleton, 1980) with constraint that present heat flow = 1.27×10^{-2} cal m^{-2} sec^{-1}, as obtained from bottom-hole measurements.

for $l(t) < h(t)$, and

$$S(t) = [\rho_m/(\rho_m - \rho_s)][l(t) - h(t)] \tag{2.28}$$

for $l(t) > h(t)$, where $-S(t)$ is uplift, ρ_m is the density of mantle, ρ_s is the density of the sediment, and $\rho_m/(\rho_m - \rho_s)$ is the isostatic loading factor.

Subsidence due to thermal contraction is given by

$$S(t) = [\rho_m/(\rho_m - \rho_s)][h(t_1) - h(t) + h(t - t_1)] \tag{2.29}$$

The distinguishing features of the Falvey–Middleton metamorphism model are that the heat flow increases up to the point of breakup and then decreases, as opposed to the other models, in which heat flow always decreases after the time of stretching. The heat flow from the metamorphism model is on the order of half that of the crustal extension model (see Fig. 2.8).

Another geologic model that may be important in some areas, particularly those like the National Petroleum Reserve of Alaska, is the low-angle thrusting model. Beaumont (1981) and Jordan (1982) have shown that low-angle thrusting with subsequent redistribution by erosion and deposition can generate a foreland basin. In the case studied by Jordan (1982), it was demonstrated that, in Wyoming, a simple elastic lithospheric flexure model with a rigidity of 10^{23} to 10^{24} N m creates a depression large enough to accommodate the sedimentary section measured in this region. Beaumont (1981) came to a similar conclusion by explaining the thickness of the sedimentary section in the Alberta foreland. The thermal history of foreland basin sediments is taken to be the simple case of constant heat flow with time, but with an obvious relationship of heat history to depth of burial.

From the foregoing discussion, we see that there are several geologic models for basin formation. The predictions of these models differ for both heat flow and subsidence as functions of time. If we are to make quantitative use of these models to predict heat flow, it is necessary to determine the optimum set of parameters for a given model for each specific well. For example, for the McKenzie crustal stretching model,

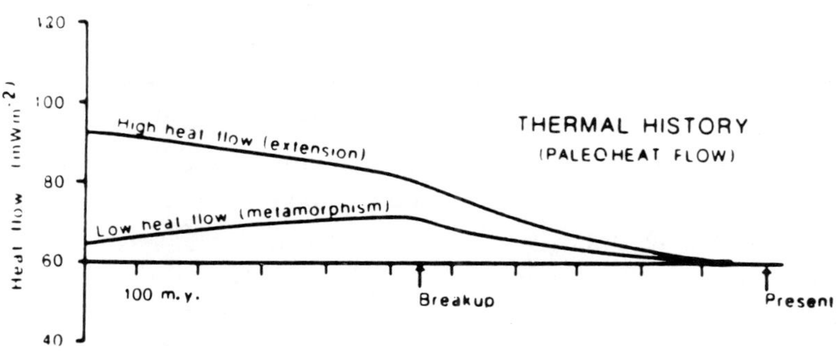

FIGURE 2.8 Comparison of heat flow resulting from McKenzie crustal extension model, derived using stretching factor of β = 1.8, and that from Falvey deep crustal metamorphism model (see Falvey and Middleton, 1981).

assuming the parametrization given in the paper by McKenzie (1978), two parameters are to be determined: the time t_0 at which crustal stretching occurred, and the parameter β which describes the magnitude of crustal stretching. The optimum values of β and t_0 correspond to those which minimize the r.m.s. difference between model-predicted subsidence and backstripped subsidence.

On the basis of these predictions for β and t_0, we find that the McKenzie crustal stretching model predicts a heat flow in the North Sea at the present time of approximately 0.8 HFU. However, on the basis of corrected bottom-hole measurements, we find that the heat flow at the present time in the North Sea is approximately 1.3 HFU. Thus, the McKenzie crustal stretching model, with the parameters given in the paper, is inconsistent with heat flow measurements. It is possible to change the parameters in the McKenzie crustal stretching model using a constraint that the heat flow predicted by the model at the present time must equal that determined from bottom-hole temperatures. This can be done by treating the time constant τ as a variable which, in addition to t_0 and β, must also be determined by the least-squares minimization. We do this by noting that the time constant is given by a^2/κ. The prefactor of the heat flow, however, can be expressed as $\kappa/a^2 \times T_1 a \times$ constant. The heat flow is proportional to $T_1 a/\tau$. We can determine the constant $T_1 a$ from the present heat flow for a chosen time constant and time of initiation of stretching. Thus, we can do a least-squares minimization simultaneously on the parameters τ, β, and t_0, subject to the constraint that the model heat flow matches that measured at the present time. The results of this calculation are shown in Figs. 2.8 and 2.9 for the subsidence and the heat flow as a function of time, respectively. The mean-squared deviation improves by 25%. This corresponds to an increase in the value of the parameter $T_1 a$ by 37% from that predicted by McKenzie. This 37% increase can occur in T_1 (the temperature at the base of the mantle), in a (the thickness of the mantle), or in some combination of the two. It is not possible to say which of these parameters should be renormalized on the basis of this simple model.

Using the McKenzie model in the North Sea, we find that the time of initial stretching of the mantle is on the order of 70–90 Ma. The stretching parameter γ (the relative crustal thinning) is in the range 0.1 to 0.3. These values agree well with the values cited elsewhere in the literature; for example, Sclater and Christie (1980) cited γ in the range 0.20 to 0.33. The predicted heat flow at the time of stretching is approximately 30% higher than the heat flow today.

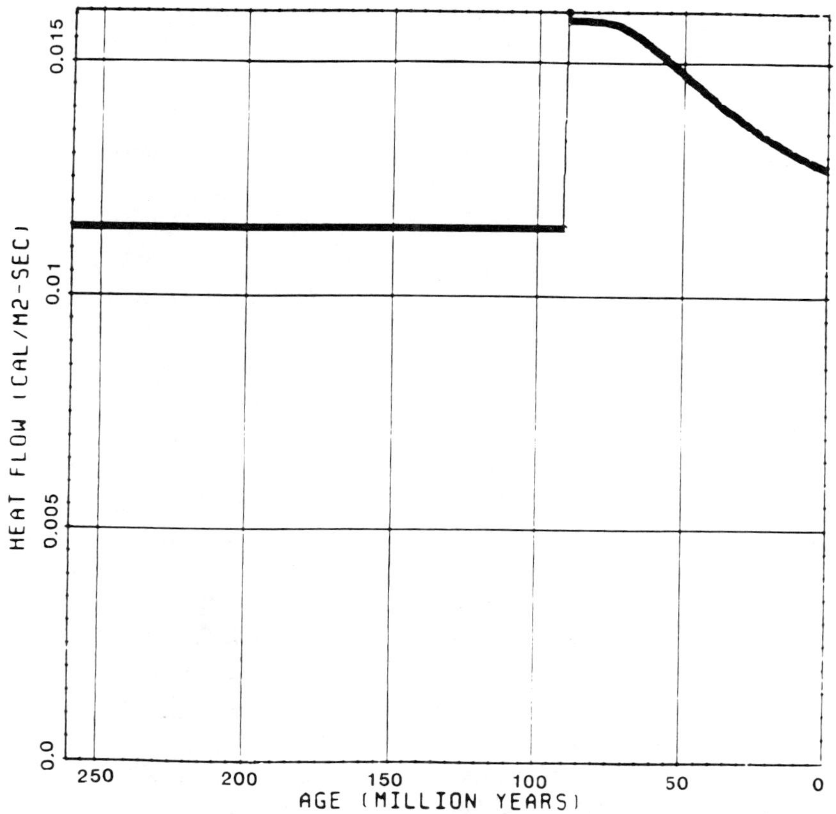

FIGURE 2.9 Heat flow resulting from McKenzie crustal stretching model.

Modifications of the least-squares technique can be applied to the other models that we have discussed in this section. It is possible to determine an optimum geologic model for a basin based on subsidence data. Unfortunately, such parameters as initial crustal thickness, average crustal density, and thermal conductivity of rocks cannot be adequately constrained solely by observed crustal subsidence. Therefore, more than one geophysical model of crustal subsidence could fit the tectonic subsidence curve, so calibration of such a thermal history requires additional constraints. For example, Royden and Keen (1980) proposed extensive subcrustal extension beneath the Labrador margin to explain initial uplift and erosion of the region. Such a model should have obvious effects on thermal indicators carried by the sedimentary sequence and should be testable. Such quantitative evaluations are necessary if geologic models are to be used as reliable hydrocarbon exploration and prediction tools away from development and COST wells.

D. Heat Flow from Vitrinite Reflectance

The fact that the various geologic models we have discussed disagree even on the sign of the derivative of heat flow at the present time makes it necessary to have independent indicators of heat flow. The most important of these is probably vitrinite reflectance. The reflectance of vitrinite is determined by the relative index of refraction of vitrinite in oil. Because the refractive index n_0 of the immersion oil is known (typically n_0 is 1.518), the reflectance is determined entirely by the index of refraction of the vitrinite, which depends on the number density of the reflecting molecules and on the oscillator strengths of the molecules. Toth *et al.* (1983) and Lerche *et al.* (1984) have shown that it is possible to obtain the initial number density multiplied by oscillator strength from the depositional surface value of vitrinite reflectance. Using the constant thus determined, the change in vitrinite reflectance with depth is determined by the change in number density with depth. This variation in number density is due to the change in number density of optically reflecting molecules due to first-order chemical kinetics, which are temperature dependent. Because we know the measured reflectances, it is possible to determine the temperature as a function of time.

The physical theory of vitrinite reflectance can be used to determine the heat flow $Q(t)$. Assume a heat flow that is characterized by several different parameters. Then, by least-squares optimization of these parameters (by comparing predicted vitrinite reflectance to those measured in a borehole), an optimized Q as a function of time is obtained (Toth *et al.*, 1983; Lerche *et al.*, 1984). For further details of the vitrinite reflectance theory and its implementation, the reader is referred to Chapter 3, where several thermal indicators are discussed. Figure 2.10 shows the use of this theory in making maps of vitrinite reflectance through time for specific formations. One point is worth mentioning here and is emphasized in Lerche *et al.* (1984). We can trust determinations of paleotemperature and paleoheat flow no further than we can trust the quality of the vitrinite reflectance data, which, most often, could be significantly improved. Errors in determination of paleoheat flow can then be ascribed both to errors in burial history evaluation and to errors in data input.

Other indicators provide measures of the variation in heat flow with time. Some of these indicators are (1) the transition of smectite to illite; (2) the variation of porosity with depth and temperature; (3) the molecular

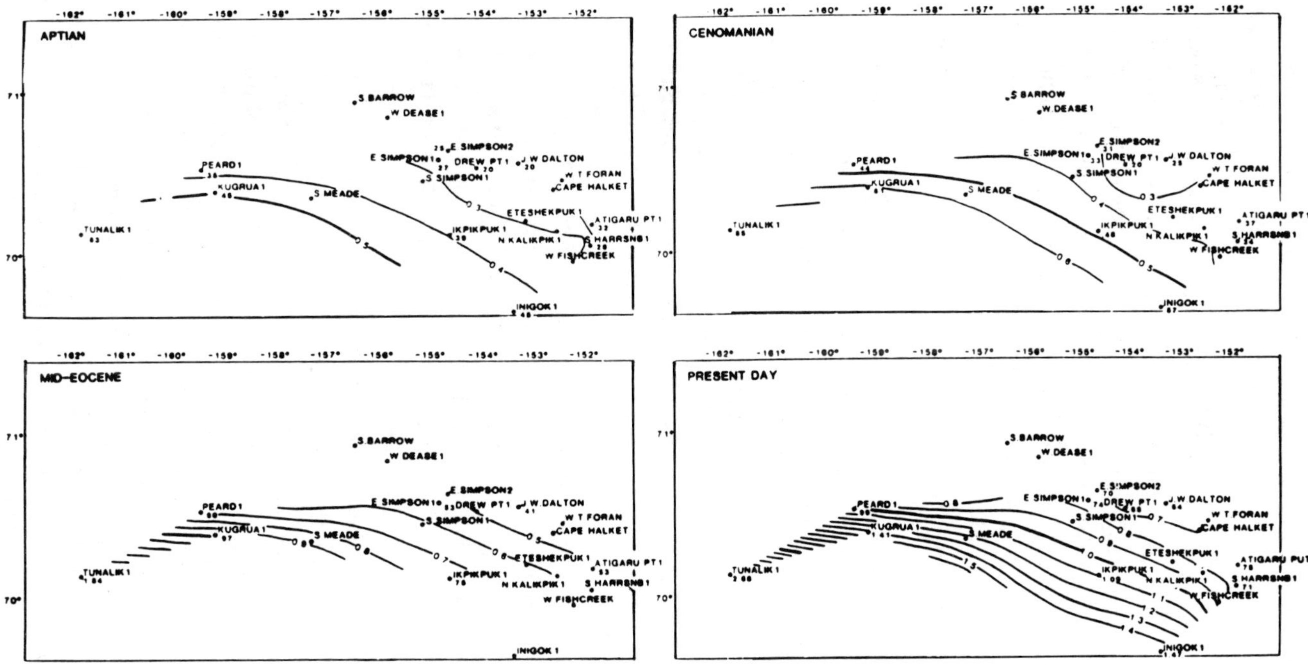

FIGURE 2.10 Contour map of vitrinite reflectance through time of Anisian sediments in National Petroleum Reserve of Alaska.

optical activity of oils as a result of asymmetric carbon atom positionings; (4) the even–odd aliphatic chain effect in oils; (5) the molecular weight distribution in oils; (6) $^{13}C/^{12}C$, $^{18}O/^{16}O$, and $^{40}Ar/^{39}Ar$ isotope anomalies with depth; (7) annealing of fission tracks in apatite, zircon, and sphene; (8) magnetic history; (9) fluid inclusions; (10) pollen translucency; and (11) biomarkers, such as sterane and hopane isomers, and sterane aromatization. Vitrinite reflectance is a generally available indicator from borehole measurements. However, this method needs to be tested against independent thermal indicators such as the annealing of apatite fission scars (Gleadow *et al.*, 1983), the variation in $^{40}Ar/^{39}Ar$ (Harrison and Be, 1983), temperature data extracted from fluid inclusions (Burruss *et al.*, 1983), and magnetic history as tied to thermal events (Middleton and Schmidt, 1982).

Thermal indicators provide knowledge of the paleoheat flow as a function of time. This information has two uses. First, a basin in which vitrinite reflectance is measured, or where some other thermal indicator is used, allows us to determine the heat flow and, hence, the thermal maturity of the acreage that is being explored. Second, for a given basin, knowing the heat flow allows us to evaluate quantitatively the geologic models that have been proposed for the origin of the basin. Thus, in principle, it is possible to determine which is an optimal geologic model for basins of a specific tectonic setting. Then, in other basins of the same tectonic setting, it is possible to use this optimal geologic model to make predictions of maturity before drilling on the basis of seismic data. Thus it is possible to rank basin maturities in the absence of borehole measurements. Therefore, we have a predictive tool for use in hydrocarbon exploration. We shall consider the thermal inversion procedure in more detail in Chapter 3.

The results of a modeling program can be divided into three classes—burial history, thermal maturity, and combined outputs.

The burial history outputs can include the burial history versus time (i.e., the depth below depositional surface as a function of time in millions of years); geohistory or total subsidence plots (Fig. 2.1) (i.e., depth below current sea level as a function of time); subsidence due to compaction; amount of subsidence due to isostatic response to sediment load; basement subsidence (Fig. 2.2) (i.e., the depth to basement with the sediments removed as a function of time); rate of sediment accumulation, subsidence, and basement subsidence as a function of time; and sea level plots as a function of time.

Thermal maturity calculations and plots can include the time–

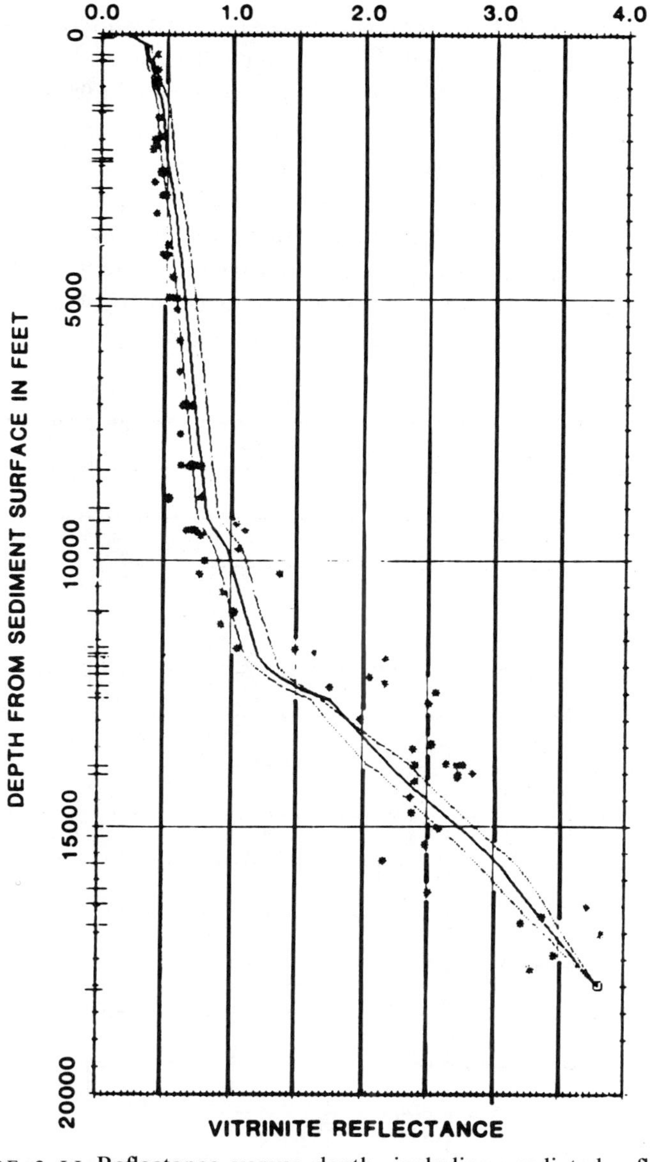

FIGURE 2.11 Reflectance versus depth, including predicted reflectance line, for Inigok 1, Alaska. Asterisks are measured vitrinite values; dark line is predicted vitrinite reflectance; dotted envelope is error envelope. Slopes of line and envelope owe their variation, in part, to rates of sediment accumulation and burial.

temperature integral (TTI) of Waples (1980) versus depth at present, vitrinite reflectance in percent versus depth at present (Fig. 2.11), vitrinite reflectance versus TTI (Fig. 2.12), and percent of kerogen converted to oil or gas for each layer and each kerogen type (types 1, 2, and 3) as a function of time. At least *seven* possible options of heat flow as a function of time could be modeled: (1) constant temperature gradient, (2) present heat flow assumed not to vary with time and determined from corrected bottom-hole and/or formation temperatures and lithologies, (3) heat flow from the McKenzie crustal stretching model with parameters determined by the least-squares optimization of model basin subsidence and back-stripped subsidence, (4) heat flow determined from the Falvey crustal metamorphism model with least-squares optimization of parameters,

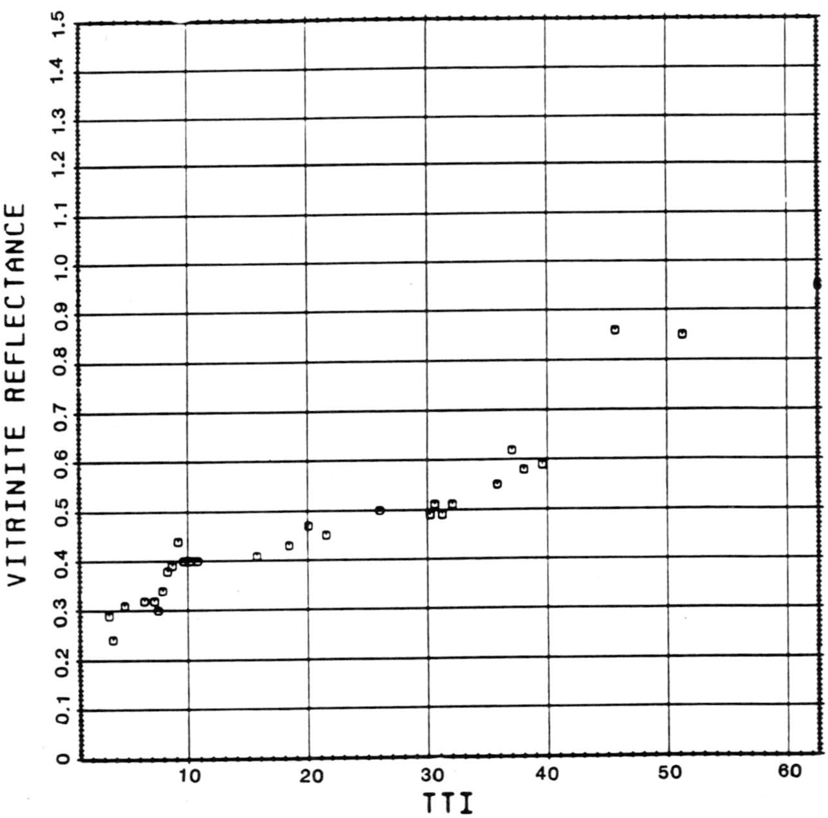

FIGURE 2.12 Cross plot of measured vitrinite reflectance versus Waples' TTI (calculated at present time assuming a McKenzie crustal stretching heat flow model with new parameters) for well A in the North Sea.

(5) heat flow determined from the Royden dike intrusion model with least-squares optimization of parameters, (6) heat flow determined from the Hellinger and Sclater (1983) two-layer extensional model, and (7) heat flow derived from vitrinite reflectance data or some other thermal indicator.

Combined thermal history and burial history calculations and outputs can include burial and/or geohistory and thermal maturation (i.e., percent kerogen converted to oil and/or gas for various geologic eras), burial history, and Waples TTI calculations. Plots of constant TTI can be produced and outlined on geohistory plots (Fig. 2.13). One can construct plots or make calculations of predicted basin subsidence based on the McKenzie crustal stretching (Fig. 2.6), the Falvey metamorphism (Fig. 2.7), or the Royden dike intrusion model as a function of time. Also, the heat flow can be calculated as a function of time based on the optimized McKenzie (Fig. 2.8), Royden, or Falvey model (Fig. 2.8) and can be

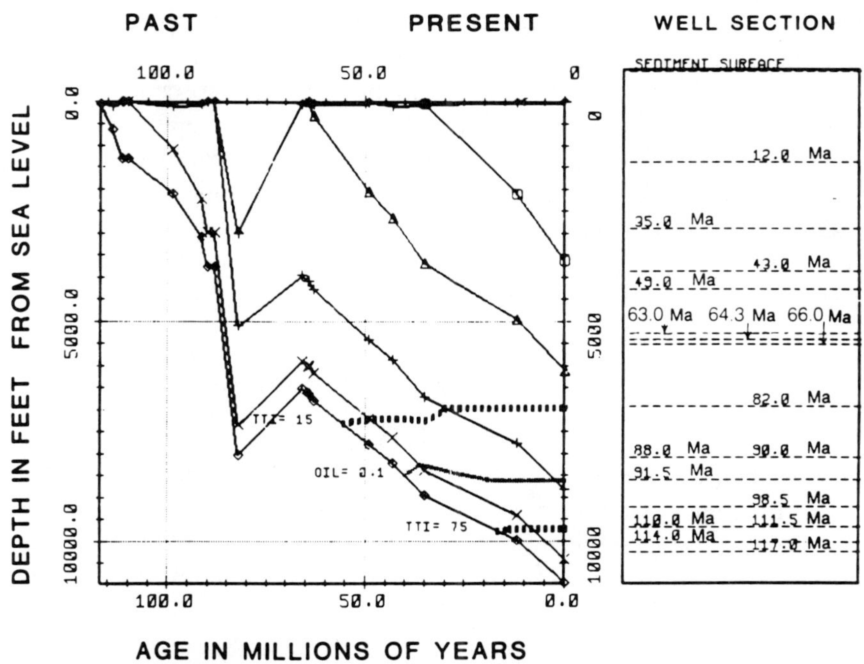

FIGURE 2.13 Machine-generated geohistory plot of Mideast well. TTI = 15 and TTI = 75 curves represent Waples' thresholds for onset and maximum generation of oil, respectively. Oil = 0.1 curve shows time and depth at which 10% of present kerogen has been converted to oil (according to Tissot and Welte, 1978).

plotted. If vitrinite reflectance data are available, one can plot both predicted heat flow as a function of time and predicted reflectance versus observed reflectance and can also obtain isoreflectance lines on the burial and geohistory diagrams.

All these different outputs can be used to determine (1) more accurate geologic models for a specific depositional basin, (2) more accurate tectonic and structural histories of a basin, (3) the sedimentary history of the basin from maps of sediment accumulation, (4) the thermal history of the basin and its effect on organic matter, and (5) the compactional history and so the possible migration history within the basin.

These outputs can be converted to maps. Oil fields can be superimposed on these maps and this information, combined with all the other mapped data, will identify hydrocarbon trends within specific geologic plays and will enable the ranking of the more prospective acreage within a basin. These different applications directly affect hydrocarbon exploration. Indirect applications include the ability to model tectonics and the physics of the crust and mantle, and to test these models using the burial and thermal histories approach. Similarly, one can model the effects of sea level variation and predict facies patterns of carbonates, clastic, evaporites, and deep-sea chemical sediments, such a metallic nodules.

Other parameters that might be extracted from burial history calculations include the thermal history of hydrocarbons from potential source rock to potential reservoir, the reservoir quality of a formation as a function of time, the acoustic impedance of the sediment as a function of depth, and fluid and gas lost from the formation as a function of time.

II. Basin Analysis of the South Mozambique Graben

In this section we illustrate how mathematical modeling can enhance the understanding of the hydrocarbon potential of a basin, given only limited data. Currently, one-dimensional burial history models assume that sediments must have been deposited, not deposited (hiatus), or eroded during each time step; meanwhile, the structural development has been largely ignored.

The primary goal is to demonstrate a basin analysis technique which marries the structural, depositional and thermal histories of an area. Such an integrated approach may greatly contribute to an exploration geologist's understanding and assessment of hydrocarbon potential of a speculative basin.

The method consists of simultaneously removing depositional layers and reconstructing structural cross sections in sequential chronostratigraphic steps back in time.

Before the technique was applied to real field data from the South Mozambique Graben, the problems of backstripping wells which penetrate faults were investigated by means of a series of tests on synthetic examples and this aspect is discussed first.

A. Backstripping Wells which Intersect Faults: Tests on Synthetic Examples

The synthetic scenario designed for the initial experiments consisted of a simple fault block which had been rotated on planar growth faults in domino style (Fig. 2.14). The burial history of this system began with deposition of three prerift layers (1, 2, and 3 in Fig. 2.14), each of which represented 5 million years. These layers were followed by deposition of two synrift layers (A and B in Fig. 2.14), also equivalent to 5 million years each. These secondary layers were of different lithology from the first three.

Pseudowells positioned across the structure were backstripped using an isostatic burial history program similar to that of Guidish *et al.* (1984). The decompaction of each chronostratigraphic layer was taken into account by using the porosity–depth functions of Sclater and Christie (1980).

After each unit had been stripped off, a new cross section was drawn (Fig. 2.14b) until the throws on the faults were restored at the surface (Fig. 2.14c). Throughout this analysis the positions of the pseudowells remained fixed.

Comparing Figs. 2.14a and c, it can readily be seen that the use of a one-dimensional burial history program to decompact and backstrip wells that cross faults presents problems. Specifically, the angles of the faults have increased, the fault traces have become curved, strata below the faults have not been properly restored, and the faults have "migrated" westward, toward the downthrown side.

These results show that the sensitivity of the pseudowell backstripping technique needed to be estimated before an actual situation could be analyzed using the one-dimensional burial history model. A sensitivity study was undertaken for a sequence of experiments similar to the initial case, where the individual effects of lithologies, spatial scale, and initial fault angles were investigated.

Two extreme lithologic cases (homogeneous shale and homogeneous sandstone), along with two variations in the present-day fault angle (40°

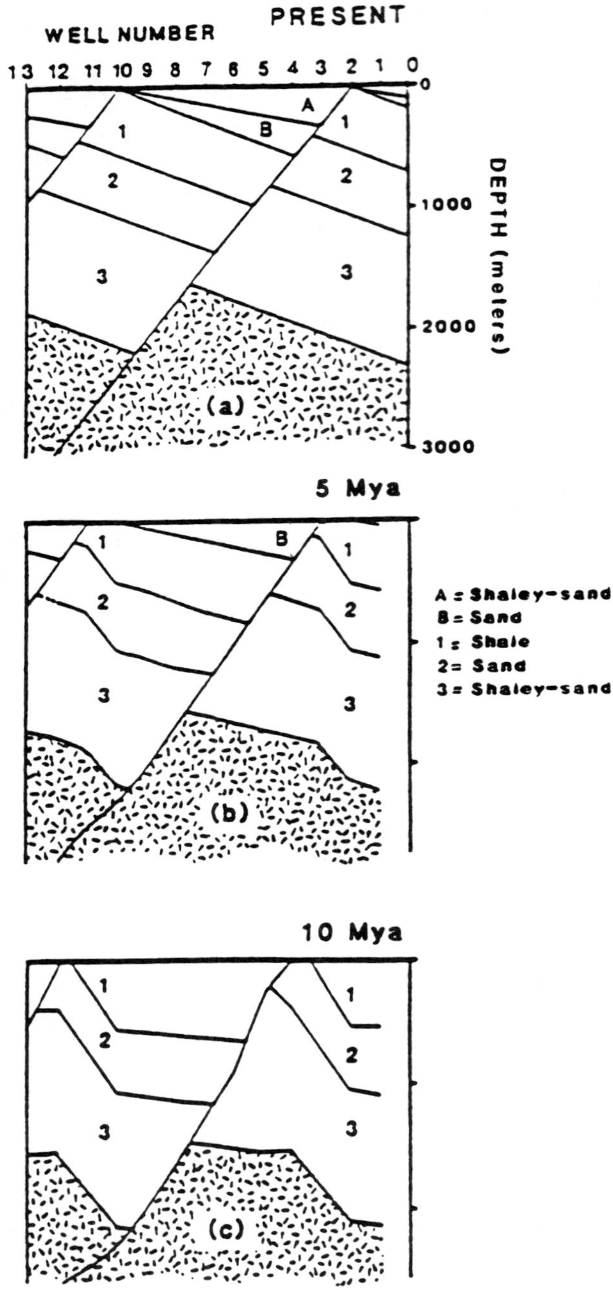

FIGURE 2.14 (a) Basic structural cross section showing the location of pseudowells used in all of the experiments. Note that in some experiments the fault angle and scale of the system vary. (b) Situation after removing layer A corresponding to the scenario 5 Mya. (c) After removal of both the syn-faulting layers A and B (10 Mya). Note that layers 1, 2, and 3 are considerably discordant. Note also the curvature of the fault trace and that sediments beneath the fault are not properly restored.

and 60°) and two different system scales (1 : 50,000 and 1 : 25,000), were backstripped and reconstructed. To summarize these results (Fig. 2.15), shallow faults are more affected by compaction than steeper faults, the angle of the fault has more variation up the fault in shales than in sandstone, and the variation up the fault is greater in the larger-scale model.

The problem of restoration of strata below the fault, noted in the first case (Fig. 2.14), was solved by treating the fault as if it were an unconformity at depth. Sediment is essentially considered to be "erod-

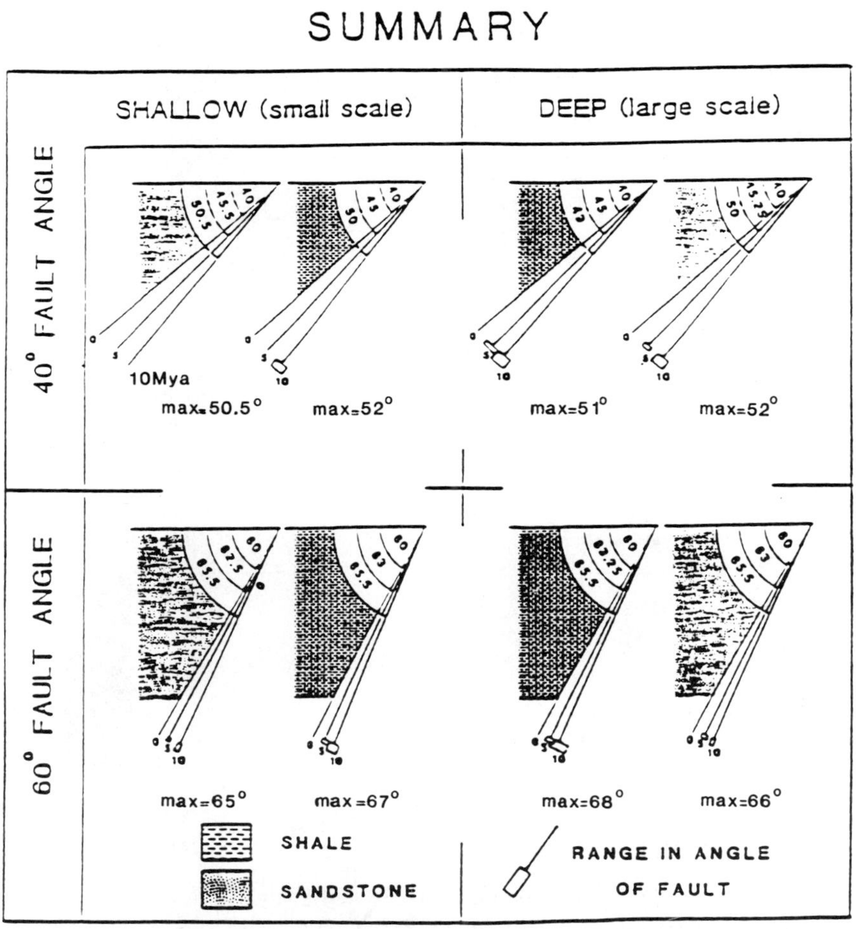

FIGURE 2.15 Summary of results of experiments considering the effects of lithology, fault angle, and depth on the backstripping of the input system shown in Fig. 2.14.

ing'' from beneath the fault, while deposition is occurring above the fault (Fig. 2.16). The values for depths to the strata are read from the burial history plots. The cross sections at time periods 5 and 10 Myr resulting from using this procedure (Fig. 2.17) were contrasted with those of a geometric solution. Comparison of the throws of the faults at different stratal levels for each solution (i.e., decompacted and undecompacted faults) at the 5-Myr time step (Fig. 2.18) shows that compaction reduces the throw of the faults and this effect decreases with depth.

Another important consideration is related to changes in the burial history after each reconstruction step, due to "contraction" of the basin.

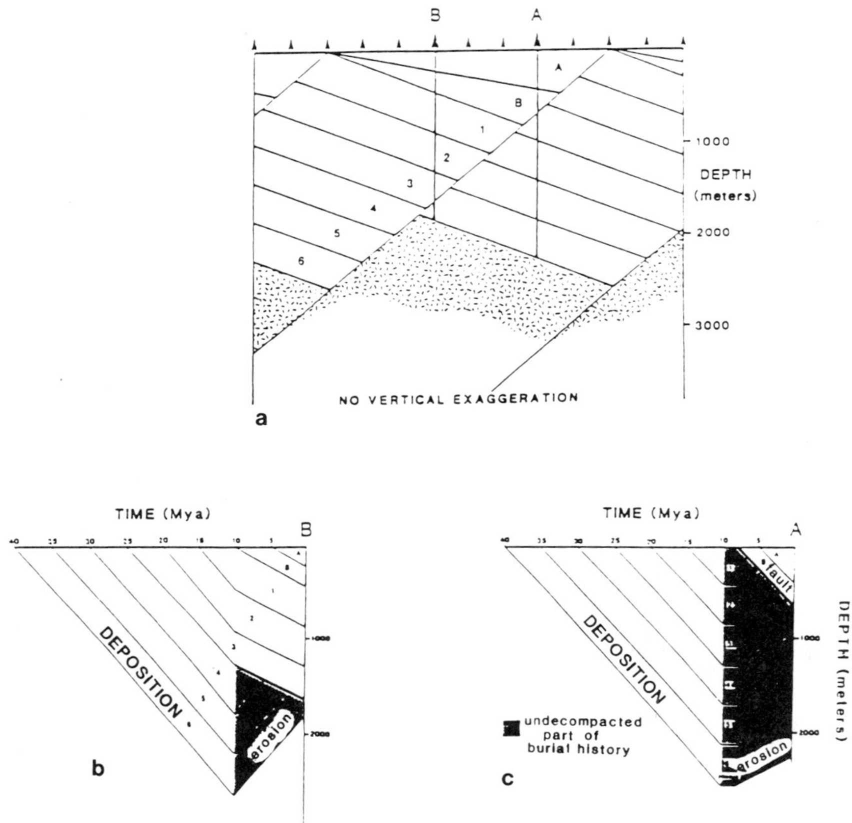

FIGURE 2.16 (a) Input system for the detailed analysis. Note positions of wells A and B. Layers 1 to 6 are prerift, A and B are synrift. (b) Burial history of well A incorporating the "erosion" below the fault solution to compensate for disturbances of bed seen in Fig. 2.14c. (c) Burial history of well B, also using the subfaulting erosion solution.

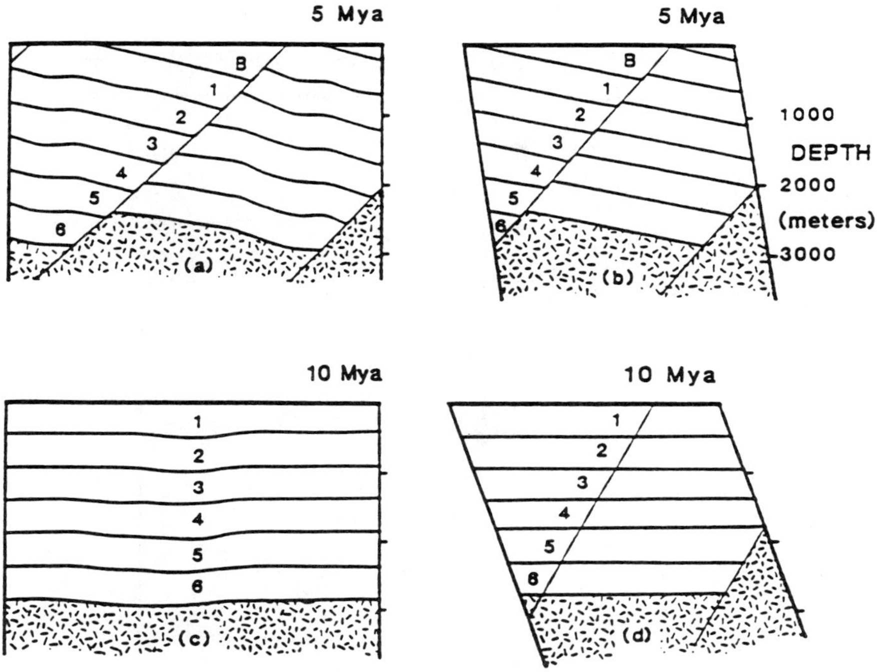

FIGURE 2.17 (a) Backstripped and reconstructed situation of Fig. 2.16a at 5 Mya. (b) Simple geometric solution (structural—not backstripped) of Fig. 2.16a at 5 Mya. (c) Backstripped and reconstructed situation at 10 Mya. Note waviness of strata compared with Fig. 2.16d. (d) Geometric solution at 10 Mya.

This is exemplified in Fig. 2.19, where well 1 burial history is radically altered from time 0 to 10 million years ago (Mya) (during the faulting). This problem is alleviated by keeping the position of the pseudowells fixed, "redrilling" each well after each layer of sediment has been removed, and then rerunning the burial history. The composite burial history of the wells in that one position is a better geologic representation than a simple "one-run" case. These constraint procedures may now be implemented on real geologic situations.

B. The South Mozambique Graben: A Case Study

1. Introduction

In this section we show how the hydrocarbon potential of a basin, the South Mozambique Graben, may be assessed by simultaneously back-

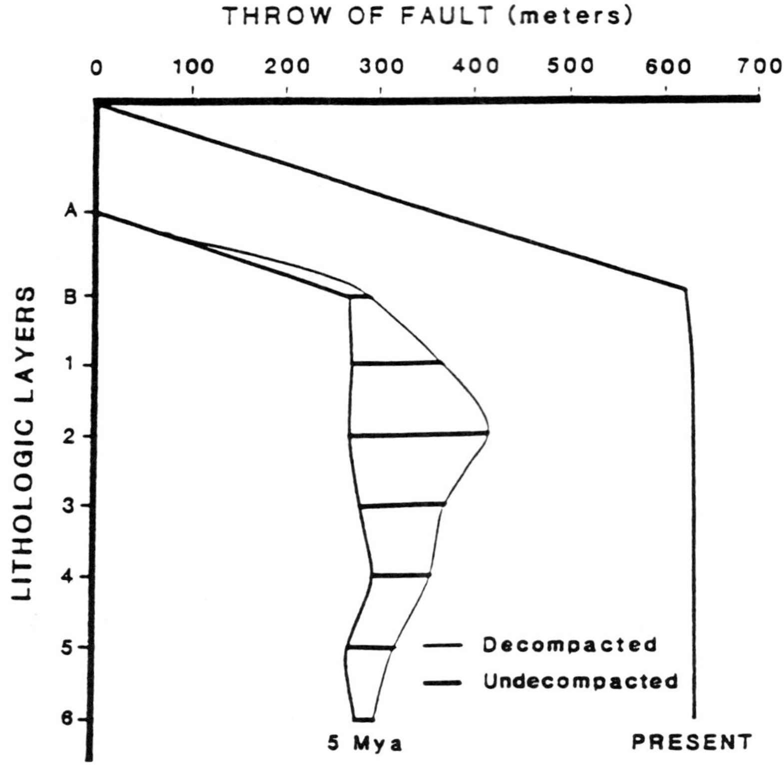

FIGURE 2.18 Comparison of the variation in throw of the fault in the undecompacted and decompacted solutions. Note that the throw is generally increased by decompaction, although this effect decreases with depth.

stripping and palinspastically reconstructing a cross section based on a seismic line perpendicular to tectonic strike.

One of the problems often encountered by exploration geologists is to make an assessment of the hydrocarbon potential of a speculative basin with limited data, often of poor quality. This study exemplifies such a situation and shows how basin modeling may improve evaluation based on the limited resources available.

The available data for offshore South Mozambique consisted of a few seismic lines, two shallow wells, and the general literature. In this study the tools applied to these data are a one-dimensional burial history program (Guidish *et al.,* 1984), a Tissot and Welte (1978) type of generation model, models of extension (Royden *et al.,* 1980; McKenzie, 1978), and some basic geological knowledge.

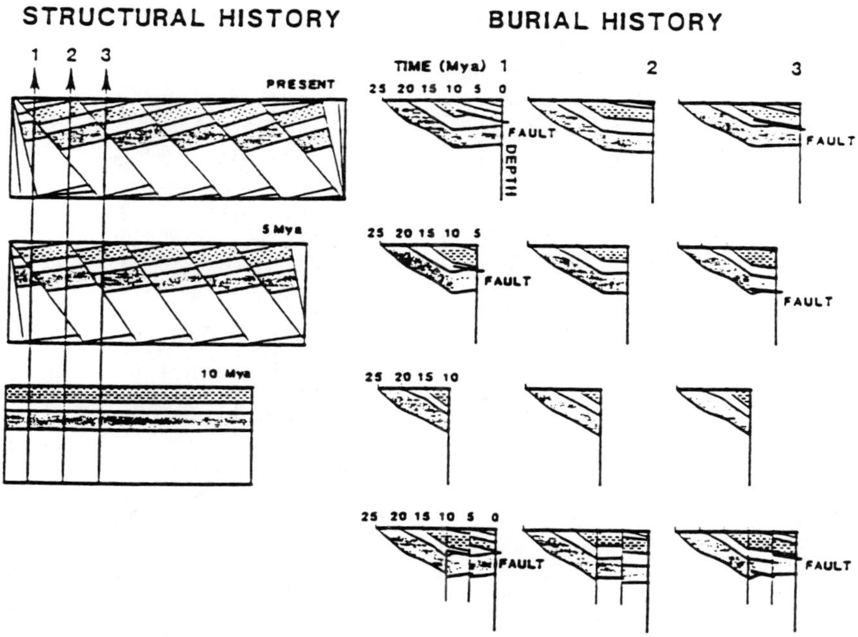

FIGURE 2.19 Variation in burial history at a location due to contraction during reconstruction. Note the changes in burial history between 5 and 10 Mya for each well. Compare with the composite burial history.

a. Geologic Setting. Initial interpretation of the data by DeBuyl and Austin (1983) revealed a V-shaped rift basin trending N–S to NNW–SSE (Fig. 2.20). The eastern border fault trends NNW–SSE, whereas the western border fault is oriented more N–S. The seismic data exhibit tilted fault blocks dipping to the east on westward-verging faults.

The tectonic setting of offshore Mozambique has not been presented coherently to date, mainly because of lack of data. Indeed, there is some debate as to the age of rifting and direction of stress fields (cf. Segoufin, 1978; Rabinowitz *et al.*, 1983; Mougenot *et al.*, 1986), although Darracott (1974) recognized that the area lies between the continental crust of Mozambique and the oceanic crust of the Mozambique channel, which separates Madagascar from the African continent. The general consensus is that Madagascar has been transported southward along a N–S-trending transform zone (Fig. 2.21). This movement has been dated on the basis of magnetic anomalies as being from 150 to 110 Ma (Segoufin, 1978) or 160 to 125 Ma (Rabinowitz *et al.*, 1983). It is also possible that initial rifting began as early as Karoo time (Permo-Triassic to Jurassic) or, more

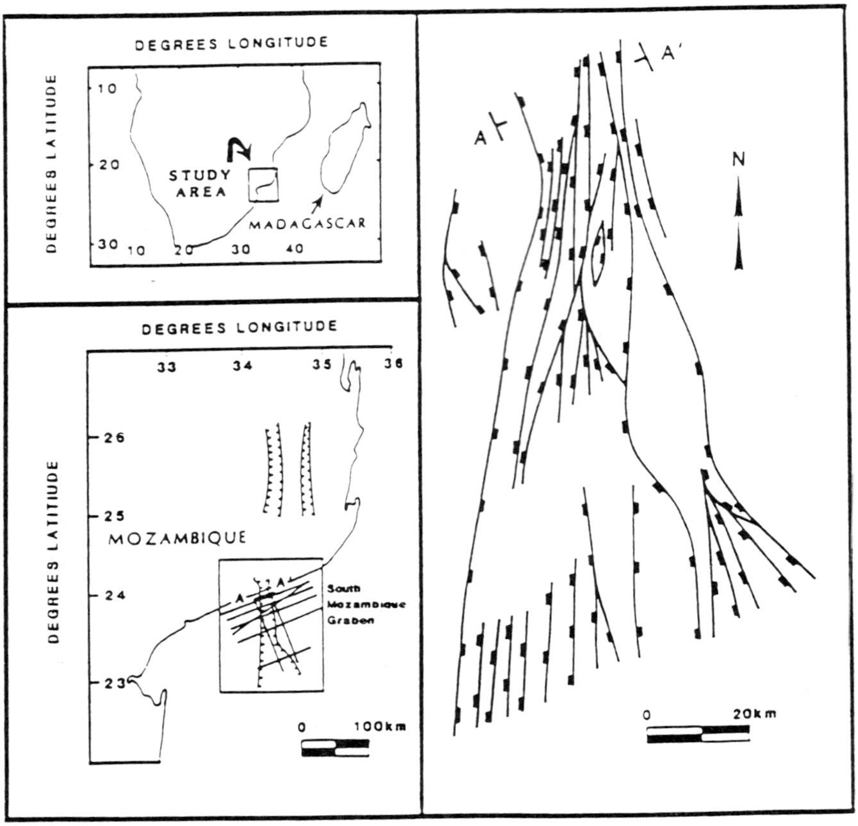

FIGURE 2.20 Location and structural trends which cut Lower Cretaceous strata in the South Mozambique Graben.

probably, is associated with the major breakup of Gondwana in late Jurassic to early Cretaceous times.

The NNW–SSE to N–S fault trends observed within the South Mozambique Graben originally may have been part of this transcurrent fault system, since flower structures are visible in deeper parts of E–W seismic profiles (Fig. 2.22). The transcurrent faults may have been reactivated later in a purely extensional regional stress field associated with the onset of Indian Ocean and South Atlantic spreading in late Cretaceous times, 100 million years before present (MYBP).

While describing the general problem of sedimentation for southern Mozambique it is useful to refer to the seismic character of the chronostratigraphic units used in this study.

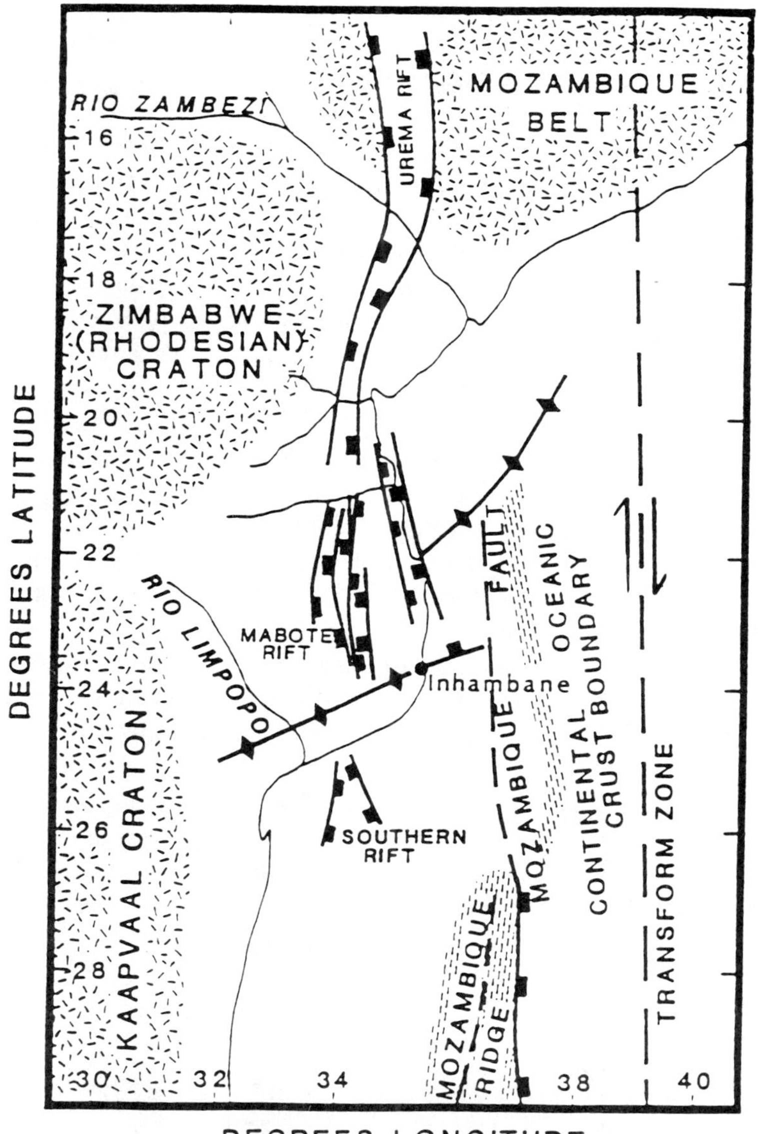

FIGURE 2.21 Main structural trends in the Mozambique region. Note the transcurrent faults in the area.

The oldest unit presumed present in the area is the Karoo (whose existence offshore has not been proved beyond doubt). This set of rocks consists of a sequence of continental deposits ranging in age from Permo-Triassic to Upper Jurassic. At the base of the Karoo is the Ecca series; coal measures, shale, siltstone, and sands. The overlying Beaufort series contains sandstones and shales and the Stormberg series is represented by volcanoclastic, lava flows and sandstones.

The Karoo unit may be present in the seismic profile, Fig. 2.22, making up part of seismic package 6. The reflectors in this package are structurally disturbed and stratigraphically discontinuous, typifying them as a synrift deposit. The age of the Karoo unit is speculative since it has never been penetrated. Reports from nearby countries with better data suggest that the Karoo was deposited during Gondwana breakup, late Jurassic to early Cretaceous times. Accordingly, this Karoo unit is taken to be 195 million years old at its base and 140 million years old at its top. This Jurassic synrift deposition is not reported onshore, probably because (1) most of the Gondwana breakup in the region now lies offshore and (2) onshore areas of this time were probably erosional highs manifested in the rock record as unconformities, as suggested by Forster (1975).

Following deposition of seismic package 6, a marine transgression begins in early Cretaceous times (Albian?) as recorded by the sands and shales of the Maputo Formation. Reflectors in seismic package 5 are discontinuous and structurally disturbed.

The Maputo Formation is overlain by the Lower Domo Shales Formation from Cenomanian to Turonian age. There is a bright doublet visible at the top and bottom of the corresponding seismic package (4), the middle being largely seismically transparent. Where visible, reflectors are discontinuous. The doublet at the top may be the mid-Turonian Domo sands.

The Domo sands doublet is truncated by an Upper Cretaceous unconformity which is overlain by the clayey silts, sands, sandy limestones, and marls of the Upper Domo Formation or Grudja Formation, depending on terminology. This unit (seismic package 3) has parallel, horizontal, continuous reflectors which may indicate a basin-type depositional setting (Flores, 1973). It is apparent from the seismic section in Fig. 2.22 that this unit was either very thin, eroded, or not deposited on the flanks of the rift.

Above the Grudja Formation lie the oolitic, algal limestones, marls and sands of the Eocene-age platform and ramp sequence of the Cheringoma Formation.

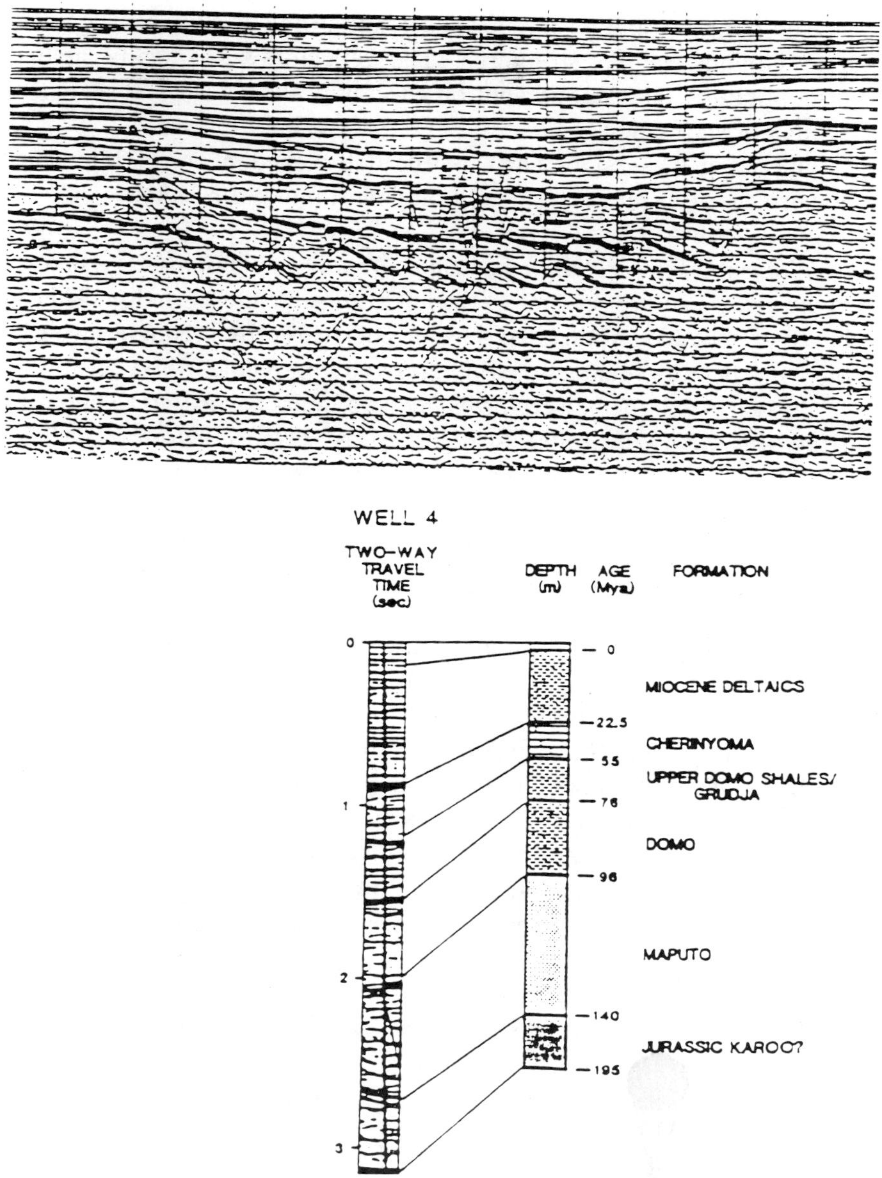

FIGURE 2.22 Interpreted seismic section trending east–west, showing flower structures at depth with more "domino" style faulting higher in the section. Note stratigraphy and time–depth conversion of well 4.

The ramp, visible on the northeast side of the seismic section in Fig. 2.22 contains slump structures. The reflectors are slightly inclined, discontinuous, and nonparallel on the ramp and faint or absent in the basinal areas.

Overlying these sediments are the deltaic sediments of Miocene to Recent age. Reflectors are gently dipping, thin, and parallel.

b. Seismic Interpretation and Time–Depth Conversion. The first step in the analysis of this basin was the initial seismic interpretation. This was based on the recognition of six seismic packages rather than following one or two reflectors regionally, which proved impossible in the rift itself.

Basic isopach maps were produced and faults correlated to determine structural trends illustrated in Fig. 2.20. A seismic section (A–A′, Fig. 2.20) perpendicular to tectonic strike was then selected. This seismic profile (Fig. 2.22) was converted to a true depth section using pseudowells 1 to 12 (Fig. 2.22). The pseudowells were located at shot points where stacked, interval velocities had been calculated. These velocities were then used to calculate the thickness of each layer. The accuracy of these thickness estimates depends on the accuracy of the velocities, which was not provided.

After obtaining a "true" section (Fig. 2.23a), the lengths of the picks were compared and the actual amount of extension as a ratio of the total extension was calculated for each time line, which may be converted into an extension rate diagram (Fig. 2.24a).

The structural interpretation was constrained by length balancing an imaginary reflector below and parallel to the deepest reflector package. The cross section is now ready for backstripping and reconstruction.

c. Backstripping and Reconstruction. The inputs to the Guidish *et al.* (1984) one-dimensional isostatic burial history program are depths to formation tops, absolute ages at these tops, layer lithologies, paleobathymetry of each layer, density of rock matrix, and porosity–depth relationships for various lithologies.

The wells backstripped in this study were pseudowells 1 to 12 on Fig. 2.22. Several assumptions are needed to develop the analysis. The lithology of the lowest layer (package 6) is presumed to be a synrift sequence of Jurassic sands.

Owing to the lack of downhole sonic and density information, the lithological porosity–depth functions were defaulted to those available in Sclater and Christie (1980). Layers of thickness 1000 m or more are taken

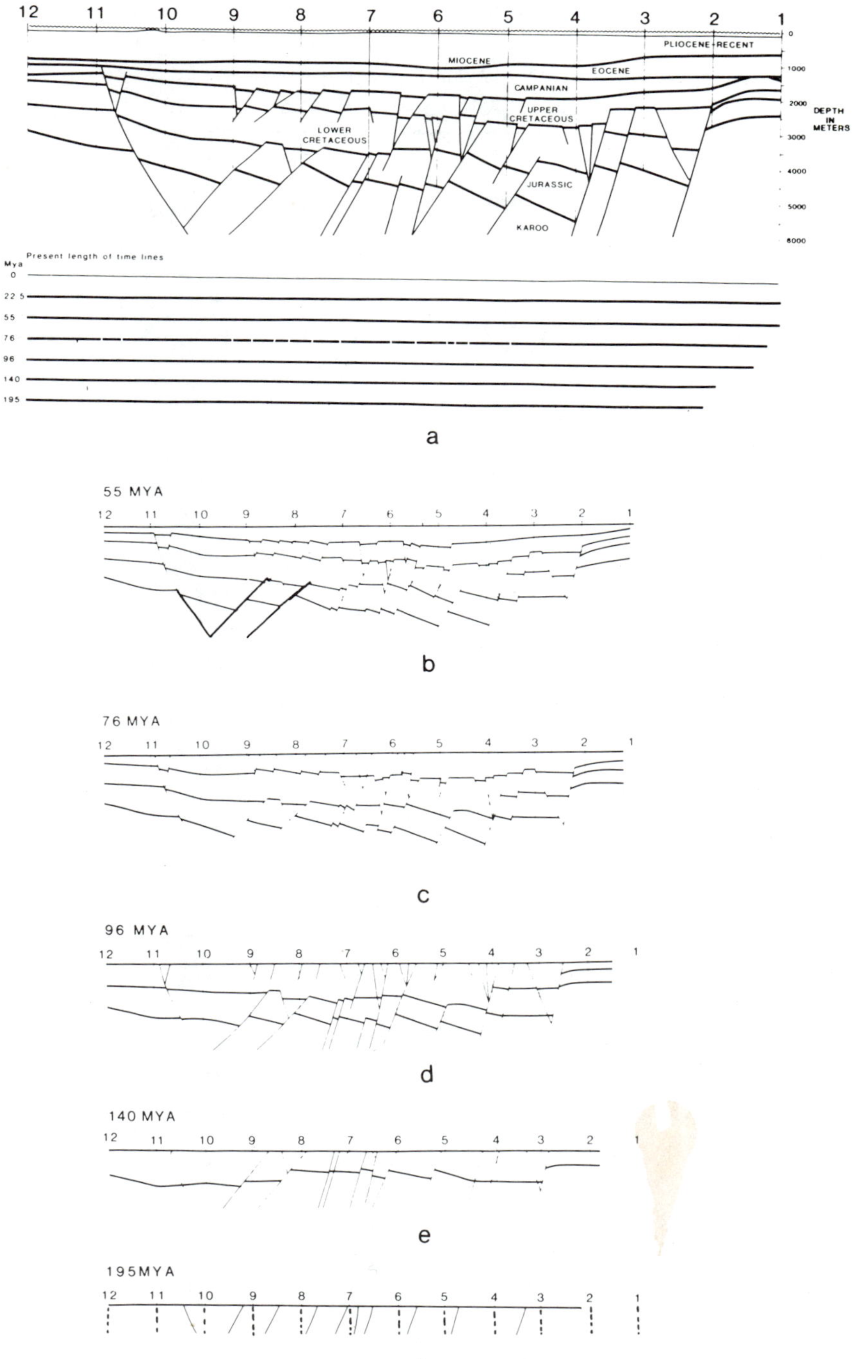

FIGURE 2.23 (a) Depth-converted section. (b) Cross section just prior to Eocene deposition 55 Mya. (c) Cross section at 76 Mya. (d) Cross section at 96 Mya. (e) Cross section at 140 Mya. (f) Cross section at 195 Mya.

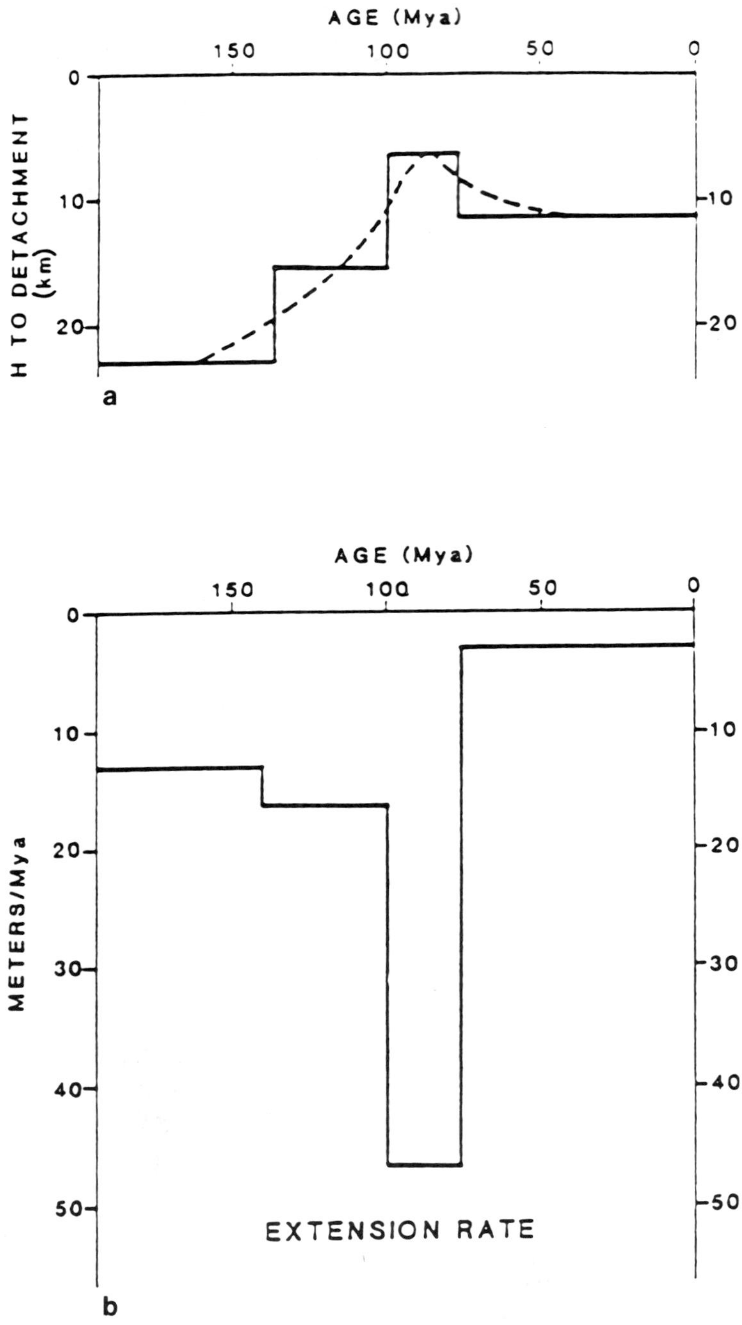

FIGURE 2.24 (a) The extension rate shown for each time period of the backstripping. (b) Variation of the depth to detachment for each time period of the backstripping.

to be monolithological. This is obviously in error; in general, however, such a working hypothesis provides an evaluation in keeping with the resolution of the bulk of the data.

More critical to the model is the estimation of absolute ages, which is precisely where the limitations of the data are hardest felt. These effects will be discussed below. Paleobathymetry estimation is again based mainly on the geologic interpretation and literature.

The backstripping program was run on all 12 wells, after which the cross section was redrawn (Fig. 2.23b). Since no faulting was evident after 55 Mya, no rebalancing was necessary after package 1 was removed.

When the second seismic package was backstripped (76 to 55 Mya), the throw on some faults had to be adjusted (Fig. 2.23b). The removal of throw on these faults slightly contracted the section such that the wells that crossed a fault had to be "redrilled." This process of backstripping, restoring, rebalancing, redrawing, and redrilling was repeated until the top of the Karoo reflector appeared as a horizontal line at the earth's surface.

Each cross section (Fig. 2.23a–f) was then analyzed for depth to detachment and length of uppermost bed. The area balancing formulas were taken from Gibbs (1983). The final lengths used in the depth to detachment calculations were between the border faults in the present-day true section, that is, within the zone of deformation. The areas were measured using a digitizer and an area-calculating program.

2. Results and Discussion of Structural and Depositional Reconstructions

When considering the variation of extension rate with time it was found that there was a slow acceleration of extension from 195 to 76 Mya, the acme of rifting being between 96 and 76 Mya (Fig. 2.24a). It was also noted that the depth to detachment seems to have shallowed from 195 Mya up to the period between 96 and 76 Mya and dropped gently ever since, as shown in Fig. 2.24b. It is significant that, even though the data are sparse and of poor quality, events of high accumulation, rapid subsidence, and quickest extension are nearly coincident at 96 to 76 Mya.

The slow extension rates from 195 to 96 Mya may be a manifestation of the early subtle, probably strike–slip, faulting, sympathetic to the southern emplacement of Madagascar. Evidence supporting this idea comes from the occurrence of flower structures in the deeper parts of the rift. The vigorous extension of the rift in late Cretaceous times resulted in the blocks being tilted in a more systematic normal domino fashion than in the previous event, probably related to the opening of the South Atlantic at this time. It may be significant that Upper Cretaceous volcanics occur at this time also (wells S-2 and S-4) (DeBuyl and Austin, 1983).

The depth to detachment in this area may represent the brittle–ductile transition zone. If so, then measured fluctuations in the depth to detachment may be an indication of the heat flow history of the region. Continuing this reasoning, the region may have experienced its highest heat flow during late Cretaceous times. This deduction is somewhat substantiated by the occurrence of Cretaceous sills and feeder dikes in the area (DeBuyl and Austin, 1983).

The fastest extension measured in the basin, some 50 m per million years (5 cm per 1000 years) is slow when compared to spreading rates of 1 to 17 cm per year for present-day opening oceans. This basin therefore does not appear to have been very active, and an extensional model to describe the heat flow and subsidence history must be applied with some caution.

3. Thermal History

Having assessed the structural and burial histories of the graben from the seismic line, some approximation to the thermal history of the basin must be made in order to apply hydrocarbon generation models in the area to complete the basin analysis. In regions with data that include downhole measurements of thermal indicators, it is possible to invert the information to obtain estimates of paleoheat flux (Lerche *et al.*, 1984). In lieu of any thermal information, the heat flow history of the South Mozambique Graben is estimated from the extensional models of McKenzie (1978) and Royden *et al.* (1980).

The geologic observations of several dikes or sills in the region, suggesting volcanic activity, favors the use of the Royden dike intrusion model. The McKenzie model does not involve any intrusive effects and is therefore probably not appropriate. However, in order to bracket the dynamic range of possible thermal histories, both the Royden and McKenzie models were used.

Another constraint to consider in refining an extension model is the heat flow history derived from depth to detachment calculations. Where the depth to detachment is known, and with the detachment assumed to be the brittle–ductile midcrustal phase boundary of greenschist to amphibolite metamorphic facies, a crude estimate of heat flux through time can be given using the equation:

$$Q/K = \mathrm{Gtz}$$

where Q is the heat flux at the surface, K the thermal conductivity, and Gtz the geothermal gradient. The variation of paleoheat flux will essentially match the shape of the depth to detachment; for example, the shallowest depth to detachment of 7 km yields the highest heat flow value of 2.8 to 3.6 HFU, taking the thermal conductivity as 5 mcal cm^{-1} $°C^{-1}$ sec^{-1} and the temperature at which the phase change occurs to be between 400 and 500°C. This would also imply a present-day heat flow value of around 1.6 to 2.1 HFU.

These models essentially predict what the heat flow must have been at a particular time after rifting of a basin, given certain parameters of that basin. Both models relate basement subsidence (tectonic subsidence) to the amount of extension and the heat flow. It is therefore possible to take the basement subsidence curves for each of the pseudowells, which are output from the burial history program (Fig. 2.25a), and find the best-fit predicted basement subsidence curve of the model that corresponds to a specific extension factor (Fig. 2.25b). Inspection of the basement subsidence curves warranted the division of the structural history into an initial rifting stage (195–109 Mya) and a later event at 96 Mya (Fig. 2.25). These two events correspond to known geological phenomena of an initial strike–slip, pull-apart basin setting, coincident with southward emplacement of Madagascar, and extension and volcanism in the late Cretaceous, respectively, as shown by seismic and well data in this study and reported independently by Mougenot *et al.* (1986).

In practice, the fitting of the observed to predicted basement subsidence curves has a degree of subjectivity. Therefore we bracket the fit to provide some idea of minimum and maximum error (Fig. 2.25b). For each well and for each of the two models there was an upper and lower limit of the possible extension factors. Each of these upper and lower extension factors was then used in the paleoheat flux equation of the model under consideration. The observed pattern of basement subsidence shows two events (Fig. 2.25a), but only the second has any direct supporting field evidence. To accommodate all possible variations, three general heat flow histories were considered (Fig. 2.26): (1) a single thermal heat pulse event at the original rifting (195 Mya); (2) a constant heat flux of 1 HFU until 96 Mya, at which time there was instantaneous rifting and an associated thermal event; (3) an initial thermal event at 195 Mya followed by a later thermal event at 96 Mya (instantaneous rifting). Such episodic rifting and subsidence has also been reported from the South China Sea by Ru and Pigott (1986).

Armed with these three possible cases and a range of extension factors for each of the events, a Tissot and Welte (1978) type of generation model

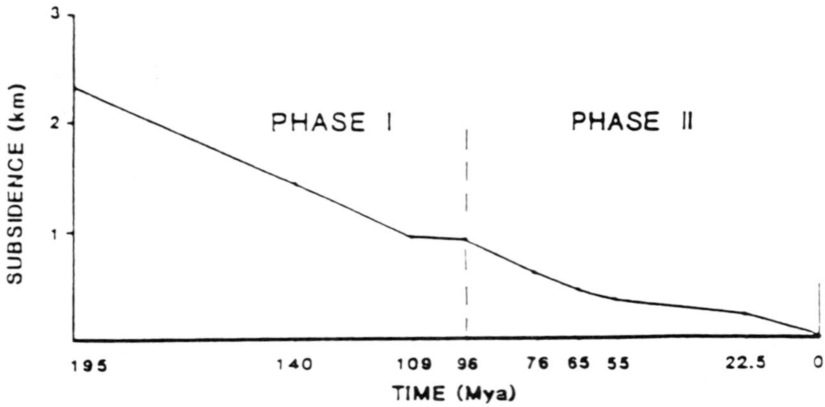

a

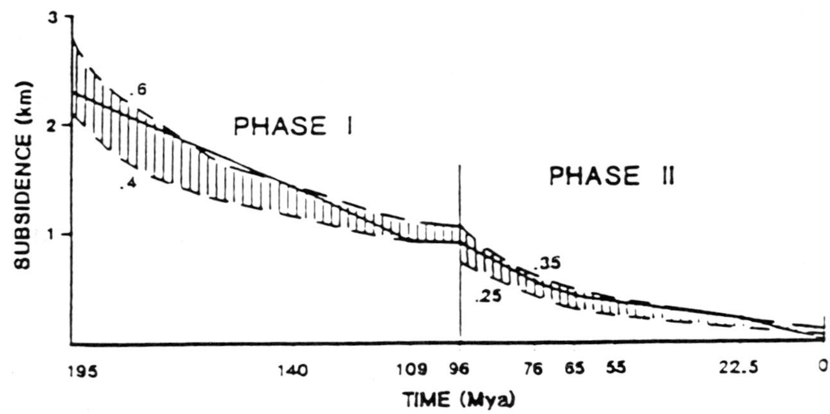

b

FIGURE 2.25 (a) An observed basement subsidence curve from well 6 shows two subsidence events. (b) Method of determining the range of extension factors from inspection of best fit of the observed curves to those predicted by the extensional models.

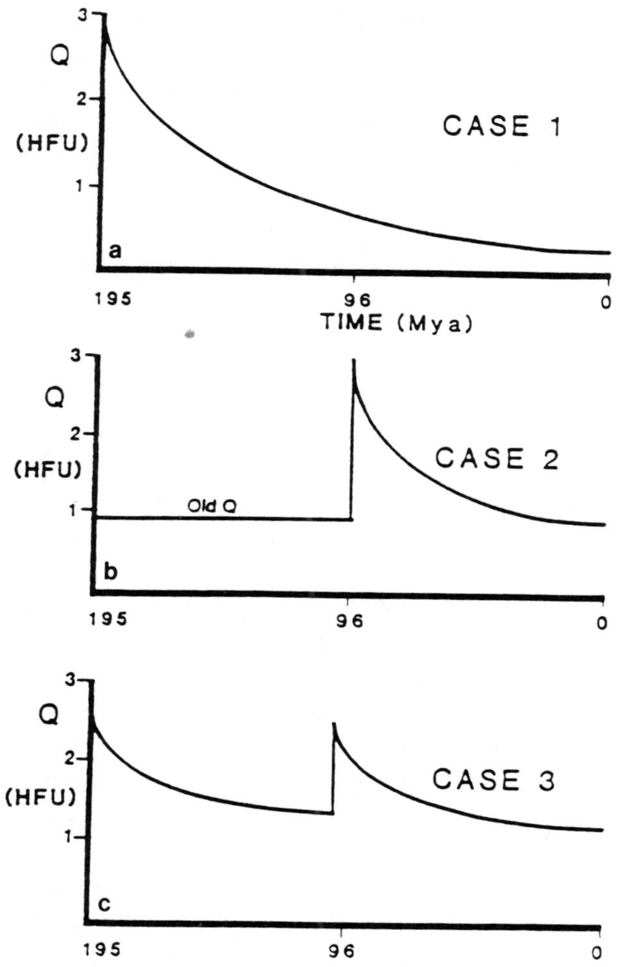

FIGURE 2.26 Three case histories: (a) an initial rifting and thermal cooling at 195 Mya; (b) a single pulse of rifting and heating at 96 Mya; (c) two rifting and thermal events at 195 and 96 Mya.

was applied to each well. The input of a heat flow curve of $Q(t)$ at time t was in the form of the equation

$$Q(t) = Q_0 \exp(\beta t)$$

where β is a variable which determines the slope and curvature of the line and Q_0 is the present-day heat flux. The extension factors taken from the models were used to formulate a β for the heat flow behavior with time in the model.

For the first two cases this technique posed no problems, as long as the present-day heat flow was known. A conservative estimate of 1.0 HFU was used for the present-day heat flux for lack of any data.

In the third case of two thermal events, it was necessary to calculate what the maximum heat flux would be at the inception of the second rifting, given the β value taken from the extension factors predicted from the observed basement subsidence curves for the first event. The maximum heat flow of the second event can be pinned by backtracking the equation from the present-day to 96 Mya with a known β as shown below:

$$Q_{R2} = Q_0 \exp[+\ \beta(T_{pd} - \mathrm{TR2})]$$

where Q_{R2} is the maximum heat flow at the second rifting event, Q_0 the present-day heat flow (1.0 HFU), T_{pd} the time from initial rifting to the present day, and TR2 the time from initial rifting to the second rifting event.

The Tissot and Welte (1978) hydrocarbon generation model is applied to all the strata. From the results of this approach we can ascertain which layers are within the oil window. Those layers may then be studied more closely for their actual source potential. The generation model predicts how many milligrams of oil are produced per gram of kerogen for each layer at each well site. This value was then plotted in the center of the formation for each well and contoured (see Figs. 2.27 to 2.30) to give maturation potential cross sections.

Such maturation potential sections were generated for each of the three cases of heat flow and for the upper and lower limits of each extension factor for both the McKenzie and Royden models of extension. In this way all the possible variations are taken into account and some degree of confidence may be placed on the results.

If there was only one rifting event at 195 Mya, Royden's model predicts that the basin is extremely oil prone (Fig. 2.27c and d) (if a source rock exists!). On the other hand, if there was only one thermal event of minimal magnitude at 96 Mya, McKenzie's model suggest that most of the basin is undermature, although still with some potential in the east, near wells 4 and 5 (Fig. 2.28a and b).

The results of the two-rifting-event case suggest that the later smaller heating event at 96 Mya has more effect on maturation of the sediment than the greater-magnitude, but earlier, rifting event (see Fig. 2.29a–d). During the initial thermal event the sediments either had not yet been

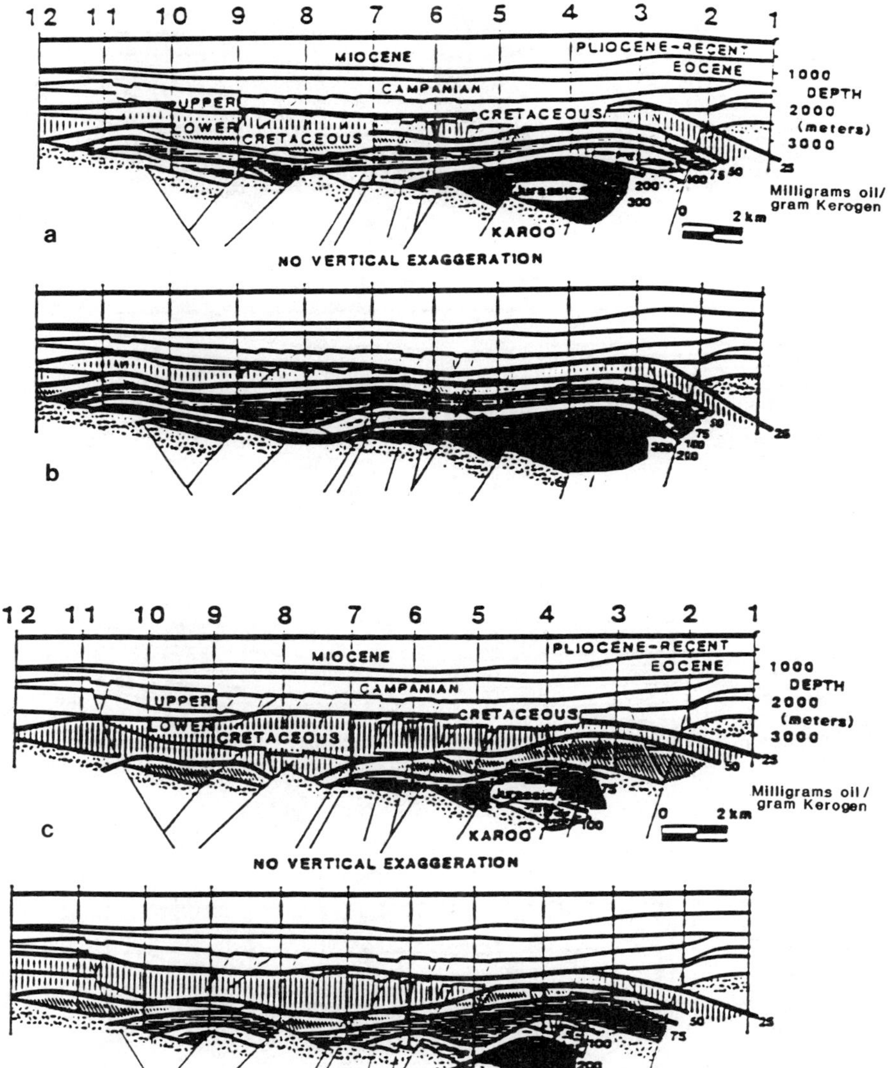

FIGURE 2.27 (a) Royden model lower limit, case 1. (b) Royden model upper limit, case 1. (c) McKenzie model lower limit, case 1. (d) McKenzie model upper limit, case 1.

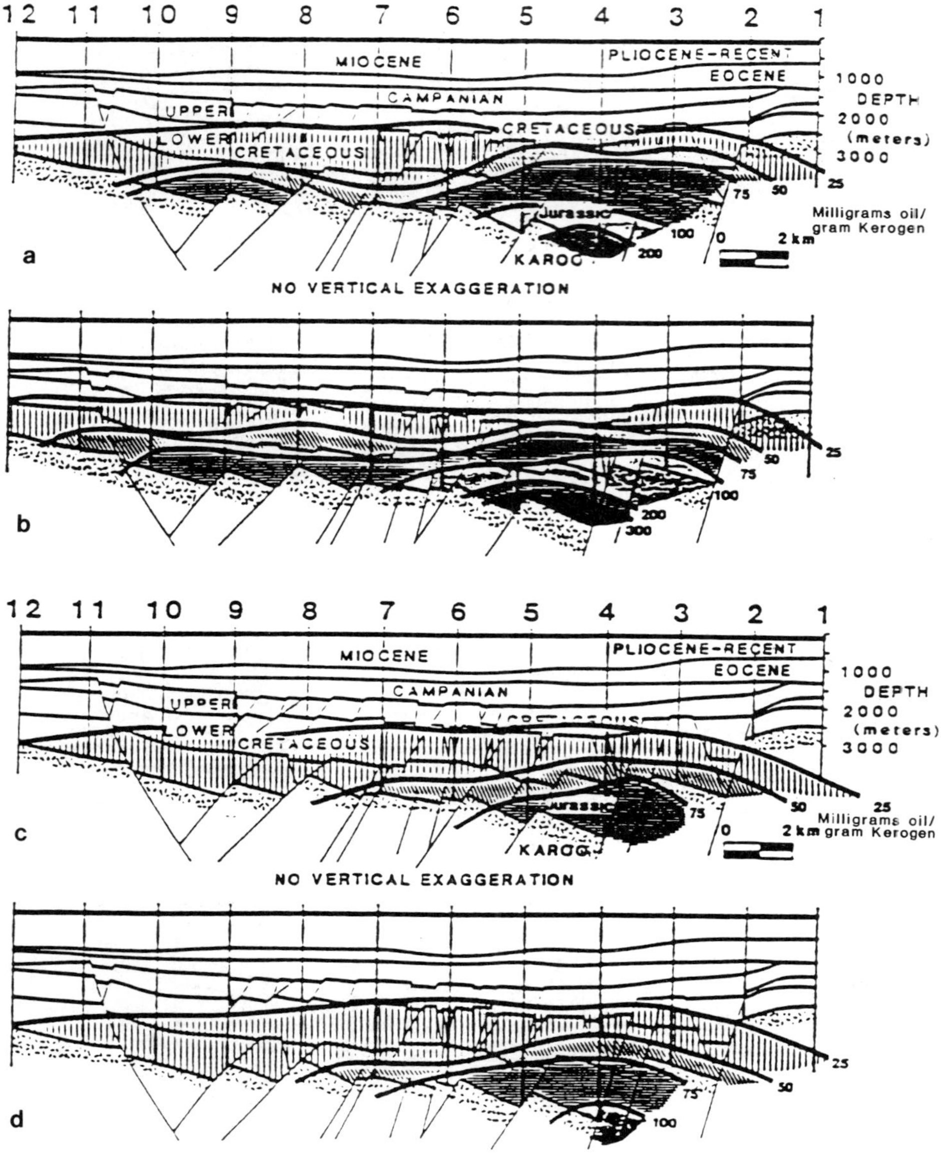

FIGURE 2.28 (a) Royden model lower limit, case 2. (b) Royden model upper limit, case 2. (c) McKenzie model lower limit, case 2. (d) McKenzie model upper limit, case 2.

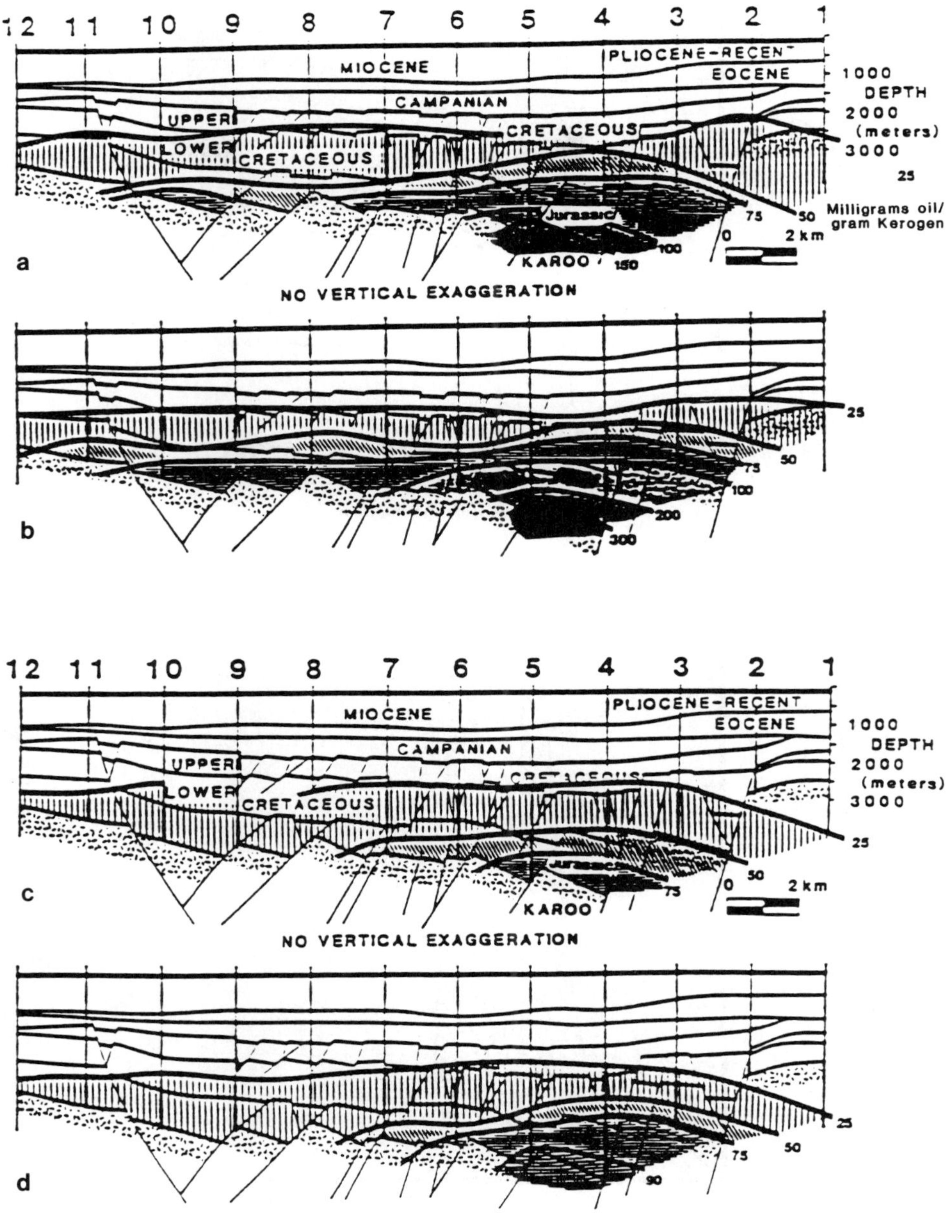

FIGURE 2.29 (a) Royden model lower limit, case 3. (b) Royden model upper limit, case 3. (c) McKenzie model lower limit, case 3. (d) McKenzie model upper limit, case 3.

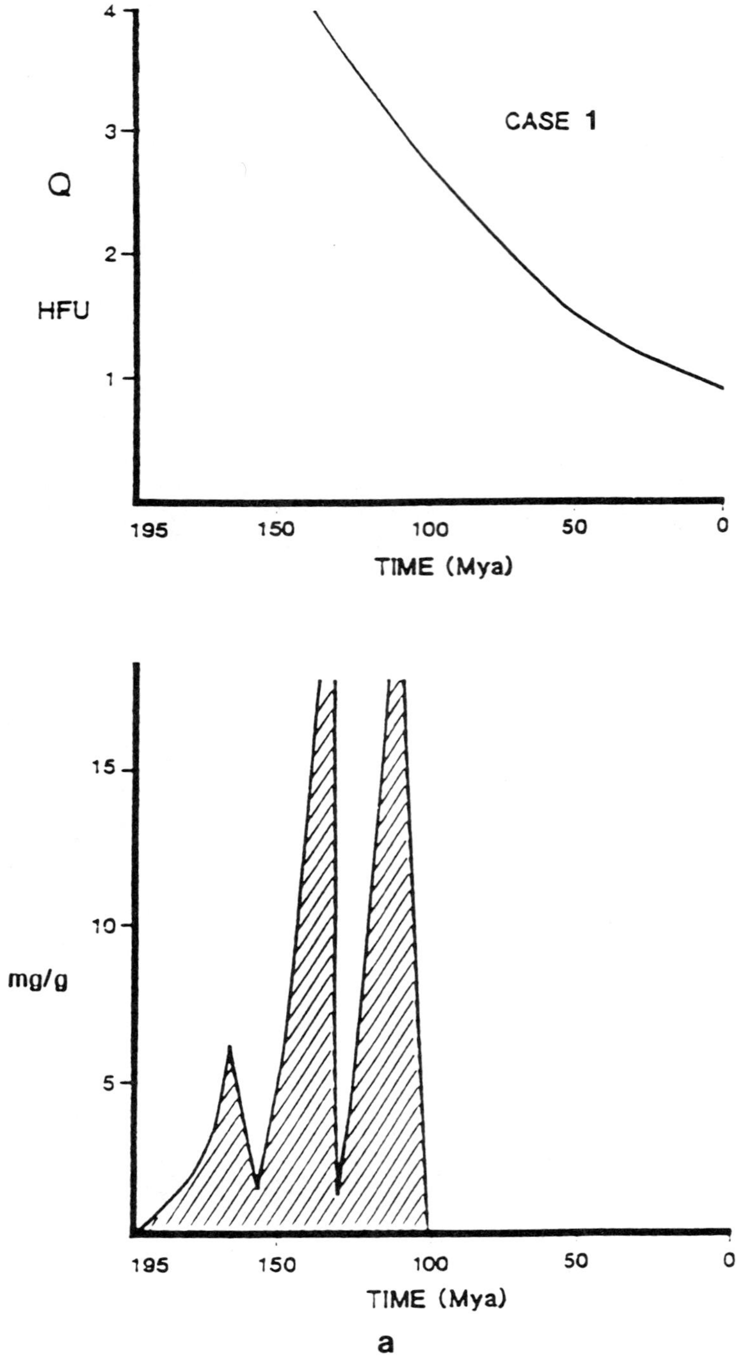

FIGURE 2.30 (a) Oil generation rate of the Royden upper limit, case 1, well 5. (b) Oil generation rate of the Royden upper limit, case 2, well 5. (c) Oil generation rate of the Royden upper limit, case 3, well 5. Oil generation values are in milligrams of oil per gram of kerogen.

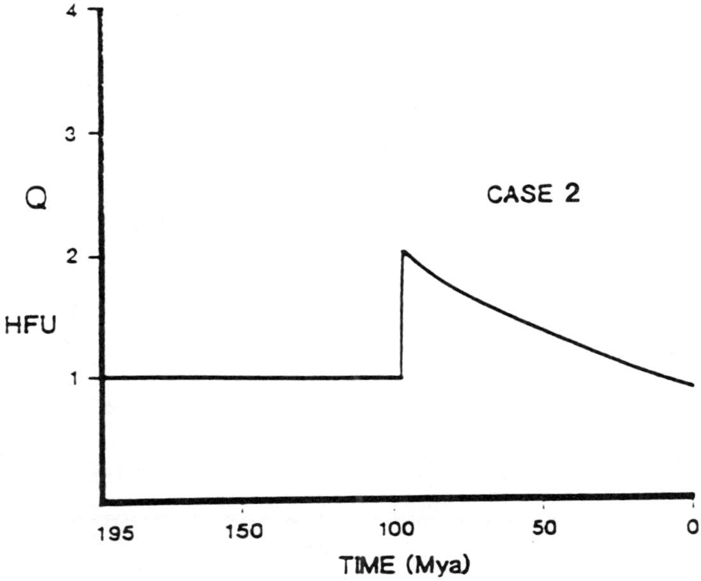

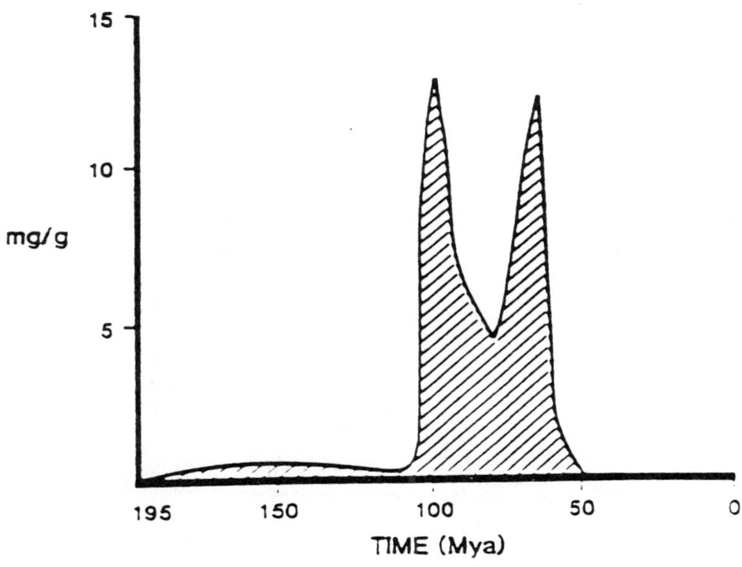

FIGURE 2.30 (*continued*)

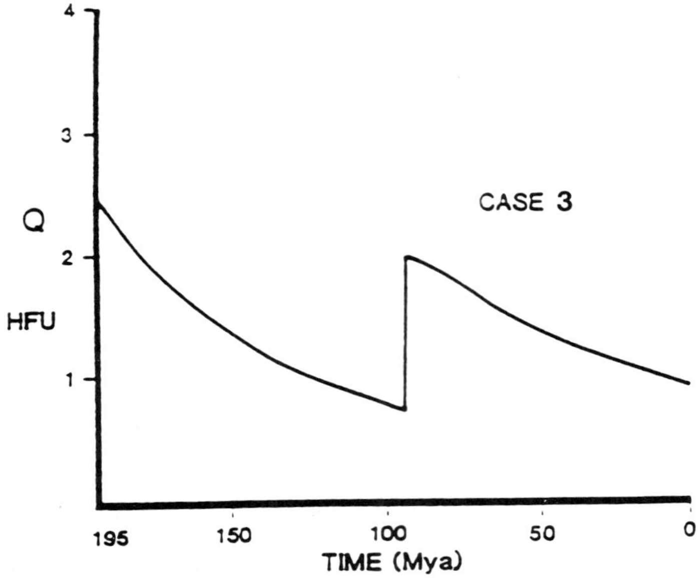

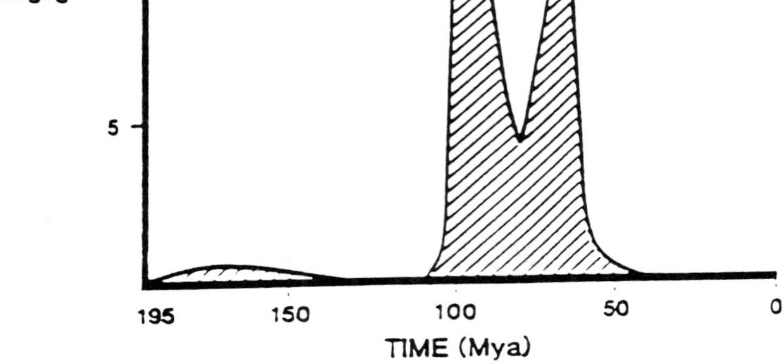

C

FIGURE 2.30 (*continued*)

deposited or were still at a relatively cool temperature near the surface. During the later event the same sediments were buried to a greater depth and so experienced higher temperatures.

The timing of generation shown for each case using well 5 (Fig. 2.30) also suggests that the second thermal event is more important for the hydrocarbon potential (Fig. 2.30b and c) of this basin since a later generation facilitates sealing and trapping. It is therefore significant that this event coincides with a raising of the brittle–ductile boundary and documented lava flows in the region.

The key is not how high the heat flow was, but rather how hot the sediments became and how long they were hot.

4. Basin Prognosis

To evaluate a basin for hydrocarbon potential we must answer five questions:

1. Does the basin have potential trapping structures? In the case of Mozambique the answer is strongly affirmative. This block-faulted type of structural regime has been exploited successfully for hydrocarbons in many areas such as the North Sea, Sirte Basin, and Gulf of Suez (Harding, 1984).
2. Does the basin have reservoirs? Most likely the synrift seismic package 6 may provide reservoir-quality sands and gravels in the form of fans at the sides of blocks.
3. Are sealing conditions present? The Lower Cretaceous and certainly the Upper Cretaceous shales are potential seals. In most cases the timing of generation from the lowermost potential sources is *after* the deposition of these shales (Figs. 2.30a and b), which would further enhance potential.
4. Is there a source rock? This is a question which may be answered definitively only with a well. However, the fact that copious shales are present (of both terrigenous and marine provenances) enhances the possibility of a high-quality organic source shale being present.
5. Has the source rock adequately matured? Based on the limits of these analyses, which should more than adequately cover the real situation, hydrocarbons will have been produced from Lower Cretaceous or Jurassic source rocks.

The advantage of the technique is in the assimilation of all aspects of the basin to tie down the relative timing of structural, depositional, and thermal events. It is important to determine if maturation of the source occurred before or after deposition of a sealing layer, or if traps were formed and sealed before maturation. The South Mozambique Graben has potential source rocks that most likely matured after sealing rock (Upper Cretaceous shales) and traps were in place (Fig. 2.30a and b).

The biggest problem encountered with an oil-probable region such as the South Mozambique Graben is potential recovery. The porosity at depths between 3000 and 4000 m is not too promising unless overpressured conditions are experienced. Up-dip stratigraphic traps may be more prolific.

The questions which cannot be answered are those of hydrocarbon migration and accumulation, although we can show that generation occurred mostly around the time of deposition of the Upper Cretaceous shale and generally after faults had become inactive. Perhaps the migration potential could be assessed using a two-dimensional fluid flow–compaction model such as that of Nakayama (1986), but the lack of more refined data makes such an endeavor of less value at the present stage of understanding of the South Mozambique Graben.

The conclusions of this application are separated into four categories. First theoretical synthetic tests are considered; then the structural and thermal conclusions are presented, followed by maturation potential and, finally, hydrocarbon potential.

a. Theoretical Testing

1. Backstripping wells which intersect faults present problems in that the faults become curved when decompacted, strata are not properly restored below faults, and faults migrate westward (down dip).

2. Sensitivity analysis of these problems revealed that shallow faults are more affected by compaction than steeper faults, the angle of the fault has more variation up the fault in shales than in sandstones, and as the scale of the system gets larger so does the apparent variation.

3. The problem of improperly restored strata below the fault was allowed for by assuming that structural "erosion" takes place below the fault.

4. The problem of changing burial history in a well due to contraction of the cross section during reconstruction was solved by "redrilling" and rerunning the burial history for each time step.

b. Tectonic and Thermal Implications

1. The structural history of the South Mozambique Graben has two rifting events recognizable on the seismic reflection profiles. The first event is of strike–slip nature, possibly sympathetic to the emplacement of Madagascar. The second event is a straight extension manifested on the seismic section as a sequence of domino-type fault blocks in the Cretaceous section. This type of structural history has recently been reported from the area just to the north by Mougenot *et al.* (1986).

2. Subsidence history also indicates two periods of extension.

3. Extension rates estimated from variation in strata lengths show an acceleration of extension at about 96 Mya.

4. Area balancing and depth to detachment calculations for each sequential palinspastic reconstruction also indicate that the depth to detachment was shallowest at 96 Mya. This may well be a tectonic/thermal event, possibly related to a shift in stress regime caused by the opening of the South Atlantic at this time.

c. Maturation Potential

1. Basement subsidence curves were coupled to extensional models of Royden and McKenzie to calculate the heat flow history of the region.

2. Two subsidence events required the use of three different heat flow histories to accommodate all possible variations with upper and lower limits.

3. Royden's model turns out to be more favorable from the maturation standpoint and geologically more appropriate in view of the abundance of sills and dikes of Cretaceous age reported in the region.

4. A source rock which can reach maturity for hydrocarbons likely lies somewhere within the Lower Cretaceous or Jurassic strata. The probability of such a source rock being present is enhanced by the high marine/terrigenous shale depositional setting for these strata.

d. Hydrocarbon Potential

1. Structural and stratigraphic traps are present due to the extensive faulting.

2. Reservoirs may be present in the synrift sands or Cretaceous rocks:

3. The Upper Cretaceous shales would suffice as an effective seal.

4. A source is unproven but likely and most probably lies in or below the Lower Cretaceous unit. Stratigraphic data would help considerably here.

5. Regardless of the model used or heat flow history configuration, the basin is expected to have mature hydrocarbons in the eastern part at depths of 3000 to 4000 m.

6. The ideal situation—relative timing of source deposition followed by faulting, then seal deposition followed by maturation of the source—seems to be the case in the South Mozambique Graben.

3

Inversion of Thermal Indicators: Theory

ONE OF THE MANY PROBLEMS confronting explorationists is the accurate prediction of source rock maturity. The problem is twofold in nature. First, an estimation of the thermal history of a basin must be made. To date, the only technique available has been geophysical modeling, an approach yielding very different thermal histories depending on the model chosen (see McKenzie, 1978; Royden *et al.,* 1980; and Falvey and Middleton, 1981). Frequently, geologic models do not agree on the sign of the derivative of heat flow with respect to time at the present. For example, the Falvey model applied to the North Sea would predict that heat flow was lower in the past than it is at present. Conversely, the McKenzie crustal stretching model would predict that heat flow was higher for approximately the past 90 Myr than at present. Thus, it is not possible to use geologic models to obtain information on heat flow as a function of time without some independent check of their validity. Second, the exact nature of the relationship of thermal history to thermal maturation is unknown. Clearly, the rule-of-thumb approach of Lopatin (1971) is inadequate (Waples, 1988). While the first-order chemical kinetics approach of Tissot and Welte (1978) has a firmer foundation, it is not completely grounded in empirical data.

The first part of this chapter relates the reflectance measurements of vitrinite collected from a well to that well's thermal history via a first-order time–temperature integral. This use of actual reflectance measurements addresses both of the problems described above. First, the maturation model (i.e., some time–temperature integral) must be capable of fitting the measured data; second, the resulting predictions of thermal histories must be reasonably consistent for wells across a basin.

The goals are (1) to summarize briefly the basic underlying theory; (2) to describe problems which arose during implementation and testing;

(3) to detail the final adopted algorithm, enabling others who are interested to gain a firm basis for their own numerical implementation of the protocol; and (4) to discuss briefly some preliminary results.

In addition, once having set up the methodology for vitrinite reflectance, that same method of inverse analysis is applicable to other thermal indicators which have their own individual chemical and physical kinetics. In Chapter 4 we show case histories of several such indicators analyzed by thermal inversion methods.

I. Overview and Basic Theory

A discussion of the theory can be found in Appendix A to this chapter. It is presented there so as not to break the general train of argument being developed and also so that the details are available to the interested reader.

The vitrinite reflectance $R(z_i)$ for each measured depth z_i in a well is first converted to a normalized VITTI index, which is a direct measure of the maturity of the sample:

$$\mathrm{VITTI}(z_i) = [R(z_i)^{1/2} - R_S^{1/2}] \Big/ \left\{ \sum_{j=1}^{N} [R(z_j)^{1/2} - R_S^{1/2}] \right\} \tag{3.1}$$

where VITTI(z_i) is the maturity index at depth z_i, $R(z_i)$ the measured vitrinite reflectance at depth z_i, R_s the reflectance of vitrinite when it is deposited (generally around 0.20, although it appears to be higher in some basins) and N the number of vitrinite reflectance measurements. Clearly, VITTI has a value of zero at the surface and a value of less than one at the deepest measured vitrinite point. Care should be taken not to confuse VITTI, the maturity index, with vitrinite reflectance.

A theoretical expression (see Appendix A) directly comparable to the VITTI index is

$$\mathrm{INORM}(i) = I^{(0)}(i) \Big/ \left[\sum_{j=1}^{N} I^{(0)}(j) \right] \tag{3.2}$$

where $I^{(0)}(i)$ is a time–temperature integral characterizing the maturity of a vitrinite sample deposited at time t_i and presently at depth z_i.

From Appendix A, the original form for the $I^{(0)}$ integrals was

$$\mathrm{TTI}(0, t_j) \equiv I^{(0)}(i) = \int_{t_i}^{0} \exp[-(T_A/T) + (T/T_D)]\, dt \tag{3.3}$$

where T_A and T_D are constants roughly analogous to an activation temperature and a doubling temperature, respectively. Each integral is taken along the burial path of a sample deposited at time t_i. The T in the above expression is the temperature along the burial path and is a function of both time and depth:

$$T(t, z) = T_{\text{surface}} + Q(t) \int_0^z dz'/K(z') \tag{3.4}$$

where $K(z')$ is the thermal conductivity as a function of depth and $Q(t)$ the basement heat flux as a function of time. [A slight technical problem arises when the heat flow varies with time because the heat transport equation then should allow for thermal diffusion of the heat flow through the sediments. However, estimates of the thermal diffusion time for sediments on the order of 15,000 ft (4500 m) in thickness yield values in the range 2–20 Myr, depending somewhat on the thermal conductivity. Accordingly, as long as the changes in heat flow take place on a time scale greater than 10 Myr, it is then an excellent approximation to regard the system as being in thermal equilibrium with the prevailing heat flux at each instant in time. In addition, only the most broad, first-order (in time) changes in heat flow are ruggedly seen in vitrinite data, lending corroborative support to the use of the balance of thermal conductivity times the temperature gradient against heat flux at each instant of time.]

The basic problem, then, is to find $Q(t)$ to minimize the differences between the theoretical INORM(i) index and the measured VITTI(i) index; that is, we wish to minimize

$$\text{SS} \equiv \sum_i [\text{INORM}(i) - \text{VITTI}(i)]^2 \tag{3.5}$$

where the summation is taken over all vitrinite samples in a well. We take as our goodness-of-fit measure MSR (mean squared residual), defined as

$$\text{MSR} = \log_{10}[\text{SS}/(i - 2)] \tag{3.6}$$

The initial algorithm for solving this highly nonlinear least-squares problem involved a steepest descent technique in a five-dimensional parameter space: T_A, T_D, and three coefficients for $Q(t)$. Preliminary results (Toth *et al.*, 1983) yielded a ballpark figure for T_D of about 400 K and a ballpark figure for T_A of less than 2×10^{-4} K. Good fits to the vitrinite data using "reasonable" $Q(t)$ functions were obtained in several wells. For many wells, however, no $Q(t)$ function was found that enabled the algorithm to fit the reflectance data.

A. Form of Paleoheat Flux Function

The original formulation of the algorithm permitted the heat flux $Q(t)$ to vary as

$$Q(t) = Q_0 + Q_1 t + Q_2 t^2 + Q_3 t^3 \qquad (3.7)$$

where Q_0 is the present-day heat flow and t is time in millions of years; $t = 0$ is the present day and t is positive back into the past.

The first observation apparent from data for several wells was that the elimination of Q_3 produced little change in the final result for $Q(t)$; that is, eliminating the Q_3 coefficient did not appreciably degrade our ability to fit the vitrinite data.

Testing continued with the Q_1 and Q_2 coefficients. Results from many wells revealed a broad range of possible solutions that fit the vitrinite data equally well. Good solutions were possible with either Q_1 or Q_2 equal to zero. The proper inference is that we do not really need two variables to define the paleoheat flux function; a single variable defining $Q(t)$ will enable us to fit the vitrinite data as well as two variables.

The final form chosen for the paleoheat flux function was

$$Q(t) = Q_0(1 + \beta t) \qquad (3.8)$$

which permits the heat flux to increase or decrease linearly in time. It could be argued that this form is inadequate in many geologic settings; for example, we could not model an igneous intrusion occurring in the middle of a well's burial history. However, examination of much vitrinite data led to the firm conclusion that any such events are buried deeply in the noise in the data and that only the most broad, first-order effects can be modeled (Figs. 3.1 and 3.2). (An alternative single-parameter formulation is $Q(t) = Q(0) \exp(\beta t)$, which manifestly keeps $Q(t)$ positive.)

B. Form of Time–Temperature Integral

Refinements to the paleoheat flux function had little impact on the broad picture of heat flux with time for the Northwest Shelf of Australia, the National Petroleum Reserve of Alaska (Fig. 3.3), and Indonesia. However, difficult problems remained with most of the North Sea wells studied. The basic symptom observed was an attempt by the algorithm to drive the heat flow to zero (and negative) in the very recent past.

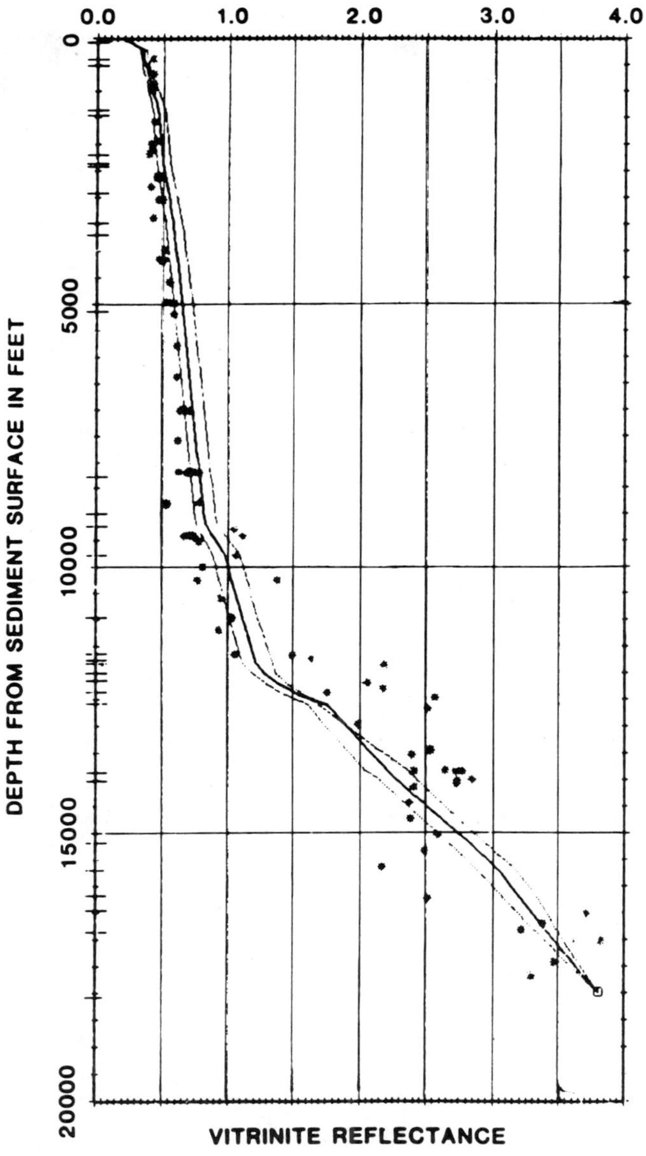

FIGURE 3.1 Reflectance versus depth, including predicted reflectance line, for Inigok 1, Alaska. Asterisks are measured vitrinite values; dark line is predicted vitrinite reflectance derived from heat flow shown in Fig. 3.2, reflectance line; dotted envelope is error envelope. Slopes of line and envelope owe their variation, in part, to rates of sediment accumulation and burial.

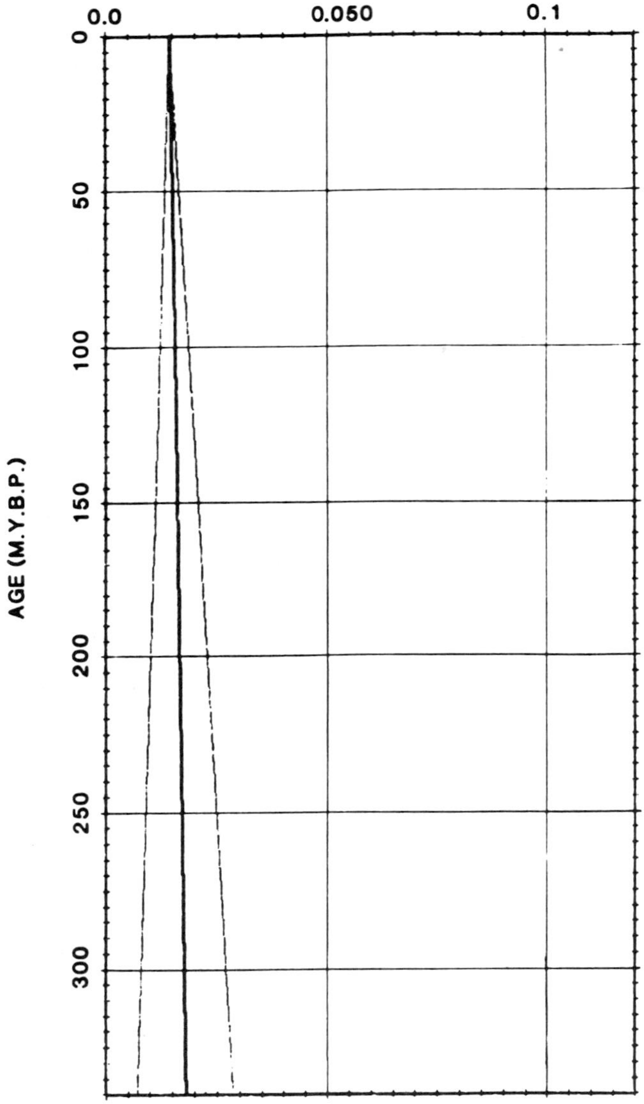

FIGURE 3.2 Heat flow versus geologic age for Inigok 1, Alaska. Dark line is best-fit predicted heat flow; dotted lines are the error envelope.

At this point, a more careful examination was made of the time–temperature integral, in particular the roles of T_A and T_D:

$$I^{(0)}(i) = \int_{-t_i}^{0} \exp[-(T_A/T) + (T/T_D)]\, dt \tag{3.9}$$

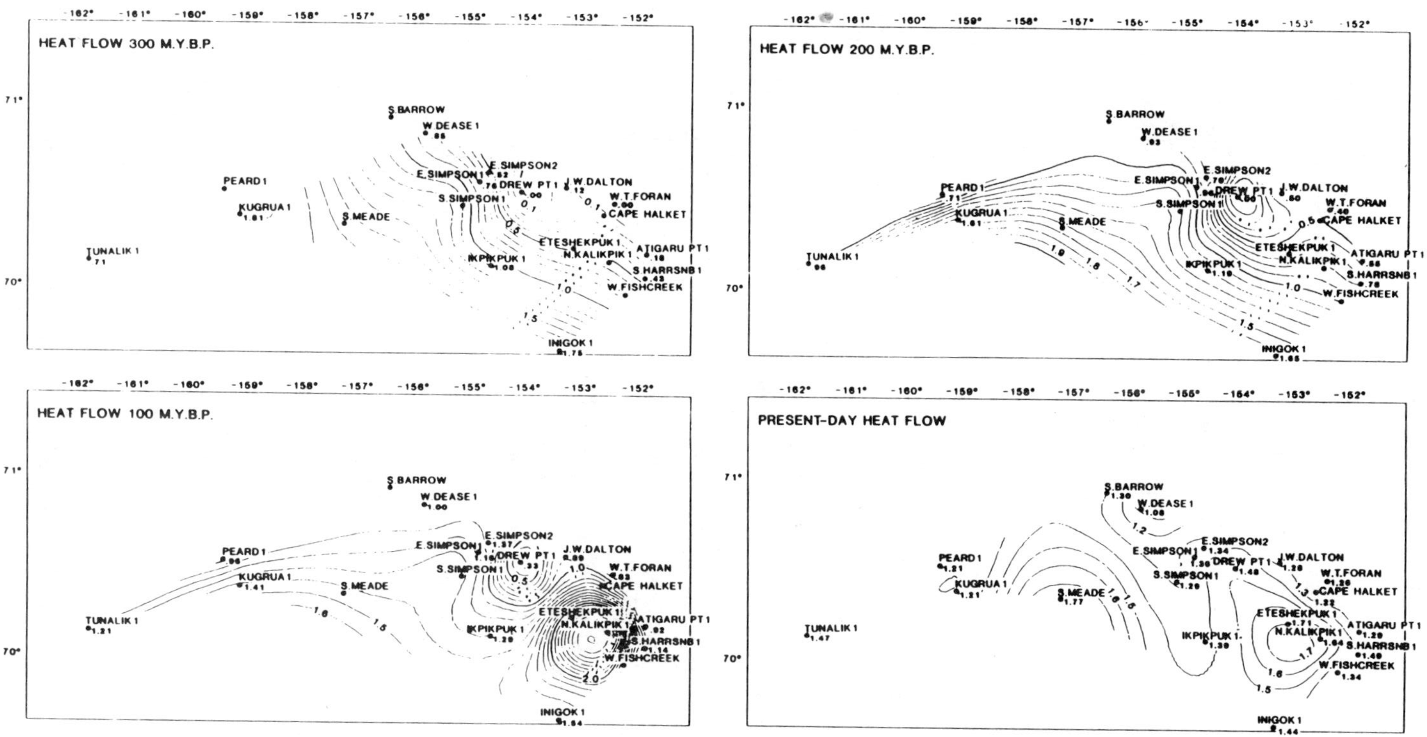

FIGURE 3.3 Contour map of heat flow through time for 21 wells from National Petroleum Reserve of Alaska. Contours are in heat flux units (HFU).

From Toth *et al.* (1983), we know that $T_A \cong 10^{-4}$ K and $T_D \cong 400$ K. The largest possible value of T_A/T occurs when T = the sediment surface temperature $\cong 280$ K. Under this circumstance,

$$T_A/T \cong 3.5 \times 10^{-7} \quad \text{and} \quad T/T_D \cong 0.7$$

Therefore, it was clear that we could safely neglect T_A/T relative to T/T_D. Additional grid searching in a T_A–T_D space for various values of β confirmed the algorithm's tendency to make T_A vanishingly small as it converged to a solution. Consequently, T_A was dropped as a parameter.[1]

Problems still remained in the North Sea data set. Several wells were examined in detail in an attempt to further understand the problem. Results from a typical well, well X, are presented here. Figure 3.4 shows the burial history of the well, and Fig. 3.5 shows the present-day depth versus vitrinite reflectance distribution. Figure 3.6 shows present-day depth versus VITTI. The six horizontal lines on this plot identify the age of the sediment in millions of years at six depths in the well section.

The vitrinite data for this well come from sediments whose ages extend only 150 Myr into the past; therefore, as shown in Fig. 3.4, the heat flow algorithm does not have to deal with the large unconformity in the Jurassic; there are no major unconformities during the period of interest. The only exceptional geologic event occurs in the Paleocene, which is marked by a period of rapid deposition; 3000 ft (915 m) of sediment were deposited in 6.5 Myr. Note that the maximum vitrinite reflectance (Fig. 3.5) is only 0.5 (the bottom-hole temperature is 70°C or 158°F at a depth of 8300 ft or 2530 m).

On Fig. 3.6, we have also added the INORM curves, that is, curves showing the predicted maturity indices for reflectance at various points in time, calculated by using a particular value of β. (In this case, $\beta = -0.01$ Myr^{-1}, indicating a lower heat flow in the past; no other value of β yielded a significantly better fit to the data.) Curve 1 is the present-day INORM curve, the final values for predicted maturity; clearly, the fit to the VITTI values is very poor.

The nature of the problem becomes clearer when one considers curves 2, 3, and 4, the INORM curves at three earlier points in time. Consider the paths of maturity for two vitrinite points: point A, deposited at 60 MYBP

[1] D. W. Waples (personal communication, 1983) has pointed out that the low value of T_A suggests that the rate-limiting step in increases in vitrinite reflectance involves breaking of van der Waals-type interactions rather than cleavage of true bonds.

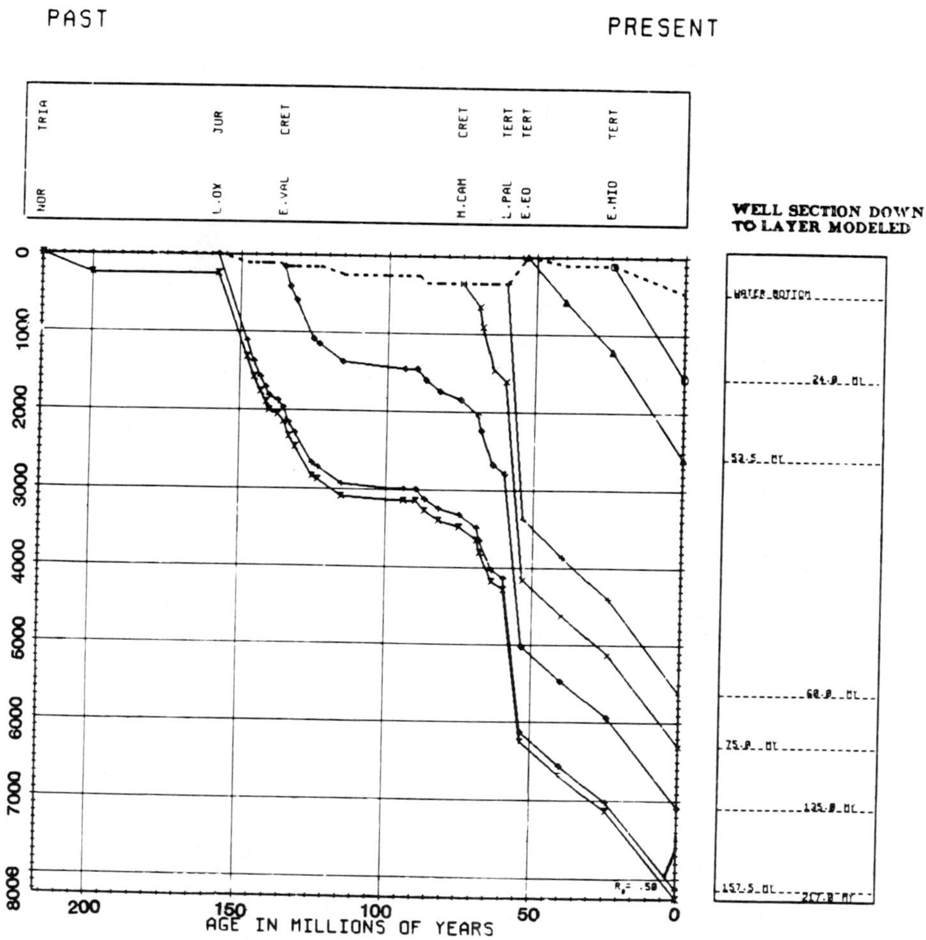

FIGURE 3.4 Geohistory plot for well X in North Sea, plotted to constant sea level datum for all flagged layers.

and at 5000 ft (1524 m) in the present-day well section; and point B, deposited at 157 MYBP and at 7700 ft (2347 m) in the present-day section. Sixty MYBP, at the time of deposition of point A, point B had already been maturing for 100 Myr and had attained a computed reflectance value of 0.55.

At this stage, we have already reached an impasse: as points A and B continue to sink and mature, the gap between the reflectance of these two points can only increase, because point B is always deeper and hotter than point A. At the present day, point B has a reflectance of 1.0 and point

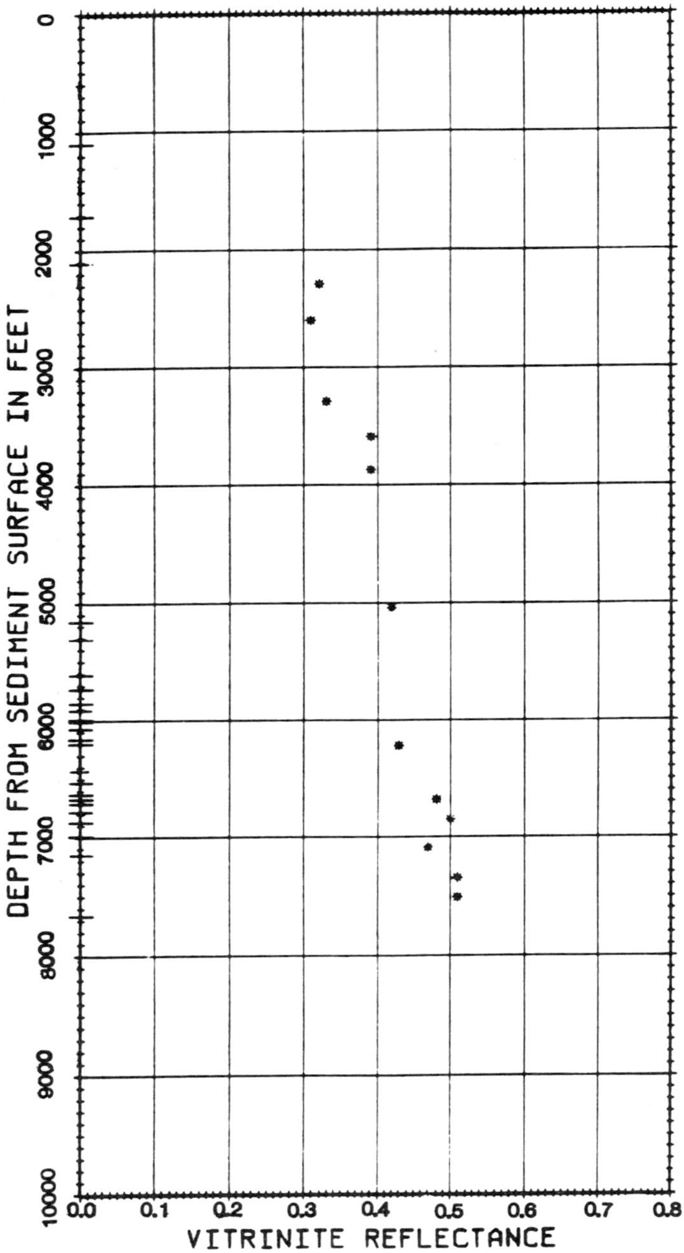

FIGURE 3.5 Vitrinite reflectance versus depth for well X in North Sea (input R_0 data).

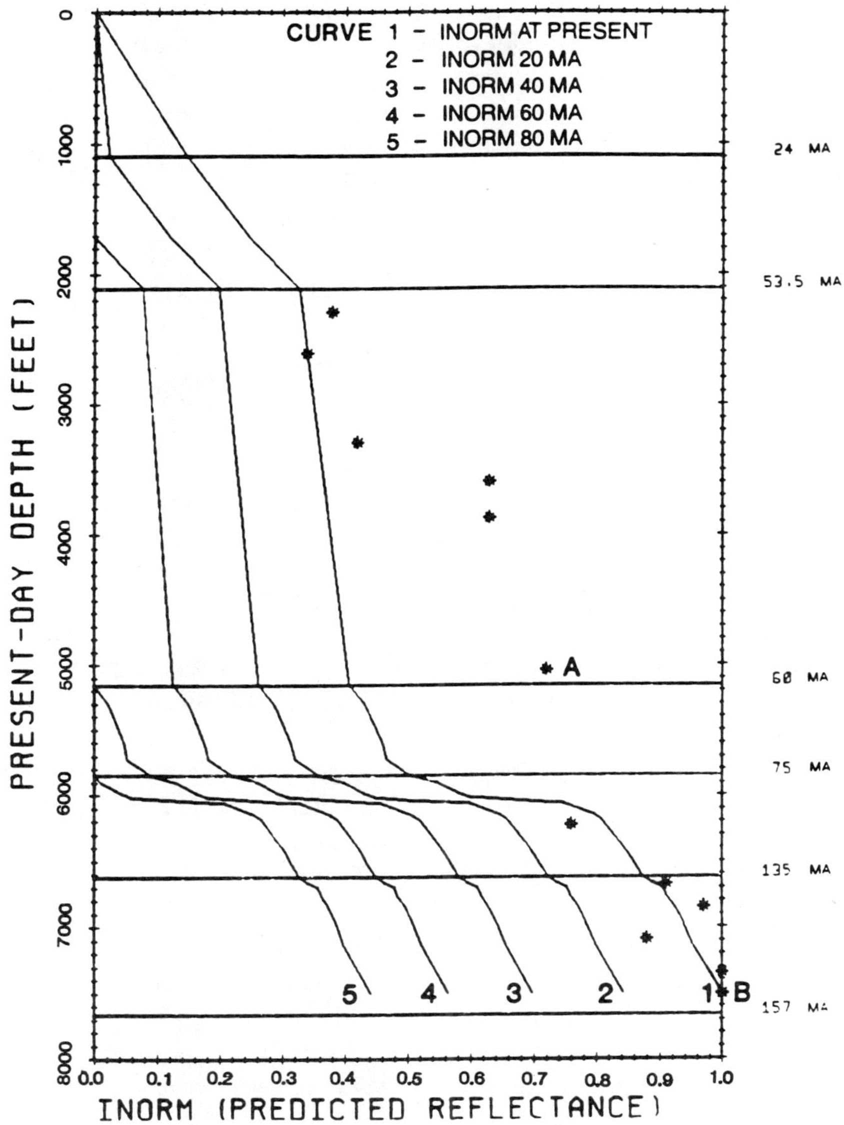

FIGURE 3.6 Predicted normalized reflectance (INORM) versus depth at five different ages for well X (North Sea), using original version of cooking integral. Asterisks are normalized maturity indices (VITTI) determined from reflectance values given in Fig. 3.5.

A a reflectance of 0.40; the gap has increased from 0.55 to 0.60. However, the actual reflectance values at these two points are 1.0 and 0.75, a gap of only 0.25; the model has "overmatured" point B relative to point A.

Two remedies were apparent. First, a higher doubling temperature would decrease the temperature dependence. When this was tried, however, the effect was very small. Second, heat flow in the past could be made even less. When the program was permitted to run free in choosing a heat flow, only a very marginal improvement was possible, even when the heat flow was dropped to zero in the recent past. The 100 Myr "jump" in maturation that point B had on point A could not be overcome.

To resolve the problem, an additional parameter, a critical temperature, had to be introduced, and the final form of the time–temperature integral became

$$I^{(0)}(i) = \int_{t_i}^{0} dt \exp[(T - T_c)/T_D], \qquad T \geqslant T_c \tag{3.10}$$

with the integrand zero in $T < T_c$. The role of the critical temperature is simple: below $T = T_c$ vitrinite does not mature.[2]

The results of grid searching for a solution in a β–T_c space for well X are shown in Fig. 3.7. A critical temperature of 285 K and a β of -0.01 produce a fit to the vitrinite data an order of magnitude better than previous attempts.

Grid searches on 20 wells from the North Sea, 21 wells from the Australian Northwest Shelf, six wells from Indonesia, and 21 wells from the National Petroleum Reserve of Alaska narrowly defined T_c to be 295 $\pm$ 10 K. Figure 3.8 shows a somewhat different situation—an Australian well whose vitrinite reflectance data suggest a higher heat flow in the past. In this case, the great majority of the well's thermal history was spent at a temperature above T_c; hence T_c cannot be determined accurately from this well, although β is narrowly defined; indeed, heat flow results for this well (and similar wells showing high heat flow in the past) are almost identical to results produced by earlier versions of the algorithm.

---

[2] D. W. Waples (personal communication, 1983) has noted that the low value of T_A, suggestive of van der Waals-type interactions, suggests a critical temperature effect because, viewed in the light of transition state theory, T_c would represent a minimum in the change in entropy between reactant and activated complex. He also noted that "The threshold for vitrinite maturation might even involve the formation of true vitrinite (starting at $R_0 = 0.35$ or 0.4%), and might thus be dependent on maturity rather than temperature as T_c indicates."

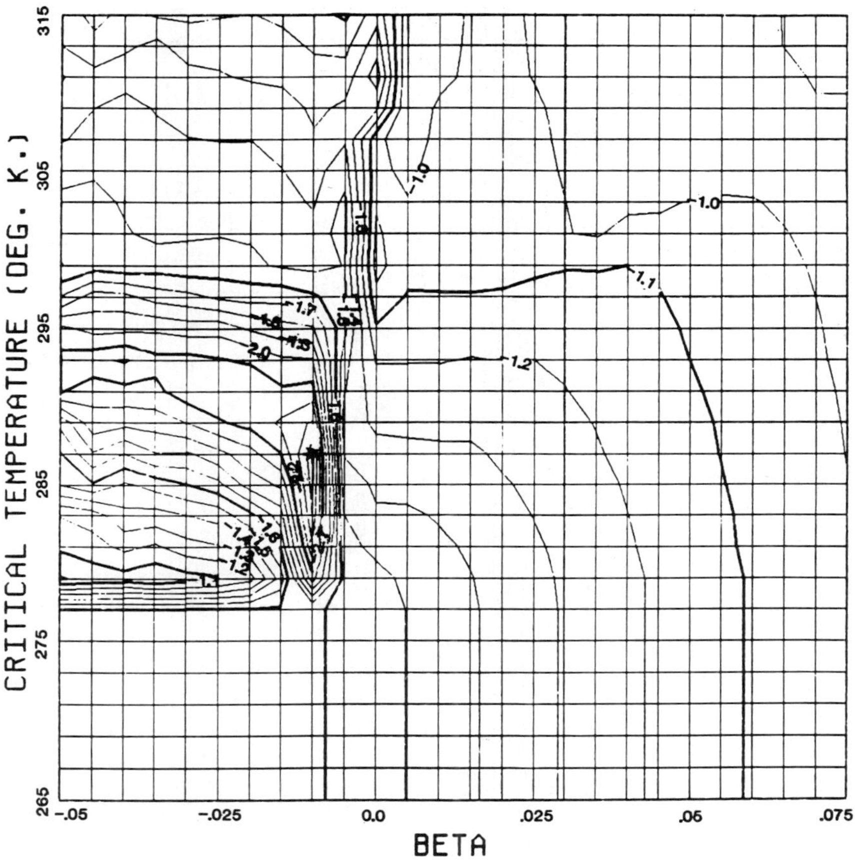

FIGURE 3.7 MSR contoured in a beta-critical temperature parameter space for well X (North Sea). Star denotes minimum of MSR.

Figure 3.9 shows the gradual maturation of North Sea well X through time. In this case, the use of T_c in the computations improves significantly the fit to the vitrinite reflectance data, and the picture is quite different from that in Fig. 3.6. Because of lower heat flow in the past, coupled with T_c, point B has reached a reflectance of only 0.2 when point A is deposited; this maturity gap slowly widens to 0.25 in the present-day section, in good agreement with the actual data.

The introduction of a fixed critical temperature made it possible to fit all sets of vitrinite data into the geohistory data bases we built for wells from Australia, Indonesia, the North Sea, and Alaska.

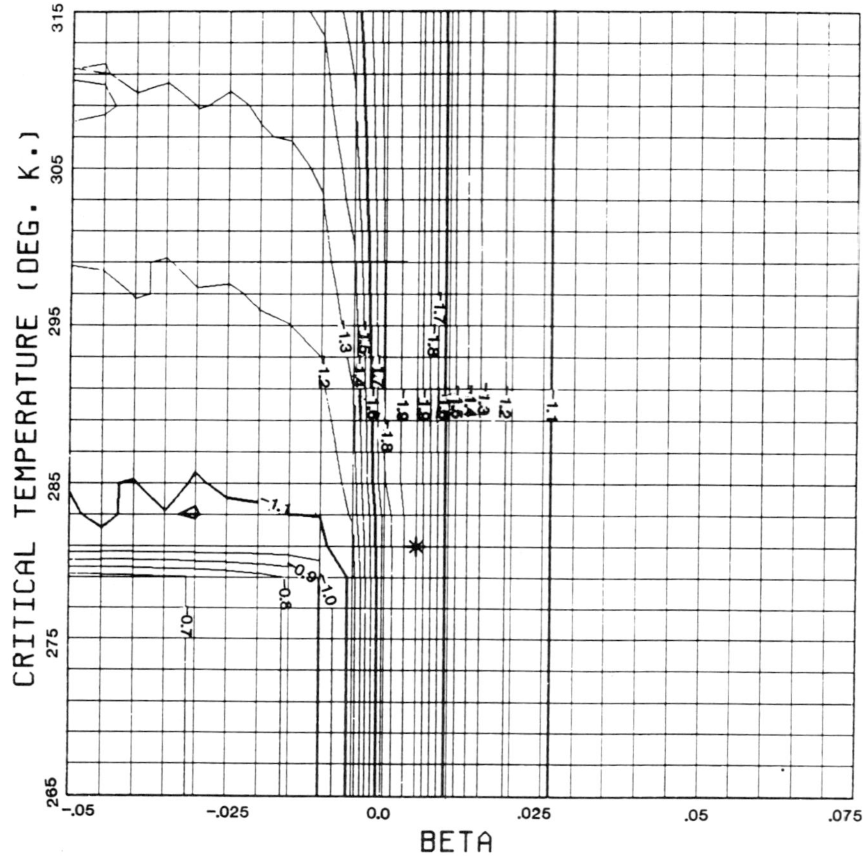

FIGURE 3.8 MSR contoured in a beta-critical temperature parameter space for Australian well A. Star denotes minimum of MSR.

C. Final Formulation and Results

The final algorithm is basically an implementation of two equations. Paleoheat flux varies linearly with time:

$$Q(t) = Q_0(1 + \beta t) \quad [\text{or } Q(t) = Q_0 \exp(\beta t)] \tag{3.11}$$

and the theoretical maturation integral possesses a critical temperature:

$$I^{(0)}(i) = \int_{t_i}^{0} \exp[(T - T_c)/T_D]\, dt, \qquad T \geqslant T_c \tag{3.12}$$

with the integrand zero in $T < T_c$.

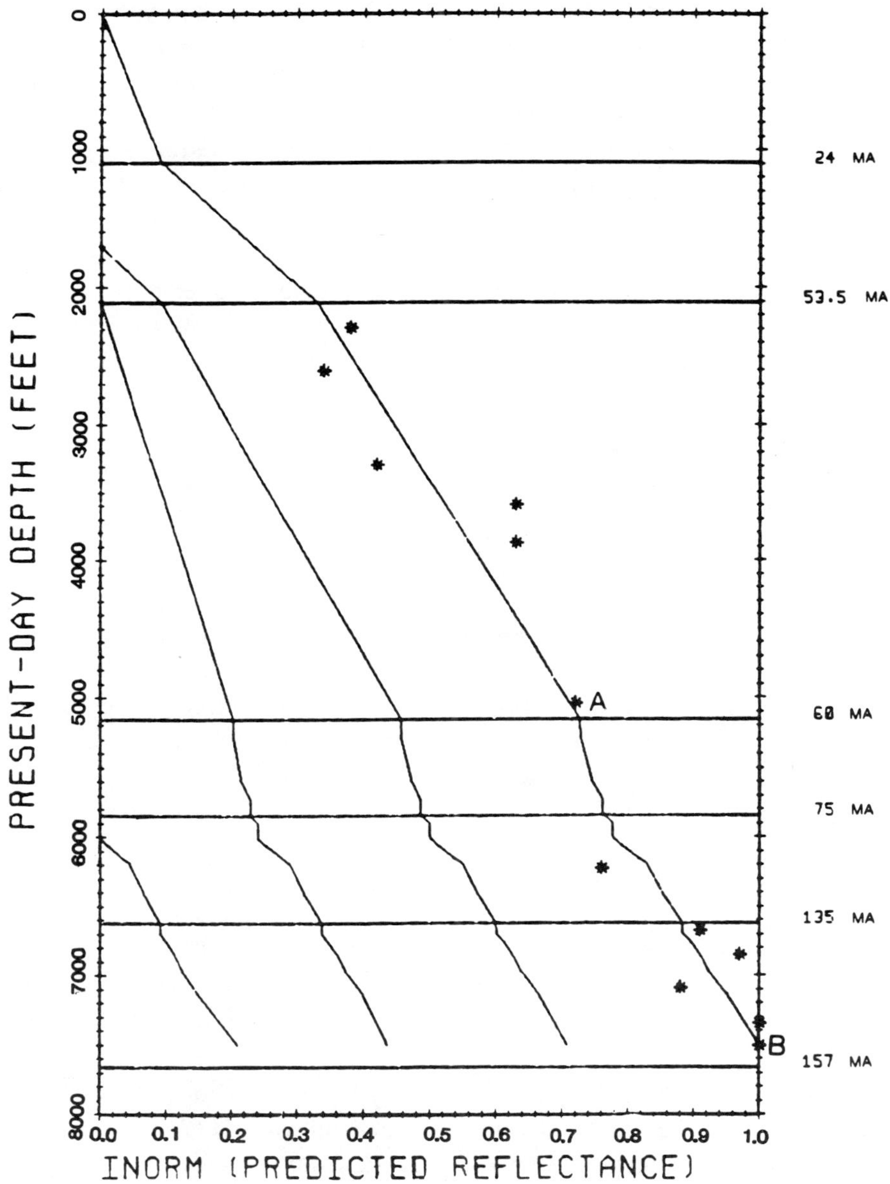

FIGURE 3.9 Predicted normalized reflectance (INORM) versus depth at four different ages for North Sea well X. Asterisks are normalized maturity indices (VITTI) determined from reflectance values given in Fig. 3.5.

Recalibration of laboratory data (see Toth *et al.*, 1983) with the final integral form led to a fix on T_D of 200 K; grid searching in a T_D–β space, with T_c fixed at 295 K, enforced the soundness of this value for T_D. With T_D and T_c fixed, it was possible to do an exhaustive trial-and-error search for the best β for each well. With the burial history parameters determined, and with T_D and T_c fixed for all wells, we have only the parameter β to be determined for each suite of vitrinite data for each well.

With the provision of high-quality vitrinite reflectance data, and with the novel inversion schemes now available (see Chapter 9 on thermal indicator tomography), we can relax the constraint that the heat flux varies linearly with time. Instead we can, for instance, take the heat flux to vary piecewise with time over selected intervals of time so that β varies with the time interval. Thus if the time intervals are labeled $t_1, t_2, \ldots,$ t_n we can write

$$Q(t) = Q(t_{n-1}) \exp[\beta_n(t-t_{n-1})] \tag{3.13}$$

in $t_{n+1} > t > t_n$, where the $\beta_1, \beta_2, \ldots, \beta_n, \ldots$ and $t_1, t_2, \ldots,$ $t_n, \ldots$ are to be determined using thermal indicator tomography as outlined later, and where $Q(0)$ is the present-day heat flux.

The form of $Q(t)$ chosen above is exponential in each time interval and the functional form allows $Q(t)$ to be continuous across each time line at $t = t_{n-1}, t_{n-2}, \ldots, t_1$. An alternative is to model the paleoheat flux evolution in the form

$$Q(t) = Q(0) \exp\left[\beta t + \sum_{k=1}^{n-1} a(k) \sin(\pi k t / t_{\max})\right] \tag{3.14}$$

where $t_{\max}$ is the age of TD and where the trend coefficient β and the oscillatory coefficients $a(k)$, are to be determined to minimize SS, as is the number n of such coefficients using a thermal indicator tomography framework.

Once the best-fitting time-dependent heat flux function has been determined, it is a simple matter to predict at any instant of time along any burial path the level of vitrinite reflectance from

$$R(z(t)) = [R_s^{1/2} + \alpha\, \mathrm{TTI}(t)]^2 \tag{3.15}$$

where

$$\alpha = \sum_{j=1}^{n} [R(z_j)^{1/2} - R_s^{1/2}] \Big/ \left[\sum_{j=1}^{n} \mathrm{TTI}(t_j)\right] \tag{3.16}$$

Current information on 300 wells sets the surface depositional value R_s at 0.2%.

Finally, one would also hope for regional consistency when looking at β values basinwide. Figure 3.10 shows a contour map of β for 20 wells in the North Sea. Note that wells grouped closely together do have similar β's. Figures 3.11 and 3.12 show the heat flow at the present day and the heat flow at 100 Ma, respectively. Note also that heat flow pattern does not increase in spatial complexity or produce "bull's-eyes" back in time.

The systematic heat flow is about a factor of 2 lower throughout the North Sea basin 100 MYBP than it is at the present. This warming trend, while contrary to much of the conventional thought about passively rifted basins (Sclater and Christie, 1980), may be related to (1) volcanic activity along the west side of England that coincided approximately with the time at which Greenland pulled away from Britain and Norway (late Paleocene), (2) massive subsidence of the North Sea basin during the early Paleocene (nearly 2000 ft or 600 m over 5–10 Myr) (Guidish *et al.*, 1984), and (3) the uplift of the United Kingdom landmass in the late Maestrichtian. A possible inference is that a hot sublithospheric upwelling

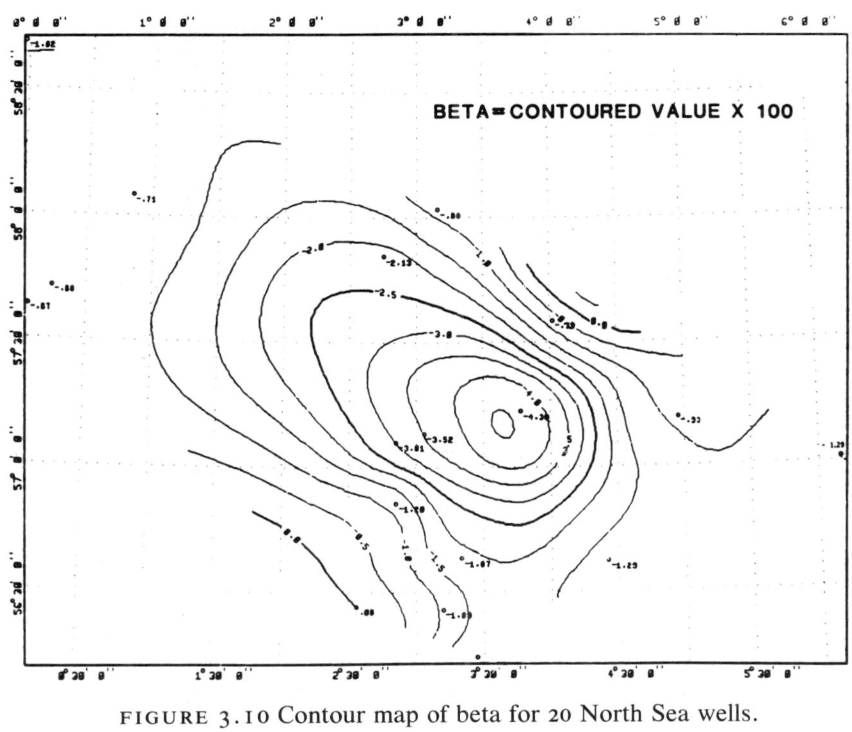

FIGURE 3.10 Contour map of beta for 20 North Sea wells.

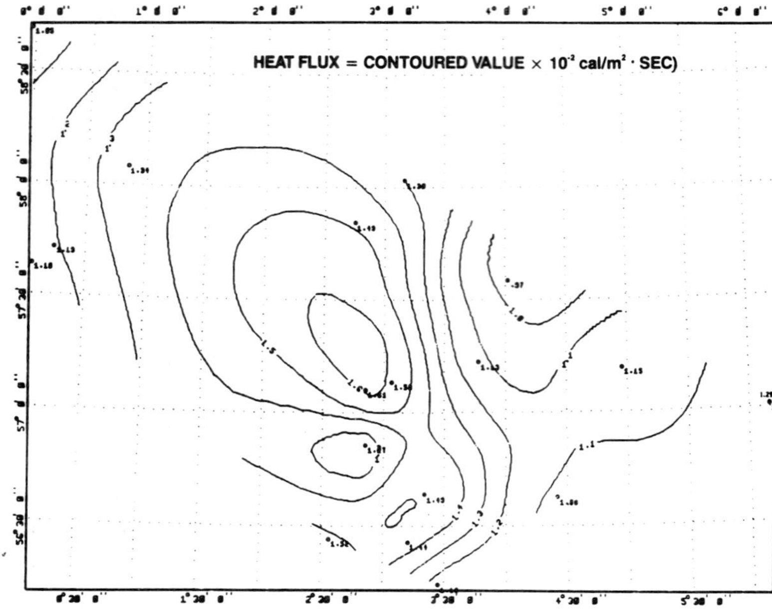

FIGURE 3.11 Contour map of present-day heat flow for 20 North Sea wells.

occurred prior to the volcanic activity. This upwelling spread out as it rose, softening the crust and hence causing the massive basement subsidence in the early Paleocene. The upwelling broke through the crust approximately 5–10 Myr later than the pull-apart. Indeed, at the northwest end of Scotland, the pull-apart was so intense that volcanic basement material was breached. In addition, it is by no means unusual to find geologic events related to plate extension occurring over a 10–20-Myr period and representing a "typical" time scale estimate for these kinds of events.

Consistency was obtained from a set of 20 Australian wells, 6 wells from Indonesia, and 21 wells from the National Petroleum Reserve of Alaska (Fig. 3.3). In the latter case, predicted vitrinite reflectance varied as a function of time, which indicates how thermal maturity also varied as a function of time and position. Again, regional consistency was found.

D. Caveats, Unresolved Problems, and Future Directions

The consistency observed in four basins argues strongly that the relationship proposed between heat flow and vitrinite reflectance is valid. Two warnings, however, should be attached to the results.

First, and most important, we can trust the results no further than we

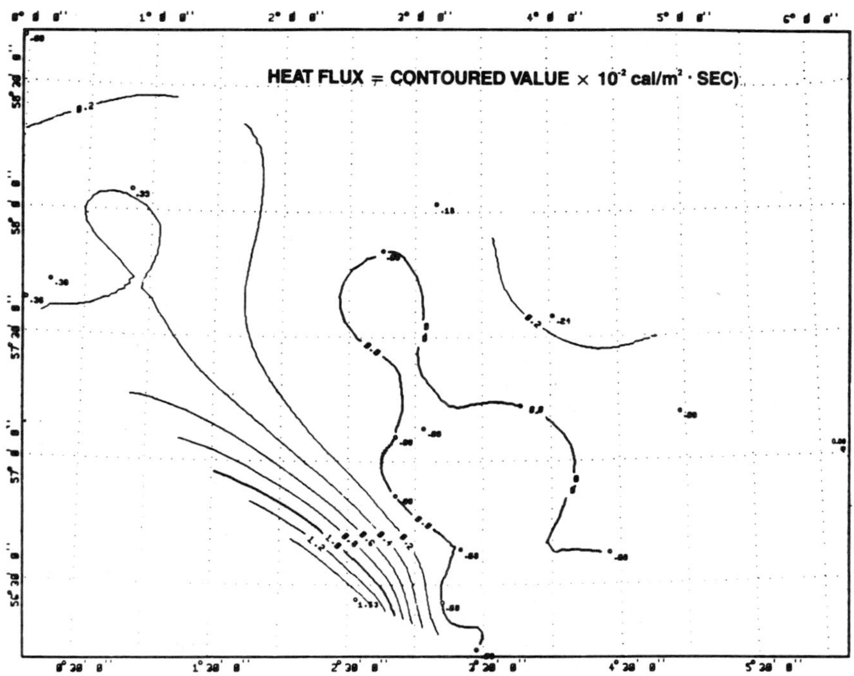

FIGURE 3.12 Contour map of heat flow 100 Ma for 20 North Sea wells.

can trust the quality of the original reflectance data. Unfortunately, there is still a considerable degree of art in the acquisition of vitrinite samples, in the measurement of vitrinite reflectance, and in the separation of primary and secondary (reworked) vitrinite. These sources of noise, coupled with the sparsity of data in many wells, lend a high degree of uncertainty to the final answer.

When accurate vitrinite data, finely sampled in depth, are available, an excellent fit of theory to observations is possible. This situation is illustrated in Fig. 3.1 for data from the Inigok 1 well in the National Petroleum Reserve of Alaska.

The second caveat is also simply stated but has more far-reaching implications: the technique presented here determines paleoheat flux from the *shape* of the reflectance curve rather than from the *magnitude* of the vitrinite reflectance data. Some attempts were made to remedy this situation. Ideally, we would like to express the relationship of the maturation integrals to the measured reflectances as

$$\alpha I^{(0)}(i) = (R_i^{1/2} - R_s^{1/2}) \tag{3.17}$$

Comparing this to our expressions for VITTI and INORM, we find

$$\alpha = \sum_i (R_i^{1/2} - R_s^{1/2}) / \sum_i I^{(0)}(i) \tag{3.18}$$

Thus, once a best value for β has been found and the $I^{(0)}$ integrals have been calculated, it is a simple matter to calculate α for a well.

Ideally, one would expect that α is a universal or at least a basinal constant. Unfortunately, the results so far show that α appears to vary slowly across a basin (see Fig. 3.13).[3] It is not clear whether a best α could be chosen, fixed in the calculations, and used to compute paleoheat flow for all wells in a basin without producing large misfits to the vitrinite data for some of the wells.

While finding a constant α is a desirable end, it should be noted that one of the most important goals—the prediction of maturity (as defined by vitrinite reflectance) at seismic stations and for wells without vitrinite data—has been accomplished. Working from contour maps of Q_0, β, and α, together with age-tied seismic data, it is possible to predict maturity at these pseudowell locations. Because β and α have behaved consistently in the basins tested, confidence can be placed in the resulting maturity predictions.

Price (1982) and Barker (1983) proposed that vitrinite reflectance values are tied only to the present-day formation temperatures. Price kindly provided his suggested relationship. When used with the vitrinite reflectance values and well temperatures for suites of wells from the North Sea, Indonesia, and the National Petroleum Reserve in Alaska, Price's proposed relationship fits the data from these three basins rather poorly, as exemplified in Fig. 3.14 for data from 20 North Sea wells, implying that some basic effect—not tying vitrinite reflectance solely to present-day formation temperature—plays a more dominant role in producing the observed vitrinite reflectance values. We suggest that the approach above provides a consistent framework for interpreting vitrinite reflectance measurements and for relating them to paleoheat flux.

One other point is in order. The use of vitrinite reflectance to derive

[3] Possible causes for this variation are (1) the lack of high-quality vitrinite data, (2) the possibility that the burial history is not isostatically controlled at all times but has an overpressured component, (3) incompleteness of the maturation theory for vitrinite, (4) insufficiently precise definition of parameters and functions entering the conversion of temperature to heat flux (such as lithology and porosity-dependent thermal conductivities), (5) need to include radioactive sources of heat in the sediments.

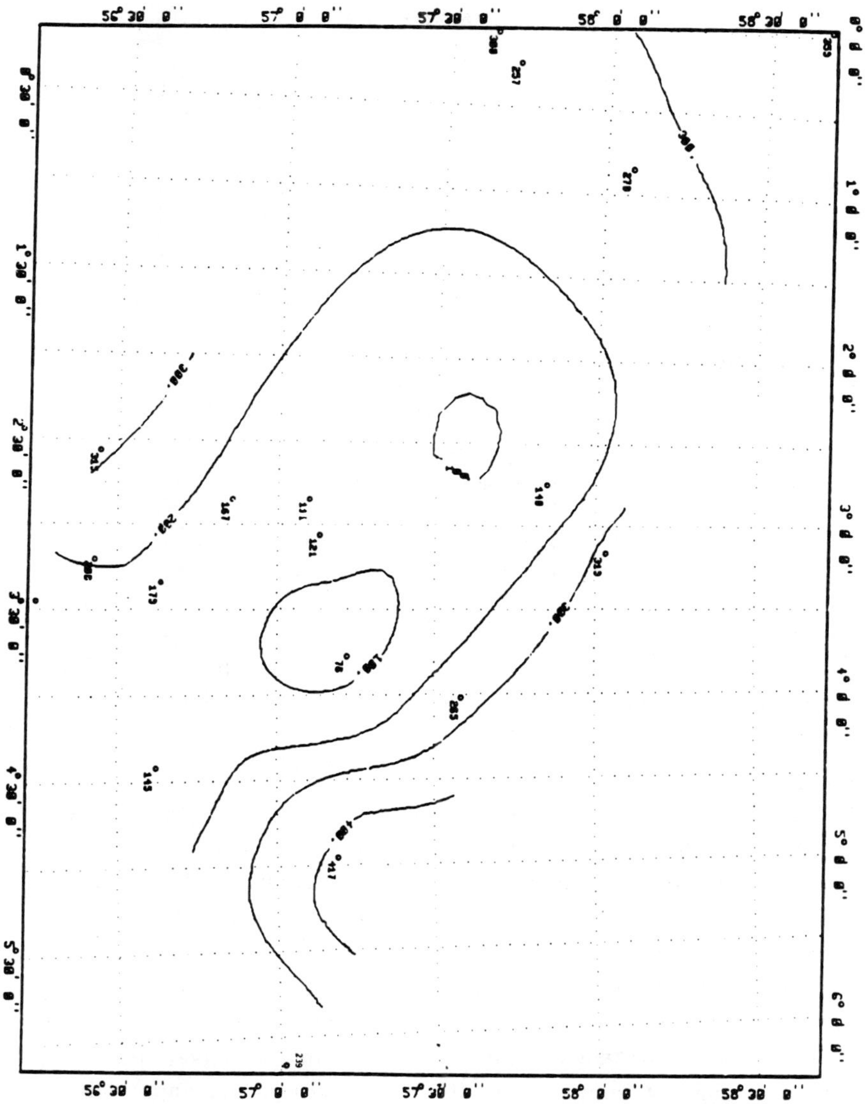

FIGURE 3.13 Contour map of alpha for 20 North Sea wells.

paleoheat flow allows us to determine the variation of temperature with time and depth for a formation. Oil generation and maturity have also been tied (Tissot and Welte, 1978; Lopatin, 1971) to the variation of formation temperature with depth and time through time–temperature integrals different from those pertaining to the evolution of vitrinite reflectance as presented here. As long as a high degree of correlation

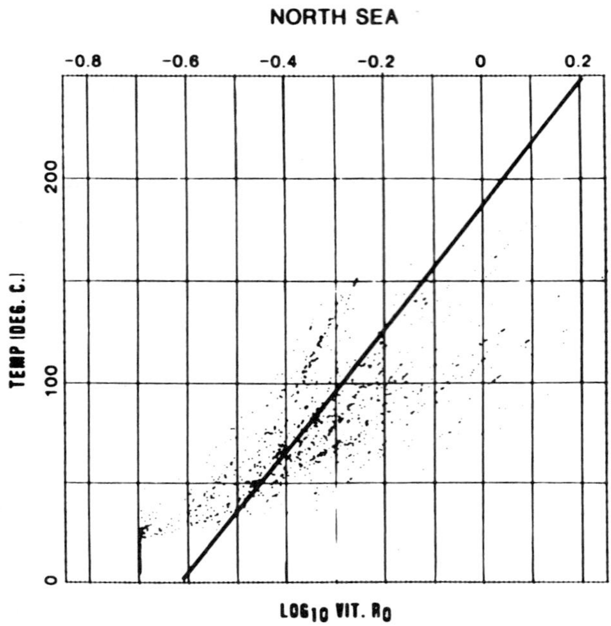

FIGURE 3.14 Comparison of vitrinite reflectance data from 20 North Sea wells and formation temperature. Solid curve is relationship suggested by Price (1982).

exists between the various time–temperature integrals, the evolution of vitrinite values acts as a good indicator of oil generation potential. The statistical degree to which the vitrinite reflectance values and time–temperature integrals proposed as oil generation indicators do correlate needs to be investigated.

It is also important to note that the thermal maturation integral invoked to monitor the increase of vitrinite reflectance with time and temperature (Lerche *et al.*, 1984) shows only a factor of 2 increase over about a 140°C dynamic range in temperature, because "typical" vitrinite reflectance measurements range from about 0.2% at the sedimentary surface to about 1–2% at depth—a factor of only about 5–10 increase. As pointed out in Lerche *et al.* (1984), the thermal maturation integral models the increase of the square root of vitrinite reflectance with time and temperature, so the thermal integral varies by only a factor of 2–3, as confirmed by Reaves and von Rosenberg (1985). This variation in the integral is not responsible for the variations we obtained in the inverted heat flow; rather the fluctuations in vitrinite reflectance data with depth, as well as uncertainties in the paleoburial history (e.g., the manner in which compaction is commonly handled is almost certainly in error when applied for all time), are the main culprits (Lerche *et al.,* 1984). Thus, poor data give poor results (Fig. 13 of Lerche *et al.,* 1984), as with the factor of 4 variation cited by Reaves and von Rosenberg (1985), but good data give good results (Figs. 1 and 2 of Lerche *et al.,* 1984). Therefore the degree of resolution and uncertainty in the paleoheat flux determined from the data inversions is precisely related to the quality of downhole thermal indicator data (Lerche *et al.,* 1984).

"Scatter" in the reflectance data is caused by reworking, difficulties in separating the different coal macerals (Castano, 1985), and suppression by hydrogen-rich kerogen (Castano, 1985; McCulloh and Fan, 1984; Price, 1985). Despite this scatter, when the thermal maturation integral was used, remarkable regional consistency in heat flow from well to well was obtained (Figs. 3, 10, 12, and 14 of Lerche *et al.,* 1984), corroborating that the "correct" variation of paleoheat flux with time had been derived.

As explained (Lerche *et al.,* 1984, appendix), the "reaction rate expressions" are tied not to the physical chemistry of a liquid (such as hydrocarbons, with a nominal doubling temperature of 10°C) but to a solid, vitrinite. This confusion of solid and liquid chemical behavior may explain why geologists have difficulty in matching vitrinite reflectance to hydrocarbon generation (Price, 1985) and other thermal indicators such as laumontite (McCulloh and Fan, 1984). Stated differently, the expressions describe the increase in vitrinite reflectance as a function of thermal history. The change in reflectance is related to the number density of the solid vitrinite surface and is, presumably, complexly related to the kinetics of bond lysis and formation in the vitrinite during maturation. The kinetics of number density change in vitrinite must not be confused with the kinetics of the individual reactions causing the change in number density. The relationship between number density change and individual chemical reactions is not obvious. Recent comparisons of various thermal indicators from the same wells check their responses to the same paleoheat flux. To date, remarkable consistency of inversions based on vitrinite reflectance, sterane isomers, and hopane isomers (see later in the book) has been found.

We also suggest that it is better to know the quantitative degree of uncertainty in the paleoheat flux, so that one can assess the dynamic range of uncertainties of a particular thermal indicator in attempts to determine, or at least bracket, thermal maturation in a particular formation.

Testing of the theoretical formulation (Lerche *et al.*, 1984) with data from more than 400 wells in many different basins has provided the firm conviction that we have reached a satisfactory measurable degree of regional consistency from well to well. However, poor data quality in general forces a treatment of paleoheat flux variation as a linear function, which if projected too far back in time can produce unrealistic heat flow values (e.g., zero heat flow in Fig. 12 of Lerche *et al.*, 1984). To circumvent this problem, a tomographic scheme is needed to treat heat flow variations so that perturbations in paleoheat flow can be measured and heat flow values at the limit of the data resolution can be better defined.

Vitrinite reflectance is insensitive to minor variations in heat flow over short periods of geologic time. Thus inversion schemes are valuable tools for evaluating the gross behavior of heat flow history and thermal maturity (Figs. 3 and 12 of Lerche *et al.*, 1984).

II. A Potpourri of Other Thermal Indicators

We have seen that the inverse scheme for relating measurements of vitrinite reflectance collected from a well to that well's thermal history provides a basis for estimating paleoheat flux. But other thermal indicators are also available in wells. The use of actual downhole thermal indicator data addresses the basic problems on a variety of fronts: first, the maturation model for each thermal indicator must be capable of fitting the measured data as well as providing a paleoheat flux with time; second, the resulting predictions of thermal histories must be reasonably consistent for wells across a basin; third, multiple downhole thermal indicators in a single well must provide the same thermal history; fourth, the geological parameters associated with a given event (e.g., an unconformity) must show other than a random pattern of variation across the basin; fifth, any physical or chemical parameters in the maturation models of a given thermal indicator must be determined and must be independent of the basin history since they are physical and/or chemical parameters, nominally not beholden to geological evolution.

The goals of this section of the chapter are (1) to outline the logic underlying the quantitative development of inverse theory applied to any thermal indicator and to provide its relation to thermal history and (2) to apply the logical pattern to a suite of thermal indicators in turn, so that one can see how the detailed physics and/or chemistry associated with a particular indicator is making its peculiar effects felt for the given indicator. The more general problems—(1) providing quantitative models and methods of determining parameters associated with a suite of geological events, such as paleo-overpressuring, unconformities, stratigraphic age determination, overthrusting, and igneous intrusions, and (2) pointing the way to an integrated approach for handling simultaneously, and on the same footing, a variety of the corollary problems and the paleoheat flux problem—will be discussed in later chapters, as will detailed case histories illustrating relevant behaviors.

In short, the problem we address is the reconstruction of both the temperature and geological history of a sedimentary basin from present-day downhole measurements of thermal indicators. By thermal indicators we mean chemical and physical properties of a sedimentary stratum whose values are determined by the thermal and burial history of that stratum. For instance, the concentration of a chemical species, whose degradation rate depends on temperature, would constitute a thermal indicator. The temperature of a sedimentary stratum depends on (1) the heat flux and (2) the burial path of the sediment, through the variation in the thermal resistance due to changes in thermal conductivity with depth and to the intrinsic subsurface burial depth of the sedimentary stratum.

A. The Logic Underlying Inversion of Thermal Indicators

In a one-dimensional burial history, variations of *all* quantities can occur only with depth z and time t. Therefore the temperature $T(z, t)$ varies only with depth z below the sediment–water interface and with time t. Since the position of a sedimentary stratum changes with time due to compaction, unconformity erosion, and so forth, we reference the ith layer by its present position z_i. If the sedimentation, subsidence, and compaction history of the basin is known precisely, the position of any stratum throughout geological time is given by the known function $z = \zeta(t; z_i)$ and the time of sedimentation of any stratum is given by the known function $\tau_i = \tau(z_i)$. Since pressure $p(z, t)$ also affects the physical and chemical evolution of the sediments, we will assume that it, too, is known precisely.

A thermal indicator TI(z) is then related to the temperature history of the stratum in which it is measured:

$$\mathrm{TI}(z) = \int_{\tau(z_i)}^{t_{\text{present}}} F(T[\zeta(t; z_i), t], p[\zeta(t; z_i), t])\, dt \tag{3.19}$$

where F is a known function.

Now suppose that we measure a number of different thermal indicators

$\mathrm{TI}_j(z_i)$, $j = 1, \ldots, N$, each at a suite of depths $z_i = 1, \ldots, M$. Can we reconstruct the temperature history $T(z, t)$? This is one part of the inverse problem which we consider in this chapter.

Note that temperature field is a function of two variables (position and time). However, the fact that temperature must obey the heat flow equation can be used to transform the problem into one with a one-dimensional unknown. Let the heat flow equation be written as

$$LT(z, t) = 0 \tag{3.20}$$

where L, the linear differential operator of the heat flow equation, is assumed to be known (that is, the thermal conductivity K, heat capacity, source terms, etc. are known). Then the temperature is completely determined by an initial condition, say that sediment is added to the basin with a given temperature, and two boundary conditions, say that the temperature at the top of the basin is always the present-day value T_s and that the heat flux through the bottom of the basin is $Q(t) = -K(\partial T/\partial z)$. Thus we can replace the problem of determining the entire temperature field with the problem of determining the heat flux $Q(t)$ at one depth in the basin. The unknowns are reduced from a two-dimensional function to a one-dimensional function. Then at any depth in the basin we have, approximately,

$$T(z, t) = T_s + Q(t) \int_0^z dz'/K(z') \tag{3.21}$$

provided the thermal recovery time through the sedimentary overburden is shorter than the time-scale marking changes in the basement heat flux.

However, when the sedimentation, subsidence, pressure, and compaction history are not precisely known, and when the causes of temperature variation are other than a heat flux solely through the basement, then a thermal indicator, measured today, also depends on these other quantities. Suppose we measure the same set of thermal indicators at a suite of depths. Can we reconstruct *both* the thermal history [through $Q(t)$] and the unknown components of the burial history and the nonbasement heat sources? This is the general inverse problem which we consider in Chapters 8 and 9.

As we shall see directly, this general sense of logic underlies all inversion schemes for all thermal indicators, although the detailed machinations change, sometimes markedly, depending on the basic physics and chemistry involved for each thermal indicator.

B. $^{40}Ar/^{39}Ar$ as a Thermal Indicator

Harrison and Be (1983) have shown that the $^{40}Ar/^{39}Ar$ age spectrum technique yields a low-temperature paleothermometer based on a simple, predictable, physical process which provides the time scale of heating. With Harrison and Be (1983), we note that the first application of the $^{40}Ar/^{39}Ar$ age spectrum technique to a variety of meteorites allowed Merrihue and Turner (1966) and Turner *et al.* (1966) to demonstrate that the ages of both crystallization and reheating could be obtained from the data. A theoretical model proposed by Turner showed that these results could be interpreted as resulting from the diffusion loss of radiogenic ^{40}Ar ($^{40}Ar^*$) during a later thermal event. When lunar material became available, many excellent examples supported a model of radiogenic ^{40}Ar gradients due to partial outgassing (Turner, 1970). More recently, several studies of terrestrial samples have demonstrated the ability of amphiboles and K-feldspars to reveal diffusion gradients of $^{40}Ar^*$ imposed naturally as a result of a reheating event or as a result of slow cooling (Harrison and McDougall, 1980, 1982; Harrison, 1981, 1982). In conjunction with known diffusion parameters, this observation allows resolution of the age of the source and later heating event. Calculations by Harrison and McDougall (1980) indicated that detrital microcline K-feldspar is capable of preserving a record of thermal events in the temperature range ~100–200°C, over geological time periods.

The loss of $^{40}Ar^*$ from microcline has previously been modeled by diffusion in an infinitely long and wide slab of finite thickness $2l$ (Harrison and McDougall, 1980). The justification for this is that the plane sheet is a suitable description of the slablike perthite lamellae whose incoherent boundaries control the effective diffusion half-width.

The solution of Fick's second law for one-dimensional flow in a solid bounded by two parallel planes at $x = \pm 1$ with initial and boundary conditions on the concentration C in the form

$$C = C_0, \qquad -1 < x < +1,\ t = 0 \tag{3.22a}$$

$$C = 0, \qquad x = +1,\ x = -1,\ t > 0 \tag{3.22b}$$

is given by (Carslaw and Jaeger, 1947)

$$C(x, t) = C_0 \frac{4}{\pi} \sum_{n=0}^{\infty} (-1)^n (2n + 1)^{-1} \exp[-D(2n + 1)^2 \pi^2 t / 4l^2] \times \cos[(2n + 1)\pi x / 2l] \tag{3.23}$$

where $C(x, t)$ is the concentration at distance x from the center of the slab after time t, D the diffusion coefficient, and C_0 the initial uniform concentration of $^{40}Ar^*$. This expression for a constant diffusion coefficient D is valid only for a square-pulse thermal history, an unlikely scenario for a sedimentary basin. However, we can accommodate a time-dependent diffusion coefficient *exactly* by replacing Dt with $\int_0^t D(t')\,dt'$. The fraction of gas, f, remaining after time t is then given by

$$f = \frac{M_t}{M_0} = 1 - \sum_{n=0}^{\infty} 8\pi^{-2}(2n+1)^{-1} \times \exp\left[-(2n+1)^2\pi^2(4l^2)^{-1}\int_0^t D(t')\,dt'\right] \qquad (3.24)$$

where $M_0 = 2lC_0$, M_t is the mass of gas in the solid at t, and M_∞ is the mass of gas in the solid after infinite time (i.e., zero concentration). For a fraction of less than 0.5, Jost (1960) has formulated a convenient approximation to equation (3.24):

$$f \cong 2\pi^{-1/2}\left[\int_0^t D(t')\,dt'/l^2\right]^{1/2} \qquad (3.25)$$

Substituting the well-known Arrhenius relationship, $D/l^2 = D_0/l^2 \exp(-E/RT)$, where D_0/l^2 is the frequency factor/grain size parameter, E the activation energy, R the gas constant, and T the absolute temperature, we can rewrite equation (3.25) in the form

$$f \cong 2\pi^{-1/2}(D_0/l^2)^{1/2}\left\{\int_0^t \exp[-E/RT(t')]\,dt'\right\}^{1/2} \qquad (3.26)$$

Thus to solve for the temperature experienced by a suite of samples, we require knowledge of the diffusion parameters for $^{40}Ar^*$ in microcline (i.e., E and D_0/l^2) and an estimate of the time scale of heating. Harrison and McDougall (1980) calculated values of D_0/l^2 from the measured ^{39}Ar release from microcline samples during stepwise gas extractions. When plotted against the reciprocal absolute temperature of the extraction step, the data below about 900°C defined an activation energy of 29 ± 2 kcal mol^{-1} and a D_0/l^2 of 6^{+14}_{-4} s^{-1}. This activation energy agrees fairly well with previous estimates of 24, 26, and 32 kcal mol^{-1} (Evernden *et al.*, 1960; Laughlin, 1969; Gerling, *et al.*, 1963) but contrasts sharply with estimates of E from homogeneous K-feldspars of 40–52 kcal mol^{-1} (Harrison and McDougall, 1980).

One way to circumvent the large uncertainty in the precise value of D_0/l^2 is to take ratios. Thus consider a suite of values f_i measured at depths $z = z_1, z_2, \ldots, z_n$ corresponding to ages $t_1, \ldots, t_n$, respectively. Then D_0/l^2 is given by

$$\frac{4}{\pi} D_0/l^2 = \sum_{i=1}^{n} f_i^2 / \sum_{i=1}^{n} \left\{ \int_0^{t_i} \exp[-E/R\ T(t')\ dt' \right\} \tag{3.27}$$

so that we can use equation (3.27) to write formula (3.26) in the form

$$\frac{f(z)^2}{\sum_{j=1}^{n} f(z_j)^2} = \frac{\int_0^{t} \exp[-E/RT(t')]\ dt'}{\sum_{j=1}^{n} \left\{ \int_0^{t_j} \exp[-E/T(t')]\ dt' \right\}} \tag{3.28}$$

which eliminates the parameter D_0/l^2, and where current depth z corresponds to age t.

We regard as a measure of the degree of agreement between observations and theory the quantity

$$X^2 \equiv [n(n-2)]^{-1} \sum_{i=1}^{n} \left\{ \frac{f(z_i)^2}{\sum_{j=1}^{n} f(z_j)^2} - \frac{\int_0^{t_i} \exp[-E/RT(t')]\ dt'}{\sum_{j=1}^{n} \left(\int_0^{t_j} \exp[-E/RT(t')]\ dt' \right)} \right\}^2 \tag{3.29}$$

Calculation of the best heat flux variation with time then proceeds in the manner described for vitrinite reflectance. For a model choice of $Q(t)$ [$\equiv Q_0(1 + \beta t)$, for example] one varies the free parameter β until X^2 takes on its smallest value, thereby providing an assessment of the time-dependent heat flux.

Having determined the best heat flux with time, an $f(t)$ value can be calculated for each depth at the present day. Let z correspond to an age t for the sedimentary layer now at depth z. Then, using the best heat flux with time, we have

$$f(t) = \left[\sum_{j=1}^{n} f(t_j)^2 \right]^{1/2} \left\{ \frac{\int_0^{t} \exp[-E/RT(t')]\ dt'}{\sum_{j=1}^{n} \int_0^{t_j} \exp[-E/RT(t')]\ dt'} \right\}^{1/2} \tag{3.30}$$

which is to be compared with observations of $f(t)$ with depth.

C. Optical Rotation as a Thermal Indicator

It is well known (see, e.g., Fenske *et al.*, 1942) that the optical rotation of cholestanes and hopanes in the sterane and triterpane families provides a measure of the degree of thermal maturity to which hydrocarbons have been exposed. Indeed, the Pennsylvania Grade Crude Oil Association has been using the optical activity of lubricating oils to distinguish Pennsylvania lubricating oils from those of other producing areas for the last 40 years or more.

The basis of the argument is relatively simple. Organic molecules are solely dextral at deposition. Under the influence of time and temperature there is a loss of dextral molecules (N_R) in favor of sinistral molecules (N_L) with the total number being conserved until, at equilibrium, the numbers of dextral and sinistral molecules are equal under racemization processes.

Because of the difference in $N_R - N_L$ the plane of polarization of optical light sent through a body containing the molecules will rotate. For a column length L of material the total angular rotation $\Delta\phi$ of the plane of light of angular frequency ω is given by (Jackson, 1975)

$$\Delta\Phi = L\omega\left[\left(\frac{1 + 2N_{Rg}}{1 - N_{Rg}}\right)^{1/2} - \left(\frac{1 + 2N_{Lg}}{1 - N_{Lg}}\right)^{1/2}\right] \qquad (3.31)$$

where g is a molecular structure factor (fixed for a given molecular weight) and where we have used the classical Lorenz–Lorentz formula for the refractive index (Jackson, 1975).

Now the stereoisomerization of dextral and sinistral molecules is usually represented as temperature-dependent first-order reactions:

$$\frac{dN_R}{dt} = -k(N_R - \tfrac{1}{2}N_0)\exp(-T_0/T) \qquad (3.32)$$

$$\frac{dN_L}{dt} = k(\tfrac{1}{2}N_0 - N_L)\exp(-T_0/T) \qquad (3.33)$$

where $N_0 = N_R + N_L =$ constant, and where T_0 is the activation temperature (related to the activation energy E through the gas constant R with $E = RT_0$).

The time-dependent solutions to equations (3.32) and (3.33) are

$$N_R = \tfrac{1}{2}N_0\left(1 + \exp\left\{-k\int_0^t dt' \exp[-T_0/T(t')]\right\}\right) \tag{3.34a}$$

$$N_L = \tfrac{1}{2}N_0\left(1 - \exp\left\{-k\int_0^t dt' \exp[-T_0/T(t')]\right\}\right) \tag{3.34b}$$

The degree of rotation per unit column length per unit angular frequency is

$$a \equiv \frac{\Delta\Phi}{L\omega} = \left[\left(\frac{1 + 2N_R g}{1 - N_R g}\right)^{1/2} - \left(\frac{1 + 2N_L g}{1 - N_L g}\right)^{1/2}\right] \tag{3.35}$$

For steranes and triterpanes taken under typical sedimentary conditions, a is normally much smaller than unity, implying that the departures of N_R and N_L from $N_0/2$ are quite small.

Then expand equation (3.35) around $N_R = N_0/2$ and $N_L = N_0/2$ to obtain

$$a \cong \frac{1}{2}\left(\frac{1 + N_0 g}{1 - N_0 g/2}\right)^{1/2} \frac{N_0 g(3 - N_0 g)}{1 - N_0^2 g^2} \exp\left\{-k\int_0^t \exp[-T_0/T(t')]\, dt'\right\} \tag{3.36}$$

In order to eliminate the dependence on the molecular structure factor g, we take the ratio of (3.36) for optical rotations on samples of ages t_i and t_j (current depths z_i and z_j). Then

$$\ln(a_i/a_j) = -k[\mathrm{TTI}_0(t_i) - \mathrm{TTI}_0(t_j)] \tag{3.37}$$

where

$$\mathrm{TTI}_0(t_i) = \int_0^{t_i} \exp[-T_0/T(t')]\, dt' \tag{3.38}$$

and the integral in equation (3.38) is taken along the burial path of the sediments. To eliminate the dependence on the rate constant k, we ratio (3.37) with respect to the sum of all measurements, that is,

$$\frac{\ln(a_i/a_j)}{\sum_{p=1}^{n} \ln(a_p/a_j)} = \frac{\mathrm{TTI}_0(t_i) - \mathrm{TTI}_0(t_j)}{\sum_{p=1}^{n} \mathrm{TTI}_0(t_p) - n\mathrm{TTI}_0(t_j)} \tag{3.39}$$

A measure of the degree of global mismatch between the observations [the left-hand side of equation (3.39)] and the predictions [the right-hand side of equation (3.39)] is then provided by

$$W^2 \equiv [n(n-2)]^{-2} \sum_{i=1}^{n} \sum_{j=1}^{n} \left[\frac{\ln(a_i/a_j)}{\sum_{p=1}^{n} \ln(a_p/a_j)} - \frac{\mathrm{TTI}_0(t_i) - \mathrm{TTI}_0(t_j)}{\sum_{p=1}^{n} \mathrm{TTI}_0(t_p) - n\mathrm{TTI}_0(t_j)} \right] \quad (3.40)$$

Since the burial history is known, the paleoheat flux and its variation with time are the only unknowns in equation (3.40). Hence we can invert equation (3.40) to determine $Q(t)$ as with the previous thermal indicators by demanding a minimum value for W^2. Once $Q(t)$ is determined, the predicted behavior of the degree of rotation at a time t along a burial path is given by

$$a(t) = \frac{1}{n} \sum_{j=1}^{n} a_j \exp\left[\frac{\mathrm{TTI}_0(t) - \mathrm{TTI}_0(t_j)}{\sum_{p=1}^{n} \mathrm{TTI}_0(t_p) - n\mathrm{TTI}_0(t_j)} \sum_{p=1}^{n} \ln(a_p/a_j) \right] \quad (3.41)$$

where the integral $\mathrm{TTI}_0(t)$ is carried out along the chosen burial path.

D. Hopane and Sterane Isomerization and Aromatization as Thermal Indicators

MacKenzie and McKenzie (1983) have presented a very detailed study of two reactions which occur in steroid hydrocarbons; one is an isomerization reaction, the other an aromatization reaction. A third reaction they investigated is another isomerization which occurs in hopane molecules and which is in many ways similar to the sterane isomerization. From MacKenzie and McKenzie (1983) we note that sedimentary steranes are thought to be derived from natural product sterols during early diagenesis (e.g., Rhead *et al.*, 1971; Gaskell and Eglinton, 1976; Dastillung and Albrecht, 1977). Almost all the natural sterols whose structures have been determined have the 8β(H), 9α(H), 14α(H), 17α(H), 20 R configuration. Different host organisms have different isomers at C-24. Since the apolar gas chromatography column used by MacKenzie and McKenzie (1983) could not resolve these isomers in any of the compounds of interest here (Maxwell *et al.*, 1980), the chiral center at C-24 is not considered further. The representation of different isomers with the α, β convention first required the molecular formula to be drawn according to a particular convention. Then α(H) implies that C—H bond at the relevant chiral center, which forms part of a ring system, points into the page. β(H) implies that the bond points out of the page, and the hydrogen lies above the page. The carbon numbering system follows an international convention. For the sterol molecule to be converted to the sterane requires the alcohol (—OH) group on C-3 be removed by dehydration and the double bond joining C-5 to C-6 be reduced chemically by the addition of hydrogen. Both processes occur relatively shortly after diagenesis, and natural product steroids without the double bond have been identified in organisms (MacKenzie *et al.*, 1982). This reduction produces steranes with only the 8β(H), 9α(H), 14α(H), 17α(H), 20 R configuration but with mixtures of isomers at C-5 (α and β) and at C-24 (R and S). The amounts of the 5β(H) steranes formed by reduction are usually small compared with those of the 5β(H) steranes. As the temperature increases, isomerization occurs at many centers (MacKenzie *et al.*, 1980). We concentrate here on that which occurs at C-20. None of the isomerization reactions proceeds at significant rates until completion of the dehydration reaction and the reduction of double bonds. The biologically inherited 20 R isomer is gradually converted to a mixture of 20 R and 20 S isomers, whose concentrations are approximately equal at equilibrium. Steroid and hopanoid hydrocarbons are found in most organic-rich sedimentary rocks with mild thermal histories (Ourisson *et al.*, 1979). The relative concentrations of the 20 R and 20 S isomers of the 5α(H), 8β(H), 9α(H), 14α(H), 17α(H) skeleton in a particular sediment can be measured by computerized gas chromatography–mass spectrometry (GC-MS). The isomerization reaction rate can be determined from the ratio y of the concentrations:

$$y = [20\ \mathrm{S}]/([20\ \mathrm{R}] + [20\ \mathrm{S}]) \tag{3.42}$$

Another of the three reactions studied by MacKenzie and McKenzie (1983) involves the removal of several hydrogen atoms and one methyl group from a C-ring monoaromatic steroid hydrocarbon (Ludwig *et al.*, 1981; Seifert *et al.*, 1983; MacKenzie *et al.*, 1981) to form a triaromatic steroid. "Unlike the isomerization reactions, there is no back reaction and the aromatization reaction is assumed to proceed to completion. All the monoaromatic species, M, are converted to triaromatic, T, steroids. The extent of aromatization, x, is given by"

$$x = [\mathrm{T}]/([\mathrm{M}] + [\mathrm{T}]) \tag{3.43}$$

The third reaction studied by MacKenzie and McKenzie (1983) is an isomerization which occurs at C-22 in hopane hydrocarbons and converts the 22 R isomer to a mixture of 22 R and 22 S. The isomerization at C-22 in C_{32} compounds was first investigated by Ensminger *et al.* (1974, 1977). As Ensminger, *et al.* (1978) pointed out, "there is a close structural resemblance between the C-20 site in steranes and C-22 in hopanes. The isomerization reactions which occur at these sites are then very similar."

On the assumption that the reactions are unimolecular and first order, MacKenzie and McKenzie (1983) obtained numerical values for the activation energy E and frequency factor A in the rate constant formula $k = A \exp(-E/RT)$. They quote the following values for the reaction rate constants:

Sterane aromatization:

$$A = 1.8 \times 10^{14}\ \mathrm{s}^{-1}, \quad E = 200\ \mathrm{kJ\ mol}^{-1}$$

Sterane isomerization at C-20:

$$A = 0.006\ \mathrm{s}^{-1}, \quad E = 91\ \mathrm{kJ\ mol}^{-1}, \quad \gamma = 1.174$$

Hopane isomerization at C-22:

$$A = 0.016\ \mathrm{s}^{-1}, \quad E = 91\ \mathrm{kJ\ mol}^{-1}, \quad \gamma = 1.564$$

where γ measures the ratio of the forward (20 R → 20 S, 22 R → 22 S) to the backward reaction.

When compounds R and S are converted into each other at rates k_1 and k_2 then for first-order chemical reactions we have

$$\frac{dR}{dt} = k_1 S - k_2 R \tag{3.44}$$

$$\frac{dS}{dt} = k_2 R - k_1 S \tag{3.45}$$

where R and S are the concentrations of compounds R and S. If neither is destroyed by other reactions, then

$$R + S = C, \tag{3.46}$$

where C is a constant.

This is basically a scheme of evolution similar to the optical rotation thermal indicator. If we assume both reactions are unimolecular and first order, then from the Arrhenius law we have

$$k_1 = A_1 \exp(-E_1/RT) \tag{3.47a}$$

$$k_2 = A_2 \exp(-E_2/RT) \tag{3.47b}$$

and

$$K = A_2 A_1^{-1} \exp[-(E_2 - E_1)/RT] \tag{3.47c}$$

In the sterane isomerization reaction $K = O(1)$, and the temperature dependence of K is small compared with those of k_1 and k_2 (van Grass *et al.*, 1982; Petrov *et al.*, 1976). Hence we can write

$$k_2 = \gamma k_1 \tag{3.48}$$

and consider γ to be constant and independent of temperature over the range with which we are concerned.

As MacKenzie and McKenzie (1983) point out, "the two isomerization reactions involve epimerization not racemization. Therefore the 20 S and 20 R forms of the sterane hydrocarbons are not mirror images of each other, and so k_1 is not necessarily equal to k_2." The same remarks apply to the isomerization of hopanes. As usual, equilibrium occurs when

$$dA/dt = O = dB/dt \tag{3.49}$$

$$\text{i.e.,} \quad k_1 B = k_2 A \tag{3.50}$$

Hence the equilibrium constant K is given by

$$K = B/A = k_2/k_1 = \gamma \tag{3.51}$$

The degree of isomerization is

$$y = (1 - R/C) \tag{3.52}$$

which, solving equations (3.46) and (3.47), yields the isomerization at time t as

$$y = \gamma(l + \gamma)^{-1}\left(1 - \exp\left\{-A \int_0^t \exp[-T_0/T(t')]\ t'\right\}\right) \quad (3.53)$$

for a sample on the sedimentary surface at $t = 0$, where $T_0 = E/R$ and where the integral in (3.53) is taken along the later ($t > 0$) burial path of the sedimentary unit. Hence, by measuring the degree of isomerization with depth and converting depth to age, we have a burial history record of the thermal history. Again, since the only unknown is the variation of paleoheat flux with time, it should be possible to invert equation (3.53) to determine the paleoheat flux.

Consider then a suite ($y_1, y_2 \ldots, y_n$) of present-day measurements of the degree of isomerization with depth ($z_1, \ldots, z_n$) and with sedimentary ages $t_1, \ldots, t_n$, respectively.

Then from (3.53) we can write

$$\ln(1 - y_i\frac{\gamma + 1}{\gamma}) = -A\ \mathrm{TTI}(t_i) \quad (3.54)$$

where

$$\mathrm{TTI}(t_i) = \int_0^t \exp[-T_0/T(t')]\ dt' \quad (3.55)$$

taken along the burial history of the sedimentary unit from deposition until the present day.

If it is thought that the rate constant A is too uncertain to be of value, we can eliminate it from the problem by taking ratios, thus

$$\left\{\frac{\ln[1 - y_i(1 + \gamma)/\gamma]}{\sum_{j=1}^{n} \ln[1 - y_j(1 + \gamma)/\gamma]}\right\} = \frac{\mathrm{TTI}(t_i)}{\sum_{j=1}^{n} \mathrm{TTI}(t_j)} \quad (3.56)$$

A measure of the degree of mismatch between observations [the left-hand side of (3.56)] and theoretical predictions [the right-hand side of (3.56)] is then provided through

$$V^2 = n^{-1}(n - 2)^{-1} \sum_{i=1}^{n} \left\{\frac{\ln[1 - y_i(1 + \gamma)/\gamma]}{\sum_{j=1}^{n} \ln[1 - y_j(1 + \gamma)/\gamma]} - \frac{\mathrm{TTI}(t_i)}{\sum_{j=1}^{n} \mathrm{TTI}(t_j)}\right\}^2 \quad (3.57)$$

Again, since the burial history paths of the sediments are known, the only unknown in equation (3.57) is the temperature along each path, and since, as before, the temperature at an instant of time is directly tied to the heat flux at that time, the fundamental unknown in (3.57) is the variation of heat flux with time. Hence by varying the parameters in the heat flux function with time, given previously, we can determine the parameters which minimize V^2. This then determines the heat flux $Q(t)$ most consistent with the observations. With this best heat flux we can then predict the isomerization ratio $y(t)$ at any time t along a given burial path from

$$y(t) = \gamma(1 + \gamma)^{-1}\left[1 - \exp\left(\left\{\frac{\sum_{j=1}^{n} \ln[1 - y_j(\gamma + 1)/\gamma]}{\sum_{j=1}^{n} \mathrm{TTT}(t_j)}\right\}\mathrm{TTI}(t)\right)\right] \quad (3.58)$$

In the case of sterane aromatization, for which no back reaction occurs, the degree of aromatization at time t for a sample deposited at $t = 0$ is then simply given by

$$x = 1 - \exp[-A \int_0^t \exp[-T_A/T(t')]\, dt'] \quad (3.59)$$

where MacKenzie and McKenzie (1983) quote the values $A = 1.8 \times 10^{14}$ s^{-1}, $T_A = E_A/R$, and $E_A = 200$ kJ mol^{-1} and where the integral is taken along the burial path of the sediment.

A line of reasoning similar to that just presented for the isomerization allows us to write

$$\ln[1 - x_i] = -A\mathrm{TTI}_A(t_i) \quad (3.60)$$

where

$$\mathrm{TTI}_A(t_i) = \int_0^{t_i} \exp[-T_A/T(t')]\, dt' \quad (3.61)$$

Again eliminate the rate constant by summing over the data points. Then a measure of the degree of misfit between observations and predictions is provided through

$$U^2 = n^{-1}(n - 2)^{-1} \sum_{i=1}^{n} \left[\frac{\ln(1 - x_i)}{\sum_{j=1}^{n} \ln(1 - x_j)} - \frac{\mathrm{TTI}_A(t_i)}{\sum_{j=1}^{n} \mathrm{TTI}_A(t_j)}\right]^2 \quad (3.62)$$

Again determine the heat flux with time which minimizes U^2. Then the prediction of the degree of aromatization at time t along a given burial history path is given through

$$x(t) = 1 - \exp\left\{\left[\frac{\sum_{j=1}^{n} \ln(1 - x_j)}{\sum_{j=1}^{n} \mathrm{TTI_A}(t_j)}\right]\mathrm{TTI_A}(t)\right\} \tag{3.63}$$

Hence we again see shaping functional forms in (3.57) and (3.62) dependent on the thermal history with the thermochemical rate constants normalized out of the problem by taking ratios.

E. Pollen Translucency as a Thermal Indicator

This method of estimating thermal maturity qualitatively was introduced and described by Gutjahr (1966) and extensively investigated by Grayson (1975) for *Carya*. By and large, a general decrease in translucency occurs with increasing depth. The pragmatic advantages of this technique, as detailed by Grayson, are

(a) Translucency measurements are made on specific taxa of palynomorphs which were incorporated into the sediments at the time of deposition. Other measurements of maturation in clastic sediments often include material secondarily derived from older sediments.

(b) The maturation characteristics of different components of dispersed organic matter in sediments differ widely. Since translucency deals with one component, it provides a more definite measurement than is possible by chemical analysis of the whole organic fraction.

(c) Only very small quantities of samples are needed.

(d) Drilling mud additives do not effect translucency measurements.

In order to convert the pollen translucency from a qualitative device into a quantitative tool for estimating thermal maturity we proceed as follows. Let monochromatic light of intensity I_0 be shone normally onto a sample pollen of thickness L. Then the amount of transmitted light, I, will be (Born and Wolf, 1964)

$$I = I_0 \exp[-La(\sigma_{sc} + \sigma_a)] \tag{3.64}$$

where σ_{sc} is the scattering cross section and σ_a the absorption cross section. Here a is a molecular scaling constant.

Since both the scattering and absorption cross sections are proportional to the number density n of the molecular component decreasing the translucency, we can combine them to write

$$I = I_0 \exp(-Lbn) \tag{3.65}$$

where b is a different constant.

The translucency τ is

$$\tau = I/I_0 = \exp(-Lbn) < 1 \tag{3.66}$$

For later convenience write

$$bn = -L^{-1} \ln\tau \tag{3.67}$$

Again take the number density of the molecular component decreasing the transparency to vary in an Arrhenius manner with

$$dn/dt = -kn \exp[-E(T)/RT] \tag{3.68}$$

where k, the reaction rate, is a constant. Again bear in mind that the pollen makes up a solid, as opposed to a single molecule, so that the activation energy $E(T)$ will have the ''standard'' solid-state format (Seitz, 1946; Smith, 1961)

$$E(T) = RT_0 - RT(T/T_D) \tag{3.69}$$

where T_0 and T_D are constants. Note further that, just as for vitrinite, the solid will have a work function energy (temperature) which must be overcome before any reaction proceeds at all. Hence equation (3.68) will describe the evolution of n provided $T > T_c$, where T_c, is the work function temperature. From (3.68) we obtain

$$n = n_0 \exp\{-k \int_0^t \exp[-T_0/T(t') + T(t')/T_D]\, dt'\} \tag{3.70}$$

or, with

$$\mathrm{TTI_c}(t) = \int_0^t \exp[-T_0/T(t') + T(t')/T_D]dt' \tag{3.71}$$

we can write

$$n = n_0 \exp[-k\, \mathrm{TTI_c}(t)] \tag{3.72}$$

where n_0 is the pollen number density at deposition and where it is understood that the integrand in (3.71) is to be set to zero whenever $T < T_c$. Since b, n_0, and k are difficult parameters to determine with precision, they can be normalized out of the problem as follows. Consider a suite of samples of pollen taken from depths $z_i, \ldots, z_n$ corresponding to sedimentary ages $t_1, t_2, \ldots, t_n$ of thicknesses $L_1, L_2, \ldots, L_n$ and with translucencies $\tau_1, \tau_2, \ldots, \tau_n$. Then from (3.67) we obtain

$$k[\mathrm{TTI_c}(t_i) - \mathrm{TTI_c}(t_j)] = \ln[L_j \ln \tau_i / L_i \ln \tau_j] \tag{3.73}$$

which normalizes the constants b and n_0 out of the problem. We can eliminate the rate constant k from equation (3.73) by summing over all samples to obtain

$$\frac{\mathrm{TTI_c}(t_i) - \mathrm{TTI_c}(t_j)}{n\,\mathrm{TTI_c}(t_i) - \sum_{p=1}^{n} \mathrm{TTI_c}(t_p)} = \frac{\ln(L_j \ln \tau_i / L_i \ln \tau_j)}{\sum_{p=1}^{n} \ln[(L_p \ln \tau_i)/(L_i \ln \tau_p)]} \tag{3.74}$$

The right-hand side of equation (3.74) is directly measurable with increasing sedimentary depth, while the left-hand side contains the thermal maturity integrals (3.71) taken along the burial tracks of the sedimentary layers (with ages $t_1, t_2, \ldots, t_n$) corresponding to depths $(z_1, z_2, \ldots, z_n)$.

The unknown is the heat flux $Q(t)$. Hence by following a prescription similar to that outlined for the previous thermal indicator measurements with depth, we can determine the heat flux variations with time. A quantitative measure of the degree of mismatch of observations and predictions is provided by the residual least-squares function

$$S^2 = [n(n-2)]^{-2} \sum_{i=1}^{n} \sum_{j=1}^{n} \left\{ \frac{\ln[(L_j \ln \tau_i)/(L_i \ln \tau_j)]}{n^{-1} \sum_{p=1}^{n} \ln[(L_p \ln \tau_i)/(L_i \ln \tau_p)]} - \frac{(\mathrm{TTI_c}(t_i) - \mathrm{TTI_c}(t_j))}{(\mathrm{TTI_c}(t_i) - n^{-1} \sum_{p=1}^{n} \mathrm{TTI_c}(t_p))} \right\}^2 \tag{3.75}$$

and the heat flux is varied until S^2 is minimized. With the paleoheat flux thus determined, we can write the assessment of pollen translucency $\tau(t)$ at time t along a given burial history path in the form

$$\tau(t) = \exp\{-\alpha \exp[-f\,\mathrm{TTI_c}(t)]\} \tag{3.76}$$

where

$$\alpha = \exp[fn^{-1} \sum_{i=1}^{n} \mathrm{TTI_c}(t_i) + n^{-1} \sum_{i=1}^{n} \ln(-L_i^{-1} \ln \tau_i)] \tag{3.77a}$$

and

$$f = \left\{ \sum_{i=1}^{n} [\ln(-L_i^{-1} \ln \tau_i)]^2 - n^{-1} \left[\sum_{i=1}^{n} \ln(-L_i^{-1} \ln \tau_i) \right]^2 \right\}$$
$$\times \left\{ n^{-1} \left[\sum_{i=1}^{n} \ln(-L_i^{-1} \ln \tau_i) \right] \sum_{i=1}^{n} \mathrm{TTI_c}(t_i) - \sum_{i=1}^{n} [\mathrm{TTI_c}(t_i) \ln(-L_i^{-1} \ln \tau_i)] \right\}^{-1} \tag{3.77b}$$

and where the integral $\mathrm{TTI_c}(t)$ is taken along the chosen burial history path to time t.

F. Fission Scar Tracks in Apatite as a Thermal Indicator

1. General Remarks

Here we follow closely, and paraphrase, the outline and rationale behind the apatite method as so eloquently expressed by Gleadow *et al.* (1983). Various environmental factors have been shown to affect the stability of fission tracks over geological time, but by far the most important and widely studied of these is temperature. Other factors such as pressure, shock, deformation, and fluids have been said to have no effect or very little additional effect compared with that of temperature (e.g., Fleischer *et al.*, 1965, 1975; Naeser, 1979). Fission tracks have also been shown to remain stable under conditions of sample weathering (Gleadow and Lovering, 1974), which is important when considering the significance of ages of detrital minerals.

All minerals used for fission track dating show track fading with increasing temperature. The sensitivity of different minerals to this thermal annealing, however, varies greatly. Of the three minerals considered by Gleadow *et al.* (1983), apatite is the most easily annealed and is the most useful for studies of thermal history. The temperatures required to anneal tracks in apatite are discussed in detail later. Tracks in zircon and sphene are much more resistant to annealing; the effective track retention temperatures over geological time are about 200 ± 50°C and 250 ± 50°C, respectively (Gleadow and Brooks, 1979; Harrison *et al.*, 1979).

In fission track annealing, as for other maturation indicators such as vitrinite reflectance, the effects of temperature and time are to some extent interchangeable. The same degree of track fading can be obtained from a short exposure to high temperatures or prolonged exposure to much lower temperatures. The usual method of studying annealing is to heat a track-bearing mineral for various times at different temperatures in the laboratory and measure the reduction in track density. A number of different studies of the annealing properties of apatite have given very similar results (Naeser and Faul, 1969; Wagner and Reimer, 1972; Nagpaul *et al.*, 1974). Heating for 1 h produced track fading in this mineral between about 250 and 350°C. Points representing the same degree of track fading appear to lie on straight lines on an Arrhenius plot (log time against inverse temperature), at least for the times between a few minutes and several months in studies in the laboratory. For 1 Myr the expected range based on extrapolation of the laboratory data would be about 50 to 150°C.

Laboratory annealing studies are nearly always performed on freshly induced tracks and it is assumed that these will represent the behavior of fossil tracks under conditions of geological annealing. However, there is growing evidence that the radiation damage of fossil tracks may undergo a gradual reorganization over long periods of time (Gleadow *et al.*, 1983). This natural ''aging'' process could result in fossil tracks having annealing properties different from those of fresh induced tracks. Thus it will be more important to understand the annealing behavior of the fossil tracks, because it is always naturally ''aged'' tracks that are annealed in geological environments.

Gleadow *et al.* (1983) showed the reduction in track density for fossil and induced tracks in an apatite heated at various temperatures for 1 h. Fossil tracks in this apatite do not begin to anneal until a significantly higher temperature (300°C) is reached compared with induced tracks (200°C). However, they then anneal more quickly to reach zero track density at the same point (375°C) as the induced tracks. Detailed arguments presented by Gleadow *et al.* (1983) suggest that the fossil track data more closely represent the behavior of tracks under geological annealing.

A more direct way to assess the fading of fission tracks in apatite under geological conditions is to study the natural annealing observed at elevated temperatures in deep boreholes (Naeser, 1979, 1981; Gleadow and Duddy, 1981). As temperature increases with depth, the apparent fission track age of apatites is observed to decrease, eventually reaching zero at the bottom of the track annealing zone. This apparent decrease in apatite age is usually referred to as the age reduction.

The relative effects of temperature and time on fission track annealing can be estimated by comparing the borehole and laboratory annealing results for fossil tracks. The track annealing zone derived in this way is significantly narrower than earlier estimates based on extrapolation from laboratory annealing data alone (Gleadow *et al.*, 1983). For geological annealing the time factor is relatively less significant than temperature. For example, changing the duration of heating by two orders of magnitude, from 1 to 100 Myr, changes the position of the annealing zone by only 25°C.

The reduction in fission track age observed in detrital apatites from an exploration well can thus be used to estimate paleotemperatures.

2. *The Areal Track Number Density Method*

To handle the problem of determining paleoheat flow from the areal number density of fission tracks we proceed as follows:

The rate of creation of fission tracks is proportional to uranium concentration (U) in the apatite sample. Since the half-life of uranium is of order 10^{10} yr while sedimentary basins are less than or of order 10^8–10^9 yr old, the rate of creation of fission tracks is constant. The rate of decay of fission tracks seems to be proportional to the number present by an Arrhenius-type law (Gleadow *et al.*, 1983). In sum, then, the total areal number density of tracks can be taken to follow the law

$$dN/dt = k_0 LU - k_1 n \exp(-T_c/T) \tag{3.78}$$

where $T_c \cong 572 \pm 50$ K, $k_1 \cong 3 \times 10^4$ yr^{-1} (Gleadow *et al.*, 1983). Here k_0 is the rate at which uranium produces tracks and L is the original track length.

For an apatite sample laid down at time $t = 0$, with zero tracks, at a later time t the areal track density N is given by solving equation (3.78) as

$$N = k_0 LU \int_0^t dt' \exp\left\{-k_1 \int_{t'}^{t} \exp[-T_c/T(t'')]\, dt''\right\} \tag{3.79}$$

where the temperature $T(t)$ varies with time along the burial history track of the apatite because of deeper burial with time and because the heat flux varies with time.

For distribution of apatite samples with depth we measure $N(z)$ directly. The current depth z is related to the deposition time t by the burial history. The concentration of uranium in each sample is determined by neutron irradiation in the laboratory (Gleadow *et al.*, 1983), as is the rate constant k_0.

It follows that the track density with depth can be used to unravel the variation of paleoheat flux with time since the temperature at any instant of time t when the apatite is at depth z is given by

$$T(t, z(t)) = T_s + Q(t) \int_0^{z(t)} dz'/K(z') \tag{3.80}$$

where $Q(t)$ is the time-variable paleoheat flux, $K(z)$ the thermal conductivity, and T_s the surface temperature.

Thus, by measuring the left-hand side of

$$\begin{aligned} N(z_i)/k_0U(z_i) &= \int_{-t_i}^{0} dt' \exp\left\{-k_1 \int_{t'}^{0} dt'' \exp[-T_c/T(t'')]\right\} \\ &\equiv \text{APTTI}(t_i) \end{aligned} \tag{3.81}$$

(where current depth z_i corresponds to sediment age t_i), we have only the one unknown $Q(t)$ entering equation (3.81).

If the rate constant k_0 for uranium production of fission tracks is thought to be not too well known, it can be normalized out of the problem by taking ratios.

In a sample of detrital apatite the areal number density of fission tracks, $N(z_i)$, for each measured depth z_i and the amount of uranium $U(z_i)$ in the sample are first converted to a normalized NU index, which is a direct measure of the maturity of the sample:

$$\text{NU}(z_i) = \frac{N(z_i)/U(z_i)}{\sum_{j=1}^{n} N(z_j)/U(z_j)} \tag{3.82}$$

where $\text{NU}(z)$ is the maturity index at depth z and n is the number of measured N and U pairs at the various depths.

A theoretical expression directly comparable to the NU index is

$$\text{INORM}(i) = \frac{\text{APTTI}(i)}{\sum_{j=1}^{n} \text{APTTI}(j)} \tag{3.83}$$

where $\text{APTTI}(i)$ is the time–temperature integral characterizing the maturity of an apatite sample deposited at time t_i and presently at depth z_i.

Each integral is taken along the burial path of a sample deposited at time t_i. The T in the above expression is the temperature along the burial path given by equation (3.80).

The basic problem is to find $Q(t)$ to minimize the differences between the theoretical INORM(i) index and the measured NU(i) index; that is, we wish to minimize

$$Y^2 \equiv \sum_{i=1}^{n} [\mathrm{INORM}(i) - \mathrm{NU}(z_i)]^2 \tag{3.84}$$

where the summation is taken over all detrital apatite samples in a well. We take as our goodness-of-fit measure MSR defined as

$$\mathrm{MSR}_Y = \log_{10}[Y^2/n(n-2)] \tag{3.85}$$

With the paleoheat flux determined by minimizing Y^2, the predicted behavior of areal track density to uranium concentration with depth z at time t along a chosen burial history path is given through

$$N(z(t))/U(z(t)) = \mathrm{APTTI}(t) \frac{\sum_{j=1}^{n} N(z_j)/U(z_j)}{\sum_{j=1}^{n} \mathrm{APTTI}(t_j)}$$

One potential drawback to the areal number density method is that it requires that *all* lengths of tracks be recorded in order to obtain an accurate measure of number. Clearly, shorter tracks are easier to overlook, leading to an anomalously low count. This, in turn, would bias the determination of heat flux in the direction of determining a higher heat flux than actually obtained, since if fewer tracks are measured rapid annealing (high heat flux) is implied.

3. The Average Track Length Method

In attempts to circumvent the potential problem of underestimating the areal number density by miscounting short fission tracks, we can use methods more beholden to longer track lengths so that a few "missing" shorter tracks will not influence the answer much. When the average track length method is used it does point in the direction of alleviating some of these problems. The detailed mathematical manipulations are recorded in part 4 of this subsection since the mean track length method requires some knowledge of the evolution of the distribution of tracks and, in the interests of brevity, it seemed appropriate to combine the detailed mathematical development.

The mean track length in samples of detrital apatite, $l(z_i)$, is measured for a suite of depths $z_1, z_2, \ldots, z_N$, corresponding to sedimentary ages $t_1, t_2, \ldots, t_N$.

Option 1. In order to circumvent any difficulties associated with incomplete knowledge of the track length at initial production, we first convert the measured mean track lengths to a normalized index AL(z_i) which is a direct measure of thermal maturity:

$$\mathrm{AL}(z_i) = l(z_i)/\sum_{j=1}^{n} l(z_j) \tag{3.86}$$

A theoretical expression directly comparable with AL(z_i) is (see subsection 4)

$$\mathrm{JNORM}(i) = L(i)/\sum_{j=1}^{n} L(j) \tag{3.87}$$

where $L(i)$ is a time–temperature integral, again characterizing the maturity of a detrital apatite sample deposited at time t_i and presently at depth z_i. The expression for L is

$$L(i) = t_i^{-1} \int_0^{t_i} \exp\{-k_1 \int_{t'}^{t_i} \exp[-T_c/T(t'')]\, dt''\}\, dt' \tag{3.88}$$

where the parameters k_1 and T have the same meaning as in the previous subsection.

Once more our task is to find $Q(t)$ to minimize the difference between the theoretical JNORM(i) index and the measured AL(i) index; that is, we wish to minimize

$$X^2 \equiv \sum_{i=1}^{n} [\mathrm{AL}(z_i) - \mathrm{JNORM}(i)]^2 \tag{3.89}$$

where the summation is taken over all detrital apatite samples in a well. Our goodness-of-fit measure is

$$\mathrm{MSR}_x \equiv \log_{10}[X^2/n(n-2)] \tag{3.90}$$

Option 2. If the length of initial tracks at production, L_{av}, is thought to be well enough known, then we do not need to normalize it out of the problem as in option 1. Instead, we can directly compare the measured length $l(z_i)$ with depth against the theoretically predicted length

$$l(i) = L(i)L_{\mathrm{av}} \tag{3.91}$$

The corresponding determination of $Q(t)$ is then through minimization of

$$W^2 \equiv \sum_{i=1}^{n} [l(z_i) - L_{av}l(i)]^2 \tag{3.92}$$

with a corresponding goodness-of-fit measure

$$\text{MSR}_w = \log_{10}[W^2/n(n - 2)] \tag{3.93}$$

One potential difficulty with the mean track length method is easily visualized by the following argument. Imagine that a large pulse of heat had been generated at some time t_0. Then all tracks produced prior to t_0 will be drastically shortened by the heat pulse event. Tracks produced after t_0 will be more tightly clustered toward the initial length at production. The average track length will then reflect more strongly the post-heat pulse tracks and will tend to overlook the pre-heat pulse shortened tracks. Thus, in a general sense, the mean track length method will tend to be biased toward cool events and/or short-time (in the past) effects since recently created tracks have not had time to be shortened. Thus the mean track length method will tend to record a *lower* paleoheat flux than actually obtained.

In attempts to overcome this potential difficulty with the mean track length method it is of interest to deal with the distribution of track lengths at each depth, which contains the greatest amount of information.

4. The Distribution of Track Lengths Method

The most important information about thermal history can be obtained from the distribution of track lengths, because the individual fission tracks are actually formed at different times. New tracks have constantly been produced throughout the history of the sample and all had a nearly constant starting length. Thus different groups of tracks have experienced different proportions of the total thermal history. The degree of shortening displayed by different tracks will reflect the range of annealing conditions they have experienced. Track length distributions thus provide a direct and powerful method of differentiating between contrasting thermal histories.

To provide a quantitative method for describing the evaluation of track length distribution in a given apatite sample and for relating the sample's thermal history to the measured track length distribution, we proceed as follows.

For a track length l, the rate of reduction of track length with time follows the law[4]

$$dl/dt = -k_1 l \exp(-T_c/T) \tag{3.94}$$

determined from the experimental results reported by Gleadow *et al.* (1983), where k_1 is a fixed rate constant and T_c a fixed critical temperature.

Let $n(l,t)\,dl$ be the number of tracks in the length range l to $l + dl$ at time t when all tracks were created of length L.

The rate of production of tracks of length l is then $k_0 U$. The rate of loss of tracks of length l is $\partial/\partial l\ (n\ dl/dt)$.

Thus the total number density of tracks of length l follows the Langevin equation

$$\frac{\partial n}{\partial t} = k_0 U\, \delta(l - L) + \frac{\partial}{\partial l}[k_1 n l \exp(-T_c/T)] \tag{3.95}$$

For a sample of detrital apatite deposited on the sediment surface at time $t = 0$ with no track density,[5] the resulting number density at time t is obtained from the solution to equation (3.95) as

$$n(l,\ t;\ L) = k_0 U \int_0^t dt' \left\{ \exp\left[\int_{t'}^t \alpha(t'')\, dt''\right] \delta\left[L - l \int_{t'}^t \alpha(t'')\, dt''\right]\right\} \tag{3.96}$$

where $\alpha(t) = k_1 \exp[-T_c/T(t)]$ and the integrals are taken over the burial paths of the sediments. Since the original track length distribution is a narrow Gaussian centered on $L_{av} = 16.3\ \mu m$ with an r.m.s. spread $\Delta = 0.9\ \mu m$ (Gleadow *et al.*, 1983), we can write the number of tracks in

---

[4] This quasi-empirical law has served rather well over the years to account in a general sense for both laboratory and field data (see, e.g., review by Green *et al.*, 1986. However, it has been suggested (Green and Duddy, 1988) that a quasi-empirical law of the form $dl/dt = -k \exp(-T_c/T)(1 - l/l_0)^{-a}$ [with $a > 0$, and l_0 the track length at production ($\sim$16.5 μm)] is more in support of data. This is a remarkable law, if true, for at $l = l_0$ it predicts an infinite rate of track reduction, independent of temperature and crystal structure; further, as $l \to 0$ it predicts a nonzero rate of reduction of zero-length tracks! In addition, the new proposed law of track length reduction does not account for crystal anisotropy effects or apatite compositional effects despite the much-vaunted claim (Green and Duddy, 1988) that such effects are paramount. We suggest that the exponential law in the text is more acceptable from a physical viewpoint at the present time, although the method given in the text can be developed for any posited law of track length reduction (Lerche, 1988c).

[5] The rather nasty problem of different track length distributions existing at deposition due to prior provenance behavior is addressed in Appendix B to this chapter.

the length range l to $l + dl$ at time t *independent* of the precise value that the original track length had as[6]

$$n(l,\ t) = \pi^{-1/2}\Delta^{-1} \int_{-\infty}^{\infty} n(l,\ t;\ L) \exp[-(L - L_{\mathrm{av}})^2/\Delta^2]\ dL \tag{3.97}$$

Use of (3.96) in (3.97) enables us to write

$$\begin{aligned} n(l,\ t) = {} & \pi^{-1/2}\Delta^{-1}k_0U \int_0^t \exp\left[\int_{t'}^{t} \alpha(t'')\ dt''\right] \\ & \times \exp\left\{-\Delta^{-2}\left[L_{\mathrm{av}} - l \int_{t'}^{t} \alpha(t'')\ dt''\right]^2\right\} dt' \end{aligned} \tag{3.98}$$

Typically, in a given sample of apatite what is plotted is the percentage distribution of track lengths observed in each range of l (frequency of occurrence)

$$P(l,\ t) = n(l,\ t)/\int_0^{\infty} n(l,\ t)\ dl \tag{3.99}$$

Using equation (3.98), we find that equation (3.99) can be written in the form

$$\begin{aligned} P(l,\ t) = {} & (\pi^{1/2}\Delta t)^{-1} \int_0^t dt' \exp\left[\int_{t'}^{t} \alpha(t'')\ dt''\right] \\ & \times \exp\left\{-\Delta^{-2}\left[L_{\mathrm{av}} - l_{\mathrm{exp}} \int_{t'}^{t} \alpha(t'')\ dt''\right]^2\right\} \end{aligned} \tag{3.100}$$

while the average length $<l(t)>$ of the percent distribution is given by

$$<l(t)> = L_{\mathrm{av}}t^{-1} \int_0^t dt' \exp\left[-\int_{t'}^{t} \alpha(t'')\ dt''\right] \equiv L_{\mathrm{av}}L(t) \tag{3.101}$$

We see then that both the frequency distribution of lengths, $P(l,t)$, for an apatite sample of age t and the average length of the frequency distribution, $<l>$, depend on the total thermal history of the sample through $\alpha(\mathrm{t}) \equiv k_1 \exp[-T_c/T(t)]$.

For a distribution of apatite samples with depth it is then possible to (1) relate the depth to the age through the burial history, (2) measure the average track length with depth (and so age), and (3) measure the frequency distribution of track lengths for a sample of a given age.

---

[6] Since $\Delta << L_{\mathrm{av}}$ we can extend the range of L integration to $-\infty$ (instead of its more proper value of zero) without introducing significant error.

At each depth z_i ($i = 1, 2, \ldots, n$) a histogram is constructed of the number of tracks whose length lies in the range l to $l + dl$. Normally (Gleadow *et al.*, 1983) dl is about 1–2 μm and the maximum track length is 16.5 ± 0.9 μm.

Essentially, then, at each depth z_i one measures the number of tracks $N(l_j, z_i)$ in the "bin" interval l_j to $l_j + \Delta l$, at depth z_i with $0 \leq l_j \leq L_{\max}$, with l_j the track length (in micrometers) at the center of the histogram bin (for 1-μm-wide bins we have $l_1 = 1/2$, $l_2 = 3/2, \ldots$). N can be converted into a probability distribution $p(l_j, z_i)$ which is a direct measure of the thermal maturity of the sample, with

$$p(l_j, z_i) = N(l_j, z_i)/\sum_{j=1}^{J} N(l_j, z_i) \tag{3.102}$$

where J is the total number of bins (for 1-μm wide bins and an $L_{\max}$ of 17 μm there are 17 bins). From equation (3.100) we have a theoretical expression, which is directly comparable to $p(l_j, z_i)$, given by $P(l_j, t_i)$. The point is that $P(l_j, t_i)$ depends on (a) the burial path followed by the detrital apatite since its deposition at time t_i, corresponding directly to its current depth z_i, and (b) the variation with respect to time of the heat flux to which the sample has been subjected. The annealing of fission tracks is sensitive to the temperature with time they record and so, therefore, to the heat flux. Note, as an aside, that the mean track length $L(z_i)$ used in the previous subsection is given by

$$L(z_i) = \int_0^\infty lP(l, t_i)\, dl \Big/ \int_0^\infty P(l, t_i)\, dl \tag{3.103}$$

Now the sensitivity of the track length distribution of each sample of detrital apatite at different depths depends on its burial history and exposure to paleoheat flux. The samples which are young, and so near the surface, will record only recent heating/cooling trends; those buried deeply will record older thermal effects—unless the tracks have been completely wiped out by a strong thermal event some time in the past and a new set of fission tracks overprinted.

In order to extract the heat flux variations with time from the distribution of track lengths, two options are again available to us.

Option 1. Suppose first we consider that the resolution on the track length distribution is *not* sufficient to enable us to do a good job of extracting the temporal variation of heat flux from a track length distribution at a single depth. Then we might anticipate that by superposing all track length distributions at all measured depths we would at least be able to obtain a gross scale of variation of heat flux. Thus our task is to find $Q(t)$ to minimize the difference between the theoretical track length probability distribution $P(l_j, z_i)$ and the measured $p(l_j, z_i)$ distribution over all depths and track lengths, that is, we wish to minimize

$$V^2 = \sum_{i=1}^{n} \sum_{j=1}^{J} [P(l_j, t_i) - p(l_j, z_i)]^2 \tag{3.104}$$

where the summation over i is taken over all detrital apatite samples in a well, and the summation over j is taken over all histogram bins. Our goodness-of-fit measure is

$$\mathrm{MSR}_V \equiv \log_{10}(V^2/[n(n - 2)J(\mathrm{J} - 2)]) \tag{3.105}$$

Option 2. It may be that we consider the track length distributions at each depth to contain sufficient well-resolved information that we can determine the paleoheat flux recorded by the apatite at each depth. Then by superposing the so-determined paleoheat flux values we can put together, in a piecewise fashion, the total evolution of heat flux over a lifetime extending from the age of the deepest measured apatite to the present day.

Thus our task is to find the heat flux $Q(t)$ which minimizes the difference between the theoretical track length distribution $P(l_j, t_i)$ at depth z_i and the measured track length distribution $p(l_j, z_i)$; that is, we wish to minimize

$$U^2(i) = \sum_{j=1}^{J} [P(l_j, t_i) - p(l_j, z_i)]^2 \tag{3.106}$$

where the summation over j is taken over all histogram bins.

For each depth z_i, our goodness-of-fit measure is

$$\mathrm{MSR}_U(i) = \log_{10}[U(i)^2/(J(J - 2))] \tag{3.107}$$

For each increasing depth z_i (corresponding to earlier and earlier depositional times) we obtain a representation of $Q(t)$ in $t_i \leqslant t \leqslant t_{i+1}$ so that we can then superpose these representations to obtain the total time behavior of heat flux throughout the well's history (Huntsberger and Lerche, 1987; Lerche, 1988a,b).

III. Summary

In this chapter on methods of inversion for thermal indicators to determine paleoheat flux variations we have provided the general logic of a suite of individual, and disparate, indicators. Each has its own particular chemical/physical process behavior, with different temperature dependences and different chemical/physical parameters. The process of inversion of such indicators is predicated on the assumption that the burial history information for the sedimentary units is known *precisely,* as are the process parameters (nominally independent of geological conditions by fiat). Thus the only unknown in the inversion of the data is the paleoheat flux evolution with time. Hence the inversion methods provided are geared to extracting that variation as sharply as possible with the minimum dependence on any other parameters of the system—which is why as many of the parameters as possible are normalized out of the inversion process.

The critical assumption for each thermal indicator is that we can tie observed quantities to an underlying known functional behavior of the quantity with temperature. Some of these relations are known precisely; some are semiempirical and subject to change as the quantitative understanding of each thermal indicator improves with further research. But any such changes in functional behavior, while changing the technical details appropriate for a given thermal indicator, will not change the basic pattern of logic presented above.

In later chapters we will examine in some detail the consequences for inversion procedures when the burial path information is itself incomplete and when the chemical and physical parameters are not too well known.

Appendix A: Vitrinite Reflectance as a Thermal Indicator

Thermal indicators provide a method of checking geologic models; they also provide an independent method of obtaining the heat flow as a function of time. Primarily we are concerned with vitrinite reflectance R as a function of depth z. It is possible to extract from the vitrinite reflectance a time–temperature integral that depends on the heat flux through geologic time. This extracted time–temperature integral has two uses. First, comparison can be made between this time–temperature integral and that obtained on the basis of the heat flow given by a geologic model. Second, the heat flow itself can be obtained directly from the extracted time–temperature integral.

This appendix is divided into three parts. First, we summarize the theory of vitrinite reflectance for a one-component system and the method of extraction of a time–temperature index from vitrinite reflectance data as a function of depth. Second, we discuss the methods of using this extracted time–temperature index (TTI) to evaluate geological models for the heat flow and to extract directly the heat flow.

Physical laws of reflection and refraction show that the fractional intensity R of monochromatic light which is normally reflected from a plane interface of vitrinite of refractive index, n surrounded by immersion oil of refractive index n_0 is given by

$$R = (n - n_0)^2/(n + n_0)^2 \tag{A1}$$

The classic Lorentz–Lorenz formula for the refractive index of a solid gives

$$n^2 = [1 + 2Ng(\omega)]\,[1 - Ng(\omega)]^{-1} \tag{A2}$$

where N is the number density of molecules and $g(\omega)$ is the molecular structure factor, which may be either positive or negative, depending on the frequency.

Thus, from equations (A1) and (A2) we see that the change in vitrinite reflectance R with depth is due to a change in N, the number density of molecules, with depth. The change in number density with depth is due to the thermochemical decrease in number density describable by a first-order reaction process as given by

$$(dN/dt)_{\text{chem}} = -k_0 N \exp[-E(T)/TR] \tag{A3}$$

where k_0 is the rate constant in inverse seconds, R the gas constant, $T = T(t,z)$ the subsurface temperature at time t and depth z, and $E(T)$ the activation energy.

It is customary to write $E(T)$ in the form

$$E(T) = E_M - RT(T/T_0) \tag{A4}$$

E_M, a constant, describes the basic molecular binding energy, and the term $T(T/T_0)$ describes the exponentiation of the reaction rate for every T_0 degrees increase in temperature (physically due to excitation of phonons and umklapp processes weakening bond strengths as the temperature rises).

If we assume all samples of vitrinite start with the same number density N_0, then we can determine the parameter $N_0 g(\omega)$ in terms of vitrinite reflectance measurements made on recently created samples (i.e., close to $z = 0$). We do this by first determining the refractive index n_* of a vitrinite sample on the surface from its value R_* of vitrinite reflectance from

$$n_* = n_0(lR_*^{1/2})/(1 \pm R_*^{1/2}) \quad \text{(A5)}$$

The quantity $N_0 g(\omega)$ is given by

$$N_0 g(\omega) = (n_*^2 - 1)/(n_*^2 + 2) \quad \text{(A6)}$$

where n_* is given by equation (A5). The quantity N in equation (A2) is then related to N_0 by

$$N(t) = N_0 f(t, t_i) \quad \text{(A7)}$$

where f is the fractional number density. Then f satisfies the differential equation

$$df/dt = \{-k_0 \exp[T(t, z)/T_0 - E_M/RT(t, z)]\}f \quad \text{(A8)}$$

The solution for the fractional number density as a function of time is given by

$$f(t, t_i) = \exp\left(-k_0 \int_{-t_i}^{t} \exp\{T[t', z(t')]/T_0 - E_M/RT[t', z(t', z(t')]\} \, dt'\right) \quad \text{(A9)}$$

With the time–temperature index TTI defined as

$$\mathrm{TTI}(t, t_i) \equiv \exp\left(-k_0 \int_{-t_i}^{t} \exp\{T[t', z_i(t')]/T_0 - E_M/RT[(t', z_i(t')]\} \, dt'\right) \quad \text{(A10)}$$

then the fractional number density as a function of time is given by

$$F(t, t_i) = \exp[-k_0 \, \mathrm{TTI}(t, t_i)] \quad \text{(A11)}$$

The fractional number density given in equation (A11) is connected to the measured vitrinite reflectance as a function of depth through equations (A7), (A2), and (A1). The small vitrinite reflectance values observed ($R \leqslant 5\%$) allow us to approximate:

$$k_0 \, \mathrm{TTI}(0, t_i) \cong [12n_0^2/(2 + n_0^2)tn_0^2 - 1)][R(z_i)^{1/2} - R_*^{1/2}] \qquad \text{(A12)}$$

In practice, we cannot measure R_* in equation (A12) because the vitrinite reflectance cannot be measured on the surface, as assumed in the derivation of equation (A12). However, we can proceed as follows. Rewrite equation (A12) in the form

$$\alpha \, \mathrm{TTI}(0, t_i) = R(z_i)^{1/2} - R_*^{1/2} \qquad \text{(A13)}$$

with

$$\alpha = \sum_j [R(z_j)^{1/2} - R_*^{1/2}] / \sum_j \mathrm{TTI}(0, t_j) \qquad \text{(A14)}$$

Then use a large suite of wells to determine the best heat flux for each well consistent with the data. In addition, since R_* is the depositional value of vitrinite reflectance, two choices exist: (a) extrapolate the vitrinite data with depth to the present-day sedimentary surface, thereby producing a frequency plot of R_* values; (b) take R_* also to be an unknown in the thermal inversion procedure given in the text of the chapter. Vary R_* for each and every well until a minimum discord is reached in predicted versus observed reflectance. Much testing on both procedures has narrowly constrained R_* at 0.2 ± 0.05.

With the value of R_* fixed we can then find the best heat flux by minimizing

$$\mathrm{SS} = \sum_{j=1}^{n} [\mathrm{VITTI}(z_j) - \mathrm{INORM}(t_j)]^2 \qquad \text{(A15)}$$

where, using equations (A13) and (A14), we have

$$\mathrm{VITTI}(z_j) = [R(z_j)^{1/2} - R_*^{1/2}] / \sum_{i=1}^{n} [R(z_i)^{1/2} - R_*^{1/2}] \qquad \text{(A16a)}$$

and

$$\mathrm{INORM}(t_j) = \mathrm{TTI}(0, t_j) / \sum_{i=1}^{n} \mathrm{TTI}(0, t_i) \qquad \text{(A16b)}$$

Equation (A15) may be used to deduce the heat flux $Q(t)$ over geologic time. This may be done in many ways, but much trial and error has led to the conclusion that least-squares minimization is the best way to proceed. Thus assume a model for $Q(t)$ [e.g., $Q(t) = Q_0 \exp(\beta t)$] and then compute the temperature as a function of time and depth and, therefore, TTI. Then compute SS. Vary the model parameters in $Q(t)$ until a minimum least-squares fit is obtained.

Several methods exist to implement the least-squares procedure. One is to search a grid for the parameters to be determined: the activation enegry E_M, the doubling temperature T_0, and the parameters characterizing the heat flux such as the rate of change at the present time and the second derivative of the heat flow at the present time. All of these parameters can be determined by finding the point in multidimensional space that minimizes the least-squares deviation. An alternative method of determining these parameters consists of a Newton–Raphson linear extrapolation technique to find a better approximation from an initial guess. This can be followed by the nonlinear or exact evaluation of the derivatives at the new trial solution and then repeating the process. Both of these techniques have their area of application.

The grid search method is most applicable in determining optimum doubling temperatures and activation energies, assuming a heat flow based on a geologic model or when we have no information about the correct values of the parameters to be determined. The output of this calculation is then a value of the least-squares standard deviation that determines the fit of the geologic model TTI to the vitrinite reflectance. Thus, it is possible to rank the heat flow predictions of geologic models.

The Newton–Raphson technique is probably most applicable where we have an estimate of the heat flow. In this case, we can choose the heat flow as a polynomial in time of fixed order, and the coefficients in the polynomial will be determined by the least-squares optimization technique outlined above.

Appendix B: Residual Track Length Distributions at Deposition

There is a rather nasty provenance problem for apatite fission scar tracks related to the thermal history and original age of the apatite *prior* to the incorporation of apatite in the deposited sedimentary layers. Thus, of the tracks measured at the present day in a given apatite sample, a certain (but unknown!) percentage can be due to tracks present prior to deposition which have been systematically shortened after burial. The remaining percentage is due to tracks created after deposition which are also systematically shortened.

For those tracks, originally of length L, produced *after* deposition (at time $t = 0$) equation (3.96) provides the number density (per gram of uranium) $n(l, t; L)$. Consider that, at deposition, the apatite had a number density of tracks (per gram of uranium) $N_0(l_0, 0; L)$ representing tracks in the length range l_0 to $l_0 + dl_0$ at time $t = 0$. At a later time t, these tracks evolve. From equation (3.95) with the right-hand term involving U set to zero, we have the remnant number density in the length range l to $l + dl$ as

$$N(l, t; L) = N_0\{l_0 \exp[-\tau(t)], 0, L\} \exp[-\tau(t)] \tag{B1}$$

where $\tau(t) = \int_0^t \alpha(t')\, dt'$.

The *total* (observed) number density of tracks per gram of uranium at time t in the sample is then

$$N_{\text{tot}}(l, t) = n(l, t; L) + N(l, t; L) \tag{B2}$$

The observable distribution of tracks in the range l to $l + dl$ for the sample of age t currently at depth z is

$$\begin{aligned} P_{\text{tot}}(l, t) &= [n(l, t; L) + N(l, t; L)] / \int_0^\infty N_{\text{tot}}(l, t)\, dl \\ &= [n(l, t; L) + e^{-\tau(t)} N(l_0 e^{-\tau(t)}, 0, L] \\ &\quad \times \left[\int_0^\infty n(l, t; L)\, dl + \int_0^\infty N(l_0)\, dl_0 \right]^{-1} \end{aligned} \tag{B3}$$

The problem, then, is to separate from $P_{\text{tot}}(l,t)$ the fission scar tracks generated and thermally developed *after* sedimentary deposition from those present at deposition and later evolved thermally.

To provide an illustration of the way this can be done, suppose first that only one source of detrital apatite is responsible for the track distributions observed at the present day and that, at deposition of each sedimentary layer, the number density of tracks (per gram of uranium) in the range $(j + 1)\Delta l \geq l_0 \geq j\Delta l$ is h_j $(j = 0, 1, 2, \ldots, N)$. Then at the later time t the number density h_j of "fossil" tracks will be in the range

$$(j+1)\Delta l e^{-\tau(t)} \geq l \geq j\Delta l e^{-\tau(t)} \qquad (j = 0,1,2, \ldots, N)$$

For brevity introduce

$$A(l, j) = 1 \quad \text{if } (j + 1)\Delta l e^{-\tau(t)} \geqslant l \geqslant j\Delta l e^{-\tau(t)}$$
$$= 0$$

Then rewrite the discrete form of equation (B3) in the form

$$P(l_j, t)\left[\sum_{k=0}^{N} h_k + \sum_{k=0}^{N} n(l_k, t; L)\right] = n(l_j, t; L) + e^{-\tau(t)} \sum_{k=1}^{N} A(l_j, k)h_k \qquad \text{(B4)}$$

where $P(l_j, t) = P_{\text{tot}}(l_j, t)\Delta l$ with $\sum_{j=1}^{N} P(l_j, t) = 1$.

Introduce the matrix $B(l,k) = 1$ and

$$R(l, k; t) = P(l, t)B(l, k) - e^{-\tau(t)}A(l, k)$$

Then we can write

$$\sum_{k=0}^{N} R(l_j, k; t)h_k = n(l_j, t; L) - P(l_j, t) \sum_{k=0}^{N} n(l_k, t; L) \qquad \text{(B5)}$$

For each value of t, we recognize equation (B5) as providing $N + 1$ ($j = 0,1,2, \ldots ,N$) linear equations for the $N + 1$ unknowns h_j in terms of the variation $n(l,t;L)$ of tracks generated and evolved *after* deposition. Suppose then that we solve equation (B5) for the h_j in terms of the n corresponding to a given time value. But we have at our disposal *all* of the remaining values of time corresponding to fission scar tracks from different sedimentary layers. Let there be M such layers. Since we have a structural form for the evolution of $n(l,t;L)$ through equation (3.96) (and lack only its magnitude) in terms of the paleoheat flux variation with time, it follows that with perfect data we can solve for $M - 1$ coefficients in the heat flux variation and for the magnitude scale of $n(l,t;L)$.

A least-squares way to approach the problem is to take

$$x^2 = \sum_{m=1}^{N} \sum_{j=0}^{N} \left\{ \sum_{k=0}^{N} R(l_j, k; t_m)h_k - [n(l_j, t_m; L) - P(l_j, t_m)\sum_{k=0}^{N} n(l_k t_m; L)] \right\}^2 \qquad \text{(B6)}$$

and to vary the h_k, the magnitude of n, and the paleoheat flux using the tomographic procedures outlined in Chapter 9, until x^2 is minimized.

For the case of many provenances supplying detrital apatite to the system, but at different times, it is clear that a generalization of the above procedure would also be appropriate. For instance, in the two-provenance case, we would have h_j set at $h_j^{(1)}$ for $t < t^{(1)}$ and at $h_j^{(2)}$ for $t > t^{(1)}$, representing the shift from the first provenance to the second at time $t^{(1)}$, and where the $h_j^{(1)}$ and $h_j^{(2)}$ $(j = 0,1, \ldots, N)$ are regarded as independent variables in the tomographic procedure. Likewise, a three-provenance model would have $h_j = h_j^{(1)}$ in $t < t^{(1)}$, $h_j = h_j^{(2)}$ in $t^{(2)} > t > t^{(1)}$, and $h_j = h_j^{(3)}$ in $t > t^{(3)}$, etc.

Thus it is possible, in principle, to recover not only the paleoheat flux variations with time from the evolved track distributions but also the original track length distributions at deposition. The pragmatic concerns of uniqueness, resolution, and precision then have to be addressed. We have not followed this point in any further depth in this book, although we are actively looking for enough downhole distributions of high-quality apatite data to implement this potentially fruitful phase of the tomographic inversion procedures.

4

Inversion of Thermal Indicators: Case Histories

IN THIS CHAPTER we report on several applications of thermal indicators to determine the thermal history in sedimentary basins.

The first example, taken from the Gulf Coast region, involves the use of three wells in close proximity and in similar geological settings with two thermal indicators: pollen translucency and vitrinite reflectance. The central points of this example are to show that (1) the requirement of local consistency of the paleoheat flux variation can be used to constrain chemical and physical parameters, nominally independent of geological conditions, and (2) with the parameters so constrained the behaviors of two very disparate thermal indicators (pollen translucency decreasing with depth, vitrinite reflectance increasing with depth) nevertheless provide a consistent paleoheat flux variation.

The second example, taken from the North Sea, involves the use of limited samples of apatite in two wells—one close to the Statfjord oil field and the other in the vicinity of the Odin field. The purpose of this exercise is to show that, despite the paucity of samples, there are roughly seventeen 1-μm "bins" for fission tracks which have been continuously produced since deposition, and it is possible to use this wealth of data at each sample depth to make an estimate of the thermal history experienced by the sediments. As we shall see, the results are not discordant with general geologic concepts related to the evolution of the North Sea.

Then, using two wells from Australia, each of which has a large suite of apatite samples with well-measured tracks, we illustrate how both the rate constant for apatite annealing (a constant nominally independent of geologic evolution) and the paleoheat flux variation can be determined more precisely by using all of the information in several modes.

The fourth example, using data from Australia and from the North Sea, illustrates how the combination of thermal indicator inversion and burial history behavior can be used in a basinal sense (44 wells in the Cooper Basin of Australia, and 29 wells in the North Sea) to attempt to tie the evolution of basement subsidence to paleoheat flux variations.

Examples to be presented in later chapters will involve the use of these and other thermal indicators both singly and multiply to determine chemical activation energies, unconformities, and stratigraphic ages.

I. Pollen Translucency and Vitrinite Reflectance: Louisiana

The notion of using pollen translucency as an interpretive tool was first introduced and described by Gutjahr (1966) and was extensively investigated by Grayson (1975) for the genus *Carya*. Gutjahr showed, by experiment, that the translucency of a specific taxon (*Quercus robur*) is affected by temperature, decreasing with exposure to higher temperatures (carbonization). In general, this agrees with the general decrease in translucency with increasing depth found by Grayson (1975). Grayson noted the pragmatic advantages of using translucency measurements: (1) Translucency measurements are made on a specific genus of a palynomorph which was incorporated in the sediment at the time of deposition. Other measurements of maturation in clastic sediments often include material secondarily derived from older sediments (that are not obviously anomalous) and/or materials that are not *closely* genetically related. (2) The maturation characteristics of different components of dispersed organic matter in sediments differ widely. Since translucency deals with one component, it provides a more definite measurement than is possible by chemical analysis of the whole organic fraction. (3) Only very small quantities of samples are needed. (4) Drilling mud additives do not affect translucency measurements. (5) Anomalous grains are easily detected. (6) The occurrence of palynomorphs is widespread. (7) The same equipment can be used as for vitrinite reflectance measurements.

In Grayson's search for a genus of pollen to use for his analysis, several properties were considered: (1) the "standard" had to have a large initial translucency so that a large range of translucency values was possible; (2) a large spore had to be available so as to be easily located and identified under the microscope; (3) it was preferable to have little ornamentation that may affect translucency and hinder the reproducibility of the measurements; (4) a medium-thick exine was to be favored for, if too thick, the exine will absorb a large amount of light and limit the translucency range and, if too thin, the exine will fold easily; and (5) the standard must occur over a long enough geological time range to be deemed useful. Grayson (1975) found that the genus *Carya* (Eocene–Recent) fulfilled many of these requirements. Measurements of translucency on *Carya* from drill cuttings were made in monochromatic light at 546 nm midway between the pores and the thin central area on the exine. Thirty grains were measured and averaged for each depth sample. The rough trend relationship between translucency and color obtained by Grayson (unpublished Amoco report) is shown in Fig. 4.1.

In order to convert *Carya* translucency from a qualitative thermal indicator to a quantitative thermal indicator we proceed as described in the following section. The same methods can be used for other palynomorphs; only the constants will change.

The pollen translucency $\tau(z_i)$ for *Carya* (a pollen genus including our present day pecans and hickories) samples from each measured depth z_i is obtained by shining monochromatic light of intensity I_0 normally onto a sample and forming the ratio

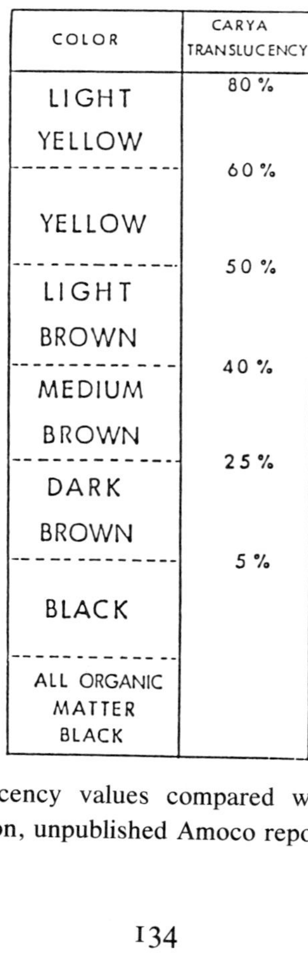

FIGURE 4.1 Translucency values compared with a qualitative color scheme. (From Grayson, unpublished Amoco report.)

$$\tau = I(z_i)/I_0 \tag{4.1}$$

where $I(z_i)$ is the amount of light transmitted through the sample from a given depth.

A theoretical expression (Chapter 3) to be compared to the translucency is given by

$$\ln \tau = -\alpha \exp[-k \, \mathrm{TTI}(t_i)] \tag{4.2}$$

where $\mathrm{TTI}(t_i)$ is a time–temperature integral for organic diagenesis, characterizing the translucency of a pollen sample deposited at time t_i and presently at depth z_i; α and k are constants. The form of the time–temperature integral is

$$\mathrm{TTI}(t_i) = \int_0^{t_i} \exp[(T - T_c)/T_D] \, dt \tag{4.3}$$

with the integrand set to zero for $T < T_c$. Hence T_c is a critical temperature below which translucency is stable. T_D is a scaling constant roughly analogous to a doubling temperature. Each integral is taken along the burial path of a sample deposited at time t_i. The T in the above expression is the temperature along the burial path; it is a function of both time and depth:

$$T(t_i, z(t_i)) = T_{\mathrm{surface}} + Q(t_i) \int_0^{z(t_i)} dz'/K(z') \tag{4.4}$$

where $K(z')$ is the thermal conductivity as a function of depth and $Q(t_i)$ is the basement heat flux as a function of time.

With some mathematical rearranging the constants α and k can be normalized out of the theoretical equation (4.2), and then the basic problem is to find $Q(t)$ to minimize the differences between quantities directly related to the theoretical and observed translucencies; that is, we wish to minimize

$$S^2 = \sum_{i=1}^{n} [P(i)_{\mathrm{th}} - P(i)_{\mathrm{ob}}]^2 \tag{4.5}$$

where P_{th} and P_{ob} are directly related to the theoretical and observed translucencies (see Chapter 3). Here the summation is taken over all pollen samples in a well. We take as our goodness-of-fit measure MSR, defined as

$$\mathrm{MSR} = \log_{10}[S^2/(i - 2)] \tag{4.6}$$

The final algorithm for determining variations of heat flux with time is basically an implementation of two equations. Paleoheat flux is chosen to vary with time in a linear fashion:

$$Q(t) = Q_0(1 + \beta t) \tag{4.7}$$

where β is the rate of change of heat flux. This heat flux variation is used together with the theoretical time–temperature maturation integral [TTI in equation (4.3)], which contains the unknown chemical constants T_D and T_c. A linear heat flux equation is chosen on the basis of past experience with the inversion of vitrinite reflectance measurements (Lerche *et al.*, 1984). We found that polynomial terms above the linear term did not produce any significantly better fit to the data. Other heat flux equations could be used, such as an exponential heat flux with time, but it is desirable to keep the model relatively simple in this first approach to see what degree of accuracy can be attained. Therefore in the present model we assume the heat flux to be linear only because of the data limitations, not because we believe that the true heat flux is linear. One drawback to the linear heat flux equation is that if the linear heat flux coefficient is negative (i.e., cooler in the past), then at a time t_* with

$$t_* = 1/\,|\,\beta\,|$$

the heat flux will first go to zero and, at earlier times, become negative. A negative heat flux is physically impossible. The correct implication is that the resolution of the data is not good enough to measure such small values of heat flux.

In the TTI calculation, the constants T_D and T_c are chemical parameters that are to be constant for any given pollen genus, regardless of any varying geologic history. So we fix T_D and T_c for a given pollen (in this case, *Carya*) and do an exhaustive search for β in each well to determine that heat flux with time which provides the best fit to the input data (most negative MSR). The parameters T_D and T_c are then varied by an exhaustive trial-and-error approach, a new β calculated (minimum MSR), and the process repeated. In this way we gradually construct a volume of solutions for surfaces of constant MSR (goodness of fit) in a space of T_D versus T_c versus β for each well. We then fix T_c and T_D at their best global solutions and do an exhaustive search for the best linear heat flux coefficient β for each well. Hence we obtain the paleoheat flux.

We can then use the *data-determined* heat flux to estimate the time and depth of source rock maturity. This may be done in one of several ways. We can plot isotemperature contours on a burial history diagram or we can plot isotranslucency contours after they have been calibrated to some "oil window" criteria.

A. Implementation of the Model and Determination of the Chemical Parameters for *Carya*

Three wells with *Carya* translucency measurements in southern Louisiana were available for this study (Pan American Miami Nos. 26 and 27, South Pecan Lake Field, Cameron Parish; and Richardson and Bass–John W. Mecon–Freeport, Humble–L.L. and E. State Lease 2414, No. 1-L, Lake Washington Field, Plaquemines Parish) (Fig. 4.2). These wells were the initial choices of Grayson, while at Amoco, to explore the pollen translucency problem. The sites were chosen partly because of their simple stratigraphy and structure, making them ideal for a study of this nature. The wells are located in deltaic shale sequences with some minor interbedded sands and are modeled as total shale because of the small percentage of dispersed sand. The wells terminate in the Miocene. High sedimentation rates in the areas (250–1100 ft of shale/Ma) suggest that the areas are overpressured, so we use a fluid flow/compaction burial history program (Cao *et al.*, 1986) with which to implement the numerical model.

A PORTION OF CAMERON PARISH, LOUISIANA

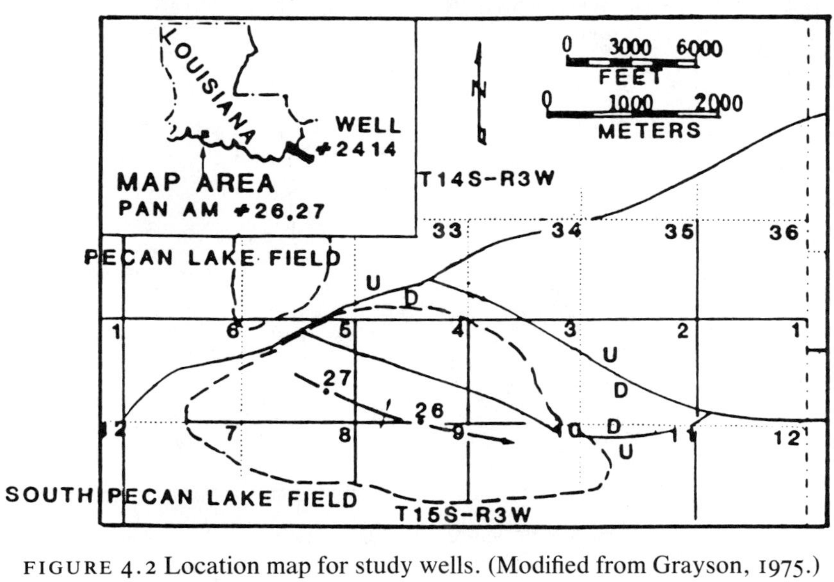

FIGURE 4.2 Location map for study wells. (Modified from Grayson, 1975.)

The choice of an appropriate burial history code is a very important aspect of determining any thermal history. To date, the calculation of thermal history assumes that the burial history of the sediments is correct. Any errors in the burial history serve to magnify errors in the calculated thermal history. Indeed, elsewhere in this volume we shall use precisely this sensitivity of the system to the burial history to estimate the ages of stratigraphic horizons and the timing and amount of erosional unconformities. With the fluid flow/compaction model we are able to model the effects of overpressured sections. Overpressuring in a section means that higher porosities can be maintained to greater depths. Hence the thermal conductivity of the section will be smaller than if the section had "normal" porosity, and so less water, in the sediment. Lower thermal conductivity results in a higher temperature gradient, so that a source rock that may have been thought to be thermally immature when no allowance is made for paleo-overpressuring may actually have reached thermal maturity. As an example, Fig. 4.3 shows the current pressure distribution and porosity versus depth for well 2414. To be noted in

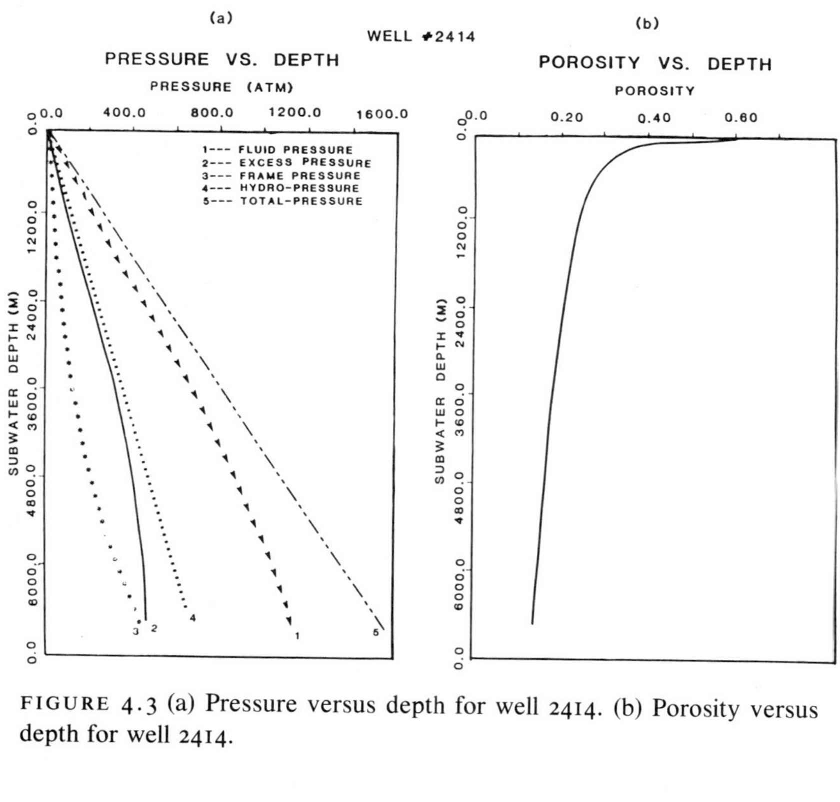

FIGURE 4.3 (a) Pressure versus depth for well 2414. (b) Porosity versus depth for well 2414.

Fig. 4.3a is the excess fluid pressure caused by the high sedimentation rate in this well (≅ 1100 ft/Ma). In Fig. 4.3b the porosity is seen to have a value of 10% at a depth of 6500 m, where the "normal" porosity of a shale would be close to zero. The effects on Pan Am 26 and 27 are similar, indicating that it is essential to use a burial history model that accounts for fluid flow in areas with such high rates of sedimentation.

With a precisely known burial history and the assumptions that the main source of heat is from the basement and the dominant mode of heat transport is by thermal conduction, we can predict the thermal history of each well using the methods described in the previous section. To invert the translucency measurements from the three wells, we need to determine the three parameters T_D, T_c, and β. Three important factors need to be considered for the inversion and thermal histories of the wells: (1) T_D and T_c must be constant for all three wells since they are chemical constants for the pollen genus *Carya* and should not depend on any geologic processes; (2) for Pan Am 27, the heat flux with time determined by the *separate* inversions of vitrinite reflectance and *Carya* translucency should be consistent; and (3) Pan Am 26 and 27 should have approximately the same thermal history because of their close proximity (1340 m; see Fig. 4.2). With these restrictions in mind, the data from the wells were inverted.

1. Results for Pan Am No. 27

Pan Am No. 27 was the first well to be examined because it contains both vitrinite reflectance and *Carya* translucency with depth. The burial history for this well is shown in Fig. 4.4. The vitrinite reflectance was inverted using methods described earlier, and the β range of acceptable values is given in Fig. 4.5. The range of β is determined for that range of MSR, the goodness-of-fit criterion, which allows the least-squares fit to deviate from the best answer by a factor of 4. This gives a $\beta_{vr} = 0.0189 \pm 0.0315$ $(\text{Ma})^{-1}$. The large range for β_{vr} is a reflection of the sparseness of the data, with only six vitrinite reflectance values recorded. (Typically, a much "tighter" range can be defined by increasing the range of the data base.) The best linear heat flux with time is shown in Fig. 4.6; the enveloping lines represent the uncertainty in the paleoheat flux determination mentioned above. In Fig. 4.7 the predicted reflectance (using the best β_{vr}) is plotted along with the input reflectance data. With this β_{vr} range in mind, the *Carya* translucency measurements in Pan Am No. 27 were inverted independently for various values of T_D, T_c, and β and a three-dimensional grid matrix of MSR (goodness of fit) values was

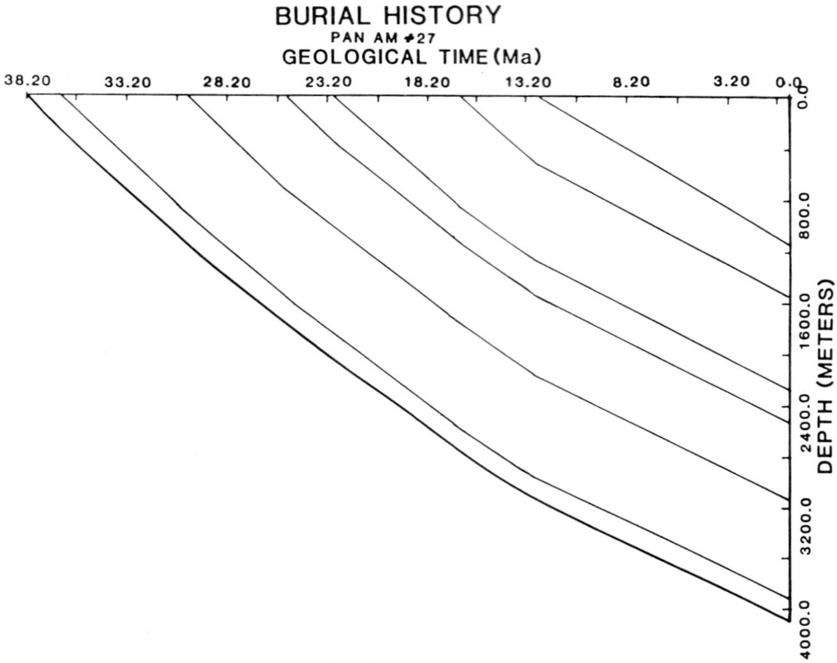

FIGURE 4.4 Burial history of Pan Am 27.

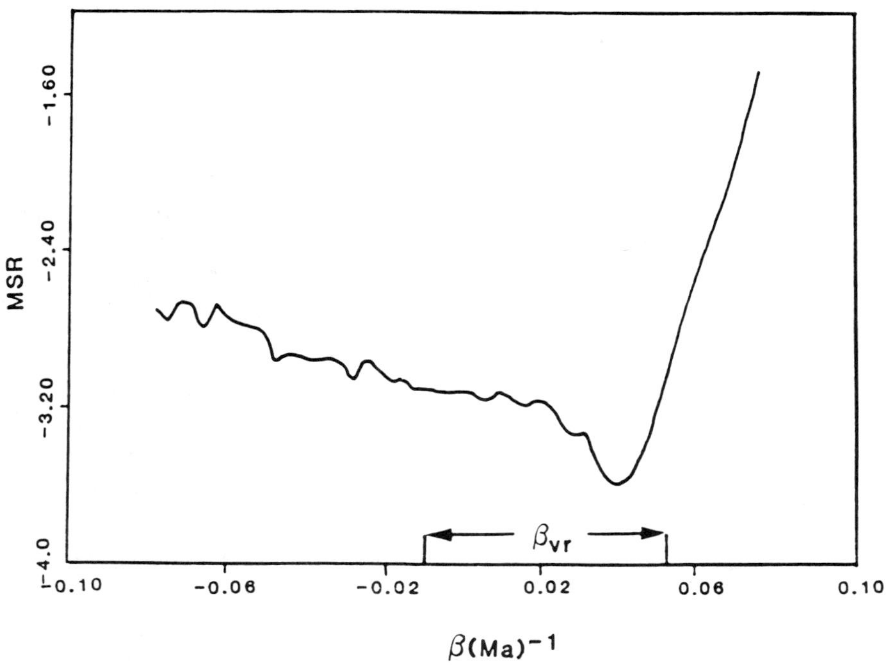

FIGURE 4.5 Linear heat flux coefficient (β) versus MSR (goodness-of-fit criterion) for the inversion of vitrinite reflectance in Pan Am 27.

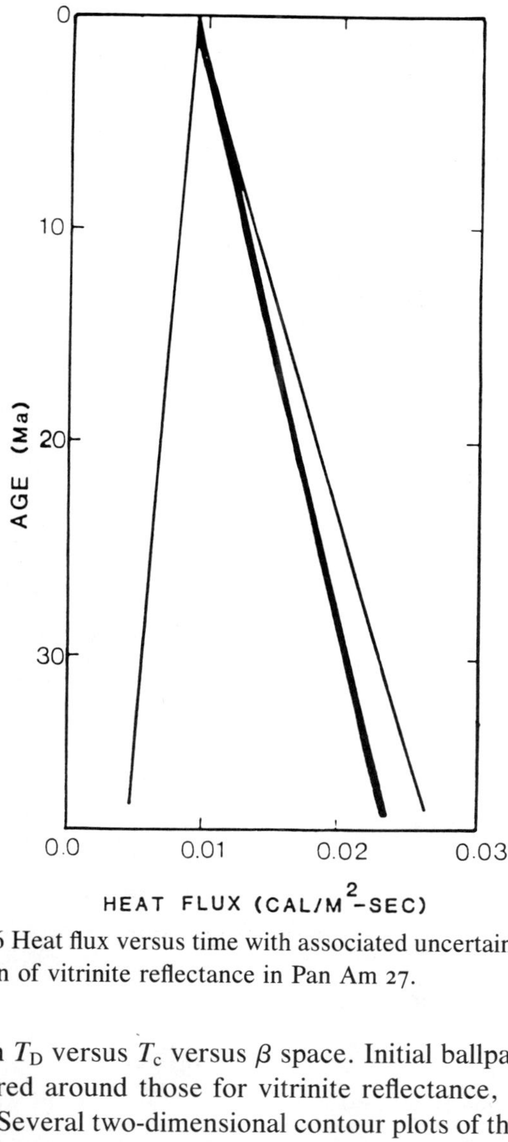

FIGURE 4.6 Heat flux versus time with associated uncertainty curves from the inversion of vitrinite reflectance in Pan Am 27.

established in T_D versus T_c versus β space. Initial ballpark guesses of T_D and T_c centered around those for vitrinite reflectance, T_D = 200 K and T_C = 295 K. Several two-dimensional contour plots of the goodness-of-fit criterion MSR were produced in T_D versus T_c space for various values of β, and from these a T_D value was estimated to be 290 ± 260 K and a T_c value to be 310 ± 40 K. The best fit is centered on the lower end of the T_c scale (273 K), hinting that this well's history may have been too hot to determine the lower boundary for a T_c value. Contour plots of the goodness-of-fit criterion MSR in higher-resolution T_D versus β space were produced for Pan Am 27 for four given values of T_c (273, 395, 305, 310 K). These four plots, stacked one upon the other, represent the three-

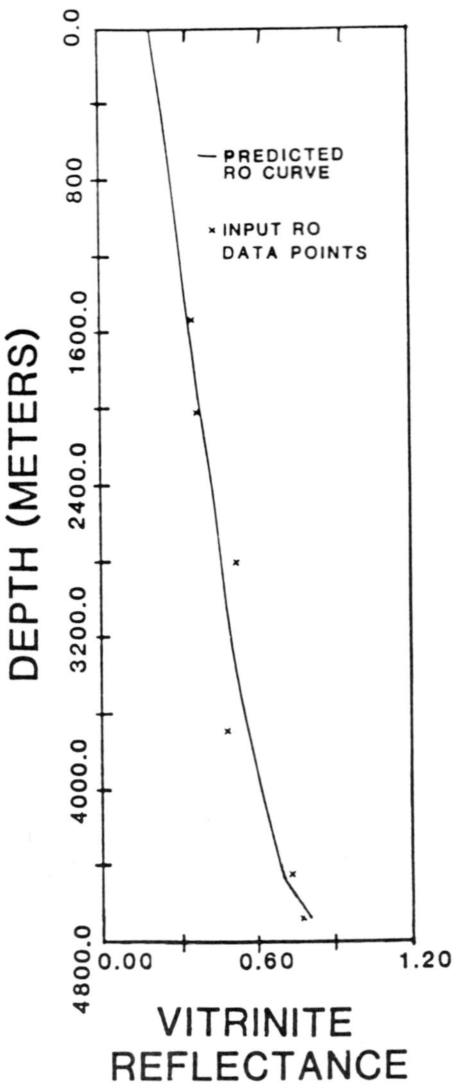

FIGURE 4.7 Predicted vitrinite reflectance with depth along with the input data for Pan Am 27.

dimensional T_D versus T_c versus β space with constant MSR surfaces contoured (see Fig. 4.8).

An allowable range for a solution is defined to be an MSR change of 0.6 from the best MSR value corresponding to a four times poorer fit. Hence, in the sequence of plots of Fig. 4.8, the −6.4 contour is an "allowable"

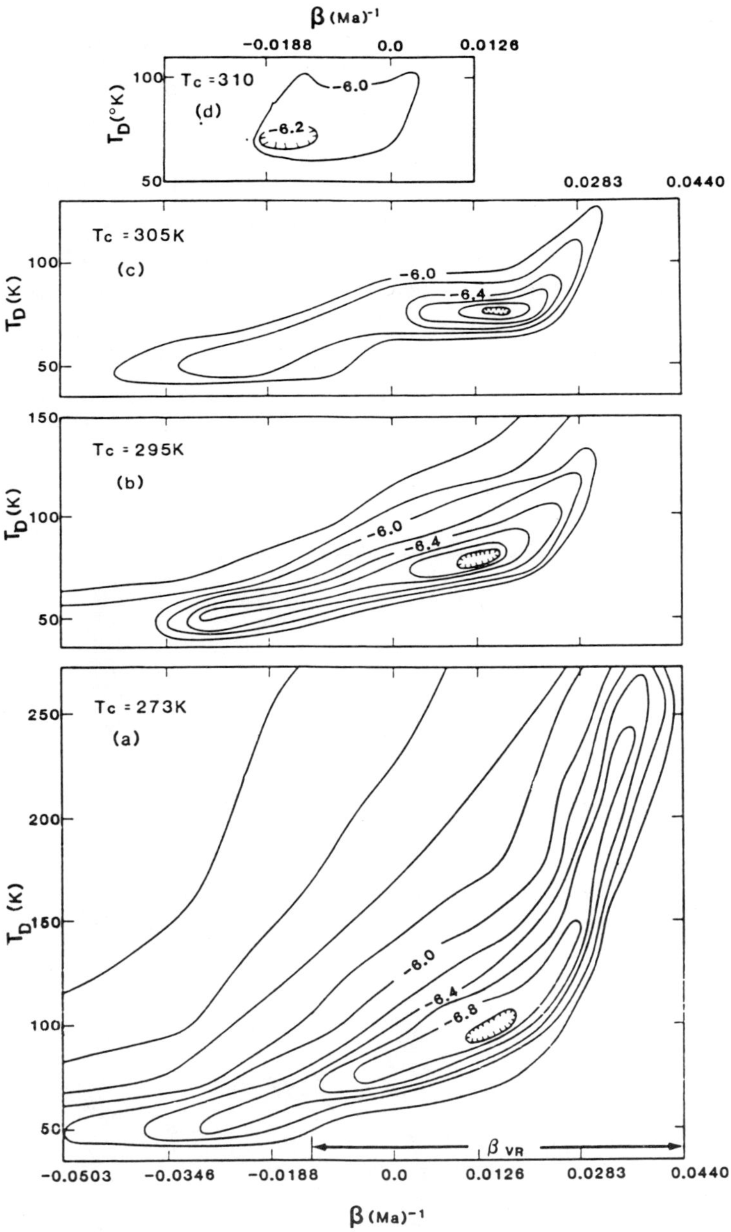

FIGURE 4.8 MSR contoured in T_D–β space for various values of T_c for Pan Am 27 (letters are referred to in text).

surface. This surface defines a solution "cone" with the widest width parallel to the T_D axis, in the $T_c = 273$ K plane (Fig. 4.8a) with $T_D = 150 \pm 100$ K. T_c plane cone slices at 295 and 305 K (Fig. 4.8b and c) also show T_D's in this range ($T_D = 75 \pm 30$ K). To be noted in the cone slice at $T_c = 310$ K (Fig. 4.8d) is the closure of the solution surface (tip of cone), indicating that no acceptable solutions occur in the T_c planes above. This can also be seen in Fig. 4.9, where MSR has been contoured in T_c versus β space with $T_D = 75$ K. An allowable range of a factor of 4 again narrowly defines $T_c = 290 \pm 20$ K. Also note in Figs. 4.8 and 4.9 that the best MSR values lie well within the β_{vr} range determined from vitrinite reflectance inversion, and the β range for *Carya* translucency inversion is predominately encompassed by the β_{vr} range. The results of the *Carya* translucency paleoheat flux determination for "allowable" values of $T_D = 100$ K and $T_c = 285$ are shown in Figs. 4.10 and 4.11.

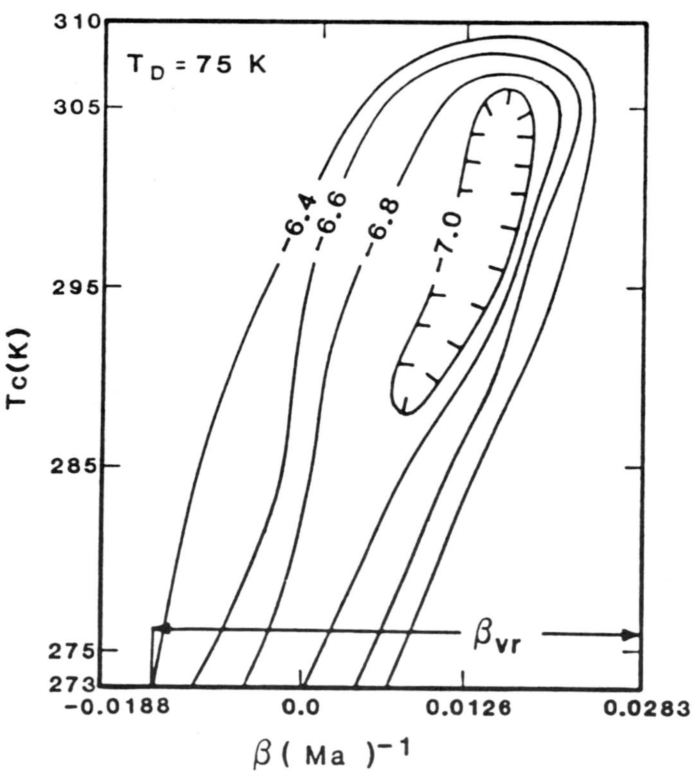

FIGURE 4.9 MSR contoured in T_c–β space with $T_D = 75$ K for Pan Am 27.

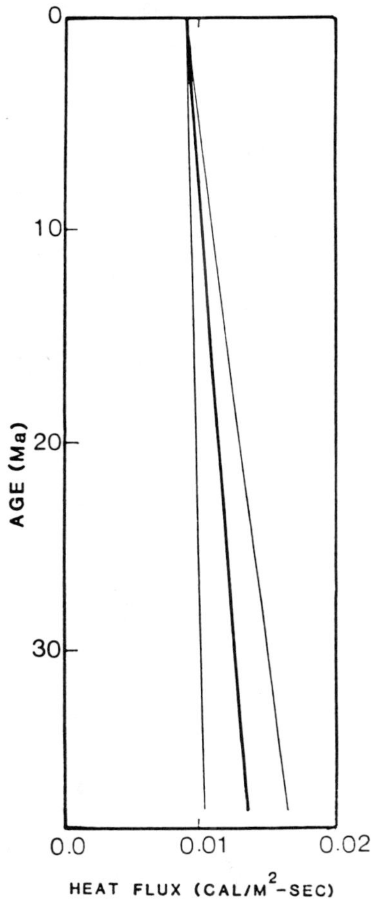

FIGURE 4.10 Heat flux versus time with associated uncertainty curves from the inversion of *Carya* translucency in Pan Am 27 with T_D = 100 K, T_c = 285 K.

Figure 4.10 shows the heat flux versus age with associated error assessments obtained by finding the minimum MSR between predicted and measured pollen translucencies (which yields the best β) and then adding 10% to this minimum value. The corresponding β's are then recorded for MSR = 1.1 × $(\mathrm{MSR})_{\min}$. (Note that this assessment is different from that used earlier for vitrinite reflectance. In the vitrinite case a larger error range was needed because of the poor quality and quantity of the vitrinite data, which did not permit the inversion procedure applied to vitrinite reflectance to obtain a unique solution. Compare Figs. 4.6 and 4.10.) In

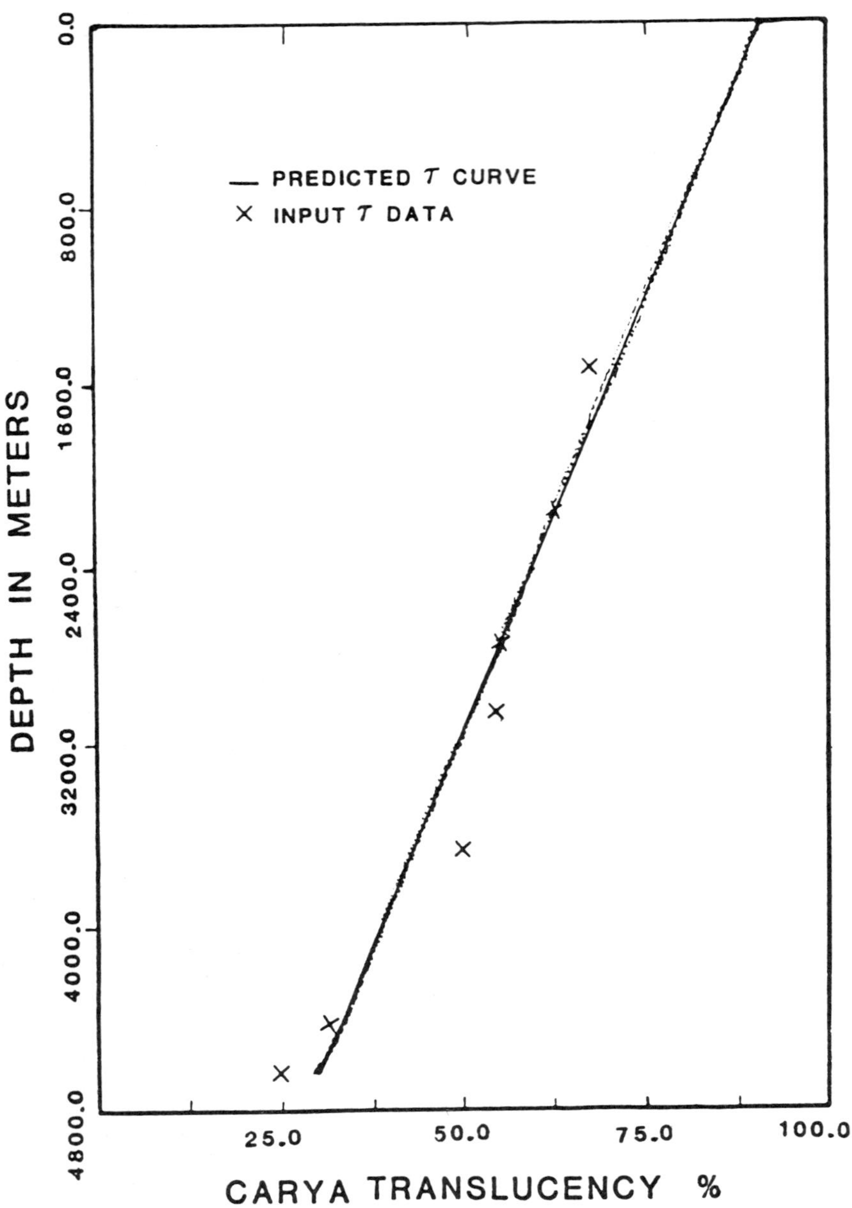

FIGURE 4.11 Predicted *Carya* translucency with depth along with associated uncertainty curves (dotted lines) and the input data for Pan Am 27. $T_D = 100$ K, $T_c = 285$ K.

Fig. 4.11 the predicted translucency with depth is plotted along with the input translucency data.

2. Results for Pan Am No. 26

Pan Am No. 26 is located 1340 m east–southeast of Pan Am No. 27 (Fig. 4.2) with a burial history (Fig. 4.12) essentially the same as that for Pan Am 27, being all shale and having roughly the same sedimentation rate (250–350 ft/Ma). Plots of contours of constant MSR (goodness-of-fit criterion) for Pan Am 26 in T_D versus β space for values of $T_c = 273$ and 305 K are given in Fig. 4.13. Solution surfaces did not show closure around any favored T_D values. On the T_c axis, planes up to 350 K were examined and the solution surfaces were still not closing. Some of the acceptable solution surface is consistent with the β_{vr} range, but this well is not useful for an accurate determination of β, T_D, or T_c.

Nevertheless, using $T_D = 50$ K and $T_c = 273$ K, consistent for Pan Am 26 and Pan Am 27, we predict the heat flux as shown in Fig. 4.14. The modeled translucency fit to the input data, using this heat flux, is depicted in Fig. 4.15. The trouble in defining a unique heat flux for this well seems

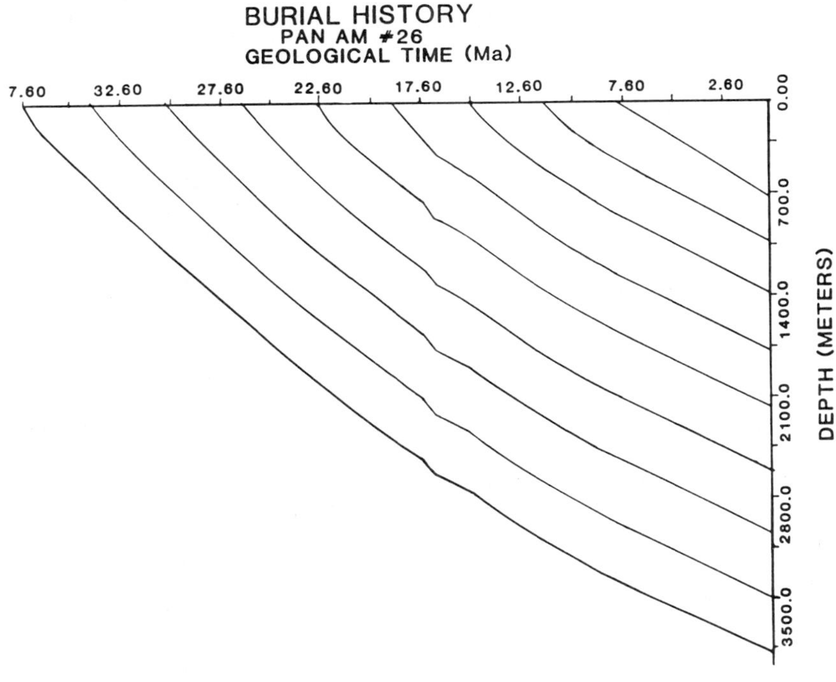

FIGURE 4.12 Burial history of Pan Am 26.

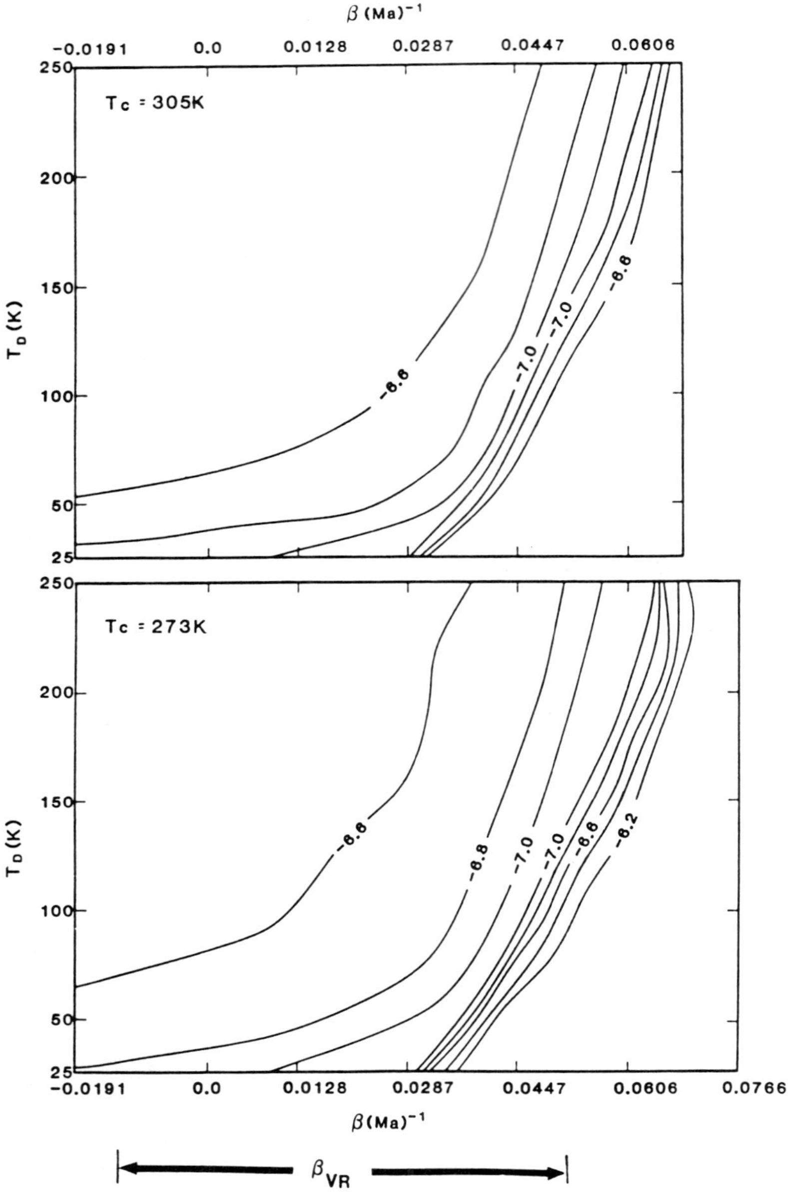

FIGURE 4.13 MSR contoured in T_D–β space for various values of T_c for Pan Am 26.

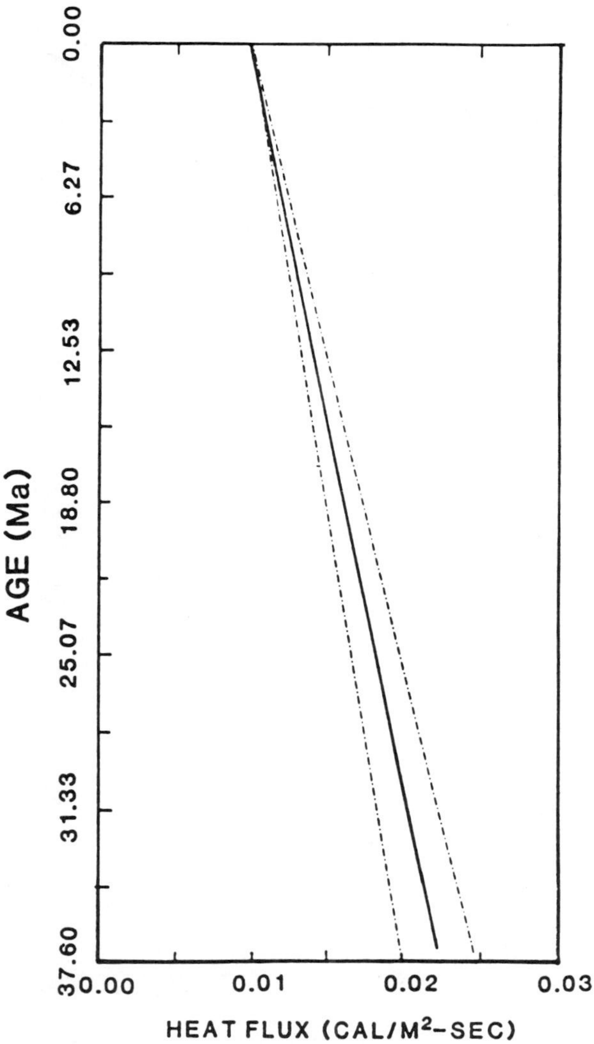

FIGURE 4.14 Heat flux versus time with associated uncertainty curves from the inversion of *Carya* translucency in Pan Am 26 with T_D = 50 K, T_c = 273 K.

to lie in the input data as shown in Fig. 4.15. Although there is an abundance of data, it is mainly dominated by translucency values of 40–50% over a depth range from 1200 to 4200 m. This "stability" of the translucency with depth is likely attributable to an incorrect burial history for the well. There may have been an increased sedimentation rate in the

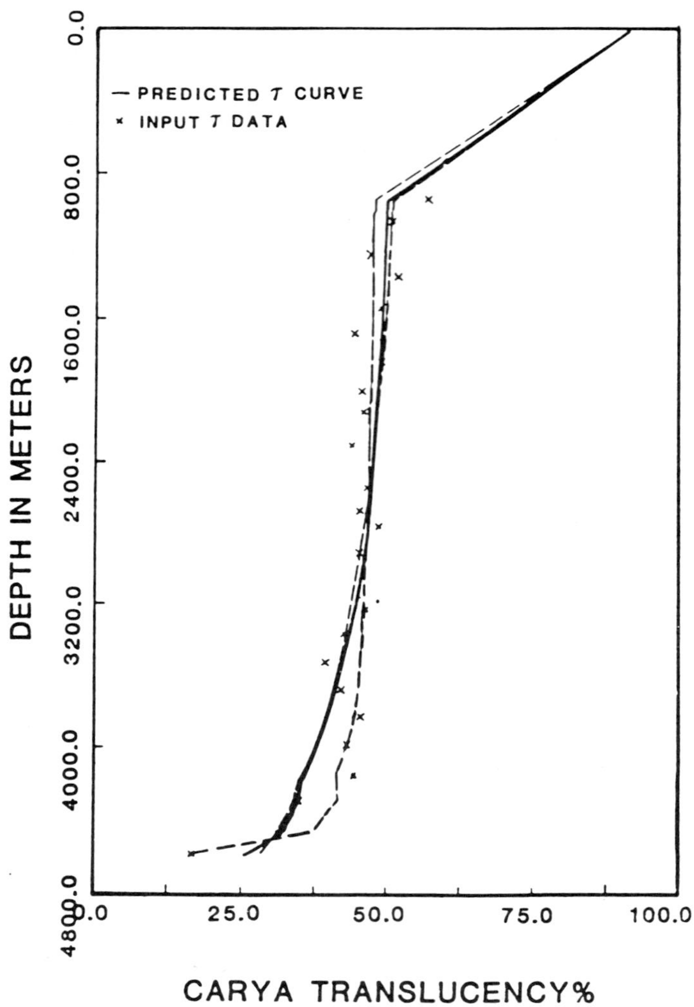

FIGURE 4.15 Predicted *Carya* translucency with depth along with associated uncertainty curves (dashed lines) and the input data for Pan Am 26 (T_D = 50 K, T_c = 273 K).

time period related to this depth or the section may have been doubled by faulting. We were not able to resolve such questions with the data available, although we did try by adjusting the sedimentation rate. Anyway, the range of the data is not sufficient to provide a *unique* solution for a thermal history.

3. Results from Well No. 2414

Well No. 2414 is about 300 miles east of Pan Am 26 and 27 (Fig. 4.2). The well's burial history is shown in Fig. 4.16. Well 2414 gives the best resolution on a T_D value as shown in Fig. 4.17. Again, allowing for an MSR fit which can be a factor of 4 poorer than the best MSR, we find $T_D = 65 \pm 35$ K in the cone slice at $T_c = 273$ K. Slices at higher T_c values show similar T_D values. Allowable T_c surfaces close by $T_c = 350$ K, giving a T_c range of 310 ± 40 K. Again, the lower limit cannot be determined in this suite of wells. Values of β overlap with those of β_{vr}.

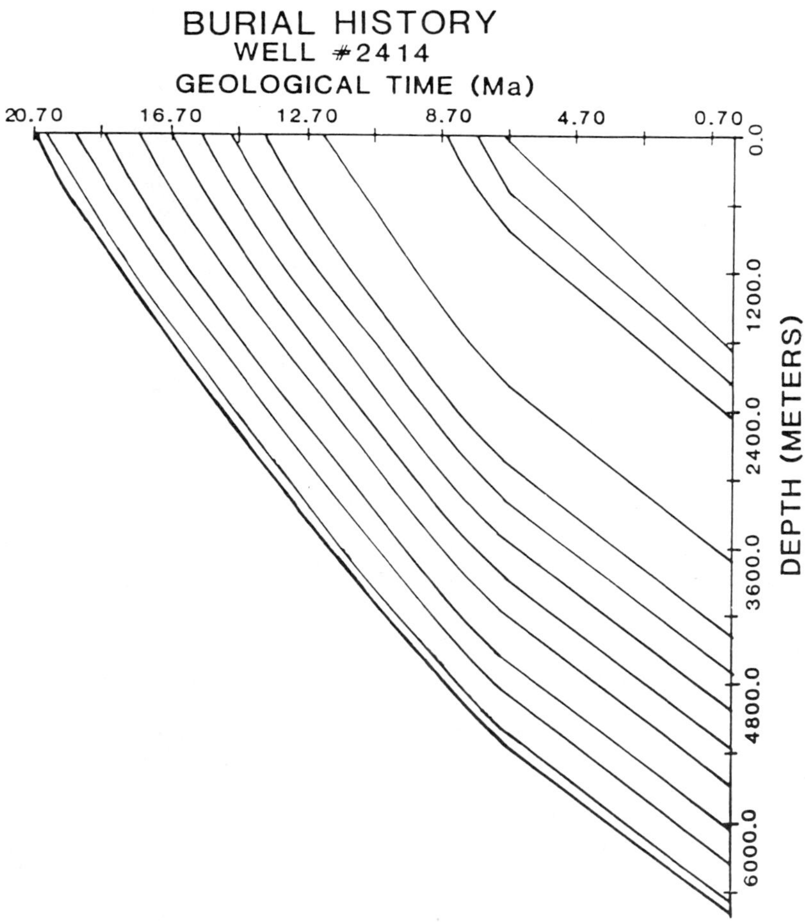

FIGURE 4.16 Burial history for well 2414.

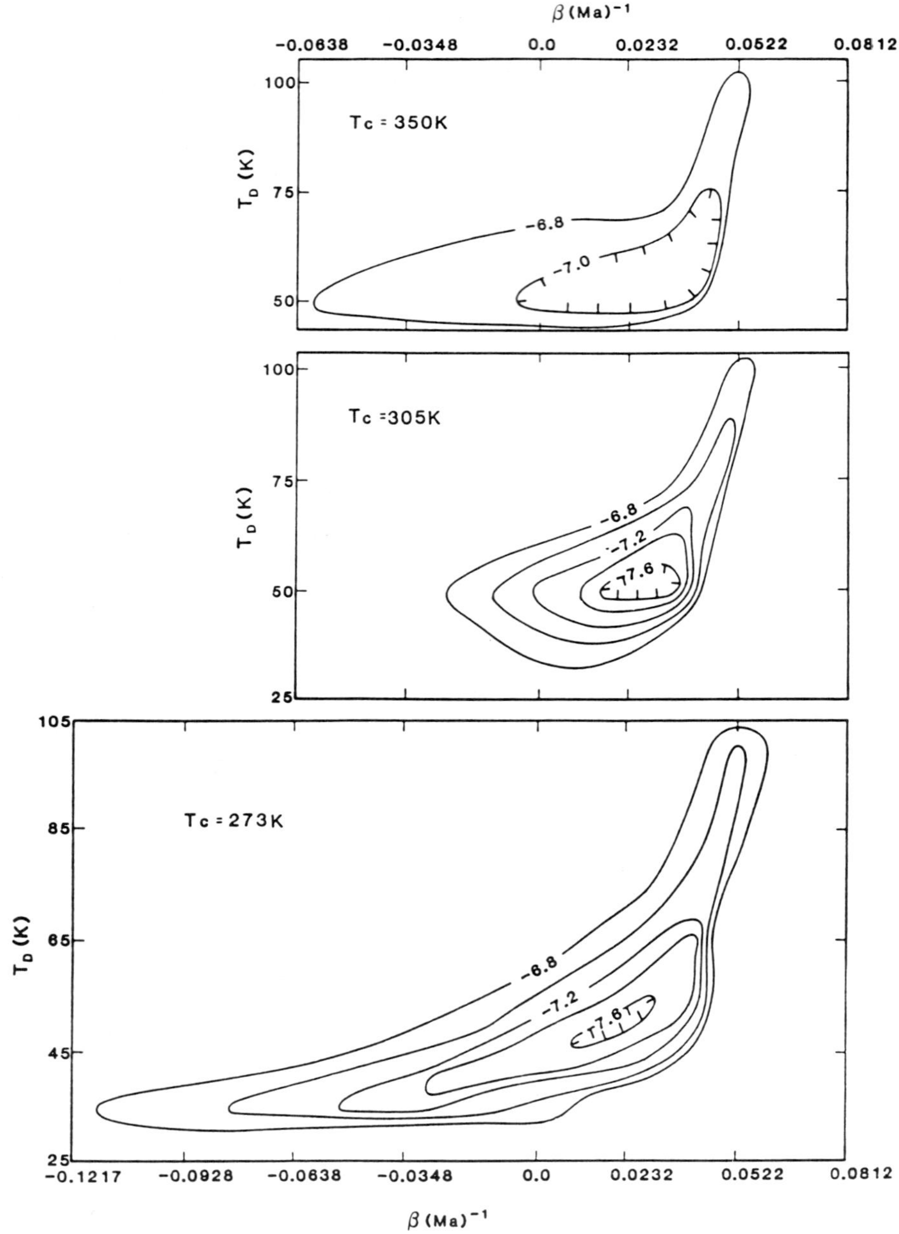

FIGURE 4.17 MSR contoured in T_D–β space for various values of T_c for well 2414.

This well is a considerable distance away from Pan Am 26 and 27 so β may vary.

The reason for the much better resolution of a solution in T_D versus T_C versus β space for this well lies with the input data, as shown in Fig. 4.18. There is an abundance of data covering a wide translucency range. Results for "allowable" values of T_D = 65 K and T_c = 273 K are shown in Figs. 4.18 and 4.19. Figure 4.19 shows the heat flux versus age with associated error assessments as calculated for Fig. 4.9. Figure 4.18 shows the predicted tanslucency with depth, using this preferred heat flux.

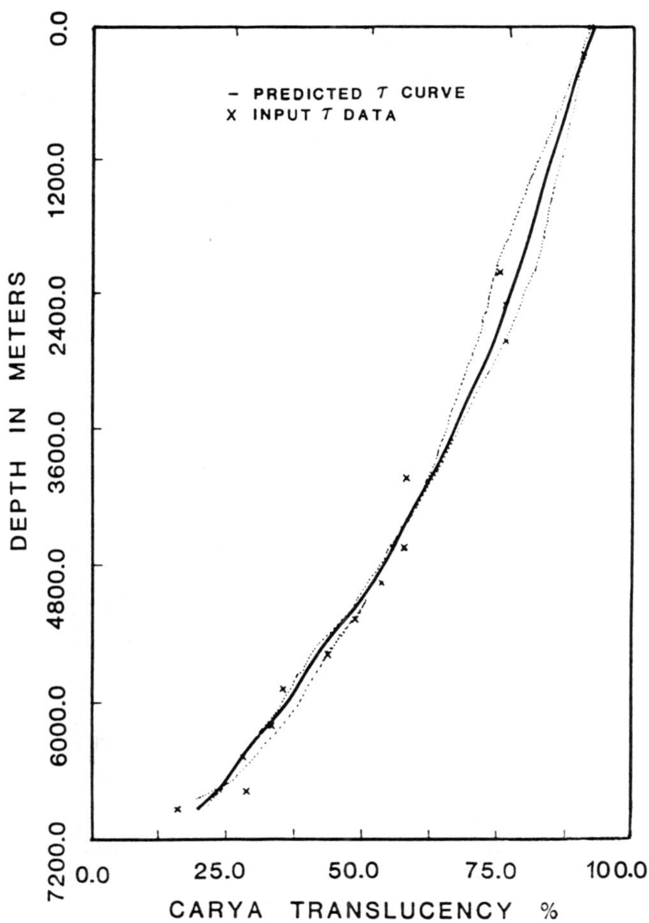

FIGURE 4.18 Predicted *Carya* translucency with depth along with associated uncertainty curves (dotted lines) and input data for well 2414. T_D = 65 K, T_c = 273 K.

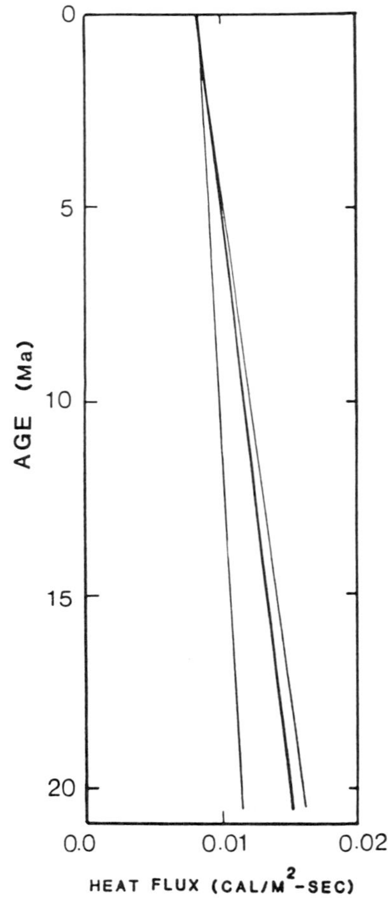

FIGURE 4.19 Heat flux versus time with associated uncertainty curves from the inversion of *Carya* translucency in well 2414 with $T_D = 65$ K, $T_c = 273$ K.

Figure 4.20 shows where the largest allowable goodness-of-fit surfaces (poorest "allowable" fits) for each well intersect the $T_c = 273$ K plane. A T_D range of 75 ± 30 K is defined at the upper end by the MSR surface of Well 2414 and at the lower end by Pan Am 27. These are both consistent with Pan Am 26. If the further constraint is placed on the Pan Am 27 surface by the β_{vr} range from vitrinite reflectance inversion, T_D is further narrowed down to $T_D = 80 \pm 25$ K. T_c is consistent for the three wells in the range defined in Pan Am 27 at $T_c = 290 \pm 20$ K.

The constants α and k can also be determined for particular pollen genera as shown in the previous chapter. Again, α and k are also constants that should not change with any geologic scenario for a given pollen genus. With an increased amount of data these constants can be determined for given genera and so simplify the inversion scheme. For the genus *Carya,* we have determined these constants consistent for the three wells to be $\alpha = 4.23 \pm 0.09$ and $k = -(2.49 \pm 0.01) \times 10^{-3}$ $(\text{Myr})^{-1}$. Given consistent α, k, T_D, and T_c we may be able to better use the inverse scheme even when few data are available. Alternatively, if no translucency data are available for a well but one has a "feel" for what the paleoheat flux was (from some other model, e.g., vitrinite reflectance inversion), translucency can be predicted with depth and time and depth of source rock maturation. Cross-checking these predictions with those from other thermal indicators will sharpen the general predictions of paleoheat flux.

We can use the pollen data to plot isomaturity contours on burial history diagrams, once the translucency has been calibrated to some "known" standard. Figure 4.21 shows isotranslucency contours plotted along with isoreflectance (vitrinite) contours on the burial history of Pan Am 27. Translucency and vitrinite reflectance mature at different rates, making a one-to-one relationship over time unattainable. With additional translucency data from source rocks we should be able to calibrate translucency to an oil window in the manner of Lopatin (1971) and later Waples (1980).

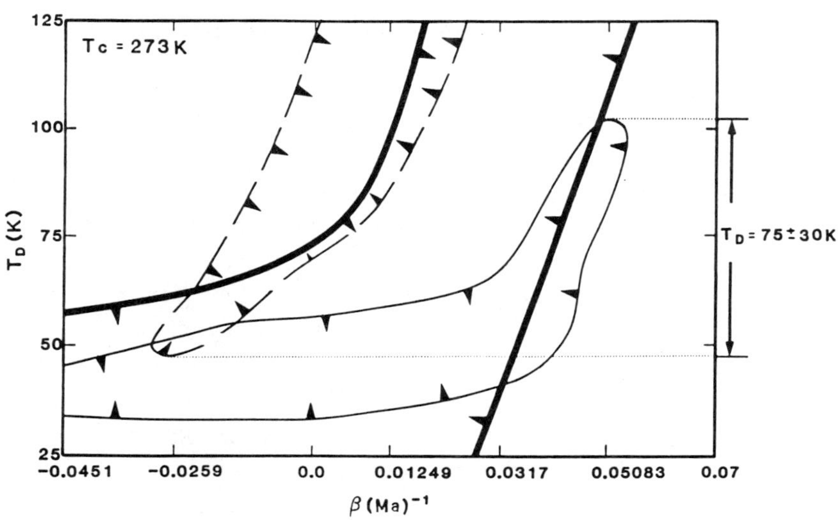

FIGURE 4.20 Allowable solution areas in T_D–β space with T_c = 273 K for the three wells, Pan Am 26, Pan Am 27, and well 2414.

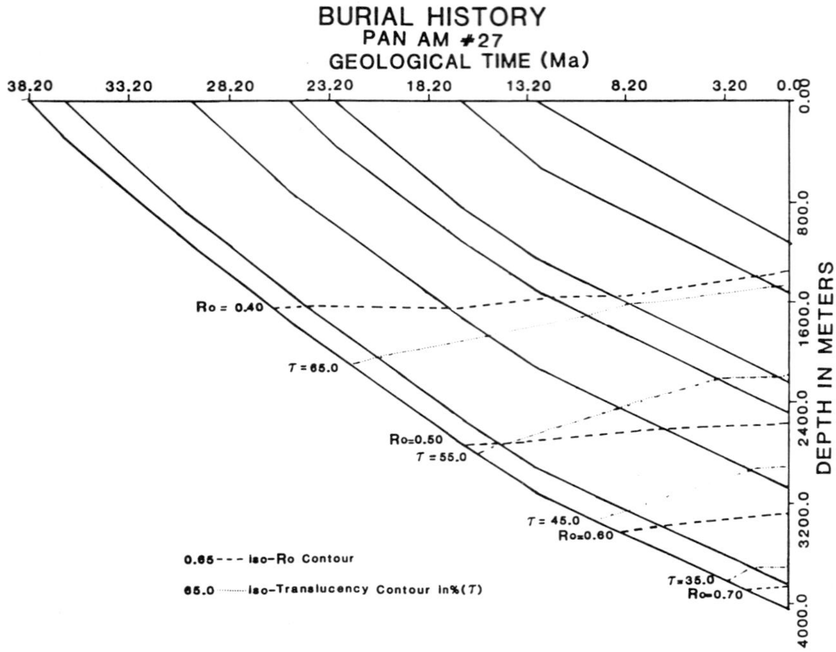

FIGURE 4.21 Burial history of Pan Am 27 with isoreflectance and isotranslucency contours (T_D = 70 K, T_c = 290 K).

II. Apatite Fission Track Distributions: North Sea

Two wells were made available in the North Sea through the courtesy of Norsk Hydro. For proprietary reasons these wells are not identified as to location and are called wells 1 and 13 respectively.

Both wells are dominated by sandy shale deposition over the last 200 Myr, with about 3000 m of total deposition in each (corresponding to an *average* rate of deposition of 15 m/Myr). In both cases a major depositional hiatus is thought to have occurred; in well 1 this hiatus (Fig. 4.22) lasted from about 170 until about 60 MYBP, while in well 13 (Fig. 4.23) the hiatus was from about 150 until 80 MYBP. Well 13 had one distribution of apatite track lengths at a depth of 2050 m (of stratigraphic age about 220 Myr), while well 1 had two distributions of apatite track lengths at 2210 m (age about 190 Myr) and 2875 m (age about 210 Myr). Thus all horizons containing the apatite samples had been exposed to nearly the total burial history of the system, including being deposited and buried before the depositional hiatuses. Present-day heat flux information

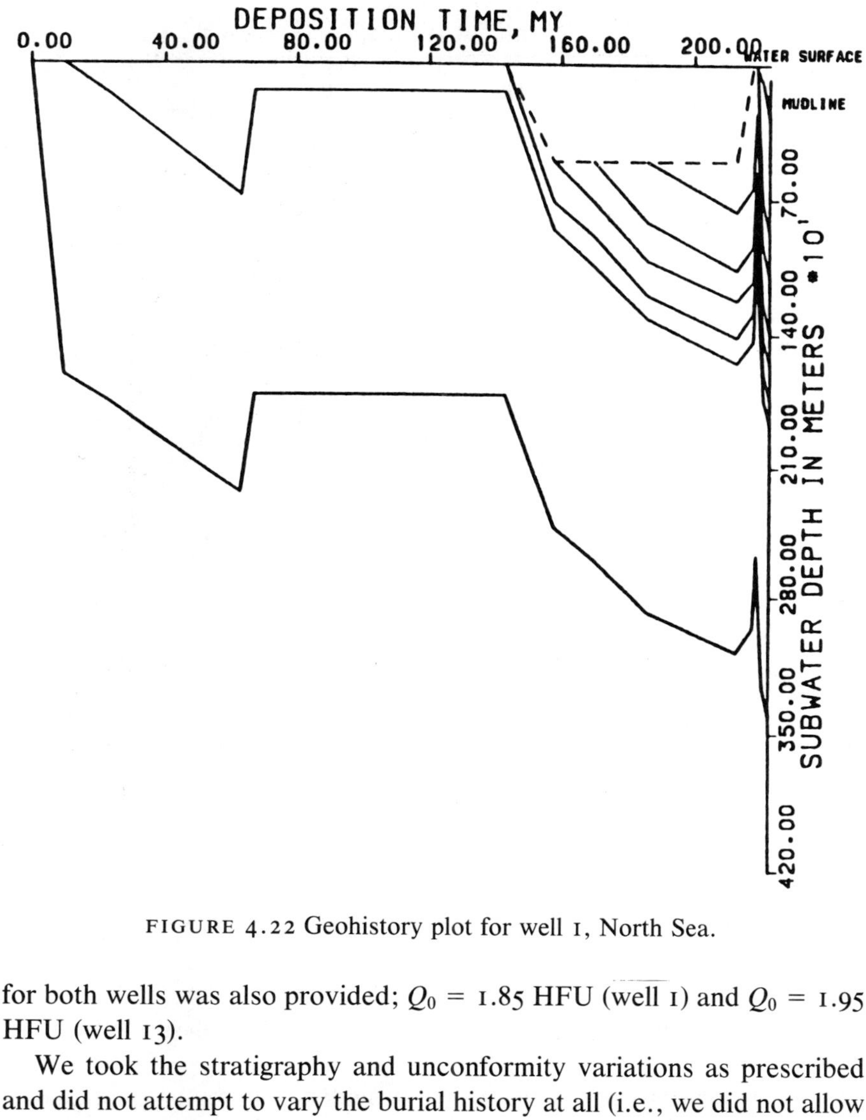

FIGURE 4.22 Geohistory plot for well 1, North Sea.

for both wells was also provided; Q_0 = 1.85 HFU (well 1) and Q_0 = 1.95 HFU (well 13).

We took the stratigraphy and unconformity variations as prescribed and did not attempt to vary the burial history at all (i.e., we did not allow deposition/erosion variations other than those specified by Norsk Hydro). In this way a precisely known burial history is available.

Armed with that information we then did two things:

1. A *dynamical* burial history inversion which allows for fluid flow, excess pressure, fracturing, and so on (see Chapter 6) and which is represented by the four geohistory curves, although curves representing burial history, pressure with depth, porosity with

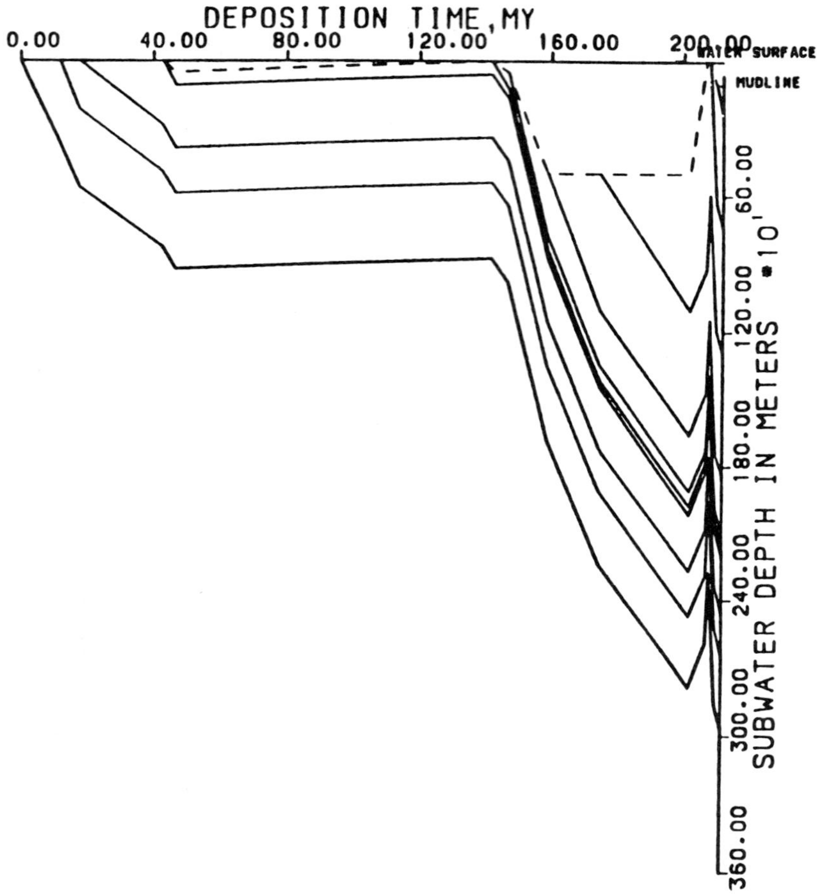

FIGURE 4.23 Geohistory plot for well 13, North Sea.

depth, permeability with depth, and rate of flow with depth can also be given.

2. A thermal inversion of the provided apatite fission track distribution. This inversion gives heat flux behavior with time (Fig. 4.24a and b).

With the heat flux behavior and the dynamical burial history we then plotted TTI values—in the sense of Waples, isotherm values with time (Fig. 4.25a and b)—and we also show an overlay of the predicted versus observed apatite track length distributions with depth (Fig. 4.26a and b). In the case of well 1, where track length distributions exist at two depths,

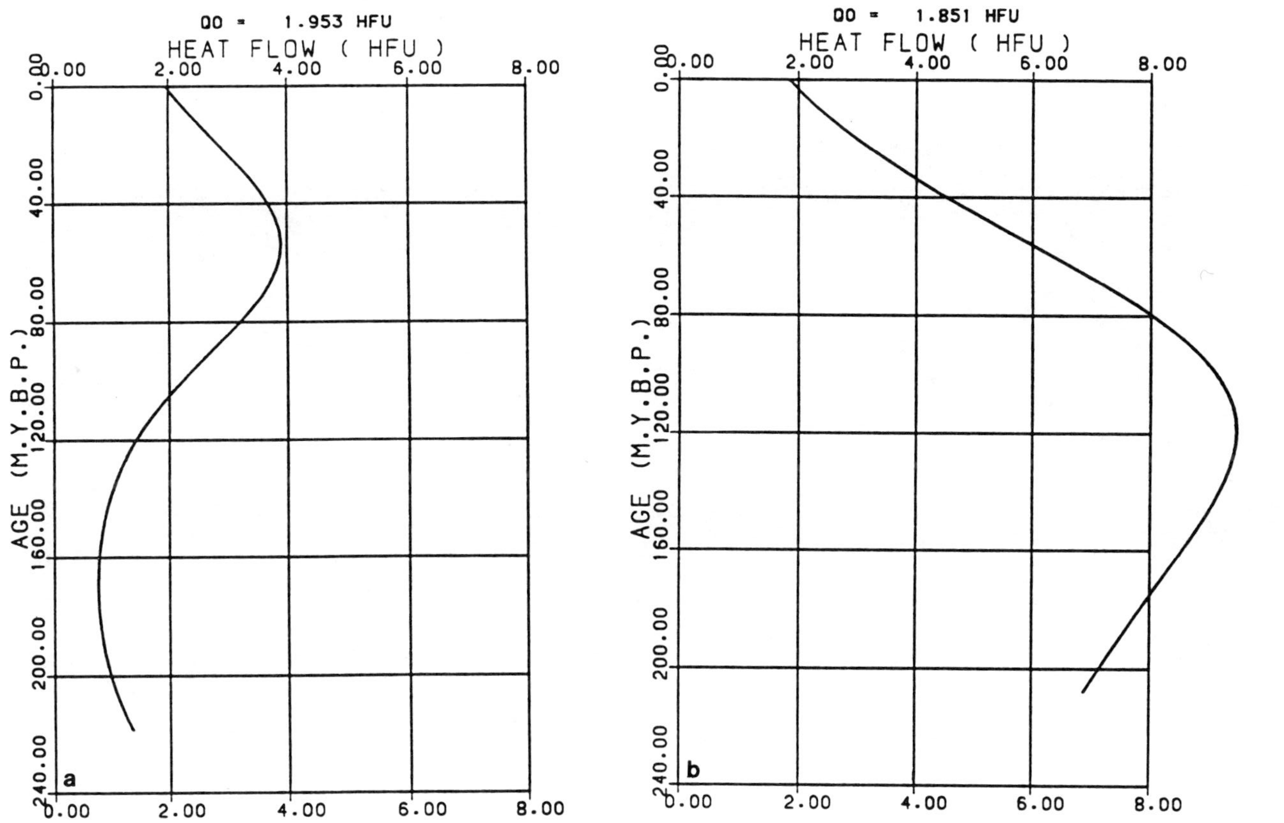

FIGURE 4.24 (a) Heat flux with time for well 1 and (b) heat flux with time for well 13, North Sea.

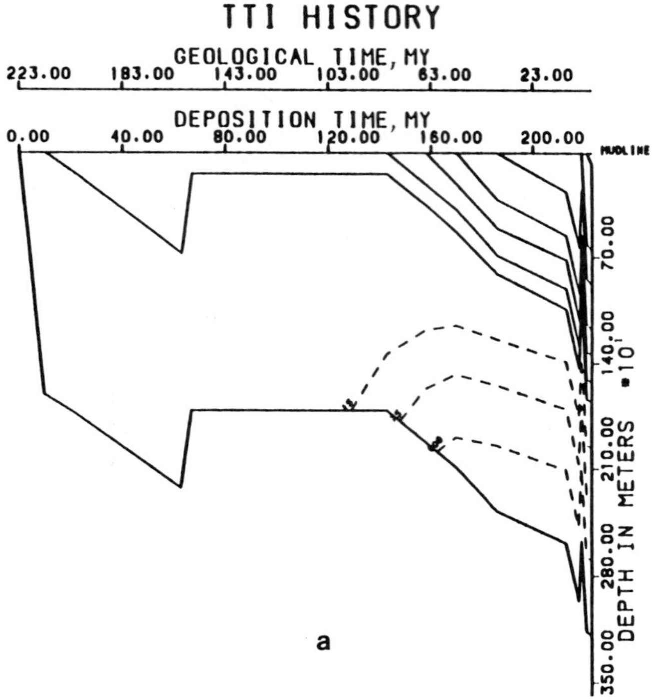

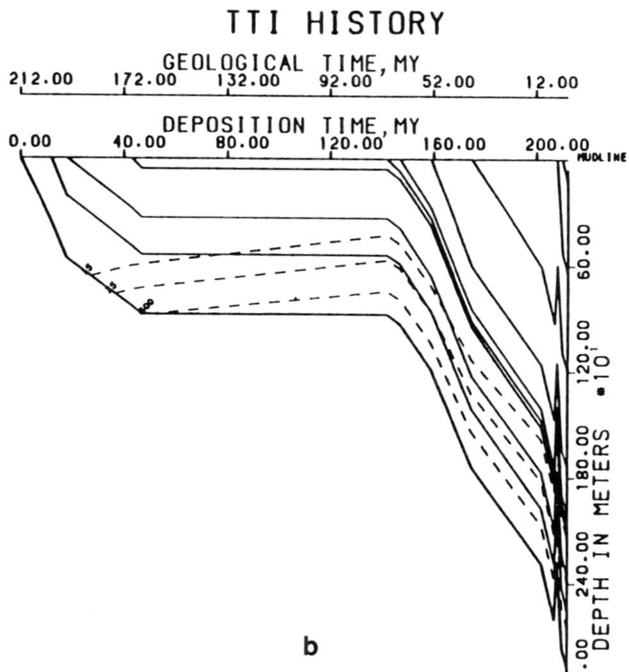

FIGURE 4.25 Overlay of iso-TTI values on burial history diagrams: (a) well 1; (b) well 13.

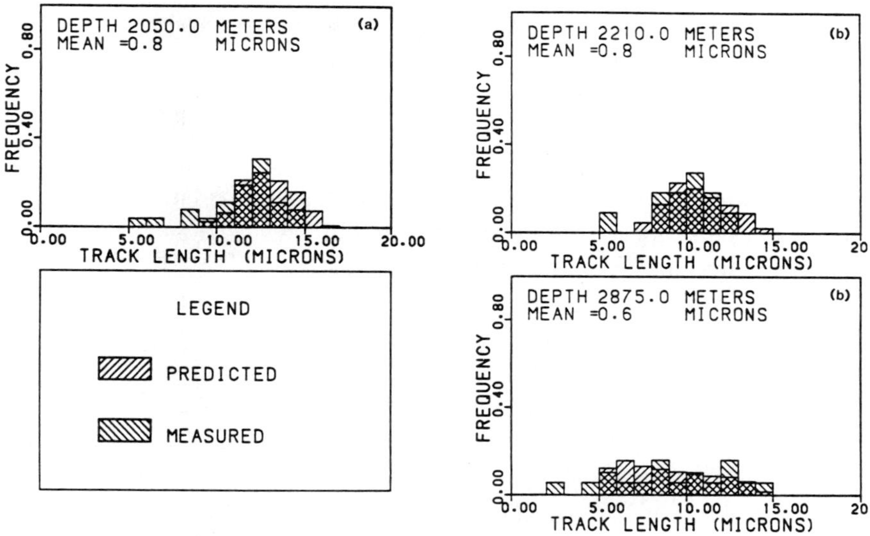

FIGURE 4.26 Predicted versus observed apatite track length distributions with depth: (a) well 1; (b) well 13.

the heat flux shows a high pulse (~8–10 HFU) about 120 ± 20 MYBP. The timing of this heat pulse and its magnitude are not discordant with the breakup scenario proposed by McKenzie (1978).

In the case of well 13, for which only one track length distribution exists, a pulse of ~4 HFU existed about 60 ± 20 MYBP, with a much cooler past (~1 HFU) around 120–160 MYBP. The high heat pulse at about 60 MYBP coincides with the massive Paleocene drop in the North Sea basin and the igneous activity recorded at about the same time.

The accuracy of these results can be seen from the predicted versus observed track length data. Having said that, it is important to qualify the results obtained by noting the intrinsic assumption that the unconformity behaviors are *precisely* specified, that is, depositional hiatuses.

In short, our concern is to exhibit the paleoheat flux behavior when all other variations are known or given, and it is assumed that such variations are immutable; that is, *all* of the track length distributions are to be accounted for solely in terms of paleoheat flux variations with time. The validity of such an assumption will be investigated in greater depth in later chapters.

With the limited data at our disposal on these two wells, we are not able to do more. Measures of excess fluid pressure with depth, or cementation and/or fracturing information on the formations, would enable improvements to be made in the dynamical burial history.

III. Apatite Fission Track Distributions: Australia

Most thermal indicators are fairly coarse in the sense that they are fairly insensitive to short-term heat flux changes and, in addition, the data are most often "noisy." Hence it is difficult to refine the resolution to paleoheat flux signatures. The latter problem is particularly severe the further back in time one attempts to "unfold" the paleoheat flux from downhole measurements of vitrinite reflectance.

In order to improve on our current capability for determining paleoheat flux, with a degree of resolution higher than we currently enjoy, we require a downhole thermal indicator which is not sensitive to migration problems, which *is* sensitive to short-term changes in heat flux, and which cumulatively records the thermal history since deposition or production. Ideally, an indicator which is continuously generated during sedimentary burial would provide differential resolution.

It has been suggested (Gleadow and Duddy, 1984) that uranium-produced fission tracks in detrital apatite are capable of recording precisely the required information.

In this subsection we examine three aspects of measured fission tracks: (i) the areal number density of tracks per gram of uranium as a function of depth, (ii) the mean track length as a function of depth, and (iii) the distribution of track lengths with depth. We will show how to invert each of these attributes in order to reconstruct the paleoheat flux and we will illustrate the degree of resolution by using apatite fission track data from two wells.

Detailed downhole detrital apatite fission track data for the two wells Kambara-1 and Curringa-1 from the Northwest Canning Basin are listed in Table 4.1 (taken from Gleadow *et al.*, 1983) for the areal track number density and uranium concentration with depth, the average track length with depth, the stratigraphic age of each depth, and the lithology for each of the wells. We also show in Fig. 4.27 the distribution of track lengths at each depth for each well. The present-day heat flux values of $Q_0 = 1.2$ HFU (for Kambara-1) and $Q_0 = 1.5$ HFU (for Curringa-1) we obtained from downhole temperature data.

Elsewhere (Guidish *et al.*, 1984) we have described in detail the isostatic burial history code which takes downhole stratigraphic data and converts them into a form detailing the burial paths of sedimentary layers with time.

Figures 4.28 and 4.29 show the resulting burial histories for Kambara-1 and Curringa-1 with selected age horizons traced through time.

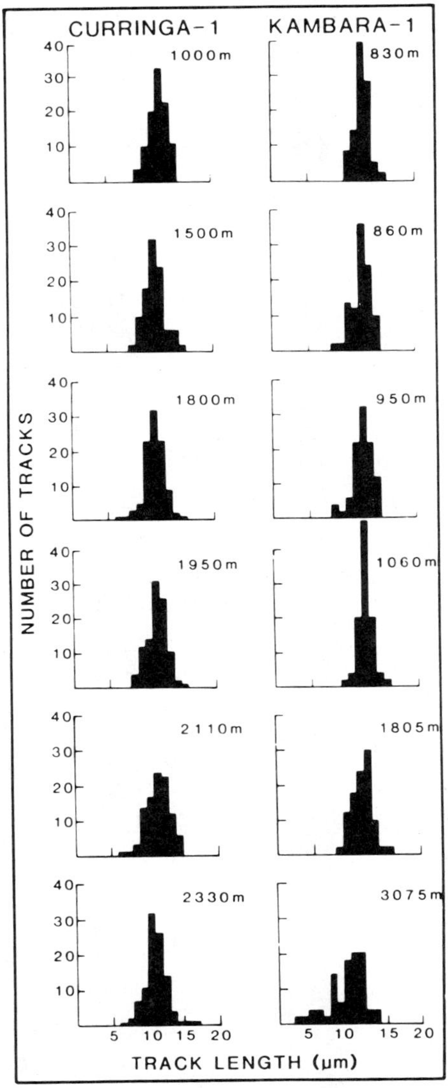

FIGURE 4.27 The distribution of fission track lengths at each depth for the Curringa-1 and Kambara-1 wells.

A. Inversion of Areal Number Density Data

Using the data presented in Table 4.1, we constructed the behavior of NU(z_i) with depth z_i for each entry in Table 4.1 for each well. We then attempted to fit, in the least-squares sense, the theoretical predicted value with depth by systematically varying β in the heat flux formula

Table 4.1 Fission Track Analytical Results—Northwest Canning Basin[a]

Sample number	Sample depth (m)	Number of grains	Neutron dose[b] ($\times 10^{15}$ n cm^{-2})	Fossil track density ($\times 10^{6}$ cm^{-2})	Induced track density ($\times 10^{6}$ cm^{-2})	Correlation coefficient	Age[c] (Myr)	Uranium (ppm)	Track length (μm)
Kambara-1									
8322-89	830	18	7.512 (3507)[d]	2.78 (971)	4.50 (786)	0.936	272 + 14	24	13.03 + 1.09 (50)
8322-90	860	7	7.512 (3507)	3.30 (848)	4.28 (550)	0.910	338 + 19	23	12.72 + 1.30 (50)
8322-90	860	3	7.512 (3507)	3.29 (362)	2.06 (114)	0.976	[677 + 74]	11	—
8322-92	945	14	7.512 (3507)	1.78 (891)	2.64 (658)	0.936	298 + 16	14	12.75 + 1.38 (50)
8322-93	1050	17	7.512 (3507)	2.18 (1297)	3.42 (1915)	0.943	281 + 13	18	12.78 + 1.06 (50)
8322-94	1160	14	7.512 (3507)	2.06 (1016)	3.12 (767)	0.971	291 + 15	16	12.48 + 1.08 (50)
8322-95	1805	15	7.512 (3507)	2.04 (1124)	3.46 (954)	0.983	260 + 12	18	11.97 + 1.43 (50)
8322-96	1838	12	7.512 (3507)	2.03 (894)	3.24 (713)	0.982	276 + 15	17	11.64 + 1.22 (50)
8322-97	2925	17	7.512 (3507)	0.605 (373)	1.78 (549)	0.933	151 + 10	9	10.06 + 2.50 (50)
8322-98	3075	10	7.512 (3507)	0.147 (42)	1.44 (206)	0.858	46 + 8	8	9.69 + 2.50

Curringa-1									
8222-16	1000	6	8.379 (3912)	2.03 (395)	2.44 (238)	0.957	403 + 34	12	12.78 + 1.25 (50)
8222-16	1000	2	8.379 (3912)	4.05 (268)	2.72 (90)	—	[707 + 87]	13	—
8222-18	1500	13	8.825 (2060)	1.91 (791)	2.96 (612)	0.982	333 + 19	13	11.84 + 1.38 (50)
8222-18	1500	2	8.825 (2060)	2.42 (178)	1.88 (69)	—	[648 + 93]	8	—
8222-19	1800	9	8.379 (3912)	2.14 (675)	3.04 (479)	0.948	344 + 21	14	11.86 + 1.42 (50)
8222-19	1800	1	8.379 (3912)	3.30 (97)	2.92 (43)	—	[542 + 99]	14	—
8222-21	1960	10	8.379 (3912)	1.16 (939)	2.18 (877)	0.995	263 + 13	10	11.65 + 1.31 (50)
8222-23	2110	14	8.379 (3912)	0.715 (839)	1.65 (968)	0.943	214 + 11	8	11.81 + 1.65 (50)
8222-24	2300	13	8.379 (3912)	0.352 (401)	1.33 (757)	0.959	132 + 8	6	11.32 + 1.36 (50)
8222-24	2300	3	8.379 (3912)	0.913 (208)	1.26 (144)	0.959	[352 + 39]	6	—

[a] From Gleadow *et al.*, 1983.

[b] Neutron dosimetry for reference glass SRM612: $B = 6.005 \times 109$ neutrons/track, $\zeta = 360 \pm 10$. Equivalent to NBS Cu dosimetry via reference glass SRM962.

[c] $\lambda_f = 7.03 \times 10^{-17}\ \text{yr}^{-1}$, $\lambda_d = 1.551 \times 10^{-10}\ \text{yr}^{-1}$, $^{235}\text{U} = 5.802 \times 10^{-22}\ \text{cm}^2$, $I = 7.253 \times 10^{-3}$.

[d] Parentheses show number of tracks counted. All track densities as for internal surfaces ($g = 0.5$).

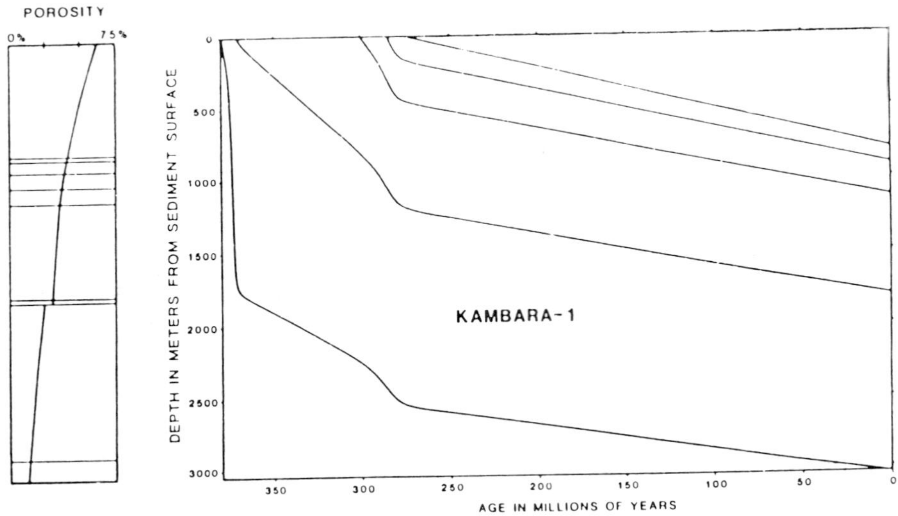

FIGURE 4.28 Burial history plot for selected age horizons in the well Kambara-1. The ordinate is in meters, the abscissa in MYBP.

$Q(t) = Q_0 \exp(\beta t)$ searching for the minimum MSR. For the values of T_c and k_1, which are needed in order to compute the theoretical time–temperature integrals, we originally chose $T_c = 573 \pm 50$ K and $k_1 = 3 \times 10^{-2}$ $(\mathrm{Myr})^{-1}$ as suggested by the laboratory experiments of Gleadow *et al.* (1983). However, not only did these choices of parameters

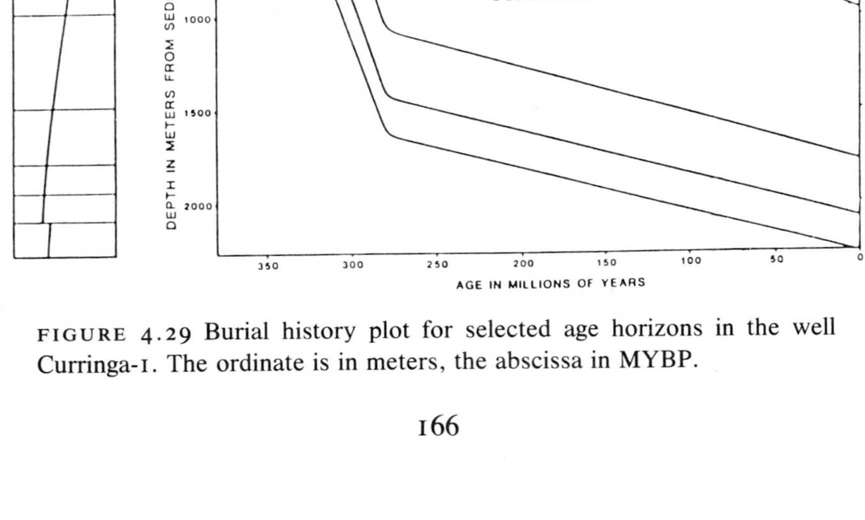

FIGURE 4.29 Burial history plot for selected age horizons in the well Curringa-1. The ordinate is in meters, the abscissa in MYBP.

lead to very poor fits of the theoretical expressions for areal number density per gram of uranium when compared to the observed values for both wells, no matter how much β was varied, but also it proved impossible to obtain any significant degree of fit of the theoretical expressions for mean track length with depth to the observations, and similarly for the track length distributions.

Two possible remedies were apparent. Either (1) the theoretical formulation was grossly in error despite the large amount of evidence based on track length reduction measurements made over the years, or (2) the critical temperature T_c and/or the laboratory-determined rate constant k_1 were not the appropriate values to use when attempting to determine fission track behavior over the course of millions to hundreds of millions of years. The argument here is that the laboratory conditions are very different in terms of pressure, ambient chemistry of surrounding materials, weathering, and so forth than the actual conditions in the geologic setting. Thus one might anticipate a different set of values to apply for T_c and k_1 in the subsurface than are determined from laboratory measurements.

An alternative point of view is to note that the theoretical expressions for areal number density, mean track length, and distribution of track lengths contain two chemical parameters, T_c and k_1, and also one geologic parameter β. Then from several wells one should be able to determine, in a consistent least-squares manner, the parameters T_c and k_1 which should at the least be basinal, if not global, constants since they refer to chemical and physical processes and not to geological variations. A similar philosophy has been followed with respect to the chemical and physical parameters entering the formulas related to vitrinite reflectance and the sterane and hopane isomer ratios. The nongeological parameters determined in this manner have indeed been proved to be global constants for vitrinite reflectance by testing on over 400 wells worldwide (Lerche *et al.*, 1984) and are consistent with constant values for the sterane and hopane isomer thermal indicators on the basis of a few wells (Armagnac, 1985).

At that time we found that we could not improve on the value $T_c = 573 \pm 50$ K advocated by Gleadow *et al.* (1983)—presumably because RT_c is an intensive energy, representative of release of stress in apatite by annealing, and so should be predominantly uninfluenced by long-term rate factors not accessible on laboratory time scales[1]. Here R is the gas constant.

[1] This situation has now improved considerably (He and Lerche, 1988; see also Chapter 9).

However, for the rate constant k_1 the story is very different. Here the rate of fission track loss on the multi-million-year time scale does not have to be closely slaved to a laboratory-determined value since the *rate* at which track annealing takes place depends on the long-term release of crystal stress in apatite. Laboratory measurements of k_1 obviously cannot mimic any stress release processes which take place on a longer time scale than about a human lifetime. Thus the rate constant for track annealing appropriate to subsurface long-term conditions must be determined from subsurface measurements.

In order to determine the most appropriate value of k_1 we therefore obtained the MSR best fit for β to the track length distributions at the observed depths for a suite of k_1 values. Plots of contours of constant MSR in a k_1–β space then provide an assessment of the best value to use for k_1.

On the basis of fitting to the shallow and deep data from the Curringa-1 well using a heat flux variation of the one-parameter form $Q(t) = Q_0 \exp(\beta t)$, on the grounds that at the shallow depths the apatite has hardly had time or resolution to determine a quadratic term in the exponential, we obtained the MSR contour plots shown in Figs. 4.30 and 4.31, respectively. We repeated the fitting procedure for every one of the five depth levels in both Curringa-1 and Kambara-1 for which we had detailed track distributions. The best k_1 values for each level are given in Table 4.2. From Table 4.2 we estimate a best k_1 given by

$$k_1 = (9.7 \pm 1.4) \times 10^{-3}\ (\text{Myr})^{-1} \tag{4.8}$$

which is a factor of 3 lower than that obtained from laboratory measurements.

For the moment we concentrate on the areal number density inversion using the rate constant k_1 given by equation (4.8).

Determining the minimum MSR by searching systematically through the space of β values for each of the two wells yields the plots shown in Figs. 4.32 and 4.33. The uncertainty in the best-fit value of β we estimate by finding the MSR values 10% higher than the minimum. The heat flux variation with time for each well is then shown in Figs. 4.34 and 4.35, while the theoretical prediction of areal track number density per gram of uranium using this best-fit β value for each well is given in Figs. 4.36 and 4.37, respectively, superposed on the datum points.

The uncertainties in fit values based on the 10% MSR rate are also shown in Figs. 4.34 and 4.35. We see that the *best* heat flux variations for

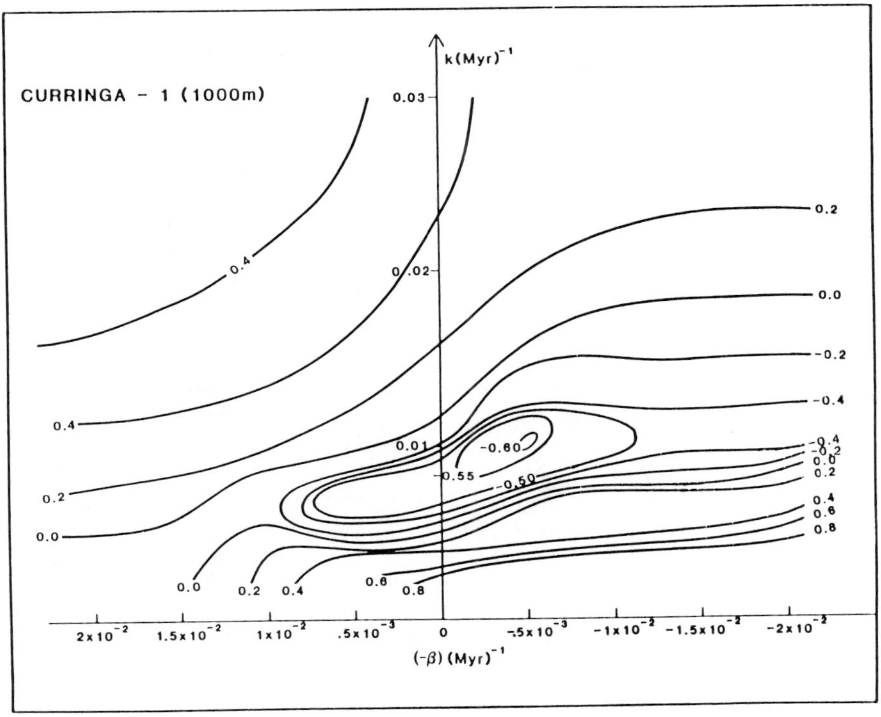

FIGURE 4.30 Plot of contours of constant MSR in k_1–β space for the track distribution data from Curringa-1 at 1000-m depth. The ordinate is k_1 in $(\text{Myr})^{-1}$ and the abscissa is β in $(\text{Myr})^{-1}$.

Curringa-1 (Kambara-1) show the heat flux decreasing in the past. It is also clear from the uncertainty in each β value that the areal number density method is too crude a measure to allow us to infer the presence of more than a one-parameter fit.

As we remarked already, a potential difficulty with the areal number density method is that if short tracks are overlooked then too low an areal number density will be recorded, leading to too high a heat flux in the past. Consider then results from the mean track length method.

B. Inversion of Mean Track Length Data

From the data presented in Table 4.1 we have the mean track lengths for each of five stratigraphic horizons for each of the two wells.

Again, we performed a least-squares fit of predicted mean track lengths versus observed track lengths for varying values of β using the heat flux from $Q(t) = Q_0 \exp(\beta t)$ and the rate constant k_1 given by equation (4.8).

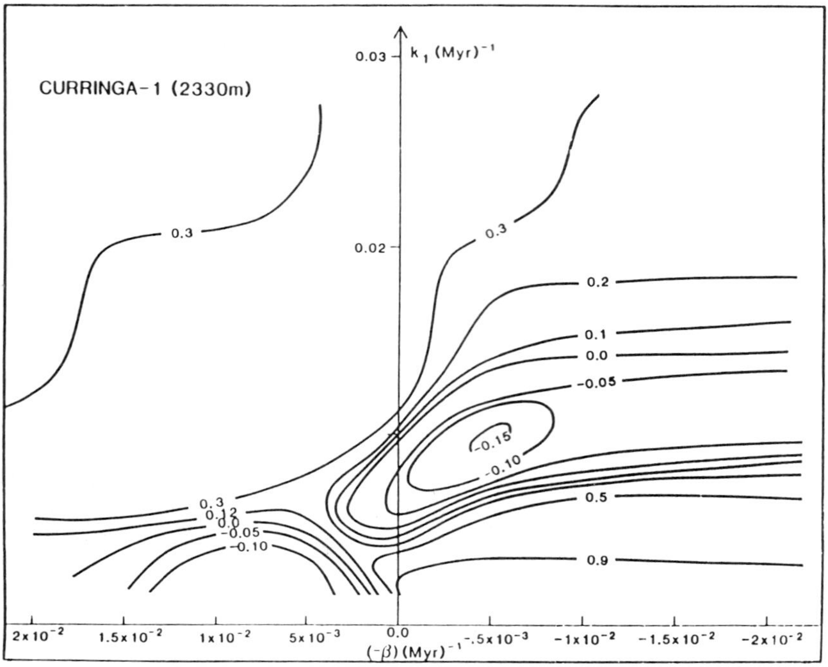

FIGURE 4.31 As for Fig. 4.30 but for the Curringa-1 track distribution data at 2330 m.

Table 4.2 Determination of Rate Constant k_1 and Best β for Individual Track Distributions[a]

Depth (m)	k_1 (Myr)$^{-1}$	β (Myr)$^{-1}$	MSR
Curringa-1			
1000	10×10^{-3}	-5×10^{-3}	−0.61
1500	3.3×10^{-3}	2×10^{-3}	−0.35
1800	6.6×10^{-3}	0	−0.45
1960	9.9×10^{-3}	-5×10^{-3}	−0.28
2110	9.9×10^{-3}	-5×10^{-3}	−0.77
2300	10×10^{-3}	-5×10^{-3}	−0.18
Average $k_1 = [8.3 \pm 1.3\ (1\sigma)] \times 10^{-3}$ (Myr)$^{-1}$			
Kambara-1			
830	6.6×10^{-3}	0	−0.20
860	6.6×10^{-3}	0	−0.29
945	20×10^{-3}	-5×10^{-3}	−0.84
1050	6.6×10^{-3}	5×10^{-3}	−0.03
1805	9.9×10^{-3}	-5×10^{-3}	−0.38
3075	16.5×10^{-3}	-1×10^{-2}	−0.30
Average $k_1 = [11.1 \pm 2.6\ (1\sigma)] \times 10^{-3}$ (Myr)$^{-1}$			
Average k_1 for both Curringa-1 and Kambara-1 is $k_1 = [9.7 \pm 1.4\ (1\sigma)] \times 10^{-3}$ (Myr)$^{-1}$			

[a] Heat flux function $Q = Q_0 \exp(\beta t)$.

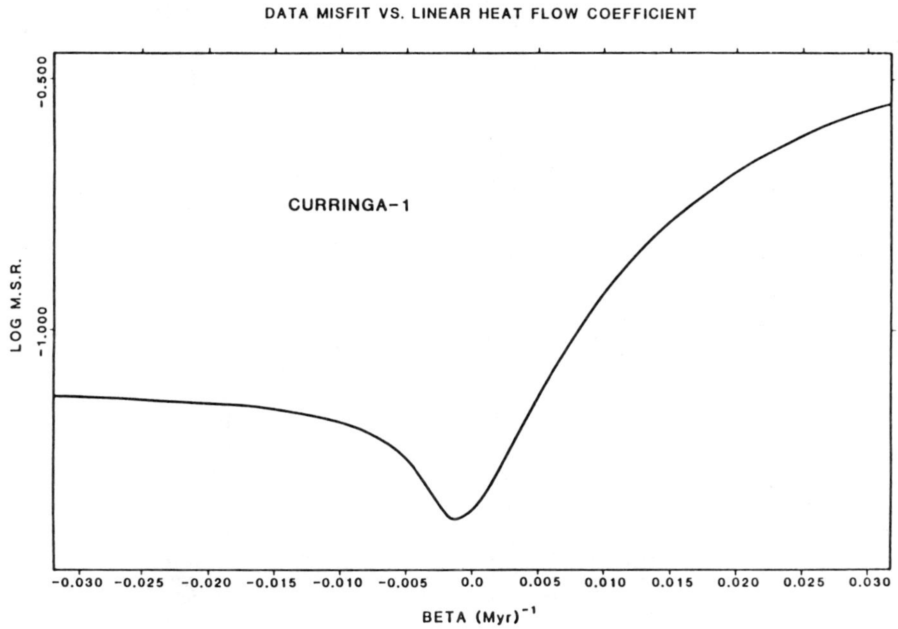

FIGURE 4.32 Plots of MSR versus β for $k_1 = 10 \times 10^{-3}$ $(\text{Myr})^{-1}$ for Curringa-1 areal track number density per gram of uranium depth data. Note the presence of a pronounced minimum in each curve at a β value of about -2×10^{-3} $(\text{Myr})^{-1}$.

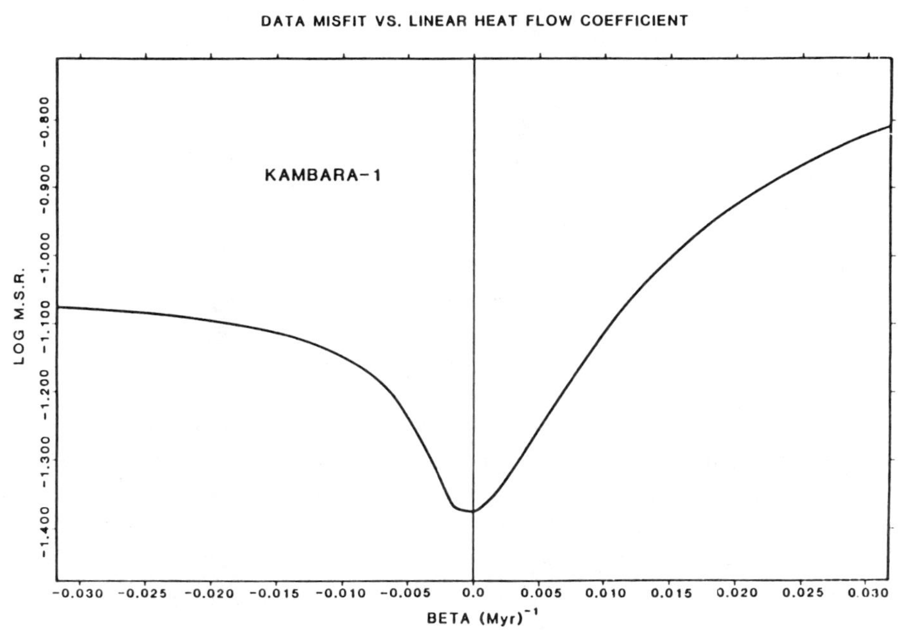

FIGURE 4.33 As for Fig. 4.32 but for areal track number density per gram of uranium depth data from Kambara-1.

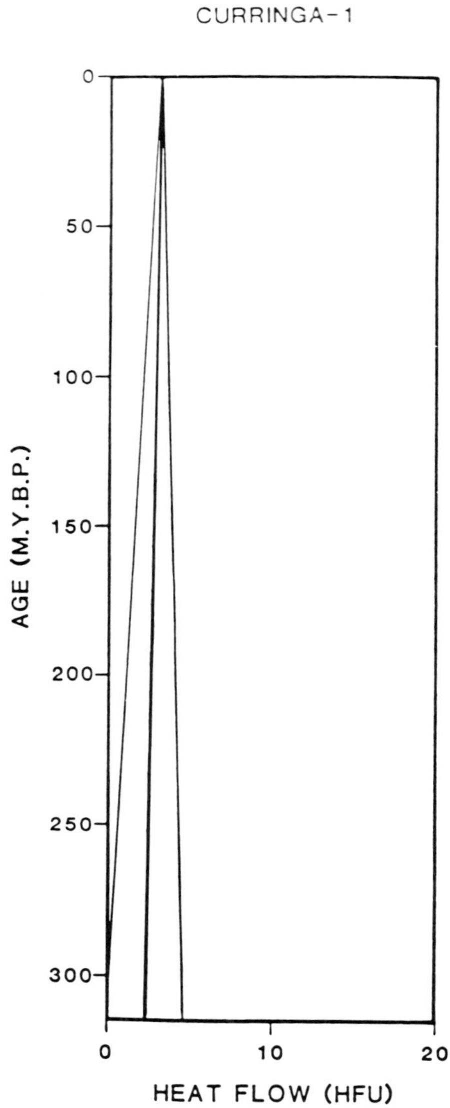

FIGURE 4.34 Plot of heat flux with time determined from the areal track density method for Curringa-1. The ordinate is time in MYBP, the abscissa is heat flux in HFU (1 HFU = 40 mW/m^2). The solid central line is the heat flux obtained using the best-fit β value; the fainter lines to either side represent error assessments based on an MSR value 10% greater than the minimum.

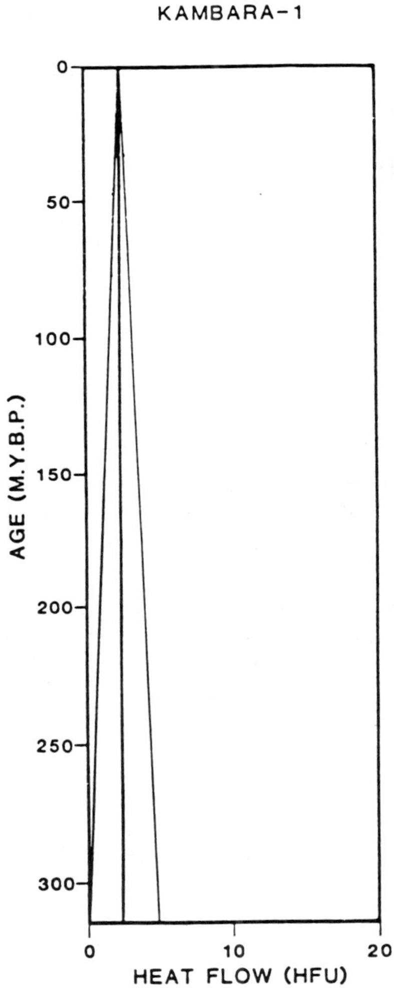

FIGURE 4.35 As for Fig. 4.34 but for areal track density data from Kambara-1.

Since we had only a few points per well, we followed option 1 and eliminated the initial track length at production from the problem.

The curves of MSR versus β are shown in Figs. 4.38 (Curringa-1) and 4.39 (Kambara-1), and the predicted mean track length behavior with depth at the present day using the best β value is shown in Fig. 4.40 for Curringa-1 and Fig. 4.41 for Kambara-1, together with estimated uncertainties. The heat flow variations with time and their uncertainties are

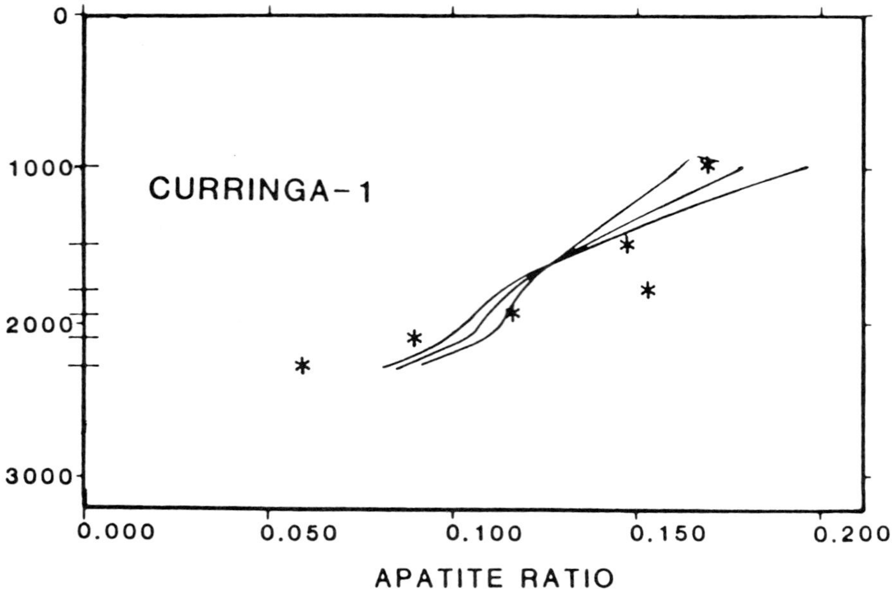

FIGURE 4.36 Theoretical prediction of areal track number density per gram of uranium with depth at the present day superimposed on the datum points for well Curringa-1. The center line is obtained using the best β corresponding to the minimum MSR; the other lines correspond to MSR values 10% greater than the minimum.

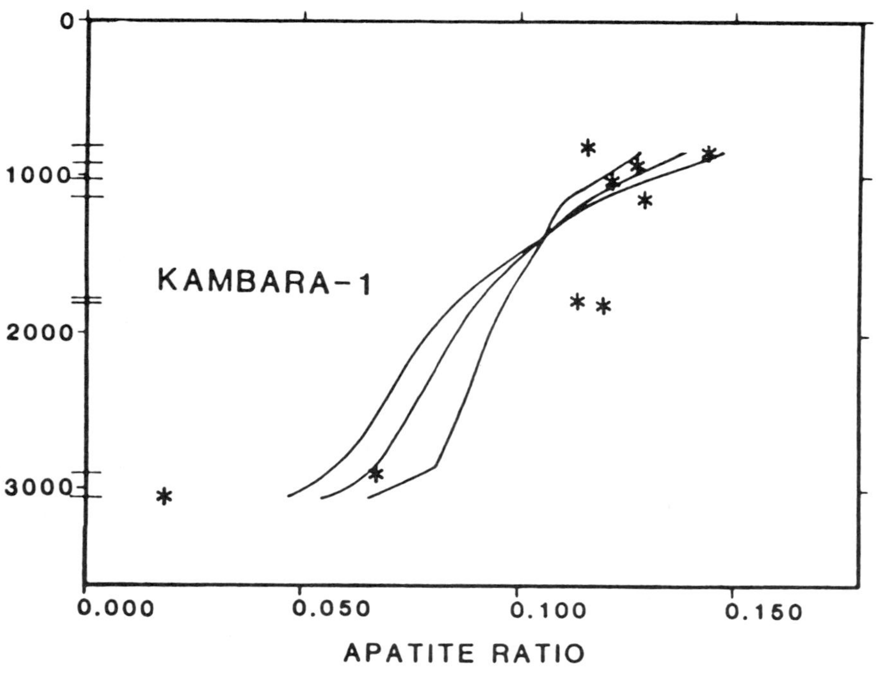

FIGURE 4.37 As for Fig. 4.36 but for areal track density data from Kambara-1.

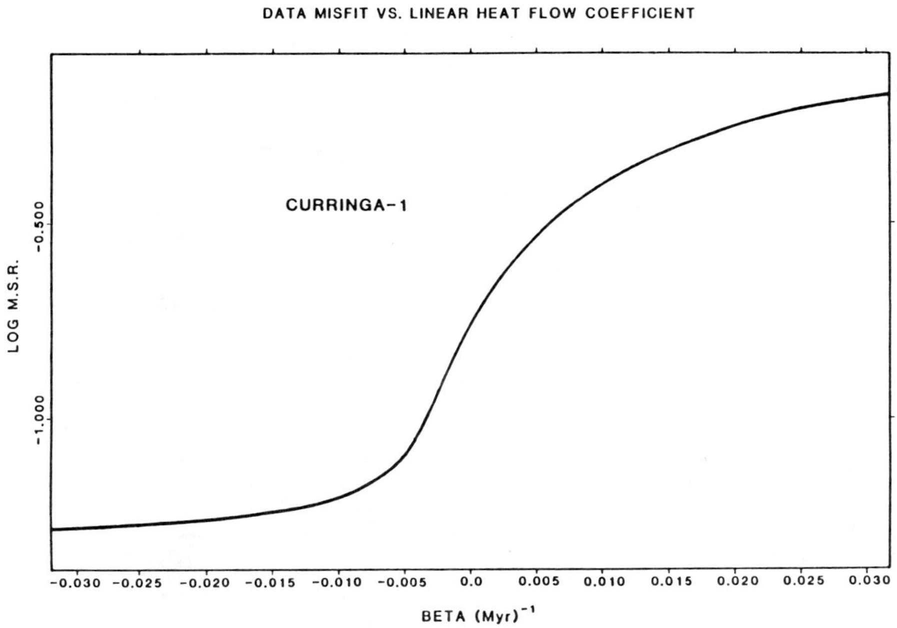

FIGURE 4.38 MSR versus β plot for the mean track length method using data from Curringa-1. Note the absence of a well-developed minimum, implying a very poor fit to the data.

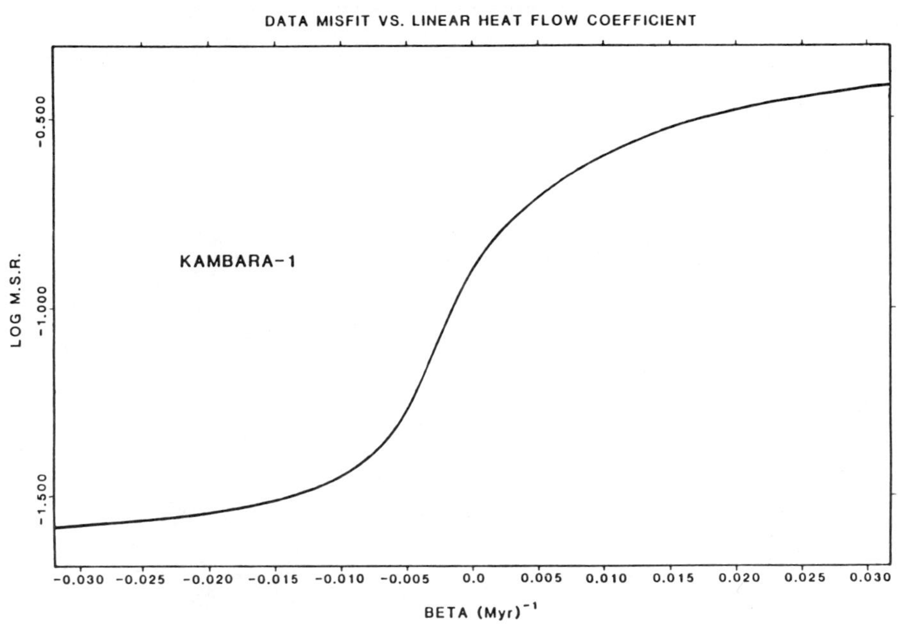

FIGURE 4.39 As for Fig. 4.38 but for mean track length with depth data from Kambara-1. Again note the absence of a well-developed minimum, implying a very poor fit to the data.

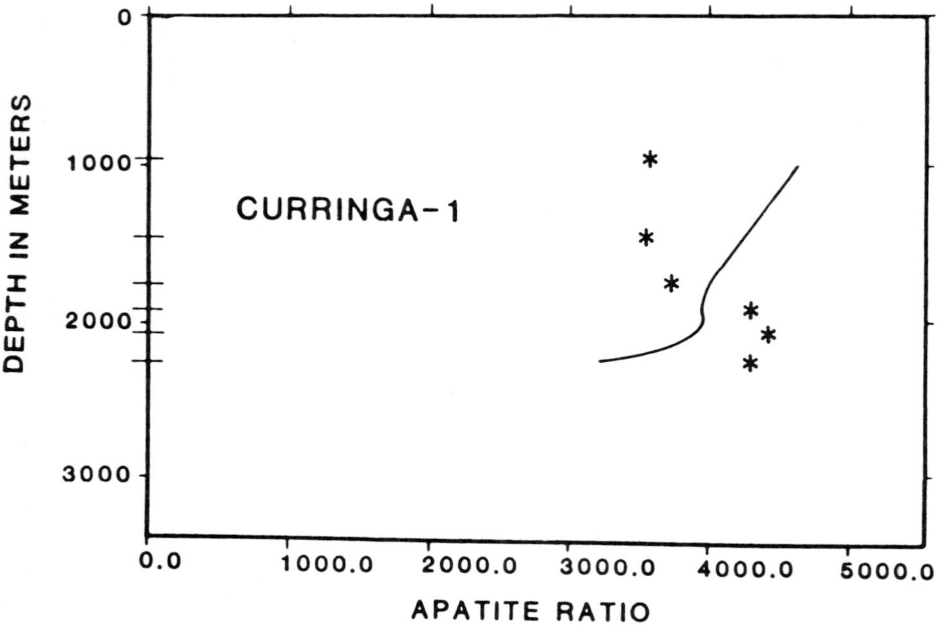

FIGURE 4.40 Theoretical prediction of mean track length with depth at the present day superposed on the observed mean track length with depth data taken from Curringa-1. The extremely poor fit is to be noted.

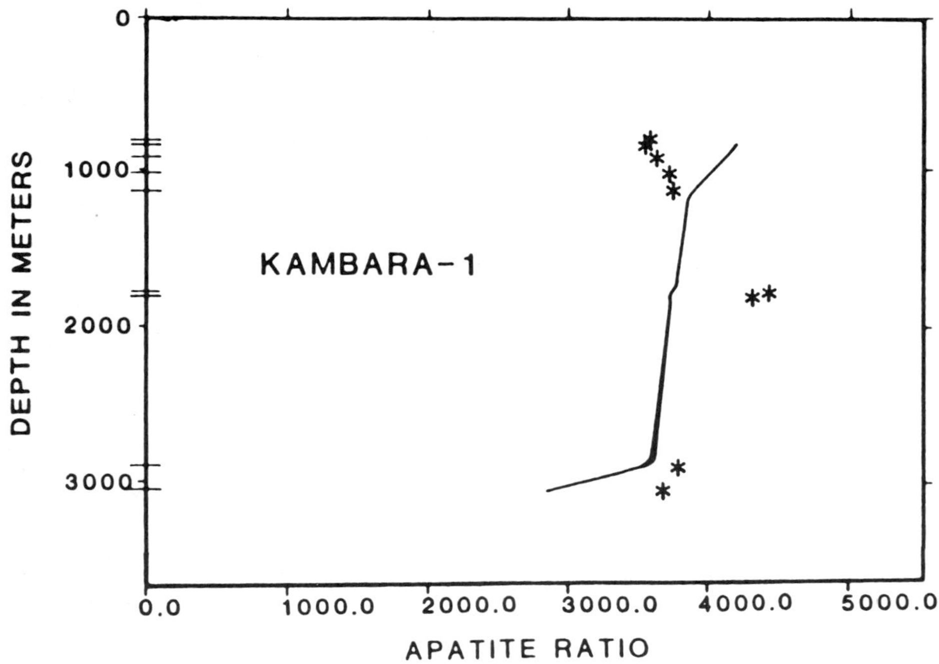

FIGURE 4.41 As for Fig. 4.40 but for mean track length data taken from Kambara-1. Again note the extremely poor fit.

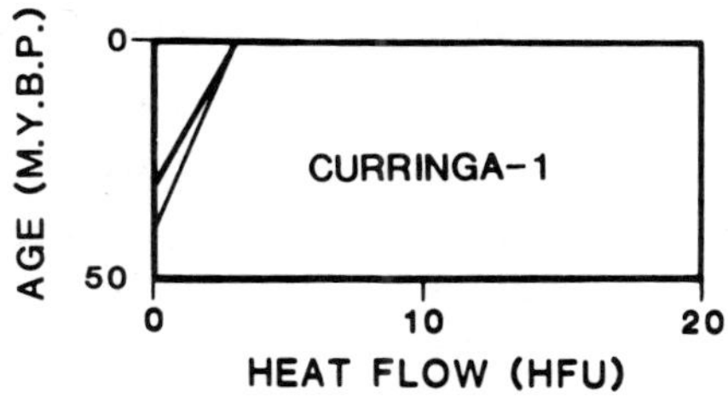

FIGURE 4.42 Plot of the best heat flux with time variation most consistent with the mean track length with depth data from Curringa-1. Note that the heat flux attempts to go to zero at about 30–50 MYBP to accommodate the recorded increasing mean track length with depth behavior.

shown in Figs. 4.42 and 4.43. The reason for this anomalous behavior is not hard to find. Inspection of the figures shows that, to within the scatter of the data, the mean track length is either constant or increasing with increasing depth. However, the one-parameter heat flux model requires that the mean track length *decrease* with increasing depth, since shorter tracks are being developed with time. The only way the inversion scheme can be consistent with the present-day heat flux and the dichotomous increasing mean track length is if the heat flux goes to zero very quickly. These anomalous mean track length data show this heat flux behavior, but note from Figs. 4.42 and 4.43 that the MSR goodness-of-fit criterion with changing β does not have a significant minimum, indicating that the mean track length data are not sufficiently well behaved for any significant measure of heat flux variation to be determined.

Despite the paucity of the data, we see that the mean track length method predicts a heat flow decreasing in the past for Kambara-1 and Curringa-1.

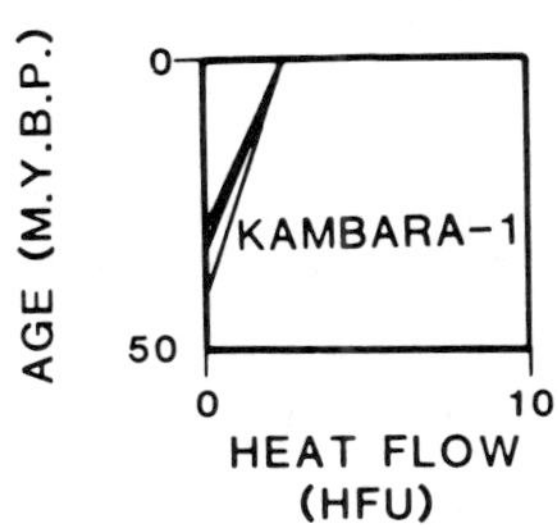

FIGURE 4.43 As for Fig. 4.42 but for mean track length data for Kambara-1. Again note the attempt to run the heat flux to zero about 30–50 MYBP to accommodate the recorded increasing, and scattered, behavior of mean track length with depth.

C. Inversion of the Distribution of Tracks with Depth

Despite having only five depths per well at which fission scar tracks in detrital apatite have been measured in Curringa-1 and Kambara-1, the actual distributions of track lengths (at 1-μm bin widths) are available (see Fig. 4.27), so that a wealth of *shape* data is available at each of the five depths. It would seem that we should be able to fit the individual distribution shapes at each depth and so determine the temporal variation of heat flux that each distribution at each depth has been exposed to.

In fact, it is precisely this wealth of shape data at each depth which enabled us to determine the effective rate constant k_1. The argument is basically simple. With a heat flux function of the form (4.7) the theoretical expression controlling the expected shape of the track length distribution at each depth depends nonlinearly on the parameters k_1 and β. Hence from five distribution shapes at five depths for each well we should be able to determine the best values of the parameters with the data. But we have two wells with which we can play the game. Each well will have a different value of β since each will have undergone a different thermal history, but the rate constant k_1 should be the same for both wells (if indeed k_1 is a chemical, or physical, constant) and not beholden to geologic effects.

To test this idea we varied k_1 and β for the shallow and deep data for both wells. Two of the resulting contours of constant k_1 in MSR–β space are shown in Fig. 4.30 for the well Curringa-1. This process was repeated for each of the layers in each of the wells, and the best k_1 values obtained are recorded in Table 4.2. From these values we obtain the best average values of k_1 as

$$k_1 = (8.3 \pm 1.3) \times 10^{-3}\,(\text{Myr})^{-1} \quad \text{(Curringa-1)}$$

and

$$k_1 = (11.1 \pm 2.6) \times 10^{-3}\,(\text{Myr})^{-1} \quad \text{(Kambara-1)}$$

Averaging these two values leads to the estimate

$$k_1 = (9.7 \pm 1.4) \times 10^{-3}\,(\text{Myr})^{-1} \tag{4.9}$$

for the value of k_1 most consistent with all of the data from both wells.

In order to obtain consistency with both shallow and deep data from both wells, we therefore fixed k_1 at the best average value given by equation (4.9) and then obtained contours of constant MSR in a β–z space. These are shown in Fig. 4.44 for Curringa-1 for track length distributions at two different depths.

With the best values of k_1 and β determined, we can then make a forward modeling prediction along each burial history track of the expected distribution of track lengths at the present day at each depth.

In Figs. 4.45 (Curringa-1) and 4.46 (Kambara-1) we have overlain the predicted track length distribution for each well on the observed distributions for each of the six depths using the best single β value consistent with all of each well's distributions. As can be seen, the predictions and observations match rather well at both shallow and deep horizons—a strong indication that the underpinning framework outlined here for determining paleoheat flux from fission scar tracks in detrital apatite by inversion is indeed basically confirmed.

One final, fairly major, point is in order. We see from the comparison of the theoretical predicted track length distributions with the observed tracks with depth, using a *single* value of β over all ages at which we have

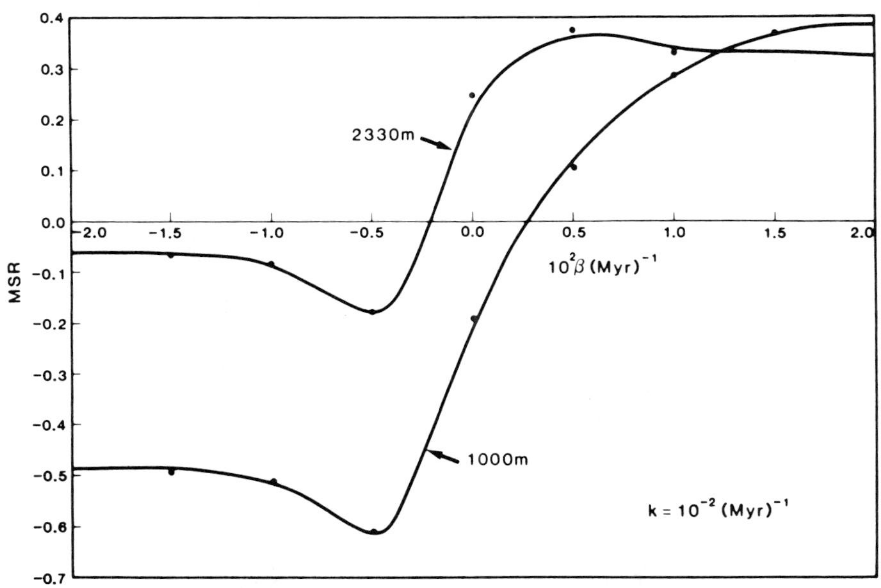

FIGURE 4.44 MSR versus β plots for the track length distribution data at the depths 1000 and 2330 m in Curringa-1. Note the presence of a pronounced minimum in both cases at $\beta \cong -5 \times 10^{-3}$ (Myr)$^{-1}$.

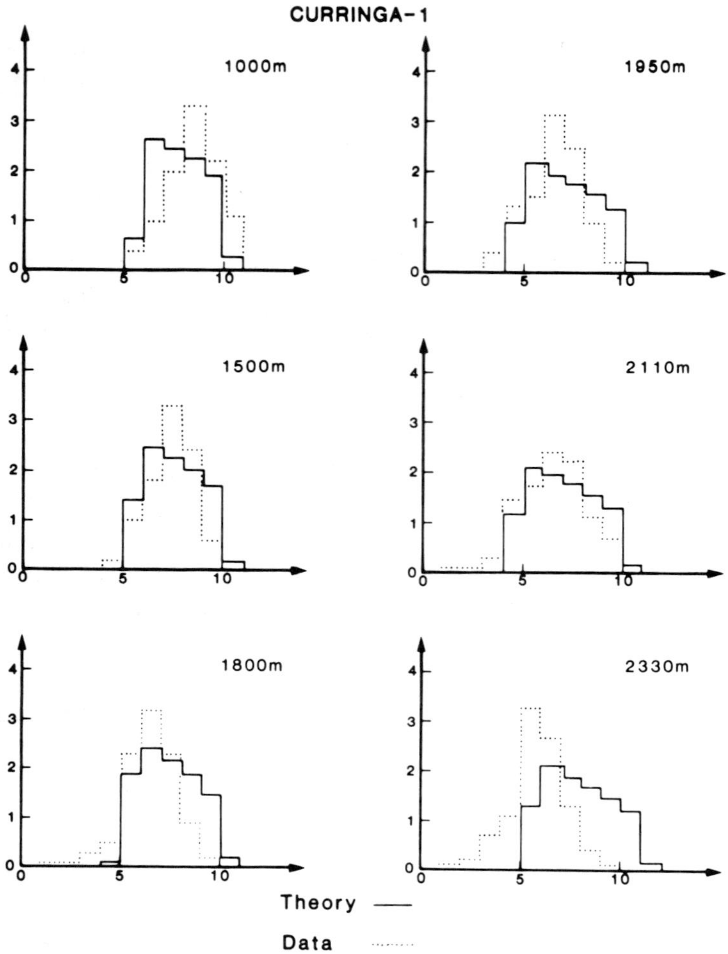

FIGURE 4.45 Theoretically predicted and observed track length distributions with depth at the present day for five depths for Curringa-1. The ordinate is number of tracks and the abscissa is track length in micrometers. The bin widths are 1 μm wide. The one-parameter heat flux variation corresponds to a best β of -5×10^{-3} (Myr)$^{-1}$.

track distributions in the one-parameter heat flux, that the fit is not perfect.

Accordingly, we wondered whether sufficient resolution was available in the track length distributions to determine a one-parameter heat flux variation in *each age range*. In this way we can find the best $Q(t)$ values for track distributions at each depth and so put together, for the first time,

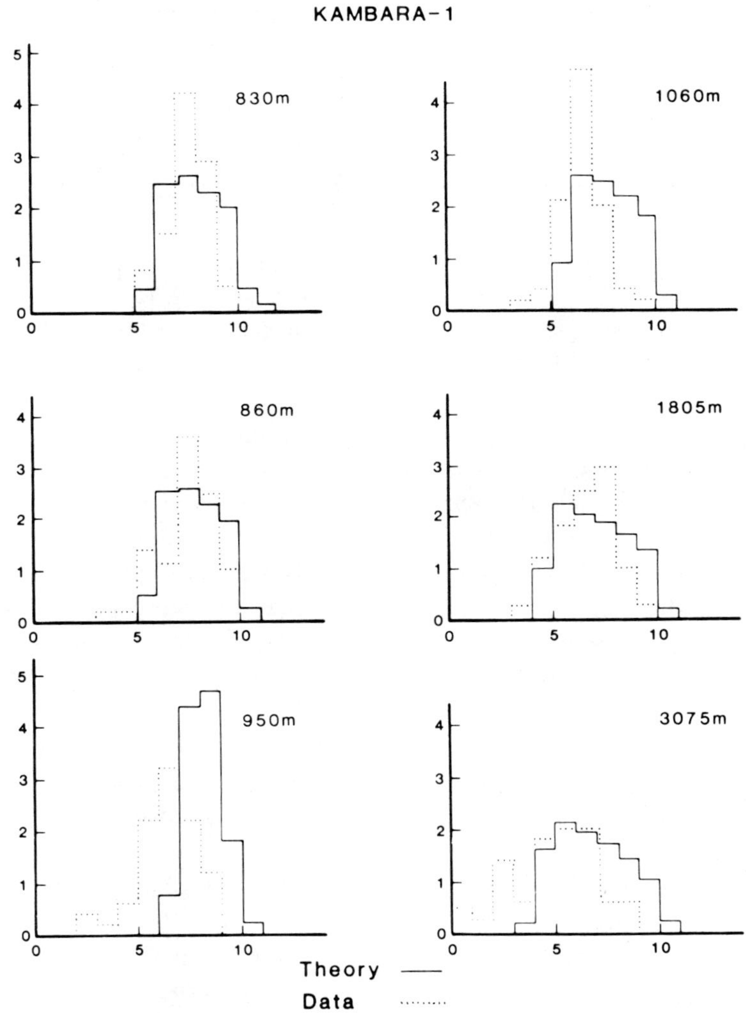

FIGURE 4.46 As for Fig. 4.45 but for track length distribution data for Kambara-1. The best one-parameter β is -5×10^{-3} (Myr)$^{-1}$.

a piecewise continuous variation of heat flux with time. The β values are shown in Table 4.2 for the two wells along with the corresponding minimum MSR values for each track length distribution. The resulting heat flux variations with time are given in Fig. 4.47 for Curringa-1 and Kambara-1 relative to the present-day heat flux values.

To within the error resolution of these $Q(t)$ values we see that the heat flux in Curringa-1 seems to have enjoyed a long slow warming trend since

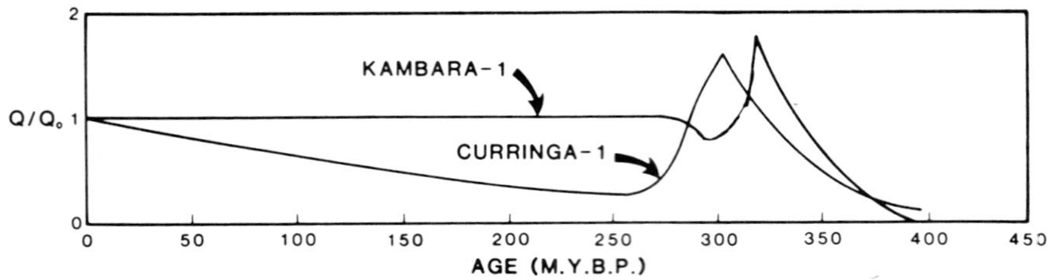

FIGURE 4.47 The best heat flux variation in time using a fixed value of β in each time interval as explained in the text and reported in Table 4.2 along with the corresponding MSR values for track length distribution data from wells Curringa-1 and Kambara-1.

the Devonian, with a thermal heating event around the early Permian to late Pennsylvanian superimposed on the warming trend.

In the case of Kambara-1 there again seems to have been a long, slow warming trend from the late Devonian until around the late Permian, when a thermal heating event took place. This was followed by a virtually unchanged heat flux since that time so that the slow post-Permian warming trend seen in Curringa-1 is not recorded by the fission tracks in Kambara-1. But both wells record some sort of thermal pulse around the 280-Myr mark.

This heat pulse is consistent with the observations reported by Gleadow *et al.* (1983) of a dolerite intrusion in the Canning Basin sometime in the early Permian to late Pennsylvanian causing fission track annealing.

We now see that we have progressed to the stage of finding three parameters, one to do with heat flux (β) and two (k_1, T_c) to do with the physical annealing process. Clearly, trial-and-error techniques are becoming unwieldly and we need a more systematic procedure. We defer development of the systematic nonlinear inversion method, called thermal indicator tomography, to Chapter 9.

We have investigated three ways of using the annealing of fission track scars in detrital apatite to determine, by inversion of downhole data, the temporal variation of paleoheat flux recorded by the differently aged apatite.

From the areal track number density method we record a heat flux variation which will be too high if short tracks are missed; from the mean track length method we record a paleoheat flux which is too low since the mean length is biased toward longer track lengths. These intrinsic biases are seen in application of these two methods to data from the two wells

Curringa-1 and Kambara-1. In addition, since neither the mean track length method nor the areal number density method uses all of the available, measured, track length information, these methods lead to a relatively large spread in the uncertainty of the paleoheat flux variations with time.

In contrast, the distribution of track lengths method for unfolding the temporal behavior of paleoheat flux uses all of the data on track length distribution at each depth (rather than just a simple moment of the distribution as do the other two methods), and the shape, and change of shape with depth, of the track length distribution provides an extremely sensitive thermal indicator.

We exploited this sensitivity to estimate from the downhole data the best rate constant k_1 for the annealing process appropriate to the multi-million-year time scale of subsurface conditions. The value obtained independently from two wells is roughly a factor of 3 smaller than that provided by short-lived laboratory experiments.

In addition, the change of track length distributional shape with depth contains a sufficient degree of sensitivity that we were able to determine not only a one-parameter form for the variation of paleoheat flux with time but also, for the first time, (albeit not too sharply) a change in direction of the heat flux variations with time.

This sensitivity to changes in direction of paleoheat flux with time is important if we are to use thermal indicators in an inverse mode to obtain short-lived (on a geologic time scale) heat pulses in sedimentary basins.

While the results reported here are encouraging, they must be viewed in a preliminary sense. The point here is that data from two wells are really not sufficient to bracket tightly the annealing rate constant k_1, as shown by the large error statements: $\delta k_1/k_1 = \pm 15\%$. Clearly, more wells are needed to reduce both the statistical and systematic errors in this determination. (See Table 4.3 for an analysis of apatite results.)

It is also clear that if the areal number density method and the mean track length method are to be used as ruggedly stable and accurate thermal indicators capable of being inverted to yield trustworthy answers, then we need to collect a number of values roughly uniformly spread with depth (considerably in excess of the six points available in this study) and we need to ensure that short track lengths are not overlooked or miscounted.

Even on the basis of the sparse data sets from just two wells, we found that the variation of the distribution of track lengths with depth is a remarkably sensitive thermal indicator that allows us to determine the

Table 4.3 Analysis of Apatite Results: Canning Basin Wells

Well	Depth (m)	Formation	Stratigraphic age (Myr)	Length L/L_0	FT[a] temperature T_L (°C)	Temperature T (°C)	FT age A (Myr)	Corrected age A_L (Myr)
Kambara-1								
8322-89	830	Poole	275	0.80	64 ± 8	48	272 ± 14	280 ± 27
8322-90	860	Grant	280–300	0.79	[66 ± 8]	49	397 ± 20	422 ± 41
8322-92	945	Grant	280–300	0.78	68 ± 7	51	298 ± 16	324 ± 33
8322-93	1050	Grant	280–300	0.78	68 ± 7	54	281 ± 13	305 ± 30
8322-94	1160	Grant	280–300	0.77	72 ± 7	56	291 ± 15	323–+ 33
8322-95	1805	Mid-Devonian sands	370–380	0.73	77 ± 6	72	260 ± 12	361 ± 28
8322-96	1838	Mid-Devonian sands	370–380	0.71	81 ± 5	73	276 ± 15	460 ± 41
8322-97	2925	Gogo	370–380	0.62	97 ± 5	100	151 ± 10	604 ± 74
8322-98	3075	Gogo	370–380	0.59	100 ± 5	103	46 ± 8	242 ± 40
Curringa-1								
8222-16	1000	Grant	280–300	0.78	[68 ± 7]	50	413 ± 26	449 ± 47
8222-18	1500	Grant	280–300	0.73	77 ± 6	61	365 ± 20	507 ± 41
8222-19	1800	Anderson/Grant	310–320	0.73	77 ± 6	66	361 ± 21	501 ± 40
8222-21	1960	Gogo	370–380	0.71	81 ± 5	68	263 ± 13	438 ± 39
8222-23	2110	Gogo	370–380	0.72	79 ± 5	71	214 ± 11	324 ± 27
8222-24	2300	Gogo	370–380	0.69	84 ± 5	81	167 ± 9	327 ± 34

[a] FT, Fission track.

physical annealing rate constant and, for the first time, to obtain an estimate of the variation of paleoheat flux beyond the simple one-parameter (heat increases or heat decreases with time) situations that have been extracted from other thermal indicators by inversion methods. Thus heating *and* cooling events (as opposed to heating *or* cooling events) in the paleodomain now can be identified using the annealing of fission scar tracks in detrital apatite. This inversion technique holds the promise of providing a fine-scale quantitative determination of paleoheat flux variations with time, provided the functional form of track length reduction with time, temperature, and length can be adequately determined, which appears to be a divergent process if Green and Duddy (1988) are correct in their latest reinterpretation of results and formulas in Gleadow *et al.* (1983).

IV. Basement Subsidence and Paleoheat Flux: Australia and North Sea

The changes of paleoheat flow and basement subsidence with time control the formation and evolution of sedimentary basins and the hydrocarbon potential of plays or prospects. Over the last few years burial history calculations have become available to assess these changes quantitatively and to classify basement motions and basinal subsidence into different types. The results can be used to control the veracity of geophysical models of basinal evolution and to predict potential areas for oil and/or gas exploration.

In this study 44 wells in the Cooper Basin, Australia, and 29 wells in the North Sea basin are used to examine relationships between subsidence and heat flux. Several types of basement subsidence can be recognized, and each type has some distinctive features which sets it apart from the others.

As we remarked in Chapter 2, the Falvey model applied to the North Sea would predict that heat flow was lower in the past than it is at present. Conversely, the McKenzie crustal stretching model would predict that heat flow was higher for approximately the past 90 Myr than at present. Thus we need some independent check of the validity of geologic models to obtain information on heat flow as a function of time. On the other hand, there seems to be little disagreement that *some* interaction of heat flux and basement subsidence must be involved in sedimentary basin evolution.

In this case study we use quantitative models to obtain some understanding about the connection between paleoheat flux and basement subsidence and, in addition, we increase our understanding of maturity of organic matter. There is the further point that if it is possible to quantify a relationship between basement subsidence and paleoheat flux variations, then in frontier areas (where we have mainly seismic information and few, if any, wells have been drilled) at least an estimate of basement subsidence can be made from seismic stratigraphy. The empirical relationships between subsidence and heat flux developed for basins with similar types of expression could then be used as a "first pass" to estimate thermal maturity in a frontier basin.

A. Thermal History Model

To determine the change of heat flux as a function of time requires that downhole thermal indicator data be inverted. For the case of vitrinite reflectance data with depth (of concern in this study) inversion of the data to determine paleoheat flux variations with time assumes that (1) the burial history of the sediments is precisely known and is isostatically controlled for all time, (2) the dominant source of heat to the sediments is from the basement, and (3) the dominant mode of heat transport is by thermal conduction. Each thermal indicator is related to an appropriate time–temperature integral which varies as a function of time and temperature with parameters that can be optimized by comparing predicted indicator values with depth to those measured in a borehole. Dominant among these parameters is the variation of paleoheat flux with time. The *simplest* formula describing heat flux change with time is

$$Q(t) = Q_0(1 + \beta t) \tag{4.10}$$

where $Q(t)$ is the basement heat flow as a function of time, Q_0 the present-day heat flow, t the time in millions of years, and β a parameter determining the linear trend of heat flow with time, to be determined from measured thermal indicator data by a least-squares method.

B. Model of Thermal Maturity of a Source Rock

From Chapter 3 we have that vitrinite reflectance change with time and depth can be obtained by

$$R(t) = \left(R_{\min}^{1/2} + \frac{I(i)}{\Sigma_j\, I(j)} \left\{ \sum_j \left[R(z_j)^{1/2} - R_{\min}^{1/2} \right] \right\} \right)^2 \tag{4.11}$$

where $R(t)$ is the vitrinite reflectance change with time t for a sedimentary layer currently at depth z, $R_{\min}$ is the depositional surface vitrinite reflectance value (currently set at $R_{min} = 0.2$ based on 300 wells), $R(z_j)$ is the measured vitrinite reflectance for a layer currently at depth z_j and of age t_j, and $I(i)$ is the time–temperature integral characterizing the maturity of a vitrinite sample deposited at time t_i presently at depth z_i, which is calculated from

$$I(i) = \int_{-t_i}^{0} \exp[(T - T_c)/T_D]\, dt, \quad \text{in } T \geqslant T_c \qquad (4.12)$$

with the integrand equal to zero in $T < T_c$, and where T is absolute temperature along the burial path. Here T_c and T_D are two thermochemical constants fixed at 295 and 200 K, respectively, based on empirical calibration against nearly 300 wells. The constant T_c is a critical temperature whose role is such that if the temperature T is less than the critical temperature T_c then vitrinite does not mature, while T_D provides a measure of the exponentiation of the reaction rate as temperature increases. A vitrinite reflectance value of $R_0 \approx 0.6$ is often considered to provide a rough rule of thumb indicating the onset of oil maturity.

C. Basement Subsidence

A local isostatic model has been used in this study to calculate basement subsidence because such a model requires the data from only one well. A regional isostatic model would require knowledge of regional tectonic movements, which is difficult to obtain in most cases (Steckler and Watts, 1978).

The basement subsidence caused by the tectonic effect is given by (Steckler and Watts, 1978)

$$Y = S[(\rho_m - \rho_s)/(\rho_m - \rho_w)] + Wd - D_s[\rho_m/(\rho_m - \rho_w)] \qquad (4.13)$$

where Y is the basement subsidence caused by the tectonic effect, S the sediment thickness, Wd the water depth at the time of sediment deposition, D_s the change in elevation of mean sea level, ρ_m the average mantle density, ρ_w the average water density, and ρ_s the average sediment density (Steckler and Watts, 1978).

D. The Parameter α

The technique presented in Chapter 3 determines paleoheat flux from the shape of the reflectance curve with depth rather than from the magnitude of the vitrinite reflectance data. Ideally we should have

$$R(t_i) = [R_{\min}^{1/2} + \alpha I(i)]^2 \tag{4.14}$$

where α is a universal constant (equation A14 of Lerche *et al.*, 1984); *provided the true heat flux is used to determine the* I(i) *integrals*.

But we can also write

$$\alpha = \sum_{j=1}^{N}[R(z_j)^{1/2} - R_{\min}^{1/2}]/\sum_{j=1}^{N} I(j) \tag{4.15}$$

which replaces the dependence of α on the poorly determined rate constants in favor of determination of α from well information. Clearly, once a best value of β has been determined and the $I(j)$ integrals calculated, it is a simple matter to calculate α for a well. Noting that the true value of α is supposed to be a universal constant, we will obtain a value close to this "truth" if and only if assumptions 1–3 are closely obeyed by the basin being modeled. Thus if the basin was significantly overpressured or fractured, formations were cemented or dissolved, the evolution was dominated by multidimensional considerations, or the like, we would violate assumptions 1, 2, and/or 3. In turn, such violations imply that the α values to be determined in a given basin will show a "spread" around some mean value and, further, that between different basins there will be a difference even in the mean values of α. For instance, a basin with rapid deposition and/or high overpressuring in the past will tend to mature the sediments more quickly so that the time–temperature integrals will, by and large, tend to be larger than they would be in a nonoverpressured basin. The corresponding α values in the overpressured basin will then tend to be lower if the basin is treated using an isostatic model for sedimentary evolution. As we shall see, such a behavior does, in fact, occur.

The input data to the study consist of present-day water depth or ground level elevation at well sites, Kelly bushing heights, sediment surface temperature, corrected bottom-hole temperature and depth, lithology, paleowater depth, ages of bases of selected layers, and vitrinite reflectance–depth pairs. Other derived parameters are determined internally by the computer in an automatic mode.

E. Results and Analysis

1. General Comments

Just about every model that has been suggested for sedimentary evolution ties the variation of basin subsidence $B(t)$ with time t to the variation of heat flux $Q(t)$. The models differ among themselves in specifying the relationship of $B(t)$ to $Q(t)$, but all models predict some relationship. In an effort to minimize the explicit role of time in such modeling procedures, what we have done here is to determine $B(t)$ from the isostatic backstripping model and $Q(t)$ from inversion of thermal indicators and then plot B versus Q. Thus if $Q(t) = Q_0(1 + \beta t)$, then

$$t = (1/\beta)\,(Q/Q_0 - 1) \qquad (4.16)$$

so that, with β determined for each well from the inversion procedure and the present-day heat flux Q_0 determined from bottom-hole measurements, by eliminating t we have that

$$B = B(t) = B[(1/\beta)\,(Q/Q_0 - 1)] \qquad (4.17)$$

Thus we can see in a B–Q plot (basement subsidence versus heat flux) for wells in a given basin how closely the B–Q behavior, obtained by superposition of many wells in the basin, is following a general trend and so providing a "universal" relation of basement subsidence with heat flux.

2. Cooper Basin, Australia

On the basis of analyzing the changes of paleoheat flux and basement subsidence for 44 wells in in the basin (all them having geologic ages ranging from 245 MYBP to the present), five basic types of response can be recognized (Fig. 4.48). Each response has features that distinguish it from the other types, as we now detail.

Type A. Here the response represents relatively rapid and large-scale subsidence of the basement, with a slow decline of the heat flux over the given time duration and early maturity of organic source material. Take well 35 as an example. From 245 MYBP to the present, the total basement subsidence was 1025 m (Fig. 4.49a), with the main subsidence period from 245 to 96 MYBP. After 96 MYBP there was just a small-scale uplift appearing at late periods. The heat flow change is from 3.8 HFU (245 MYBP) to 1.6 HFU (present), for an average decline rate of 9×10^{-3} HFU/Myr (Fig. 4.49b). Organic material deposited at 245 MYBP reaches thermal maturity ($R_0 = 0.6$) at 86 MYBP (Fig. 4.49c), so that it took 159 million years to mature.

Type B. In contrast to type A, this type of basement has relatively

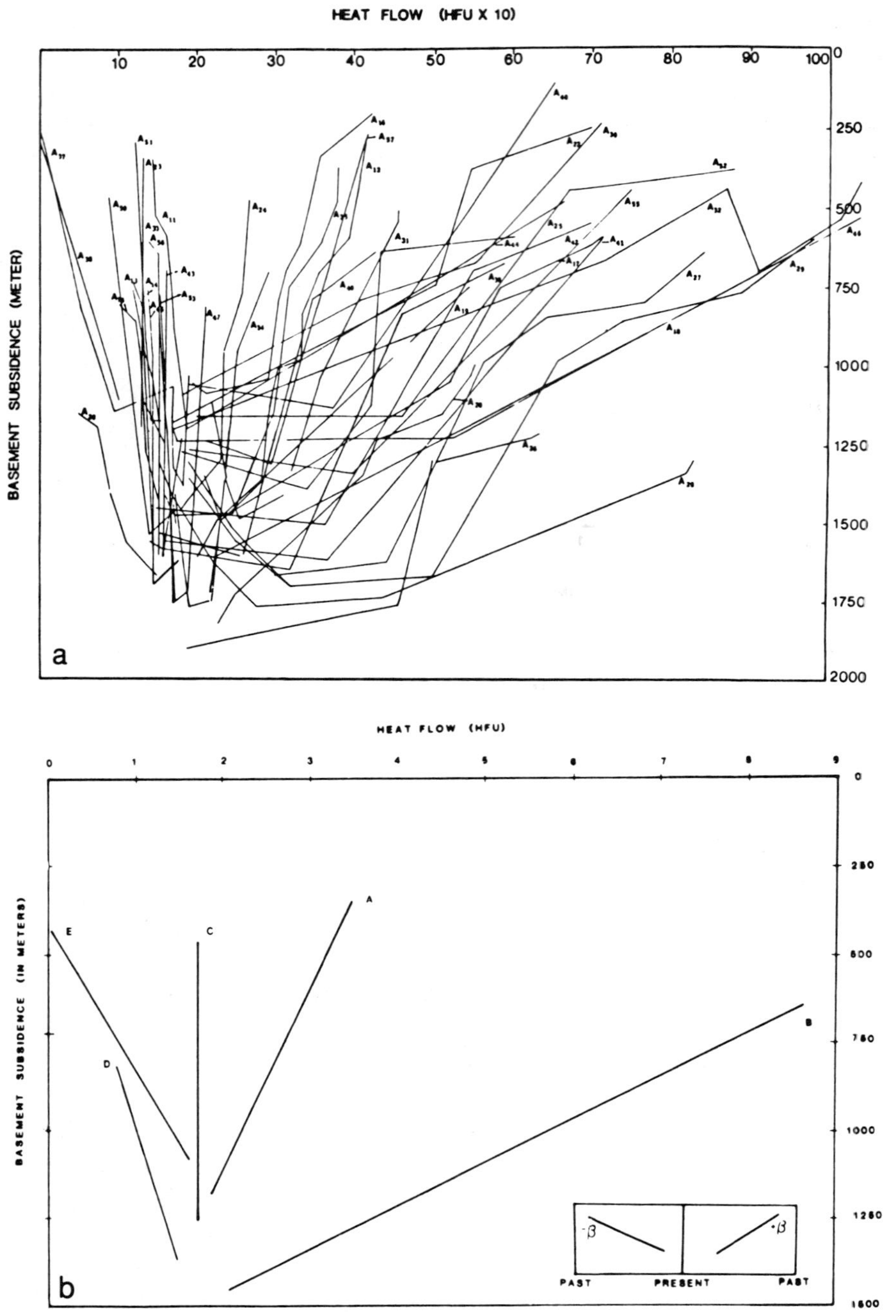

FIGURE 4.48 (a) Relationship between paleoheat flow and basement subsidence, Cooper Basin, Australia based on 44 wells. (b) Schematic types of response.

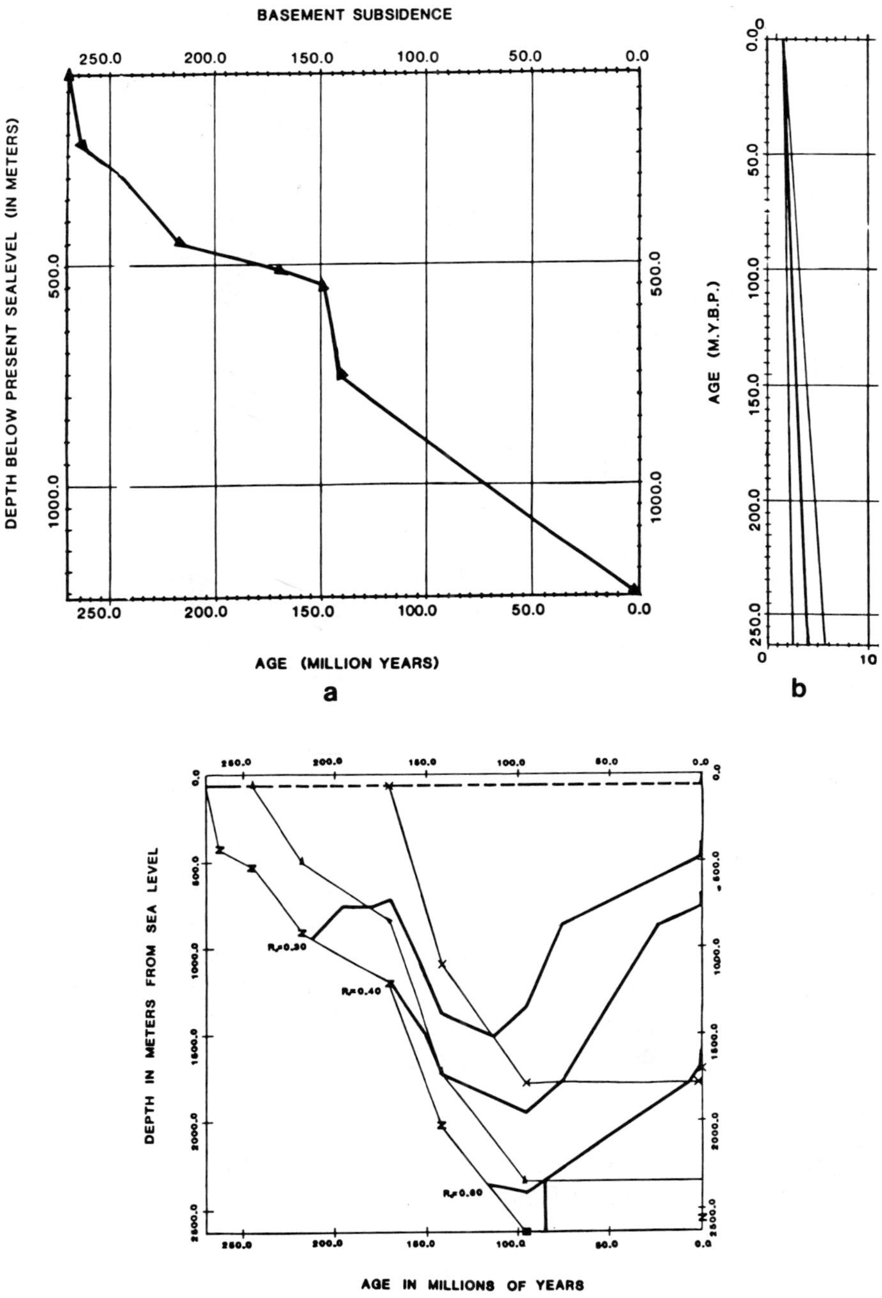

FIGURE 4.49 For well 35 in the Cooper Basin: (a) basement subsidence vs. geologic age; (b) heat flow change with time; (c) burial history with maturity contours.

slow subsidence, rapid decline of the heat flow, and later maturity. For instance, in well 18 the total basement subsidence during the past 245 million years was 785 m (less than that in well 35) (Figure 4.50a); the heat flow change was from 8.1 HFU (245 MYBP) to 2.2 HFU (present) for an average decline rate of 2.4×10^{-2} HFU/Myr (considerably more than in well 35) (Fig. 4.50b). Organic material deposited at 245 MYBP reached maturity at 76 MYBP, so that it took 169 million years to mature (more than that in well 35 by 10 Myr) (Fig. 4.50c). Although both the initial heat flow and present-day heat flow are higher in well 18 than well 35, because of the relatively slow subsidence and rapid heat flow decline the sediment temperature stays relatively low, leading to a comparable time for thermal maturity to be reached.

Type C. This type of response we label "transitional." An apparent feature of the basement motion is that during a long period, the heat flow change is very small. In such a case, the value of β calculated by the computer is zero, implying that for a long period the heat flow is stable. For this type of basement, the maturity of organic material depends mainly on the initial heat flow and the amount of the basement subsidence.

Generally speaking, higher initial heat flow and/or a larger amount of subsidence for a given time duration makes for earlier maturity, but some care must be exercised with such generalizations. For example, in well 23 the initial heat flow is 1.3 HFU, the total basement subsidence is 963 m, the onset of maturity was at 45 MYBP, and it took some 200 million years to reach maturity. However, in well 43 the initial heat flow is 1.6 HFU (higher than in well 23), the total basement subsidence is 891 m (less than in well 23), but the onset of maturity was at 25 MYBP, and it took 220 million years to reach maturity (more than in well 23). The main reason for this apparent dichotomy is that there was a relative uplift of the basement in well 43 after 96 MYBP, so that, although it has a higher initial heat flow, because of the uplift the sediments cooled somewhat, leading to a little longer time being needed to reach maturity. This example illuminates the point that besides the thermal history, burial history is also a rather important element for oil/gas generation.

Type D. Both type A and type B responses have higher heat fluxes in the past, while the type C response has a heat flux comparable to the present-day value (to the point where resolution limits the ability to say whether the heat flux was changing at all). The type D response corresponds to a heat flux in the past significantly lower than the present-day value. With a linear (in time) heat flux modeling function, the

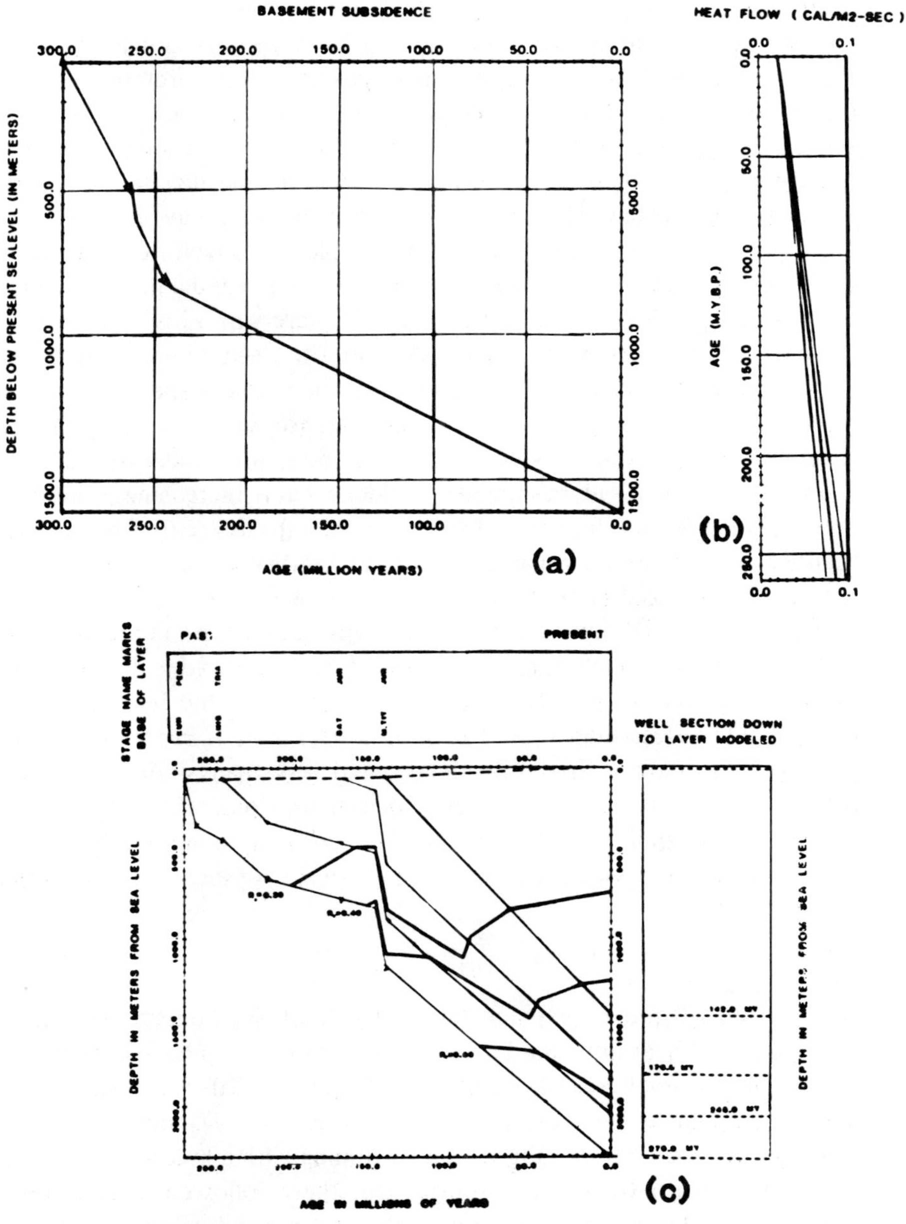

FIGURE 4.50 For well 18 in the Cooper Basin: (a) basement subsidence vs. geologic age; (b) heat flow change with time; (c) burial history with maturity contours.

mathematical implication is that the heat flux will be zero at time $t = 1/\beta$ and be negative at earlier times. From both geological and physical viewpoints such results are absurd. A more correct interpretation of such anomalous results is that before the time $\sim 1/\beta$ both the heat flux and sedimentary temperature were so low that there was not enough energy to mature any vitrinite which then existed. As the heat flux increases the vitrinite starts to mature. For instance, in well 50 we have a β value of -2×10^{-3} (Myr)$^{-1}$, so on the basis of a simple linear heat flux with time, we have a zero heat flow value 500 million years ago and a negative value at earlier times. This result just means that there was a low heat flow in the past and resolution has been lost around the 500-million-year mark. About 265 MYBP the basement began to subside rapidly and, at about the same time, magmatic activity became intense in the Cooper Basin (Kantsler *et al.*, 1983), which made the heat flow increase rapidly. Thereafter, both rapid basement subsidence and increasing heat flow compensated for the low heat flux and slight subsidence at times earlier than Permian. Vitrinite began to mature at 245 MYBP, and at 90 MYBP the organic material passed the maturity line of $R_0 = 0.6$.

Type E. The distinction between the type E and type D responses is that type E has much longer periods of low heat flow history. For example, in well 38 at about 200 million years ago the heat flow was extremely low (less than 10% of its present-day value). From 175 MYBP the basement began rapid subsidence, and less than 5–10 MYBP the organic material reached the maturity threshold. Comparing well 50 with well 38, we see that well 38 took a much longer time to mature. The main reason for this behavior is the much longer period of low heat flow in the past in well 38.

3. *North Sea Basin*

We inverted vitrinite reflectance data and well lithology data to determine the changes of heat flow and basement subsidence from 64 MYBP to the present in the North Sea basin. Most wells in the North Sea basin turned out to be similar in response to types D and E. Types D and E make up about 90% of the 29 wells (Fig. 4.51). Beginning with a long period of low heat flow and relatively slow subsidence, there followed a later rapid subsidence and increasing heat flow with a correspondingly late maturity time. Take well 20 as an example. Until about 21 million years ago the heat flow was very low (Fig. 4.51a), and for a long time (~250 MYBP until around 60 MYBP) the extent of the subsidence was small (Fig. 4.52b; note that the depth is in feet). Thereafter the basement activity began to intensify and was accompanied by rapid subsidence and increasing heat

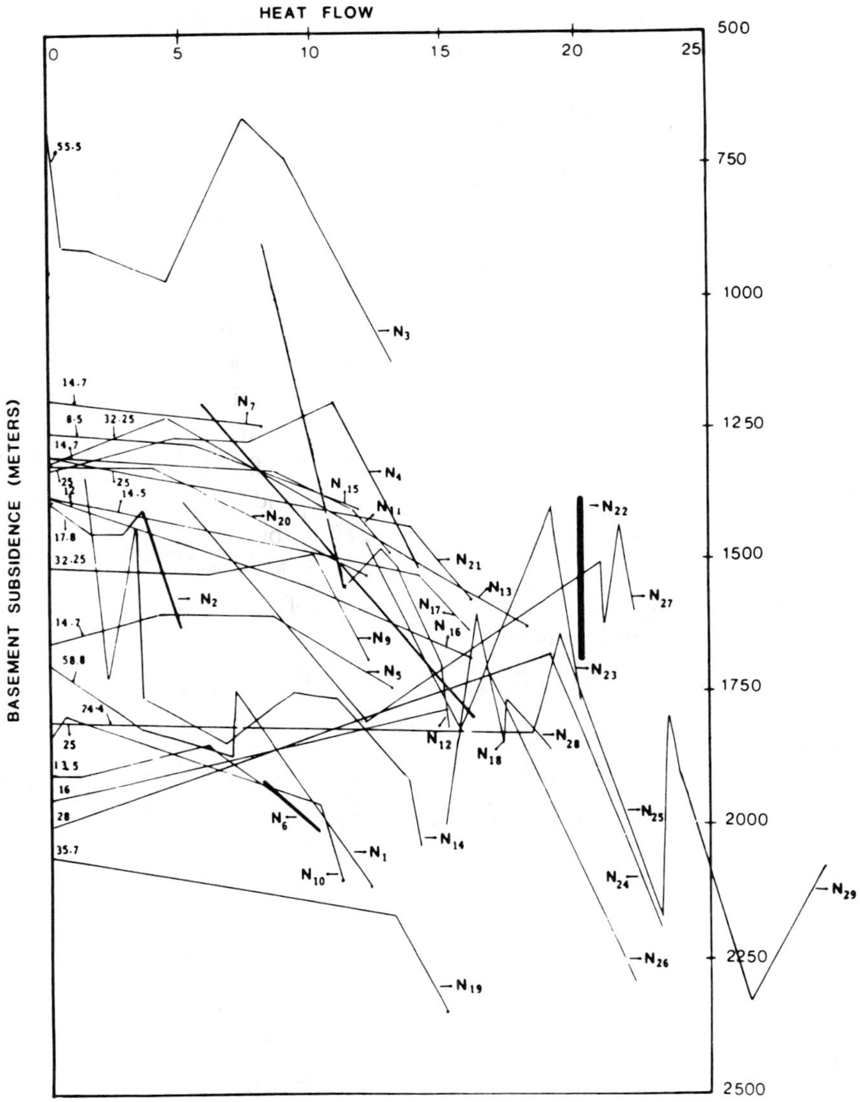

FIGURE 4.51 Relationship between heat flow and basement subsidence in the North Sea basin, based on 29 wells.

flow caused by magmatic activity (P. A. Zeigler, 1975). Even so, organic material deposited at 250 MYBP is still immature at present (Fig. 4.52c). Because of the long period of low heat flow and slow subsidence, there just has not been enough energy accumulated to mature the organic material.

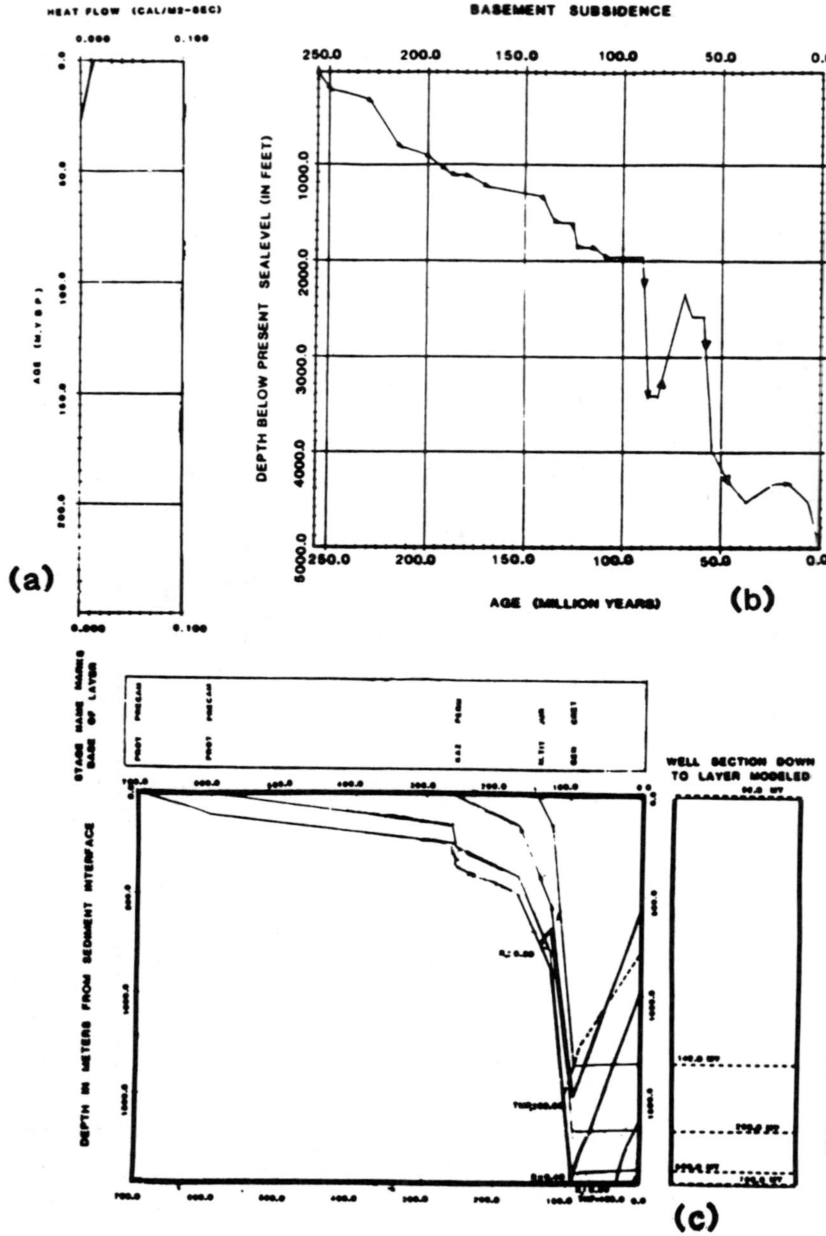

FIGURE 4.52 For well 20 in the North Sea basin: (a) heat flow change with time; (b) basement subsidence vs. geologic age; (c) burial history with maturity contours.

F. Summary

From the work reported here on 44 wells in the Cooper Basin, Australia, and 29 wells in the North Sea basin it would seem that:

1. Burial history inversion and thermal history inversion present a quantitative method for assessing the evolution of the basement subsidence, the heat flow, and the thermal maturity of organic material. Classifying the relationship of basement subsidence to paleoheat flux variations in a basin has uncovered responses different from what preconceived models of basinal evolution might have indicated.
2. We label the responses of types A, B, and C "high heat flow" types of basement. In the results from 44 wells, the majority of the wells would classify the Cooper Basin as belonging to the A, B, or C type; the results from 29 wells indicate that the North Sea basin belongs to the D or E type.
3. The areas that have a large amount of basement subsidence and a slow heat flow decline in the past (or a rapid heat flow increase) are the most favorable for hydrocarbon generation.
4. The generic class of thermal cooling models of basement subsidence, which have been in vogue for the last few years, do not represent a universal response. Indeed, the situation in which an increasing heat flow is accompanied by a rapidly subsiding basement is not encompassed by such popular models, and this type of situation is not an isolated type of response as is shown by the results from many wells in both the Cooper Basin and the North Sea basin.
5. The absolute amounts of the heat flow changes with time do not control in a dominant fashion the amount of basement subsidence, although a clear trend of responses in the forms of types A through E is readily discernible. There must be other causes which control the absolute amounts of subsidence or uplift of a basin.
6. The ranges of the α values have obvious distinguishing characteristics in the two basins. The α values for the 29 wells in the North Sea basin yield a mean of $\alpha = 160 \pm 25$ (Fig. 4.53a), while the 44 wells in the Cooper basin provide a mean value of $\alpha = 650 \pm 50$ (Fig. 4.53b). Within the framework of an isostatic burial history, and with assumptions (a), (b), and (c) in force, these results suggest that there is indeed a constant α value for each particular basin,

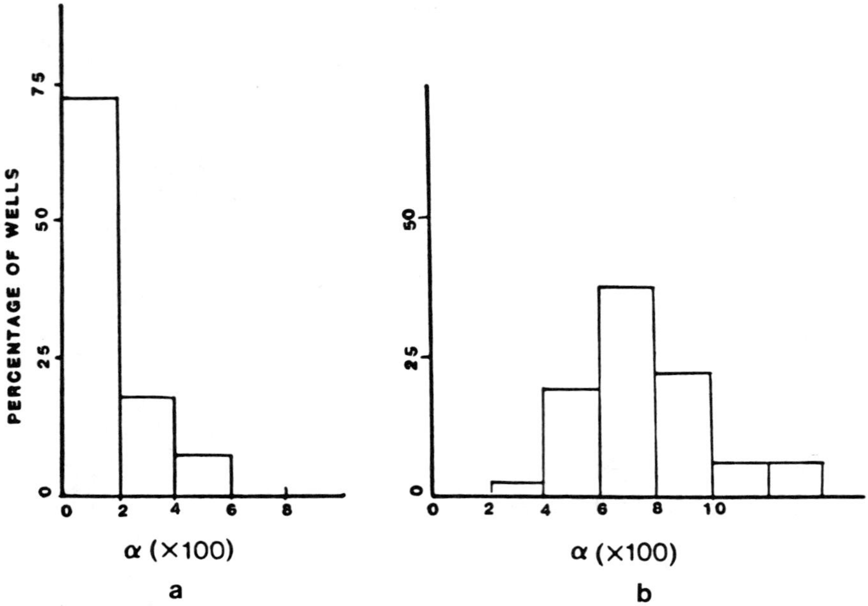

FIGURE 4.53 Plot of frequency distribution of α values in (a) the North Sea basin and (b) the Cooper Basin, Australia.

providing firm corroborative support for the underlying tenets of the basic inversion theory posited for vitrinite reflectance as a thermal indicator. But because of the lack of precision of the data and the intrinsic assumptions of the isostatic model, it still seems to be difficult to choose a constant α value valid for *all* basins and then to compute the paleoheat flow for all wells in a particular basin without producing large discrepancies in the predicted versus observed vitrinite data for some of the wells. Further, the discordance between the two mean α values for the Cooper Basin and the North Sea basin (650 versus 160) implies that at least some of the assumptions (a), (b), and (c) need to be reexamined. It would, for example, be of interest to repeat the work reported using a dynamical fluid-flow/compaction model of sedimentary evolution so that overpressuring with time (strongly suggested in the North Sea but likely lacking in the Cooper Basin) and the associated enhanced thermal maturity are accommodated. We speculate that such a modeling procedure, which better represents basinal evolution, would produce a closer accord of the α values for the two basins, as well as improve the empirical relationships uncovered here.

5

Hydrocarbon Generation Models

THE PRODUCTION OF LIQUID and gaseous hydrocarbons from organic matter by thermal degradation processes was put on a firm footing in the 1960s (see, e.g., Philippi, 1965). Much earlier it was already known that biogenic processes could also produce degradation of organic material (see, e.g., Hunt, 1979, and references therein). But knowing that processes exist which lead to hydrocarbon generation and quantifying of those processes are very different matters.

Two major difficulties are apparent at the outset of any attempts to provide quantitative kinetic models.

First is the sheer magnitude of the task. Organic matter (mainly insoluble kerogen) is composed of biological polymers (lipids, proteins, amino acids, etc.) with molecular weights in the multimillion range; by way of contrast, methane (CH_4) has a molecular weight of 16. We are invited to construct a complete kinetic reaction scheme which accounts correctly for the detailed breakdown of organic matter to methane.

Second is the sheer time scope of the task. We know from isotopic measurements, stereochemistry, and gas chromatography–mass spectrometry that the production of hydrocarbons in the earth occurs on time scales ranging from fractions of a megayear to hundreds of megayears, in a temperature regime in the rough range 0–200°C, and in an evolving environmental "smog" of sedimentary pores containing water laced with varying and variable amounts of ions, molecules, and minerals. We are invited to construct a chemical kinetic framework which allows correctly for all of these factors, as well as for surficial reaction chemistry on grain surfaces.

Faced with these two major problems, as well as a host of important but subsidiary difficulties, two major thrusts of research have evolved over the last 20 years or so in attempts to bypass or at least constrain and bracket the major problems. Because it will be a very long time indeed before a complete thermokinetic behavior is available (if it ever is), pyrolysis experiments in the laboratory have been used, in conjunction with theoretical concepts, to try to determine at least the dominant controlling reaction kinetics in two areas: (1) in attempts to discover the order of the chemical reactions (first-order, second-order, etc.) and (2) in attempts to discover the sequence of chemical reactions (series, parallel, catalytic, etc.).

The simplest model assumes that oil generation can be modeled by a single first-order reaction of the Arrhenius type,

$$dM/dt = +k(1 - M), \quad \text{where } k = A \exp(-E/RT) \tag{5.1}$$

with M the mass of oil produced per unit mass of kerogen, t the time, E the activation energy, R the gas constant, T the absolute temperature, and A the frequency factor (occasionally we also write $E/R = T_0$, the activation temperature). {Note, as an aside, that if temperature is expressed in degrees Celsius, $T = 273 + C$, then for $|C| << 273$ we have $k = A \exp[-T_0/(273 + C)] \cong A \exp(-T_0/273) \times \exp[CT_0/(273)^2]$. Write $T_D = (273)^2/T_0$, when k takes the form

$$k = A' \exp[(C - T_c)/T_D] \tag{5.2}$$

where $A' = A \exp(-T_0/273 + T_c/T_D) =$ constant. Then we have the solution for M in the form

$$\ln(1 - M) = -A' = \text{TTI}(t) \tag{5.3}$$

where

$$\text{TTI}(t) = \int_0^t \exp\{[C(t') - T_c]/T_D\} dt' \tag{5.4}$$

which we recognize as providing a basis for the Lopatin (1971) rule-of-thumb estimate for hydrocarbon maturity popularized by Waples (1980).}

On the other hand, such simplistic models—and the determination of the best values of A and E by matching predicted behavior to observations (see, e.g., van Heek and Juntgen, 1968; Connan 1974)—are known not to be very relevant. Oil is not a single molecular substance, no gas production is allowed for, and different types of kerogen, from provenances ranging from terrestrial to marine, produce different oils. In addition, there has persisted an annoying irreconcilable difference between the activation energy of 40–70 kcal/mol required from laboratory pyrolysis and the range of 10–20 kcal/mol needed to form oil under geological conditions (Sweeney *et al.*, 1987).

Only a quantitative approach, allowing a computation of the amount of oil and gas generated in any place in the basin as a function of time, can provide information on (1) the amount of oil and gas generated per square kilometer in each source rock and in every part of the basin; (2) the timing of hydrocarbon formation, timing of the development of impermeable seals, and timing of folding and faulting, so traps can be identified; (3) the amounts of oil and gas that have been expelled from the source rock into the porous reservoirs and how much is still retained in the source rock; and (4) the evaluation of the ultimate oil and gas reserves of a sedimentary basin. However, because many of the intermediate steps in petroleum and gas generation from kerogen degradation are unknown, such a scheme has to be phenomenological; that is, parameters have to be determined by least-squares fitting to basin and laboratory data. Such a mathematical model of petroleum generation accounting explicitly for geologic time was first introduced by Tissot (1969) and is discussed fully by Tissot and Espitalie (1975). As sketched in Fig. 5.1a, the idea is to simulate the formation of oil from kerogen in six parallel, first-order, reactions (first stage) and the formation of gas (CH_4) from oil in a single reaction (second stage). The model is based on kerogen degradation and uses the general scheme of first-order chemical kinetics. Kerogen is a macromolecule composed of polycondensed nuclei bearing alkyl chains and functional groups—the links between nuclei being heteroatomic bonds or carbon chains. As the burial depth and temperature increase, the bonds are successively broken, roughly in order of increasing rupture energy. The products generated are heavy heteroatomic compounds, carbon dioxide, and water, then progressively smaller molecules, and finally hydrocarbons. At the same time, the residual kerogen becomes progressively more aromatic and evolves toward a carbon residue.

To apply the phenomenological model of Tissot and Welte (1978) to the prediction of change in hydrocarbon content of rocks with respect to time and temperature, it is necessary to assume first-order chemical reaction kinetics for the breakdown of kerogen to hydrocarbons as governed by

$$-dx_i/dt = k_i x_i \tag{5.5a}$$

$$k_i = A_i \exp[-E_i/k_B T(t)] \tag{5.5b}$$

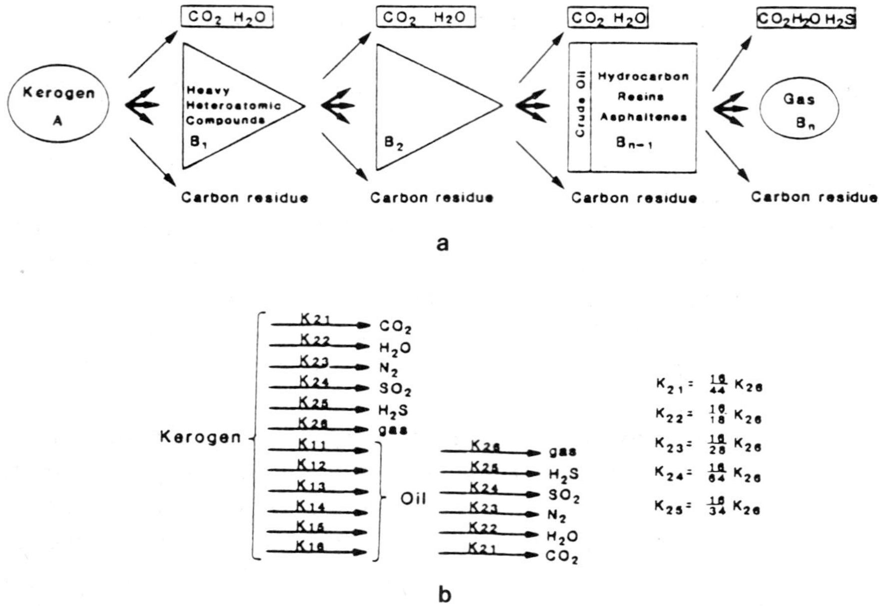

FIGURE 5.1 (a) General scheme of kerogen degradation under thermocatalytic cracking. (b) Two detailed kinetic frameworks of oil and gas production. The Tissot (1969) scheme is produced by setting K_{21} through K_{25} to zero and by ignoring K_{26} in the direct transformation of kerogen to oil and gas.

where x_i is the number density of component i, k_i is the reaction rate for this reaction, k_B is Boltzmann's constant, A_i is the reaction rate in the limit of high temperature, E_i is the activation energy, and $T(t)$ is the temperature as a function of time acting on the element x_i.

In the Tissot and Welte scheme, three types of initial kerogens are considered. The three types differ in their hydrogen-to-carbon ratio. Loosely speaking, these three types can be considered type 1 (algal) with a high petroleum potential, type 2 (amorphous) with an intermediate petroleum potential, and type 3 (woody) with a low petroleum potential. Tables of the activation energies (E_i), the reaction constants (A_i), and the initial values (x_i) for each kerogen type have been given by Tissot and Welte (1978, Table V.4.1).

To calculate the quantity of hydrocarbons produced by potential source rocks, it is necessary to use the above calculations and measurements or estimates of the percentage of organic material in those source rocks. Nakayama and Van Siclen (1981) detailed the methodology of this system to work out a hydrocarbon budget for a basin. In order to make predictions about the hydrocarbon history of a basin, it is also necessary to determine the temperature as a function of time. Tissot and Welte (1978) assumed that heat flux was constant and used constant geothermal gradients in their models. However, by allowing the heat flux to vary with time, their model is improved, provided the variation of heat flux with time can be determined.

Presumably, what should be done at our present stage of understanding is to let the frequency factors and activation energies be "free" parameters and determine the values most consistent with observational data in a globally optimized least-squares sense, using the best burial history, fluid flow, and thermal histories currently available. In addition, a sensitivity study should be undertaken to determine the degrees of resolution, precision, uniqueness, and stability of any chemical parameters so obtained. On the other hand, major problems still remain: there is no guarantee that the kinetic framework proposed is the correct phenomenological description (e.g., why have only six channels for oil production and only one channel for gas production?). Two extreme end-member lines of development can be carried forward to provide different quasi-phenomenological hydrocarbon generation schemes: (1) we can modify the Tissot type of sequence of reaction kinetics; (2) we can modify the number of channels of the reaction kinetics.

For instance, as sketched in Fig. 5.1b, an obvious modification adds the major gaseous products from both kerogen degradation and oil cracking. Such a modified model is represented as follows. Hydrocarbon generation is simulated in two stages: kerogen degradation and oil cracking. In the kerogen degradation stage:

1. Total kerogen degradation for type n is given by

$$\frac{dX_n}{dt} = -X_n\left(\sum_{i=1}^{6} K_{1i} + \sum_{j=1}^{6} K_{2j}\right) \tag{5.6}$$

where X_n is the concentration of kerogen, K_{1i} is the reaction constant for oil with bond type 1, and the K_{2j} are the reaction constants for different gases (CH_4, H_2, N_2, etc.).

2. Direct gas production is given by

$$dU_j/dt = K_{2j} \times X_n \tag{5.7}$$

3. Direct oil production is given by

$$\frac{dY_i}{dt} = K_{1i} \times X_n - Y_i \sum_{j=1}^{6} K_{2j} \tag{5.8}$$

The amount of oil type i is Y_i, and total oil is

$$Y = \sum_{i=1}^{6} Y_i.$$

In the oil cracking stage, for gas type j from oil cracking:

$$\frac{dU_j'}{dt} = K_{2j} \sum_{i=1}^{6} Y_i \tag{5.9}$$

Total gas is

$$G = \sum_{i=1}^{6} (U_j + U_j') \tag{5.10}$$

with gas of type j being U_j.

The value of K_{1i} in this model is as in Tissot's model. The values of K_{2j}, the reaction constants for gas production, are poorly known. If the reaction constants of different gaseous compounds are associated with the relevant molecular weights (the lighter a compound, the less energy it needs to react and so the higher its reaction constant), then the reaction constant is inversely proportional to a compound's molecular weight; for example, $K(CO_2) = 16/44\ K(CH_4)$.

Again, it would seem appropriate not to prejudge the values the chemical constants are to have but rather to let data determine the values.

Yet one can question the sense in retaining a fixed number of reactions for degrading kerogen (six in the Tissot type of model). The charge can be leveled that the number of chemical parameters to be determined can be increased systematically until the reaction kinetics are sure to agree with any set of measurements. Arguments for simplicity (Occam's razor) suggest that other types of reaction kinetics (with fewer parameters but greater flexibility) might be more appropriate. Braun and Burnham (1987), Sweeney *et al.* (1987), and Burnham *et al.* (1987) have provided both laboratory pyrolysis data and an application to the Uinta basin which, they suggest, are better understood in terms of a distributed activation energy model (an infinite number of channels for hydrocarbon production but with an energy weighting per channel).

In its simplest form, the distributed model writes a kerogen mass, K, loss equation under thermochemical kinetics in the form

$$dK/dt = -\alpha(T, E_0,\sigma)K \tag{5.11}$$

but, instead of α being the rate constant, k, appropriate to a single kinetic reaction, what is written is

$$\alpha(T, E_0, \sigma) = A \int_0^\infty dE \exp(-E/RT)P(E; E_0, \sigma) \tag{5.12}$$

that is, the reaction kinetics are distributed over energy with a channel probability $P(E; E_0, \sigma)$. The most common assumption (Braun and Burnham, 1987) on the channel probability is that

$$P\alpha \exp[-(E - E_0)^2/\sigma^2] \tag{5.13}$$

where E_0 and σ are constants so that α takes the form

$$\alpha = A \int_0^\infty dE \exp[-E/RT - (E - E_0)^2/\sigma^2] \tag{5.14}$$

The distributed activation energy kinetics then depend on a frequency factor A, a mean activation energy E_0, and a spread in activation energies σ, together with the assumption that the probability is gaussianly distributed in activation energy. Bruan and Burnham (1987), Sweeney *et al.* (1987), and Burnham *et al.* (1987) report good fits to laboratory pyrolysis data with such a phenomenological model and further suggest that the chemical parameters derived from such laboratory fitting are also appropriate in geological applications. While not yet as well developed as the Tissot type of model for hydrocarbon generation, the distributed activation energy model provides an interesting alternative.

The appearance of at least three model types for hydrocarbon generation may seem disquieting, for it suggests that we still have some distance to go before we can unravel the correct quantitative scheme for hydrocarbon production. But this very variation can be used to advantage in assessing hydrocarbon proneness of a basin. The obvious strategy is to invoke *all* of the models of hydrocarbon generation in putting together the development of a sedimentary basin. Each will predict different amounts and rates of hydrocarbon (both oil and gas) production at different stages in the basin's evolution. But the important point is that these individual assessments of hydrocarbon generation provide a quantitative measure of the uncertainty and irresolution of the timing and amount of hydrocarbon production. This uncertainty can then be invoked in an exploration economic sense to assign ranking probabilities (Nind, 1981).

In addition, some qualitative controls are available to limit the degree of uncertainty of hydrocarbon models. For example, since the van Krevelen diagram (Fig. 5.2) indicates a systematic loss of hydrogen and oxygen in maturing kerogen, measurements of the location of a partially matured kerogen sample set a qualitative limit on its total maturity. This

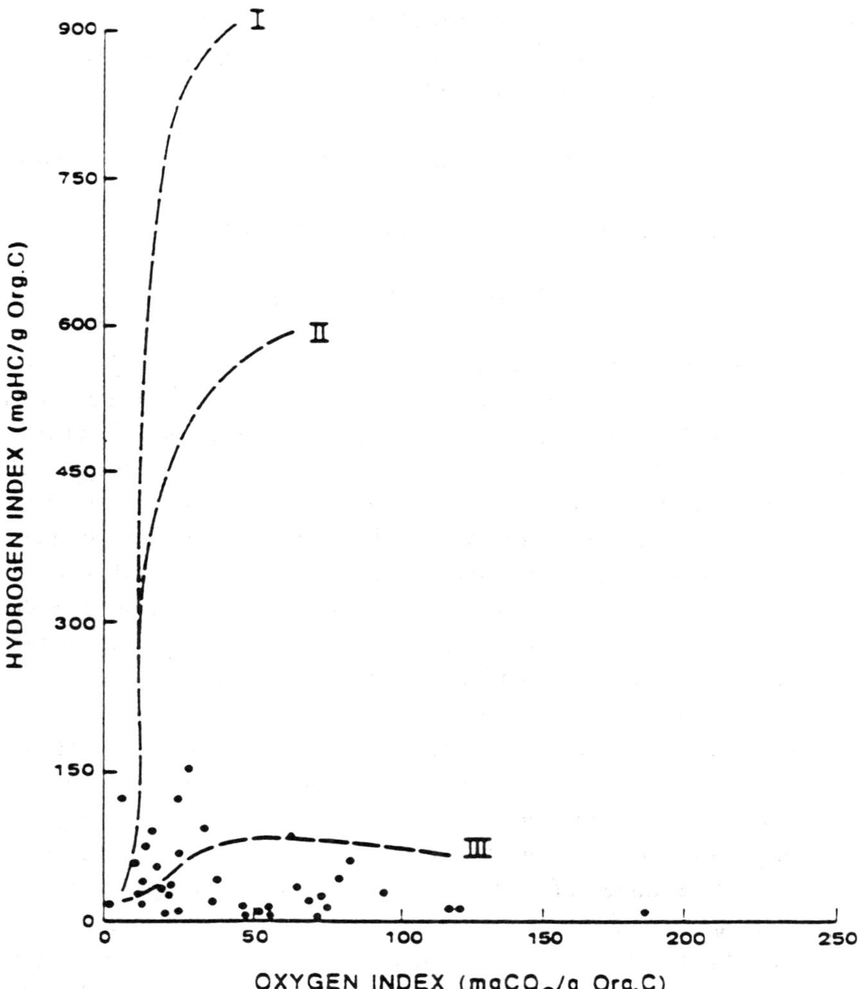

FIGURE 5.2 A van Krevelen diagram of hydrogen index to oxygen for organic samples taken from the Pinedale Anticline area, Wyoming. The dashed lines provide the general trend paths expected for type I, II, and III kerogens.

can be compared, in a rough sense, with modeled hydrocarbon generation. Thus if measurements indicate low H/C and O/C ratios, indicating high maturity, but the particular hydrocarbon generation model has hardly started to produce, it is unlikely that that model is appropriate. Again, gas chromatography–mass spectrometry measurements of the composition of oils indicate whether they were produced early in the basin's history or are more recent (Hunt, 1979; Waples, 1984).

A generation model not conforming to these rough rule-of-thumb restrictions is unlikely to be a highly probable reaction kinetic pathway. Other models would then be preferable.

But by whatever technique invoked, three factors are apparent. First, any model of *total* amount of hydrocarbon *generation* depends on the amount of kerogen per unit volume of rock (the "richness" or "leanness" of a potential source rock), the burial path of the sedimentary units, and the thermal history of the sediments. Second, the generation of hydrocarbons does not have to take place solely *in situ* with the kerogen. Migration of hydrocarbons (e.g., water-soluble isoalcohols) in solution, in suspension, or as a separate phase does not remove the hydrocarbons from their thermal bath. Generation to lighter hydrocarbons and to gas can carry on during migration or even after emplacement in a reservoir accumulation site. Third, while kerogen typically makes up less than a few percent of rock mass (Hunt, 1979), generated hydrocarbons influence the fluid flow and fluid pressure, because they form a mobile phase invading pore volume, with a different capillary surface tension than water, and can act to microfracture their very generation and/or accumulation sites and thereby become lost to the system.

Some of the components of these three factors are roughly insensitive to changes in the absolute amounts of hydrocarbons generated, being dominated by other considerations (e.g., burial history is surely dominated by total deposition rate), whereas others (e.g., a rock close to its fracture limit) can be affected significantly by a small change in the amount of hydrocarbon generation.

Thus there is a sensitive interplay, as well as a nonsensitive component, between dynamical and thermal evolution of a sedimentary basin and hydrocarbon production. We really do need to improve on our ability to model hydrocarbon generation in a more fundamental manner than is done by currently available quasi-phenomenological models.

6

Fluid Flow and Compaction: Time-Dependent Models

THIS CHAPTER deals with the evolution of porosity, fluid flow, and overpressure in sedimentary basins. A discussion is given of:

1. The effects of deposition, compaction, and erosion in modifying the flow of subsurface fluids.
2. The effects of buoyancy, capillarity, and relative permeability and of thermal expansion and convective heat transfer on fluid flow both directly as driving forces and indirectly in terms of pressure generated from the conversion of solid kerogen to liquid oil and gaseous hydrocarbons, processes influenced by the local temperature and hence by the local heat budget.
3. The effect of fractures in modifying the flow of fluid and the effect of fluid pressure in generating fractures.
4. Solubility and also chemical cementation and dissolution, which change both the subsurface porosity and fluid migration pathways and provide reservoir seals and regions of paleo-overpressuring and which are, in turn, influenced by the migration pathways of the subsurface fluids.

At deposition, shales have a porosity of order 70–75% while sands typically have depositional porosities of order 50%. The porosities of shales and sands in deep cores (say 5000 m, to fix ideas) may be typically 5–20%. Therefore during the process of burial a volume of fluid roughly comparable to the sediment volume has somehow left the system. This is such a large factor that the movement of subsurface fluids must play a dominant role in any attempt to understand the evolutional history of geological processes including, but not limited to, petroleum formation and migration, generation of overpressuring, cementation and dissolution of sedimentary rocks, fracture formation and subsequent cementation or dynamical closure, reservoir formation and seals, and formation of ore deposits. Compaction-driven flow drives petroleum migration, especially for primary migration from source rocks into carrier beds (Athy, 1930b; Bonham, 1980). It has been argued that compaction-driven and gravity-driven ground water flow drive secondary migration within carrier beds (Rich, 1921; Athy, 1930b; Toth, 1980). Overpressuring in basins is also caused by sediment compaction (Dickinson, 1953; Rubey and Hubbert, 1959; Brederhoeft and Hanshaw, 1968) and may be modified by aquathermal pressuring. Graf (1982) has suggested that high pore pressure gradients concentrate subsurface brines by reverse osmosis across shaley strata. Ore deposits may have formed (1) slowly during basin evolution (Noble, 1963; Jackson and Beales, 1967; Dozy, 1970) or (2) suddenly out of overpressured basins (Sharp, 1978; Cathles and Smith, 1983) by discharge of subsurface brines (Barnes, 1979). Cementation and dissolution processes within sedimentary rocks have also been related to chemical transport by fluids (Milliman, 1974; Hayes, 1979; Harris *et al.*, 1985), while sediment diagenesis has also been related to subsurface fluid flow by Sibley and Blatt (1976) and Wood and Surdam (1979).

Despite the importance of subsurface fluid flow for geological problems, little quantitative modeling has been done on a basinwide scale. Sharp (1976) and Sharp and Domenico (1976) developed a one-dimensional model of vertical compaction-driven flow and heat transfer, which they applied to relatively rapidly subsiding basins. Their analysis, however, does not consider lateral flow, which may be more important than vertical flow in many sedimentary basins (Magara, 1976a). Cathles and Smith (1983) used a two-dimensional model to predict the thermal effects of sudden dewatering of overpressured basins, but, as Bethke (1985) has remarked, "it is not clear whether their method is applicable to general studies of compaction-flow, and they only briefly discuss their calculation technique." The organization of this chapter is as follows: First we provide a general discussion of the processes to be included and how they cause fluid flow. This is followed with the mathematical details of a two-dimensional model of compaction-driven flow in accreting media. The model formulates flow in a set of unitized coordinates which remove completely difficulties usually encountered with the moving (in time and space) sediment–water boundary, as well as allowing for continuous vertical and lateral deformation of deposited sediments. The advantages are that more accurate and complete modeling of compaction-driven flow is possible than heretofore. Two numerical examples of flow are presented (1) to illustrate the comparison of similarities and differences with isostatically compensated models of sediment evolution and (2) to provide general insights into the effects of variable depositional lithologies, erosion, and subsequent deposition on the development of subsurface fluid movement, porosity and overpressure.

I. The Deposition–Compaction Process

A. General Remarks

Although the process of hydrocarbon migration is not well understood, the water expelled from the sediment as a result of compaction is a probable transporting medium of the hydrocarbons generated within the source rock. Shale contains most of the water in the subsurface and, despite shale's relative impermeability, fluid expulsion is implicit in the observed decrease of porosity with depth.

The flow that occurs in the temperature–depth interval in which hydrocarbons are generated from kerogen is of particular interest. Temperatures of 60 to 150°C commonly occur at depths of 2000 to 5000 m. While a source rock is undergoing burial in this depth zone, available flow may carry hydrocarbons from source to reservoir. *Migration potential* is defined here as the volume of flow that occurs during the period of hydrocarbon generation. Flow through a source rock during an interval of hydrocarbon generation extends upward through the pores of the overlying rock and may eventually pass through a reservoir rock. Thus, the flow history of a source must be considered in conjunction with the flow history of overlying reservoirs to evaluate the migration potential in a given stratigraphic section.

B. Deposition and Compaction

The effective overburden is the immersed weight of the overlying granular matrix, the effective vertical stress is the component of the sum of forces at grain contacts, and the excess pressure is the pore pressure in excess of hydrostatic. In general, the overburden is supported by a skeleton of particles and by fluid pressure in the enclosed pore volume, so that the effective overburden stress is equal to the sum of effective stress and fluid pressure.

At the instant an increment of sediment is deposited, the excess pressure is increased through the entire underlying column by the immersed weight of the increment. This is evident since time is required for the expulsion of fluid and, until compaction proceeds, effective stress cannot increase. Compaction proceeds as fluid escapes from the rock, allowing frame pressure to increase until it can support the effective overburden. Finally, as excess pressure approaches zero, flow ceases.

As deposition proceeds, however, overburden can accumulate at a greater rate than can be compensated for by the escape of pore fluid from a thick shale. Thus, significant excess pressure can develop that may easily reach a magnitude associated with "overpressured shales." The development of excess pressure retards compaction, resulting in both a greater porosity and a greater permeability (all other processes controlling permeability being regarded as held fixed) than are normal for a given depth, which, combined with the excess pressure gradient, implies a relatively high rate of flow for fluids in shale at a given depth.

Since excess pressure increases with the rate of deposition, rapid burial at the site of a delta, or beneath a thrust sheet, can be expected to develop excess pressure in underlying thick shales even though they had previously been hydrostatically pressured. This can lead to a "rejuvenated" expulsion of pore fluid, which may then find its way to potential reservoirs. Thus gradients of excess pressure may delineate flow patterns which point to current accumulations of petroleum.

If fluid pressure is indeed hydrostatic, then excess pressure is zero and the effective overburden is entirely supported by effective vertical stress. The effective stress, being equal to overburden, can then be related to porosity, thus defining a compaction function applicable in the immediate region and, possibly, to other shales similar in composition and geologic age.

If equivalent porosities are then found in overpressured regions at greater depths, it is often assumed that the effective stress is also equivalent and that the additional overburden is supported by excess pressure. The excess pressure can then be estimated from shale porosity when porosity is known as a function of depth. The accuracy of this assumed equivalence is unknown. Excess pressure can also be estimated by inferring the porosity from equivalence in sonic velocity, resistivity, or density.

II. The Basic Model

The fundamental underlying physical law to be used is that the mass density of a fluid is conserved. Thus, the rate of compaction of sediments is balanced by a net flow of fluid out of the locally compacting volume.

The flow velocity, in turn, is related to the gradient of the excess pressure by means of Darcy's law, about which we shall have more to say in a moment. The excess pressure may be produced by

1. the actual overburden less that fraction of the overburden supported by the matrix pressure of the rock, to which we may add other sources of excess pressure—such as
2. that caused by thermal expansion,
3. that caused by hydrocarbon generation and also by differential buoyancy,
4. that caused by capillary pressure, and
5. that caused by smectite transforming to illite (which transformation releases copious amounts of water)—although there is still considerable debate as to whether this transformation creates excess pressure.

A. Basic Equations

The relation of fluid flow velocity to the excess pressure gradient through Darcy's law is surprisingly subtle. Basically, Darcy's law requires that the mass flow of water per unit area, F, driven by an excess pressure gradient $\partial p/\partial l$ across a rigid, unmoving, noncompacting rock of length l be such that

$$F = -\rho K \mu^{-1} \, \partial p/\partial l \tag{6.1}$$

where K is the rock permeability in the direction l, ρ the density of water, and μ the viscosity. But in the time-dependent situation, not only is the rock moving downward towards basement as a result of compaction of the underlying rock but also a single unit thickness of rock at a given time does not maintain its thickness as it moves; that is, there is differential compaction. In order to apply Darcy's law, we must find a frame of reference in which the rock can be considered not to compact or move, write down the fluid flow equations in that frame, and then express the answers in the "physical" frame where the rock *does* move and compact.

To handle this problem the physical depth coordinate z, measured from the basement upward to the sediment–water interface, is replaced by the fully compacted coordinate ζ, where at any instant of time the transformation between ζ and z is

$$z = \int_{\zeta_b(x,\, t)}^{\zeta} [1 + e(\zeta', x, t)] \, d\zeta' \tag{6.2}$$

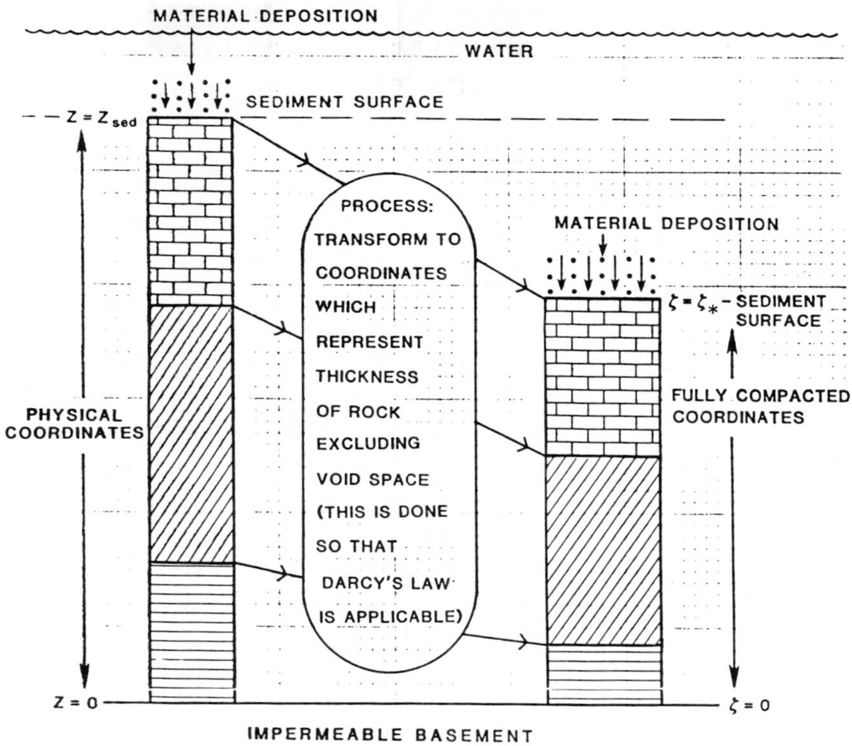

FIGURE 6.1 Sketch of the transformation of coordinates from physical depth z to the fully compacted ζ.

where the ζ coordinate is defined relative to basement, $\zeta_b(x, t)$, at each lateral location and where e is the void ratio[1] (see Fig. 6.1). Similarly, the surface $\zeta = \zeta_*(x, t)$ and the surface void ratio $e_*(x, t)$ can depend on lateral position as well as time, and they, too, are to be specified by the model maker.

We take the effective stress σ to be made up of two components: (1) the frame pressure p_f, defined as the part of the effective overburden that can be supported by grain contacts transmitting stress without further compaction so that $p_f = p_f(e)$, and (2) a negative component equal to any increase in the excess pressure generated by other effects such as hydrocarbon pressure or thermal expansion. Therefore, we can write

$$\sigma = p_f(e) - p$$

[1] The connection between void ratio e and porosity ϕ is $e = \phi/(1 - \phi)$ or $\phi = e/(1 + e)$.

where p is the generated pressure for which a model dependence must be assigned (but more on this point later). Empirical laws for different materials relate frame pressure to void ratio (see, e.g., Sharp and Domenico, 1976; Bethke, 1985). For example, for shales $p_f(e) \propto e^{-a}$ with $a \approx 3$. The point is that $p_f(e)$ is considered to be a *known* function of the void ratio for each lithology. For the case of compaction alone the relation between frame pressure and void ratio is for primary consolidation. The changes in the frame pressure with void ratio law introduced by, say, chemical cementation and dissolution, thermal expansion, fracture production, etc., we defer for the moment. Equally, when a model is provided for the generated pressure term, we can consider p as a known function of quantities which enter the equations.

For the moment, then, we can take σ to have a prescribed dependence on appropriate variables and, in particular, we shall write σ in the form above so that the effects of frame pressure and generated pressure are clearly defined.

At any lateral and vertical position the excess pressure p is given by

$$p = g \int_{\zeta}^{\zeta_*(x,\, t)} [\rho_s(\zeta',\, x,\, t) - \rho_f(\zeta',\, x,\, t)]\, d\zeta' \\ - [p_f(e) - p_f(e_*)] + p(\zeta,\, x,\, t) - p_*(x,\, t) \tag{6.3}$$

where we have subtracted the values at the sediment–water interface since the excess fluid pressure must be zero there, and where the particle density ρ_s may vary with both time and position since the lithology of the sediments deposited varies with time and basinal position.

Since the accumulating sediments are compressed solely in the vertical direction, in the ζ–x frame we can use Darcy's law so that vector flow $\rho_f v$ is given through

$$-\rho_f v = K_z \rho_f [\mu(1 + e)]^{-1} \frac{\partial p}{\partial \zeta}\, \hat{\zeta} + K_x \rho_f \mu^{-1} \frac{\partial p}{\partial x}\, \hat{x} \tag{6.4}$$

where $\hat{\zeta}$ and $\hat{x}$ are unit vectors in the ζ and x directions, respectively, and where K_z and K_x reflect the difference in permeability due to the anisotropic nature of the compacting material. When combined with the requirement that fluid be conserved, equation (6.4) provides the two-dimensional fluid migration equation in the ζ–x coordinates:

$$\frac{\partial}{\partial t}(\rho_f e) = \frac{\partial}{\partial \zeta}\left[K_z \rho_f (\mu(1 + e))^{-1} \frac{\partial p}{\partial \zeta} \right] + \frac{\partial}{\partial x}\left(K_x \rho_f \mu^{-1} \frac{\partial p}{\partial x} \right) \tag{6.5}$$

Equation (6.5), when combined with equation (6.3) relating excess fluid pressure to void ratio, is the basic fluid flow equation requiring solution.

B. Transformation to a Stable Coordinate System

Two problems are evident with equations (6.3) and (6.5). The first problem is that the depositional surface $\zeta = \zeta_*(x, t)$ evolves with time, so we have to keep adjusting the upper surface boundary condition since solutions must have $e = e_*(x, t)$ on $\zeta = \zeta_*(x, t)$. However, moving boundaries are notoriously difficult to handle, both analytically and numerically. We would prefer to avoid this requirement. The second problem is that not only is the sedimentary surface changing with time but so also is the value of the void ratio on the surface, since the lithology of the accumulating sediments changes; that is, on $\zeta = \zeta_*$ we have $e = e_*(x, t)$ and the surface void ratio $e_*(x, t)$ changes with time and lateral basinal position. We would prefer to avoid this problem, too, since it can lead to difficulties similar to those generated by the moving boundary position.

The way to bypass *both* problems is the following. Define a fractional vertical coordinate u through

$$u = [\zeta - \zeta_b(x, t)]/[\zeta_*(x, t) - \zeta_b(x, t)] \tag{6.6}$$

such that on $\zeta = \zeta_b$, $u = 0$ and on $\zeta = \zeta_*$, $u = 1$, and also define the fractional void ratio V through

$$V = e(\zeta, x, t)/e_*(x, t) \tag{6.7}$$

In terms of these variables, equation (6.5) can be written in the form

$$\begin{aligned}
&\frac{\partial V}{\partial t} + \frac{V\,\partial \ln(\rho_f e_*)}{\partial t} + \frac{\partial V}{\partial u}\left[-u\frac{\partial \ln(\zeta_* - \zeta_b)}{\partial t} + (\zeta_* - \zeta_b)^{-1}\frac{\partial \zeta_b}{\partial t}\right] \\
&= (\zeta_* - \zeta_b)^{-2}(\rho_f e_*)^{-1}\frac{\partial}{\partial u}\left[K_z\rho_f/\mu^{-1}(1 + e_* V)^{-1}\frac{\partial p}{\partial u}\right] \\
&\quad + (\rho_f e_*)^{-1}\left\{\frac{\partial}{\partial x}\left(K_x\rho_f\mu^{-1}\frac{\partial p}{\partial x}\right) - \frac{\partial}{\partial x}\left[K_x\rho^{-1}\mu_f(uA + B)\frac{\partial p}{\partial u}\right]\right. \\
&\quad \left. - (uA + B)\frac{\partial}{\partial u}\left(K_x\rho_f\mu^{-1}\frac{\partial p}{\partial x}\right) + (uA + B)\frac{\partial}{\partial u}\left[K_x\rho_f\mu^{-1}(uA + B)\frac{\partial p}{\partial u}\right]\right\}
\end{aligned} \tag{6.8}$$

where

$$A(x, t) = \frac{\partial \ln(\zeta_* - \zeta_b)}{\partial x}, \qquad B(x, t) = (\zeta_* - \zeta_b)^{-1}\frac{\partial \zeta_b}{\partial x}$$

In terms of u, x variables the excess pressure can be written

$$p = g(\zeta_* - \zeta_b) \int_u^1 (\rho_s - \rho_f)\, du - [p_f(e_* V) - p_f(e_*)] + p(u, x, t) - p(1, x, t) \tag{6.9}$$

while the physical depth z is given through

$$z = (\zeta_* - \zeta_b) \int_0^u [1 + e_*(x, t)\, V(u', x, t)]\, du' \tag{6.10}$$

Note that if ζ_b and ζ_* are independent of lateral position x, if ζ_b is independent of time t so that ζ_b can be set to zero, and if $K_x = 0$, then equation (6.8) reduces to the one-dimensional fluid migration equation. The boundary conditions on equation (6.8) are:

1. On the sediment surface $\zeta = \zeta_*$ (correspondingly $u = 1$), we require $V = 1$.
2. On the basement surface $\zeta = \zeta_b(x, t)$ (correspondingly $u = 0$), we require that there be *no* fluid flow across the surface $\zeta = \zeta_b(x, t)$, that is, we require

$$\frac{\partial p}{\partial u}\left[K_x(1 + e)^{-1}\left(\frac{\partial \zeta_b}{\partial x}\right)^2 + K_z\right] - \frac{\partial p}{\partial x} K_x(\zeta_* - \zeta_b)\frac{\partial \zeta_b}{\partial x} = 0 \tag{6.11}$$

There are advantages to having a two-dimensional formulation of the fluid migration equation. First, the fact that the migration of fluid under compaction is *not* restricted to vertical motion but includes lateral flow of fluid means that a modeling of primary migration paths is possible. Second, the fact that both the basement geometry $\zeta_b(x, t)$ and the depositional surface geometry $\zeta_*(x, t)$ are permitted to be freely specified by the model maker means that we can follow the flow of fluid sediment accumulating at the margin of a basin and filling seaward.

The price that is paid for converting to *fixed* boundaries in the fractional coordinate system and fractional void ratio is that the evolution of the surface void ratio e_* and the evolution of the depositional surface, $\zeta_*(t, x)$ both appear in the evolutionary equation (6.8) for the fractional void ratio and are represented by the second and third terms, respectively, on the left-hand side of equation (6.8). This price is worth paying on two counts: first, we gain an enormous simplification in dealing with fixed boundaries; second, the underlying topological structure of the equation is maintained (i.e., the equation is *not* converted to one of higher order, which, mathematically speaking, would have had extremely unpleasant repercussions).

Similarly, we can include varying lithological sequences being deposited both in time and in lateral position (i.e., the surface void ratio e_* is a function of both lateral position and time—directly reflecting facies changes), so that the changes in fluid migration with time developed by successively advancing and retreating shorelines, say, can be modeled with ease.

Also, we can allow generation pressure P_g to vary in space and time—as would occur, for example, with pressure generated by hydrocarbons—so that the basic two-dimensional model can include a large variety of factors and effects which are thought to influence oil and gas migration.

III. Two Simple Numerical Models

The purposes of the two models presented here are (1) to compare and contrast sedimentary behavior during burial when fluid flow is allowed for with the corresponding burial history under the assumption of instantaneous adjustment of the sediments to isostatic burial (see, e.g., Guidish *et al.*, 1984, 1985), and (2) to demonstrate simply the interesting patterns of porosity and pressure behavior which result when different lithologies are juxtaposed in a depositional setting and when erosion and subsequent further deposition also take place.

The simplest geometrical situations we have been able to devise to illustrate these two points as clearly as possible deal with "layer-cake" deposition so that the deposited sedimentary layers remain forever parallel layers. We also suppose that the basement surface remains flat. This geometric situation permits the two model cases to exhibit the salient points in a pristine manner, without obscuration by modifying factors such as tectonic uplift or laterally variable deposition rates and facies change.

The numerical coding of the fluid flow/compaction equations (6.8) and (6.9) has been done using an implicit method. The resulting code has been tested for veracity against two analytic models which solve the fluid flow equations exactly and which are presented elsewhere (Glezen and Lerche, 1985).

Since Bethke (1985) has noted that the fluid flow equations had not previously been known to have any exact analytic solutions, it would seem that this is the first time such solutions have become available to use as templates for gauging the accuracy of numerical codes. Here we do not present any details of coding procedures, which have certain novel aspects of their own, since our main interest is in developing an appreciation of the basic dynamical processes.

A. Comparison with Isostatic Burial History

To fix ideas, a 1200-m layer of shale with a depositional porosity of 75% is deposited at a constant rate of 120 m/Myr for a 10-Myr period. Under isostatic burial conditions, at each instant of time the layer adjusts to an equilibrium balance between frame pressure and effective overburden. Thus the behavior of porosity with depth for the shale layer is constantly adjusted as though the shale had infinite permeability. In order to set up a familiar framework as a template to use in gauging the differences between behavior on the basis of an isostatic burial history and behavior on the basis of self-consistent fluid flow evolutionary equations, we have used the burial history method of Guidish *et al.* (1984, 1985). That particular approach specifies the porosity with depth variations for each lithology (in lieu of being able to determine the variation *ab initio,* which is the goal of the fluid flow effort) based on present-day measurements in the North Sea. The default functions used in that paper are taken from Sclater and Christie (1980). In the fluid flow case the rule connecting shale frame pressure and porosity used for illustration is as given in the Appendix to this chapter. The isostatic result is that at the end of the 10-Myr depositional period the thickness of the shale layer is about 800 m, as shown in Fig. 6.2 by the curve marked A. Since no further deposition occurs, the shale layer maintains its 800-m thickness for all time after deposition ends. In contrast, the solution of the fluid flow equation, allowing for a finite shale permeability (with a permeability–porosity law as given in the Appendix), yields a shale thickness shown by the curve marked B in Fig. 6.2. The initial rapid sedimentation is faster than the fluid can escape from the shale so that the shale thickness remains closer to its 120-m/Myr increasing thickness. At the end of the 10 Myr of deposition the shale layer has nearly its full depositional thickness of 1200 m. Thereafter, as time proceeds, fluid permeates out of the shale layer and, over about another 10–15 Myr, the shale rapidly thins to about 1000 m thick. But since the permeability of the shale decreases as the porosity decreases, it becomes harder for fluid to escape. As a consequence, after the first rapid loss of water there is a long slow period of adjustment, lasting some 70–80 Myr for the numerical values chosen, in which fluid slowly oozes out of the shale as it approaches its equilibrium thickness.

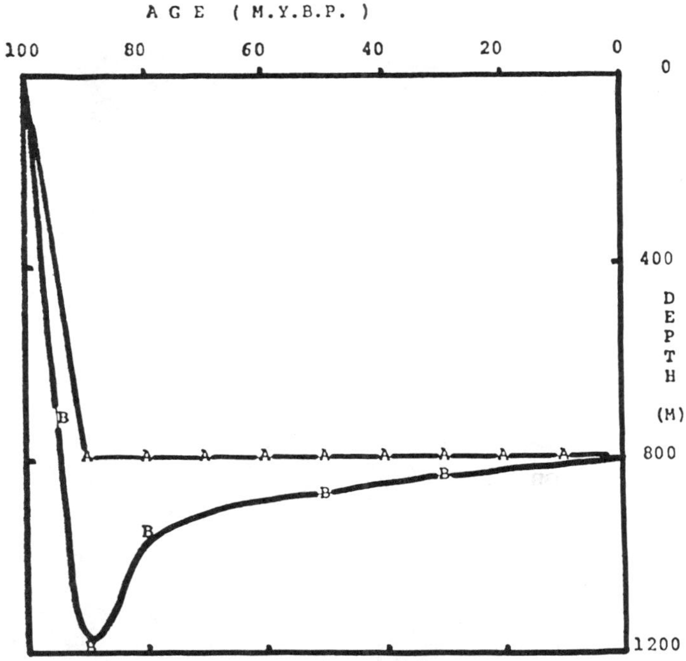

FIGURE 6.2 Shale thickness with time for deposition at 120 m/Myr for 10 Myr. Curve A represents the shale thickness computed using isostatic burial history; curve B represents the shale thickness computed using the fluid flow/compaction equations given in the chapter.

Figure 6.3 shows the porosity with depth behavior at the instant (10 Myr) that deposition stops. The curve labeled 1 is the porosity with depth law resulting from solving for the balance of frame pressure against excess overburden and so represents isostatic burial (infinite shale permeability). The curve marked 2 shows the porosity with depth when the finite permeability of shale is allowed for. From Fig. 6.3 we see that the porosity allowing for fluid flow is larger at depth than that obtaining under isostatic conditions. We note that there is a rapid loss of porosity with shallow depth for curve 2 followed by a slow decline, representing the fact that the permeability of the underlying shale is smaller than that of the near-surface shale so that it is more difficult to remove fluid. Hence the porosity is maintained in excess of the isostatic case.

Figure 6.4 shows the evolution of porosity with depth as time proceeds. Curve A represents the porosity with depth just at the end of the depositional period (10 Myr), curve B at the end of 20 Myr, curve C at

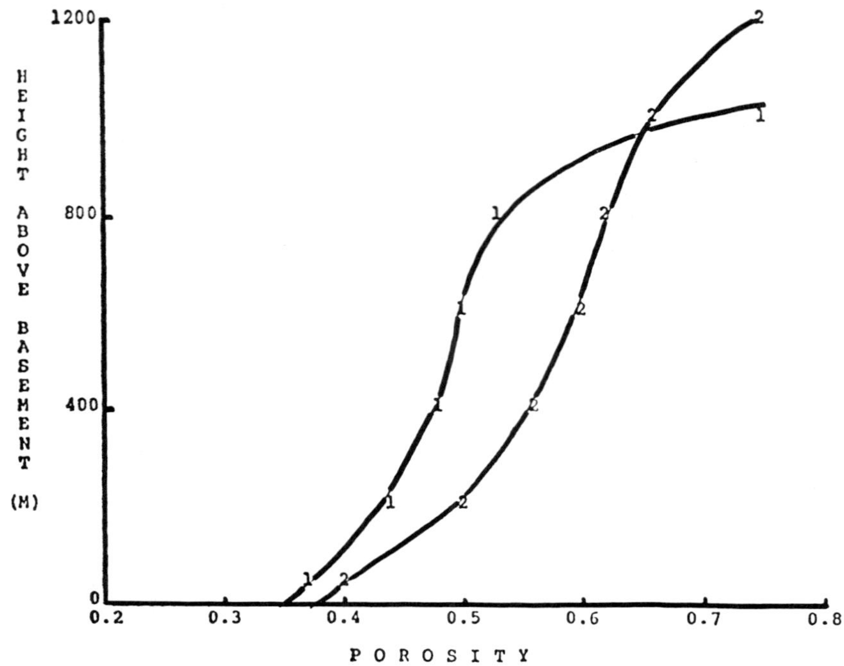

FIGURE 6.3 Porosity with depth behavior for the shale of Fig. 6.2 at the instant (0 Myr) deposition ceases. Curve 1 is the behavior assuming isostatic burial history; curve 2 is the behavior using the fluid flow/compaction equations.

the end of 50 Myr, and curve D at the end of 300 Myr. We note that the compaction of the shale (as far as the porosity with depth law is concerned) is not complete until nearly 300 Myr have passed and that the final shale thickness is some 600 m—half of its original deposited value. We also see that there is then an extremely rapid decrease of porosity with depth, dropping from a surface value of 75% to 35% on the order of 100 m. For comparison, the equivalent drops are on the order of 1000 m (curve B, 10 Myr) and 700 m (curve C, 50 Myr) so that, as presented in Fig. 6.2 in a different way, there is a relatively rapid efflux of fluid from the shale followed by a long slow oozing of fluid.

Figure 6.5 illustrates the evolution of excess fluid pressure with time. Curve A shows the excess fluid pressure with depth just at the end of the depositional period (10 Myr), curve B at the end of 50 Myr, curve C at the end of 300 Myr, and curve D at the end of 1000 Myr. Comparison of Fig. 6.5 with Fig. 6.4 demonstrates that the excess fluid pressure takes considerably longer than the porosity (by nearly a factor of 3) to reach

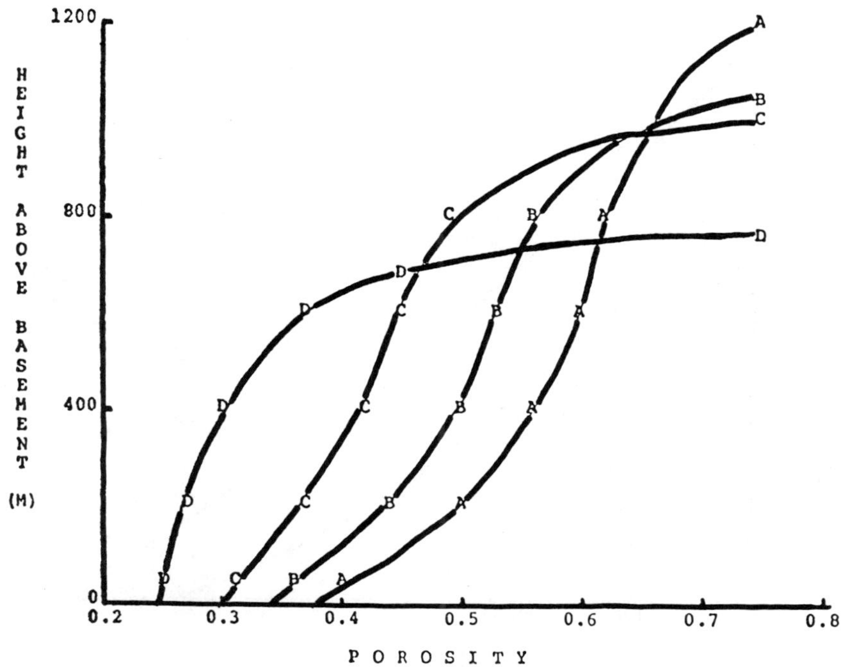

FIGURE 6.4 Evolution with time of porosity with depth for the shale of Fig. 6.2. Curve A is at 10 Myr, curve B at 20 Myr, curve C at 50 Myr, and curve D at 300 Myr.

equilibrium. The reason is that the excess fluid pressure is nonlinearly dependent on the porosity and permeability so that, even though the porosity has nearly completed its evolutionary approach to equilibrium, a very small porosity in excess of the isostatic equilibrium value corresponds to a relatively large excess fluid pressure (the same basic principle is invoked in hydraulic rams). Since the shale's permeability is now very low, it takes a very long time indeed for the remaining amount of excess fluid to ooze out of the shale and hence reduce the excess fluid pressure to zero.

Three specific lessons are to be learned from this simple example:

1. The self-consistent solution of fluid flow compaction models corresponds to a different end state than obtained by specifying a priori an approximate behavior for porosity with depth and then using that specification in an isostatic equilibrium burial history model.

2. There is a rather rapid loss of near-surface porosity quite

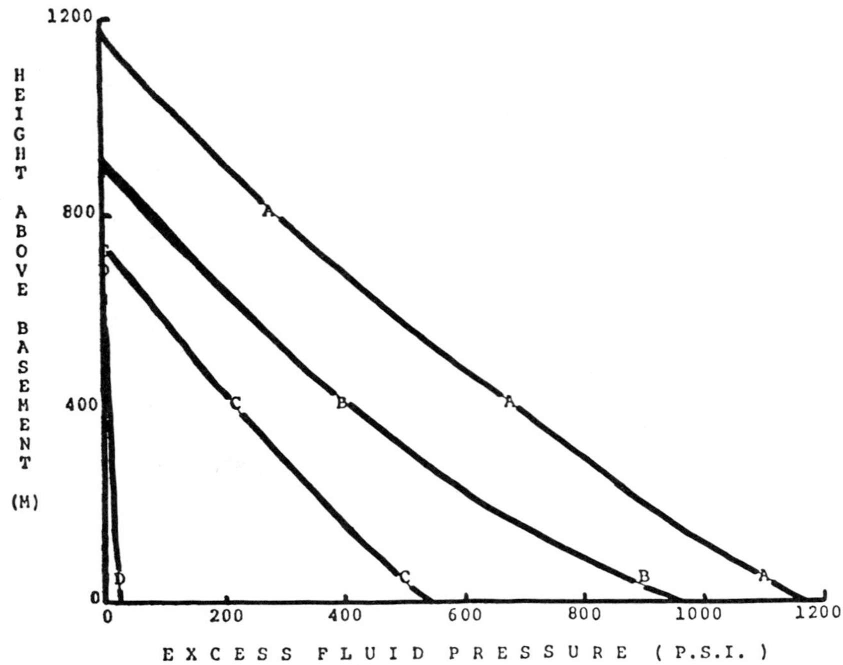

FIGURE 6.5 Evolution with time of excess fluid pressure with depth for the shale of Fig. 6.2. Curve A is at 10 Myr, curve B at 50 Myr, curve C at 300 Myr, and curve D at 1000 Myr.

soon after deposition ceases followed by a long period of slow readjustment and oozing of fluid as the shale becomes more impermeable.

3. The relaxation of excess fluid pressure with depth takes a considerably longer period of time (a factor of 3 for the numerical example given) than the relaxation of porosity with depth.

More general inferences based on the detailed results presented here are given later.

B. Variable Lithology, Deposition, and Erosion

In order to illustrate the interesting evolutionary patterns of porosity and excess fluid pressure that can arise within sedimentary basins—in particular the occurrence of overpressured sands as has been observed (e.g., Hunt, 1979, figure 6-17)—a simple mixed lithology depositional/erosional model was constructed using the rules connecting frame pressure with porosity and permeability with porosity given in the Appendix.

First 50 m/Myr of shale with a surface porosity of 75% is deposited for a period of 10 Myr; then this partially compacted shale is eroded at the rate of 25 m/Myr for the next 10 Myr. Sand is then deposited, with a surface porosity of 49%, at the rate of 50 m/Myr for the next 10 Myr, following which shale is again deposited for the next 10 Myr at 50 m/Myr. Deposition then ends and the evolution of the system is followed for a further 60 Myr. The depositional sequence and rate of deposition (erosion) with time are as sketched in Fig. 6.6.

Figures 6.7 through 6.11 illustrate how both the porosity (Figs. 6.7a through 6.11a) and various pressures (Figs. 6.7b through 6.11b) vary with depth as time proceeds.

Inspection of the figures labeled (a) indicates several new phenomena not uncovered by the simple shale layer deposition model given previously. Figure 6.7a shows the usual rapid loss of shale porosity in the first few tens of meters (from a surface value of some 75% to 60%) followed by a much slower decline with depth, basically paralleling in

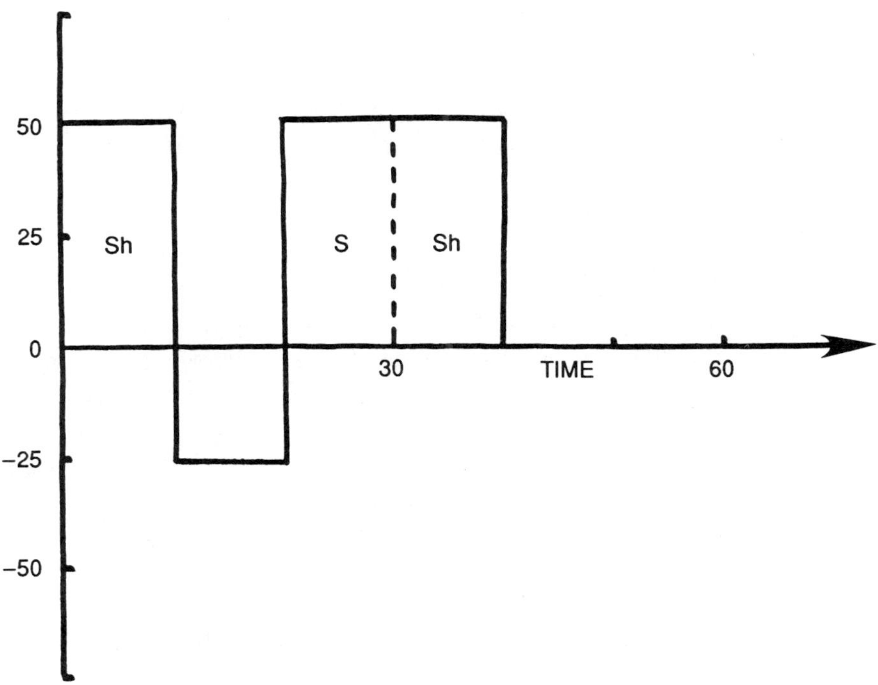

FIGURE 6.6 Rate of deposition/erosion of the shale–sand–shale sequence discussed in text. The ordinate is in units of meters per million years, the abscissa in units of million years. Shale is identified by the symbol Sh, sand by S.

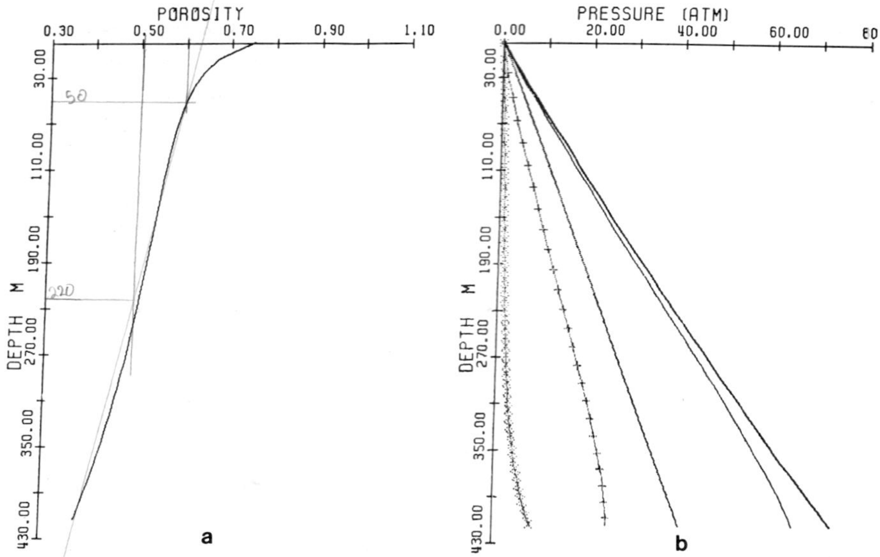

FIGURE 6.7 (a) Variation of porosity with depth at the end of 10 Myr when 500 m of initial thickness of shale has been deposited. (b) Variation with depth of lithologic pressure, fluid pressure, hydrostatic pressure, frame pressure, and excess fluid pressure at 10 Myr. The curves are identified as follows: lithologic pressure (heavy black line), fluid pressure (lighter line), hydrostatic pressure (hash mark line), excess fluid pressure (dotted and crossed line), frame pressure (fuzzy dotted line).

behavior the situation already seen. At the end of 20 Myr, after erosion has taken place, Fig. 6.8a shows that the shale's surface porosity is now some 44% and that porosity exhibits merely a steady decline with depth, reflecting the fact that the material was eroded from the surface downward. The sand layer which is then deposited (with surface porosity 49%) in the next 10 Myr has a permeability some six orders of magnitude larger than that of the underlying shale. Hence, as shown in Fig. 6.9a, the overlying sand is close to equilibrium and has a roughly linear decrease of porosity with depth, in accordance with a roughly linear increase of sand frame pressure with decreasing porosity. Fluids from the underlying shale which make their way into the overlying sand can then escape rapidly. The porosity behavior with depth of the underlying shale at this instant of time is another story. It shows first an increase from about 30 to 34% with increasing depth followed by a decrease at greater depths as also shown in Fig. 6.9a. What has happened is that the effective overburden produced

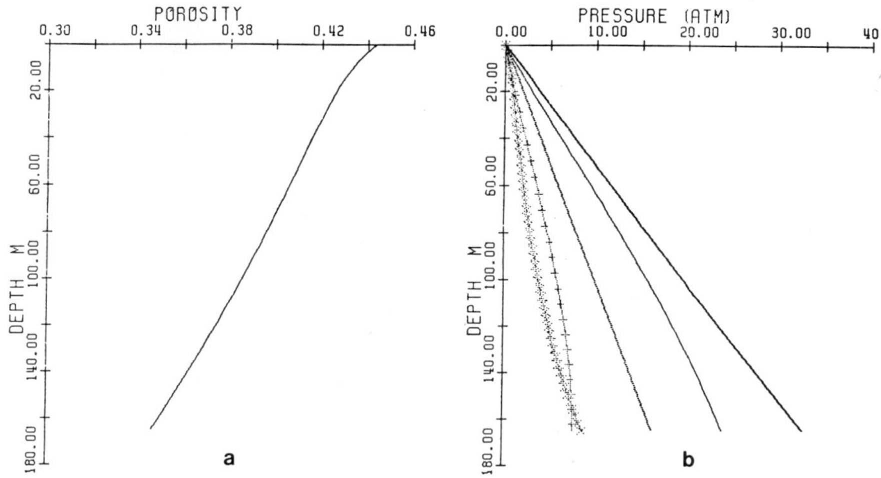

FIGURE 6.8 (a) Variation of shale porosity with depth at 20 Myr following 10 Myr of erosion at 25 m/Myr. (b) Variation with depth of lithologic pressure, fluid pressure, hydrostatic pressure, frame pressure, and excess fluid pressure at 20 Myr. The curves are identified as follows: lithologic pressure (heavy black line), fluid pressure (lighter line), hydrostatic pressure (hash mark line), excess fluid pressure (dotted and crossed line), frame pressure (fuzzy dotted line).

by the sand has compacted the top layer of shale, thereby reducing the shale's permeability. Fluid in the underlying shale is then "trapped" and can escape only at the slow oozing rate permitted by the topmost shale layers. Hence the porosity is higher in the middle of the shale layer than near the top. The deeper decrease in shale's porosity is a consequence of the shale being compacted against the impermeable underlying basement, forcing fluid upward through the shale.

In Fig. 6.10a we see the result of adding more shale (at the rate of 50 m/Myr for 10 Myr) to the system. Now the sand body is trapped between two relatively impermeable shale layers and, with no lateral connectivity of the sand body to another, the sand body cannot remove its fluid content except at the rate allowed by the shale permeability. Hence the fluid remains trapped and the sand porosity holds nearly constant at about 35%. The shale underlying the sand retains its increased porosity as shown in Fig. 6.10a, while the overlying shale again shows the usual initial rapid decrease in porosity with depth followed by a slower decline.

Deposition now ceases and the system evolves for the next 60 Myr. As sketched in Fig. 6.11a at the end of a total of 100 Myr the overlying shale

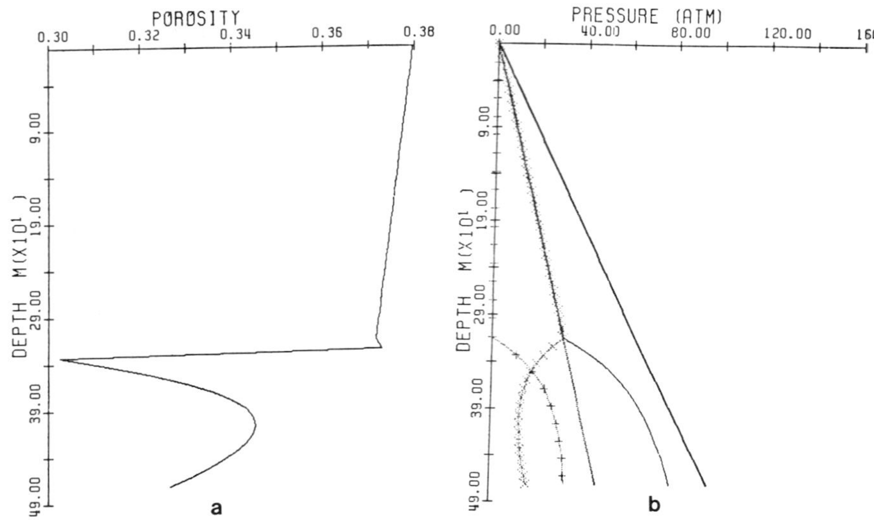

FIGURE 6.9 (a) Variation of shale and sand porosity with depth at 30 Myr following 10 Myr of sand deposition at 50 m/Myr. (b) Variation with depth of lithologic pressure, fluid pressure, hydrostatic pressure, frame pressure, and excess fluid pressure at 30 Myr. The curves are identified as follows: lithologic pressure (heavy black line), fluid pressure (lighter line), hydrostatic pressure (hash mark line), frame pressure (fuzzy dotted line), excess fluid pressure (dotted and crossed line).

layer has now assumed the "classic" profile of a very rapid decrease (from 75 to 40% porosity) in about 100 m followed by a slow decline to about 30% porosity at 300 m. The trapped sand layer has an even greater problem than it did at earlier times in displacing its contained fluid because the permeability of the overlying shale has been decreased so that the loss rate of fluid from the sand is also decreased. The sand stays at close to 36% porosity. The shale underlying the sand layer has had time to remove some small amount of its entrapped fluid so that its porosity no longer shows the "hump" so characteristic of rapid deposition at earlier times—but the effect is small, cutting the porosity only from a peak of some 34% to 30%.

The curves of Figs. 6.7b through 6.11b track with depth the lithologic overburden, hydrostatic pressure, total fluid pressure, excess fluid pressure and frame pressure at different times (the curves are identified in the figure legends).

In Fig. 6.7b (after 10 Myr of shale deposition at 50 m/Myr) we see that the overburden is nearly completely supported by fluid pressure with a

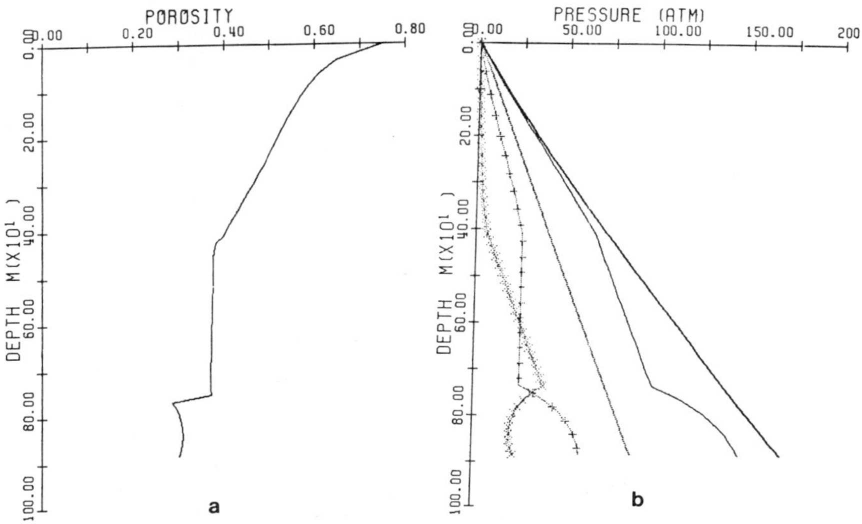

FIGURE 6.10 (a) Variation of porosity of shale, sand, and shale again with depth at 40 Myr following 10 Myr of shale deposition on top of the sand at 50 m/Myr. (b) As for Fig. 6.7b but at 40 Myr.

very small frame pressure contribution, representing the fact that deposition has occurred much more rapidly than fluid can penetrate through the shale so that not much fluid has yet escaped from the system. After erosion Fig. 6.8b (20 Myr) shows that the fluid overpressure has been diminished and that the frame pressure is comparable to the fluid pressure. The deposition of a layer of sand (high permeability) as shown in Fig. 6.9b (30 Myr) causes the system to develop overpressure only in the underlying shale. The sand overburden is totally supported by the sand's frame pressure. The increased overburden produced by the sand compacts the shale a little, forcing an increase in the fluid pressure, so that the support of the shale is now provided more by fluid than by frame pressure.

Further deposition of a relatively impermeable shale layer on top of the sand (Fig. 6.10b, 40 Myr) now causes the sand to become overpressured since it can no longer release its fluids to the surface as the overlying shale acts to seal the system more tightly. Note that the excess fluid pressure is nearly constant with depth in the sand because of the high permeability.

Both the underlying and overlying shale layers remain overpressured. Deposition now ceases and the system evolves. At the end of 100 Myr after initial deposition, Fig. 6.11b captures the slow evolution of the

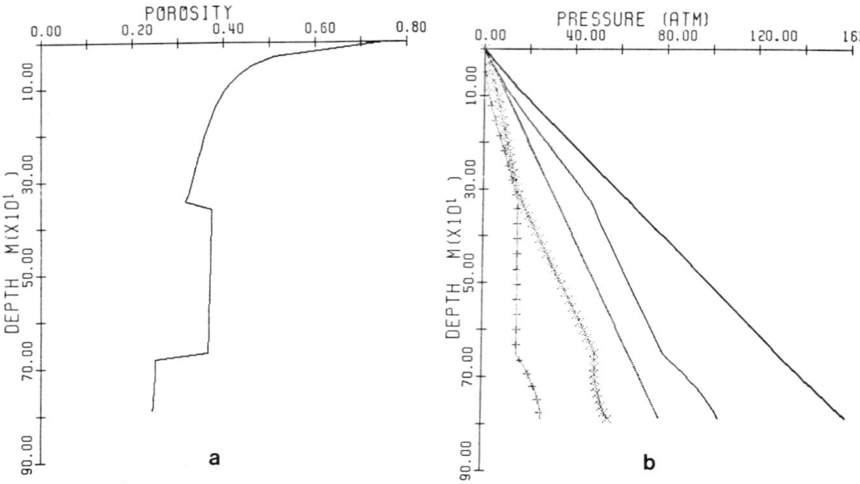

FIGURE 6.11 (a) Variation of porosity of the juxtaposed shale/sand sequence of Fig. 6.10 at the end of a further 60 Myr, for a total elapsed time of 100 Myr. (b) Variation with depth of lithologic pressure, fluid pressure, hydrostatic pressure, frame pressure, and excess fluid pressure at 100 Myr. The curves are identified as follows: lithologic pressure (heavy black line), fluid pressure (lighter line), hydrostatic pressure (hash mark line), frame pressure (fuzzy dotted line), excess fluid pressure (dotted and crossed line).

system. The degree of fluid overpressuring has been cut down, the frame pressure increases, and the system slowly relaxes, heading toward its final equilibrium condition of zero excess fluid pressure.

The lessons to be learned from this simple example are:

1. The existence of overpressured sands, as observed (Hunt, 1979, Fig. 6-17), requires that the sand body be isolated (in a connecting sense) so that no "draining" of the excess fluid pressure take place and requires that an overlying, relatively impermeable, sealing material be present.

2. The development of overpressure in shales takes place either when shale deposition is so fast that there is not enough time for fluid to escape or when a shale, nearly at equilibrium conditions, is subject to a rapidly deposited overburden of any lithology (sand in our numerical example).

3. The time scale for evolution of the depositional sequence toward equilibrium is controlled not only by the intrinsic time scales at which fluids can leak out of individual component members of the

composite system but also by the juxtapositional order of the various lithologic units as they are deposited and by the degree of erosion that has taken place. For example, Fig. 6.8 indicates that the shale layer after erosion is closer to equilibrium than it was in Fig. 6.7 prior to erosion in the sense that the excess fluid pressure has been diminished. But this closeness is then upset by the further deposition of a sand layer.

From a technical point of view, by using novel mathematical transformations, this section has shown how to bypass both of the fairly severe numerical problems which arise from (1) the presence of a moving sediment–water boundary as deposition or erosion takes place and (2) application of Darcy's law to a system which is subsiding, deforming, and differentially compacting.

The two simple numerical examples presented of solutions to the fluid flow/compaction equations were designed with two purposes:

1. To illustrate the major departures (from the ad hoc assumption of isostatic burial history) of self-consistent models of sedimentation which take account of the different permeabilities and so a finite speed of fluid flow relative to accumulating sediments. The major conclusion here is that the relaxation of excess fluid pressure and of sedimentary thickness with time to their equilibrium conditions occurs on a timescale of order 50–100 Myr, comparable to the age of sedimentary basins. These effects then play a major role in influencing the development of structures, fracturing, fluid migration, thermal maturity, and cementation and dissolution.
2. To illustrate the evolution with time of both porosity and fluid overpressuring with depth. This evolution is influenced by four major factors: the rate of deposition/erosion of lithologic layers; the intrinsic permeabilities of the various layers; the sequence of juxtaposition of the deposited lithologies; and the degree of connectivity, or isolation, of the various sedimentary units—in particular of the highly permeable materials.

The fundamental reasons for modeling the fluid history of an accumulating sedimentary sequence are that (1) the development of structures, fractures, and the thermal maturity of the basin are influenced to no small extent by the fluids and (2) the movement of pore water is associated with the generation and primary migration of hydrocarbons.

Pore fluid motion is further influenced by the presence of fractures, faults, seals, deposition and erosion, tectonics, and stratigraphic sequences. Other influences on fluid migration include the development of overpressured zones, the presence of evolved gas, and the development of diapirs.

The development of a model for the migration of pore fluid in a basin has several requirements:

1. A balance between simplification and inclusion of detailed behaviors for all relevant factors. Excessive simplification restricts the applications, while detailed inclusion of all relevant factors complicates the model to the point where its effective application is frustrated.

2. A numerical solution structured so as to attain a sufficient degree of accuracy, sustained stability, and acceptably short computer time.

3. A sophisticated graphics output to display a multiparameter system in two spatial dimensions that varies with time.

Burial history and thermal history are dependent on fluid history including the generation and migration of hydrocarbons. Burial history may include the movement of pore fluid through the sediments as well as depth of burial, temperature, and hydrocarbon generation as functions of time. In turn, the development of excess pressure, of fractures, and of fluid movement is influenced by the associated temperature history and hydrocarbon generation. In short, burial history and fluid history are not two independent processes but are different aspects of the same process. Ultimately, the merging of these two approaches into a unified framework provides an extremely powerful tool for better assessing and risking the hydrocarbon potential of basins.

We now treat the interaction of fluid and thermal flow, hydrocarbon generation and gas pressure effects, fracturing, and chemical dissolution and precipitation. As we shall see, each process introduced can be fit simply into the basic fluid flow equation given above and, with the use of the fractional vertical coordinate system which cleanly bypasses the otherwise serious numerical problems associated with moving boundaries, a better understanding of the ruggedly stable components of the various processes can be achieved.

IV. Thermochemical, Hydrocarbon Generation, and Saturation Considerations

We develop expressions for two sources of generated pressure: hydrocarbon generation and thermal expansion. The amount of hydrocarbon generation is a response to the thermochemical breakdown of kerogen. Assume that biogenic gas does not play a significant role in the accumulation of the gas associated with petroleum. For hydrocarbons to be trapped in the subsurface, sealing by shale or evaporite is necessary as well as good-quality reservoirs. Sealing capacity is related to the size of pore throats, which in turn is related to the porosity. At shallower depths where biogenic gas production is active, the porosity of shale is usually large, hence the sealing capacity tends to be low. Therefore, there would be little potential to retain gas even if biogenic gas had been generated in the surrounding area.

We illustrate the pattern of development using the first-order chemical reaction for the breakdown of kerogen following an Arrhenius equation scheme modified from Tissot and Welte (1978) and Tissot and Espitalie (1975). Thus from Chapter 5 we can write

$$dX_i/dt = -Ck_iX_i \tag{6.12}$$

$$Ck_i = \sum_{j=1}^{6} A_j \exp[-E_j/R_GT] \tag{6.13}$$

where X_i is the concentration of kerogen of type i, Ck_i the reaction constant, E the activation energy, R_G the gas constant (1.986 cal K^{-1} mol^{-1}), A the frequency factor, T the temperature, and t the time, and i, j denote kerogen type and reaction type, respectively. Here up to six reaction types can be allowed for. Figures in Chapter 5 show the general scheme of active kerogen breakdown to oil, carbon dioxide (expressed by CO_2 in the following context), and natural gas (expressed by gas in the following context). The reaction constants can be related to each other as follows.

Assuming that the production of gas and CO_2 from oil can be represented by a single reaction with $E = E_g$ and $A = A_g$, the total rate constant for this case is

$$Ck^{(i)}_{g2T} \equiv Ck^{(i)}_{g2} + Ck^{(i)}_{D3} = A^{(i)}_g \exp[-E^{(i)}_g/R_GT)] \tag{6.14}$$

Since the molecular weights of kerogen and oil are much heavier than those of gas and CO_2 (16 and 44, respectively) we consider that the rate constant from kerogen to gas or CO_2 should be close to the rate constant from oil to gas or CO_2. Namely,

$$Ck_{D1} = Ck_{D2} = Ck_{g1} \tag{6.15}$$

and the direct channel breakdown of active kerogen of type i to CO_2 should be in proportionate representation to the initial fractional amount X; that is, we write

$$Ck_{D1} = X_e^{(i)}/(\sum_j^6 X_j^{(i)})Ck_{g2T} \tag{6.16}$$

where the third reaction type of kerogen is assumed to correspond to that of oil to gas generation as treated in Tissot and Welte (1978). Finally, it is anticipated that the ratio $Ck_{g2}^{(i)} : Ck_{D3}^{(i)}$ is roughly proportional to the amount of each molecule available. Four hydrocarbon molecules are consumed to produce gas (CH_4), whereas only two oxygen molecules are needed for CO_2. A typical type I kerogen has a hydrogen : oxygen ratio of about 18 : 1, a type II kerogen a ratio of about 7 : 1, and a type III kerogen a ratio of 2 : 1 (Hunt, 1979). Therefore the estimated values of $Ck_{g2}^{(i)} : Ck_{D3}^{(i)}$ are 9 : 1, 7 : 2, and 1 : 1 for type I, II, and III kerogens, respectively.

In addition to this, we assume that the rate constant of the reaction from kerogen to bitumen is equal to that from bitumen to oil. We write

$$Ck_A^{(i)} = Ck_o^{(i)} \tag{6.17}$$

From the above discussion, we can determine all the rate constants if the values of frequency factors ($A_j^{(i)}$, activation energies ($E_j^{(i)}$), and genetic potentials ($X_j^{(i)}$) are given for each type of kerogen.

A. Generation Pressure

The addition of mass Δm_0 to the fluid in a time Δt attributed to oil generation (O) is

$$\Delta m_0 = \Delta t \, dO/dt \tag{6.18}$$

Also, the additions of mass to the fluid by gas and CO_2 generation are

$$\Delta m_g = \Delta t(dG/dt)SS_g e_L \tag{6.19}$$

$$\Delta m_D = \Delta t(dD/dt)SS_D e_L \tag{6.20}$$

where SS is the solubility of gas, e_L is the void ratio of total liquid phase, t is time duration, and subscripts g and D denote gas and CO_2, respectively. Therefore, the total mass increase in liquid phase is

$$\Delta m_L = \Delta m_0 + \Delta m_g + \Delta m_D \tag{6.21}$$

The change in void caused by this liquid mass is

$$\Delta e_L = (\Delta m_L/\rho_L)(1 - C_p\, PR) \tag{6.22}$$

where ρ_L is the density of liquid, C_p the liquid compressibility, and PR the pressure.

The increase of density in the liquid phase is

$$\begin{aligned}\Delta\rho_L &= (\rho_L e_L + \Delta m_0 + \Delta m_g + \Delta m_D)/(e_L + \Delta e_L) - \rho_L \\ &= (\Delta m_0 + \Delta m_g + \Delta m_D - \Delta e_L \rho_L)/(e_L + \Delta e_L)\end{aligned} \tag{6.23}$$

In addition, from the definition of compressibility in general we can write

$$C_p\, \Delta P_L = -(\Delta V/V) \tag{6.24}$$

where ΔP_L is the pressure increase by liquid generated, V the original volume of liquid, and ΔV the volumetric change by liquid generated. Here the weight should not change before and after compression by liquid generated:

$$\rho_L V = (\rho_L + \Delta\rho_L)(V + \Delta V) \tag{6.25}$$

Rearranging this we have

$$(\Delta\rho/\rho) = -[\Delta V/(V + \Delta V)] = -(\Delta V/V)/(1 + \Delta V/V) \tag{6.26}$$

Substituting equation (6.25) into (6.23), we obtain the formula for the pressure increase by liquid generated ($\Delta\rho_L$):

$$\Delta P_L = (1/C_p)(\Delta\rho_L/\rho_L)/(1 + \Delta\rho_L/\rho_L) \tag{6.27}$$

On the other hand, the additions of mass to the gas phase (not in solution) by gas and CO_2 generations are

$$\Delta m_g' = \Delta t(dG/dt) - \Delta m_g \tag{6.28}$$

$$\Delta m_D' = \Delta t(dD/dt) - \Delta m_D \tag{6.29}$$

Considering the increases in terms of concentration of gas and CO_2 in gas phase ($\Delta G'$ and $\Delta D'$), the equations become

$$\Delta G' = \Delta m_g'/M_g \tag{6.30}$$

$$\Delta D' = \Delta m_D'/M_D \tag{6.31}$$

where M is the total mass of molecules (16 g/mol for CH_4 and 44 g/mol for CO_2) and g and D denote gas and CO_2, respectively. Using van der Waals equation, the mass of gas occupies volume

$$\Delta e_g = \mathrm{fwv}(P, T, \Delta G') \tag{6.32a}$$

$$\Delta e_D = \mathrm{fwv}(P, T, \Delta D') \tag{6.32b}$$

where fwv represents van der Waals function connecting pressure, volume, and temperature. Calculate the increase in number density for gas (ΔN_g) and CO_2 (ΔN_D):

$$\Delta N_g = (\Delta G' - N_g\, \Delta e_g)/e_g \tag{6.33}$$

The increases in pressure in the gas phase within the time interval Δt due solely to the increase in number densities are

$$\Delta P_g = (1/2) f_g\, \Delta N_g kk\, T \tag{6.34a}$$

$$\Delta P_g = (1/2) f_g\, \Delta N_g kk\, T \tag{6.34b}$$

where f_g and f_D are the numbers of degrees of freedom of a gas molecule and a CO_2 molecule, respectively ($f_g = 6.452$, $f_D = 6.667$), kk is Boltzmann's constant ($= 1.38 \times 10^{-16}$ erg K^{-1}), and T is the temperature (K). The total vapor pressure is then given by

$$\Delta P_v = \Delta P_g + \Delta P_D + \Delta P_W \qquad (6.35)$$

where ΔP_W is the vapor pressure of water, as given in Haas (1978).

Finally, the total generation pressure (P_G) is given by a simple summation of pressure increases in liquid phase [equation (6.27)] and in gas phase:

$$P_G = \Delta P_L + \Delta P_V \qquad (6.36)$$

B. Effects of Temperature Variations

At a given position $x \equiv (\zeta, x)$, the subsurface temperature may vary for a variety of reasons. In addition to continued burial, there can be a change in the basement heat flux with time, and there can be a change in fluid flow which will convect heat away from, or toward, a particular subsurface position. Thermal expansion of the pore fluid, increasing the partial pressures of natural gas and carbon dioxide, and raising the liquid pressure will retard compaction and, at the same time, increase the rate of fluid flow. The thermal expansion of the granular matrix and the corresponding increase in pore volume only partially accommodate the increase in pore fluid volume.

The effects can be described as follows. Due *solely* to temperature variations, there will be an increase in void ratio of the rock at position x in the amount

$$1/e(\partial e/\partial t)_{\text{temp}} = \beta_T(\partial T/\partial t)_x \qquad (6.37)$$

where β_T is the logarithmic coefficient of thermal expansion of the rock at position $x \equiv (\zeta, x)$ at time t.

This increase in pore volume has two effects: it lowers the number density per unit volume of each component species, and it changes the partial pressures of the natural gas and carbon dioxide. The changes in partial gas pressures due to temperature variations are described through

$$(\partial p_g/\partial t)_{x,\text{temp}} = \tfrac{1}{2} f_g k N_g (\partial T/\partial t)_x \qquad (6.38a)$$
$$(\partial p_D/\partial t)_{x,\text{temp}} = \tfrac{1}{2} f_D k N_D (\partial T/\partial t)_x \qquad (6.38b)$$

while the change in number density of each gaseous species obeys

$$\frac{1}{N_g}\left(\frac{\partial N_g}{\partial t}\right)_{x,\ \text{temp}} = -\beta_T\left(\frac{\partial T}{\partial t}\right)_x = \frac{1}{N_D}\left(\frac{\partial N_D}{\partial t}\right)_{x,\ \text{temp}} \tag{6.39}$$

The change in fluid mass density ρ_L due to the change in temperature is then

$$1/\rho_L(\partial\rho_L/\partial_t)_{x,\text{temp}} = -\beta_L(\partial T/\partial t)_x \tag{6.40}$$

where β_L is the logarithmic coefficient of thermal expansion of the fluid at position x at time t. Since a change in temperature does not alter the total mass of liquid occupying the pore space, the fractional shift in e_L due solely to changes in temperature is given by

$$1/e_L(\partial e_L/\partial t)_{x,\text{temp}} = \beta_L(\partial T/\partial t)_x \tag{6.41}$$

Since the total change in pore volume must be completely occupied by the fluid, natural gas, and carbon dioxide, it follows that

$$\left(\frac{\partial e_g}{\partial t}\right)_{x,\text{temp}} + \left(\frac{\partial e_D}{\partial t}\right)_{x,\text{temp}} = (e\beta_T - \beta_L e_L)\left(\frac{\partial T}{\partial t}\right)_x \tag{6.42}$$

Again, on the assumption of equilibrium between natural gas and carbon dioxide in the gas phase, we have

$$\frac{1}{e_g}\left(\frac{\partial e_g}{\partial t}\right)_{x,\text{temp}} = \frac{1}{e_D}\left(\frac{\partial e_D}{\partial t}\right)_{x,\text{temp}} \tag{6.43}$$

Note that no statement has been made as yet about the causes for the changes in temperature with respect to time at a given position—such changes arise when one considers the evolution of temperature and fluid flow simultaneously, together with the possible variations of basement heat flux with time.

C. Two-Dimensional Heat Transfer

The flux of heat Q is to be related to the temperature gradient by means of the thermal conductivity. This law also refers to heat flowing through a noncompacting, nonmoving rock. Therefore, it is suitable to apply the ζ-coordinate system to heat transfer problems.

The amount of heat H stored in the material at position ζ is

$$H = T\,(\rho_L C_L e_L + \rho_s C_s) \tag{6.44}$$

where T is the temperature, ρ the density, C the specific heat, e the void ratio, and subscripts L and s stand for liquid and solid, respectively. The change of heat at a point (ζ, x) is due both to thermal diffusion and to convection since the fluid moves:

$$\frac{\partial H}{\partial t} = \frac{\partial}{\partial \zeta}\left[\kappa_z(1 + e)^{-1}\frac{\partial T}{\partial \zeta} + \frac{\partial}{\partial x}\left(\kappa_x \frac{\partial T}{\partial x}\right)\right] + \frac{\partial}{\partial \zeta}\left[\frac{\rho_L e K_z C_L T}{\mu(1 + e)}\right]\frac{\partial P}{\partial \zeta} + \frac{\partial}{\partial x}\left(\frac{\rho_L e K_x C_L T}{\mu}\right)\frac{\partial P}{\partial x} \qquad (6.45)$$

where κ is thermal conductivity, K is permeability, and z and x denote vertical and horizontal directions, respectively.

The first and second terms on the right-hand side of equation (6.44) express the amount of heat transferred by diffusion, and the third and fourth terms represent the heat transferred by fluid movement, defined here as "convection."

For boundary conditions, we take the temperature to be given at the sedimentary surface and a conductive balance of heat at the basement, i.e.,

$$\frac{\kappa_z}{1 + e}\frac{\partial T}{\partial \zeta} + \kappa_x \frac{\partial \zeta_b}{\partial x}\frac{\partial T}{\partial x} = Q\left[1 + \left(\frac{\partial \zeta_b}{\partial x}\right)^2\right]^{1/2} \quad \text{on } u = 0 \qquad (6.46)$$

where Q is the heat flow at basement and the other variables are the same as before. The heat flow (Q) is a function of the location (x) and the time (t) and must be prescribed.

This type of equation is basically the same as that for the fluid flow; therefore, a conversion to the fractional vertical coordinates is also appropriate.

Since the accumulated sediments are compressed solely in the vertical direction, in the ζ–x frame we can use Darcy's law so that the vector liquid flow, $\rho_L v$, is given through

$$-\rho_L v = \frac{K_z}{\mu(1 + e)}\rho_L \frac{\partial p}{\partial \zeta}\hat{\zeta} + \frac{K_x \rho_L}{\mu}\frac{\partial p}{\partial x}\hat{x} \qquad (6.47)$$

where $\hat{\zeta}$ and $\hat{x}$ are unit vectors in the ζ and x directions, respectively, and K_z and K_x reflect the difference in permeability due to the anisotropic nature of the compacting material. When combined with the thermal expansion and the mass increase to the fluid from the thermochemical evolution of kerogen in ζ–x coordinates, we can write the two-dimensional fluid migration equation in the form

$$\frac{\partial}{\partial t}(\rho_L e_L) = \frac{\partial}{\partial \zeta}\left[\frac{K_z \rho_L}{\mu(1+e)}\frac{\partial p}{\partial \zeta}\right] + \frac{\partial}{\partial x}\left(\frac{K_x \rho_L}{\mu}\frac{\partial p}{\partial x}\right) - \rho_L e_L \beta_L \frac{\partial T}{\partial t} + \frac{\partial m_L}{\partial t} \tag{6.48}$$

Equally, the thermal balance equation takes on the form

$$\frac{\partial H}{\partial t} - \frac{\partial}{\partial \zeta}\left[\frac{K_z e_L C_L T}{\mu(1+e)}\frac{\partial p}{\partial \zeta}\right] - \frac{\partial}{\partial x}\left(\frac{K_x}{\mu} e_L C_L T \frac{\partial p}{\partial x}\right)$$
$$= \frac{\partial}{\partial \zeta}\left(\frac{\kappa_z}{1+e}\frac{\partial T}{\partial \zeta}\right) + \frac{\partial}{\partial x}\left(\kappa_x \frac{\partial T}{\partial x}\right) + Q(x, \zeta, t)\delta[\zeta - \zeta_b(x, t)] \tag{6.49}$$

where κ_z and κ_x reflect the differences in thermal conductivity in the vertical and lateral directions, respectively, due to the anisotropic nature of the compacting material.

D. Microfracturing

Excess pressure P is the difference between the pore pressure and the hydrostatic pressure P_e at the point, and pore pressure is expressed as overburden pressure P_{ob} minus effective stress σ. Therefore, we write

$$P = (P_{ob} - \sigma) - P_e \tag{6.50}$$

The effective stress consists of two parts: the frame pressure P_f, which is the part of the effective overburden that can be supported by grain contacts transmitting stress without further compaction, and a negative component generated by the hydrocarbon generation process, P_G. Therefore, P should be expressed as follows:

$$P = (P_{ob} - P_e) - (P_f - P_{f*}) + (P_G - P_{G*})$$
$$= g\int_z^{z_*} \frac{\rho_s + e\rho_w}{1+e}\,dz' - g\int_z^{z_*} \rho_w\,dz'$$
$$- (P_f - P_{f*}) + (P_G - P_{G*}) \tag{6.51}$$

where g is the acceleration due to gravity (= 980 cm/s^2), ρ_s and ρ_w are the densities of sediment matrix and water, respectively, and $*$ denotes the value at the surface. Using the relation in equation (6.2), we have

$$P = g\int_\zeta^{\zeta_*}(\rho_s - \rho_w)\,d\zeta' - (P_f - P_{f*}) + (P_G - P_{G*}) \tag{6.52}$$

Furthermore, it is assumed that empirical laws for different lithologies relate frame pressure P_f to void ratio e and void ratio e to vertical permeability K_z as

$$e/e_{*sh} = (P_f/P_{f_*})^a \quad \text{for shale} \quad (6.53a)$$
$$e = e_{*ss} - c(P_f - P_{f_*}) \quad \text{for sandstone} \quad (6.53b)$$
$$K_z/K_{*sh} = (e/e_{*sh})^{b_{sh}} \quad \text{for shale} \quad (6.53c)$$
$$K_z/K_{*ss} = (e/e_{*ss})^{b_{ss}} \quad \text{for sandstone} \quad (6.53d)$$

where a, c, b_{sh}, and b_{ss} are constants and $*$, sh, and ss denote the value at surface and the values for shale and sandstone, respectively.

But there is a physical limit of internal pore pressure that subsurface rocks can contain. Du Rouche (1981) proposed that the maximum pore pressure P_{max} should be a function of overburden pressure P_{ob} and Poisson's ratio of the rock, τ, in the form

$$P_{max} = P_{ob}\,\tau/(1 - \tau) \equiv fP_{ob} \quad (6.54)$$

where typical values for τ are 0.2, 0.4, and 0.3 for sandstone, shale, and carbonate, respectively (Du Rouche, 1981). Empirical observations suggest the following form for shale (Lerche *et al.*, 1984):

$$P_{max} = 0.85P_{ob} \quad (6.55)$$

Either way we can determine the maximum value of pressure for each point. At the instant of fracture the initial value of fracture permeability is determined. The evolution of fracture permeability is to be modeled as being proportional to the cube of the fracture width and to approach infinity as the excess pressure approaches the effective overburden. Thus, the excess pressure is prevented from exceeding the excess overburden. Let the fracture permeability K_f be given by

$$K_f = \frac{K_{F*}e_{min}(e_f/e_{f*})^3}{2(P/P_{ob} - 1)^2}[\text{sgn}(P - fP_{ob}) + 1]\left(\frac{1}{f} - 1\right)^2 \quad (6.56)$$

where $\text{sgn}(P - fP_{ob}) = +1, 0$, or -1 according as $P > fP_{ob}$, $P = fP_{ob}$, or $P < fP_{ob}$, respectively. Here e_{f_*} is the value of the fracture void ratio at the initial instant of fracture. The factor $[\text{sgn}(P - fP_{ob}) + 1]/2$ turns K_f on and off as the excess pressure P crosses the fracture limit determined by the overburden fP_{ob}.

The factor $(1/f - 1)^2/(P/P_{ob} - 1)^2$ causes the fracture permeability K_f

to approach infinity as the excess pressure P approaches the limiting overburden pressure P_{ob}. This will limit the increase in excess pressure, honoring the physical limitation that pore pressure cannot exceed overburden pressure. The factor $(e_f/e_{f_*})^3$ represents the modeling of the effect of fracture width ω on permeability K_f, where $K_f \propto \omega^3$. The initial fracture permeability $K_{F_*}(e_{min})$ is scaled to provide a smooth transition from interstitial permeability to fracture permeability.

The evolution of the fracture void ratio e_f is to be determined by the net effect of divergence of flow, thermal expansion, and hydrocarbon generation within the region of fracture. The fracture permeability is, then, a function of the increasing width of fracture, depending on the assumed spacing between fractures (fracture density).

When the effective stress again starts to increase, due either to the addition of overburden or to the decrease in pore pressure attributed to the loss of fluid through the fractures, we expect the fractures to begin to close. The closure of the fractures may be due to precipitation of dissolved minerals or to mechanical compaction. In either case, we assume closure of the fractures corresponds to the point at which effective stress is again increased to the maximum evolved effective stress (frame pressure).

In due course, the drop in pore pressure and corresponding increase in effective stress will tend to close the fractures, but only when the effective stress exceeds the previous maximum will intergranular compaction be resumed. However, the decrease first of fracture permeability and then of intergranular permeability is likely to result in a second cycle of fracture generation.

In summary, fracture evolution is simulated by its effects on formation permeability. For pore pressure P_p lower than a critical fraction f of the lithologic overburden pressure P_{ob}, fractures are closed; for $P_p \geqslant fP_{ob}$ the formation permeability is modeled by

$$K_f = K + K_s\,(1.0 - P_c/P_p)^2 \qquad P_p \geqslant fP_{ob} \tag{6.57a}$$

$$K_f = K \qquad P_p < fP_{ob} \tag{6.57b}$$

where K is the permeability in the absence of fractures, K_s is the permeability of sand, P_p is the total pore fluid pressure (hydrostatic plus excess), and $P_c = fP_{ob}$ is a critical pressure (Eaton, 1969; Palcianskas and Domenico, 1980), normally taken to be around 0.8–0.9 of the lithology overburden. We use 0.85 as a default value.

E. Cementation and Dissolution

We simulate the effects of cementation and dissolution indirectly in terms of the evolution of formation permeability with

$$K = K_p \qquad \text{in } t < T_1 \qquad (6.58a)$$

$$K = K_p\left[1.0 - \left(\frac{t - T_1}{T_2 - T_1}\right)C(-D)\right] \qquad \text{in } T_2 \geqslant t \geqslant T_1 \qquad (6.58b)$$

$$K = K_p[1.0 - C(-D)] \qquad \text{in } t > T_2 \qquad (6.58c)$$

where $C(D)$ is the fractional degree of cementation (dissolution), T_1 (T_2) is the time at which either cementation or dissolution starts (ends), and K_p is the formation permeability immediately prior to cementation (dissolution) onset.

It can, of course, be argued that this is an extremely primitive method of accommodating cementation and/or dissolution processes. One might suggest that use of the dependences of mineral concentrations on solubility, temperature, pressure, activity, mobility, etc. would be a much preferred path to follow. While such an approach is, indeed, preferable in principle, from a pragmatic point of view it adds a complexity to the basic problem which is not justified by data availability. The point is that one would have to track mineral concentrations for all times and depths of interest, including allowing for variable depositional amounts, saturation in mixed concentrations of varying minerals, ion replacement effects in phases, etc. Such a program of research would be a major undertaking which is hardly justified by the imprecision of our knowledge of the constitutive equations or of the input data. We believe that the simple, primitive rule above is a good method of providing an estimate of the dominance, or otherwise, of diagenetic effects on the flow of fluid in sedimentary basins.

F. Porosity, Permeability, and Thermal Conductivity

The porosity calculation for a shale and sandstone mixture in the model is as follows: Given the P_f value, the system calculates the void ratios for shale and sandstone; then the ratios are weighted according to their lithology:

$$e = V_{sh}e_{sh} + (1 - V_{sh})e_{ss} \qquad (6.59)$$

where e is the calculated void ratio for the mixture of shale and sandstone, V_{sh} is the volumetric fraction of shale to the total rock volume and subscripts sh and ss represent the values for shale and sandstone, respectively. Porosity ϕ is simply derived from the void ratio.

$$\phi = e/(1 + e) \tag{6.60}$$

If values are assigned for the "effectiveness" factor for cementation (Cmf), the void value is multiplied by Cmf so that the cementation process is performed.

To calculate vertical permeability (thermal conductivity) K_z, a random distribution of shale and sandstone particles is assumed (Lewis and Rose, 1970).

$$K_z = K_{sh}^{V_{sh}} K_{ss}^{(1 - V_{sh})} \tag{6.61}$$

where K_{sh} and K_{ss} represent the permeabilities (thermal conductivities) of pure shale and sand, respectively, and are parameters for calculation.

G. Buoyancy and Migration

Since we have a system composed of three separate phase flows in the real world, the fluid flow equation is solved three times, corresponding to water, oil, and gas-phase flows. However, solving takes much computational time and is uneconomic. We need to keep a simple structure for the model, so we make the following assumptions to approximate fluid migration:

1. The total void change is calculated assuming the whole fluid consists of water.
2. However, the amount of flow is calculated in three separate phases; the rate proportional to

$$q_w : q_o : q_g = \frac{Kk_w}{\mu_w} \Delta P_w : \frac{Kk_o}{\mu_o} \Delta P_o : \frac{Kk_g}{\mu_g} \Delta P_g \tag{6.62}$$

where K is permeability, k is relative permeability, μ is viscosity, and w, o and g denote water, oil, and gas, respectively. For the pressure difference for each phase (ΔP_o, ΔP_g, ΔP_w), the buoyancy effect is taken into account as

$$\Delta P_o = \Delta P + g(\rho_w - \rho_o)\,\Delta\text{Dep} \tag{6.63a}$$
$$\Delta P_g = \Delta P + g(\rho_w - \rho_o)\,\Delta\text{Dep} \tag{6.63b}$$
$$\Delta P_w = \Delta P - g(2\rho_w - \rho_o - \rho_g)\,\Delta\text{Dep} \tag{6.63c}$$

where g is the acceleration of gravity, ΔP the difference in excess pressure, ρ the density, and ΔDep the difference in physical depths.

3. The capillary pressure P_c is calculated by

$$P_c = 2\tau_s(1/R_{c1} - 1/R_{c2}) \tag{6.64}$$

where τ_s is the interfacial tension as a constant variable (the default values may be 70 dyn/cm for oil and 30 dyn/cm for gas) and R_c is the pore throat diameter. Furthermore,

$$R_c = \tfrac{1}{2}\,\mathrm{COEF}\,d_m \tag{6.65}$$

where d_m is the grain diameter and COEF the coefficient for the pore throat. If we assume packing of equal-sized spheres, COEF is theoretically related to porosity ϕ as follows (Nakayama, 1986)

$$\mathrm{COEF} = \mathrm{RRAA}\,\phi^2 + \mathrm{RRBB}\,\phi \tag{6.66}$$

where the constants are RRAA = 1.92, RRBB = 0.0882. Capillary pressure is always exerted from the smaller pore throat (R_{c1}) to the larger pore throat (R_{c2}).

4. After calculating the amount of each fluid that has moved into or out of each volume element, the flow rates are adjusted so that the total amount remaining equals the void calculated in the previous step 1.

5. The irreducible water cannot be replaced by any oil or gas; therefore, oil and gas saturations do not exceed (1 − Swirr), where Swirr is the irreducible water saturation given as an input variable.

Appendix

A. Explicit Functions, Examples

Units are centimeters, seconds, grams, and dynes.

1. Void ratio is a function of frame pressure:

Shale: $e/e_* = (p_f/p_{f_*})^A$, where A may equal $-1/3.5$

Sand: $e = e_* - C(P_f - P_{f_*})$, where C may equal 0.7×10^{-9}

2. Permeability is a function of void ratio:

Shale: $K/K_* = (e/e_*)^B$, where B may equal 4.0
Sand: $K/K_* = (e/e_*)^B$, where B may equal 6.4

	K_*	e_*	p_*
Shale	10^{-12}	3	6.9×10^4
Sand	0.275×10^{-6}	0.612	1

B. Parameters Defined

t ≡ time, in seconds

z ≡ vertical coordinate, elevation above basement, increasing upward to the seafloor

ζ ≡ transformation of the vertical coordinate, representing the thickness of rock, exclusive of void space, increasing upward from the basement:

$$z = \int_0^{\zeta} (1 + e)\, d\zeta$$

$\zeta_*(t, x)$ ≡ value of ζ at the surface of the sediment accumulating at the seafloor

u ≡ normalized coordinate representing elevation above the basement relative to elevation of the seafloor above the basement in the ζ coordinate: $u = (\zeta - \zeta_b)/(\zeta_*(t, x) - \zeta_b(x, t))$

x ≡ lateral coordinate of true length in a frame of reference that employs ζ as the vertical coordinate

e ≡ void ratio, the ratio of void/grain volume

$e_*(t, x)$ ≡ void ratio at the surface of the sediment accumulating at the seafloor; changes with time and lateral position as composition of sediment changes with time and lateral position

V ≡ fractional parameter representing the void ratio relative to the initial void ratio e_* at seafloor: $V = e/e_*(t, x)$

K_z ≡ permeability perpendicular to x, as commonly defined relative to a fixed volume of rock (cm^2), flow rate = $(K/\mu)(\partial p/\partial \zeta)(1 + e)^{-1}$

K'_z ≡ permeability with respect to the ζ coordinate and incorporating viscosity μ: $K'_z = K_z(e)[\mu(1 + e)]^{-1}$

K_x ≡ permeability, parallel to x as above

K'_x ≡ permeability including effect of viscosity: $K'_x = K_x/\mu$
p ≡ fluid pressure in excess of hydrostatic, (dyn/cm^2), dependent on effective stress and overburden
P ≡ total fluid pressure (dyn/cm^2)
σ ≡ effective stress (dyn/cm^2), the sum of vertical forces between grains, across a unit area, dependent on void ratio; since compaction is nearly irreversible, an increase in excess pressure or decrease in overburden can cause a decrease in effective stress with no corresponding decrease in void ratio
μ ≡ absolute viscosity expressed as a function of temperature, salinity, and pressure (dyn s/cm^2)
ρ_f ≡ density of pore fluid (g/cm^3), expressed as a function of temperature, salinity, and pressure
ρ_s ≡ average grain density of the rock (e.g., 2.65 g/cm^3 for quartz); may vary with time and position as composition of accumulating sediment changes
g ≡ acceleration due to gravity (cm/s^2)
T ≡ temperature (K)
F ≡ mass flow per unit area

7

Applications of Fluid Flow, Compaction, and Thermal Indicators to Single-Well and Basinal Settings

I. The Navarin Basin COST No. 1 Well

For the Navarin COST No. 1 well, most of the detailed geological, geophysical, and geochemical data are reported in Geological and Operational Summary, Navarin basin COST No. 1 Well, Bering Sea, Alaska, by Turner *et al.*, (1984) (OCS Report MMS 84-0081).

Navarin basin COST No. 1 well, completed on October 22, 1983, was the first well drilled in the remote Navarin basin, Bering Sea, Alaska (Fig. 7.1). The Navarin basin, a Tertiary sedimentary basin on the outer continental shelf of the Bering Sea, is nearly 483 km (300 miles) west of mainland Alaska and covers an area about the size of Maine, 83,400 km^2 (32,215 square miles). The northwest boundary of the basin lies within the disputed area along the international border between the United States and the Soviet Union. Shelf water depths range from less than 97 m (320 ft) to 152 m (500 ft) at the shelf break. The basin is delimited by the northwest boundary continental shelf break on the southwest and by shallow Mesozoic basement rock on the southeast, northeast, and northwest (Steffy *et al.*, 1985).

The Navarin basin COST No. 1 well is located in the east central part of the basin, reaching a depth of 4,999 m (16,400 ft). The sedimentary section penetrated by the COST well is composed of Neogene and Paleogene clastic sedimentary rocks deposited in a marine environment and Cretaceous clastic sediments deposited in fluvial, floodplain, and deltaic environments. The Cretaceous and Tertiary sedimentary rocks are separated by a diachronous angular unconformity. In the interval between 3,901 m (12,800 ft) and 4,587 m (15,048 ft), rocks of late Cretaceous age

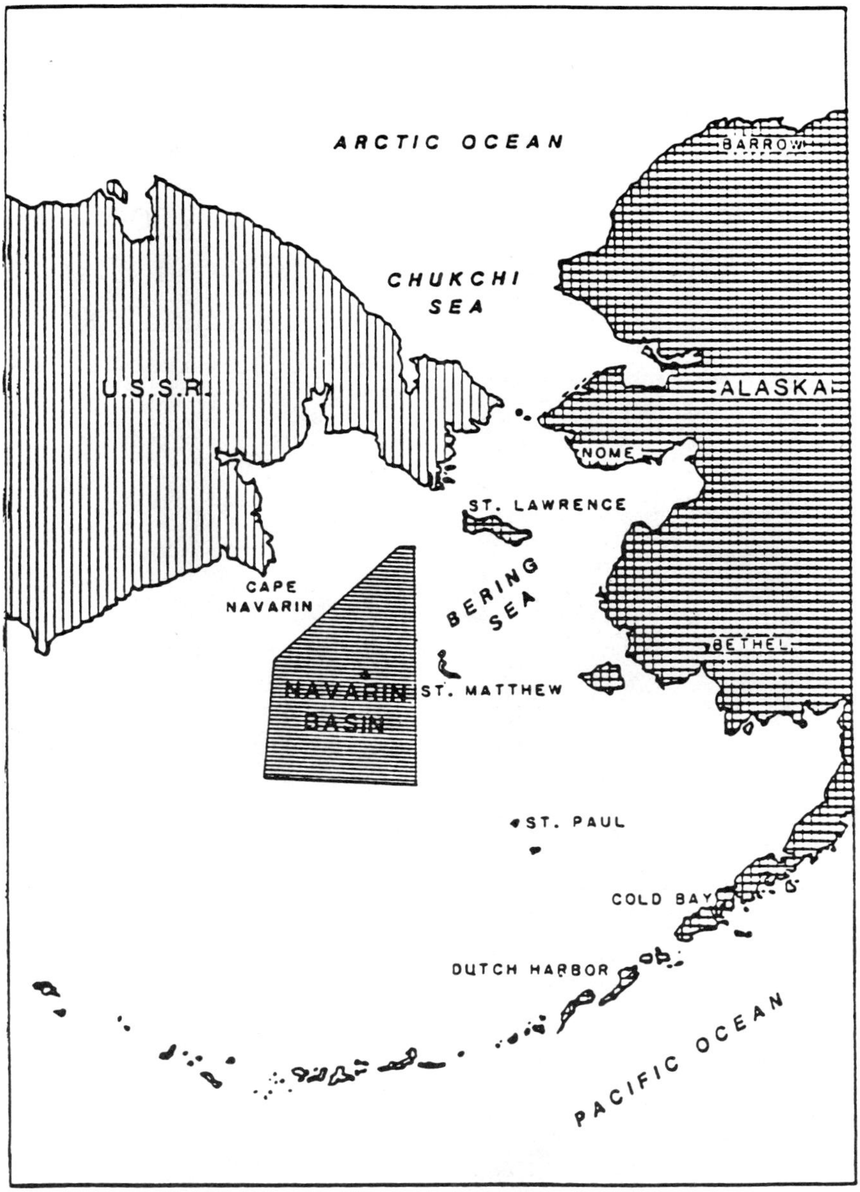

FIGURE 7.1 Location of the Navarin basin COST No. 1 well.

are intruded by numerous basaltic and diabasic sills. Intrusion probably occurred during early Miocene and late Oligocene time (Turner *et al.*, 1984). The penetrated section is divided into nine zones, designated A through I on the basis of lithology, deposition environments, diagenetic alteration, and well log characteristics (Fig. 7.2). Lithologic zone A

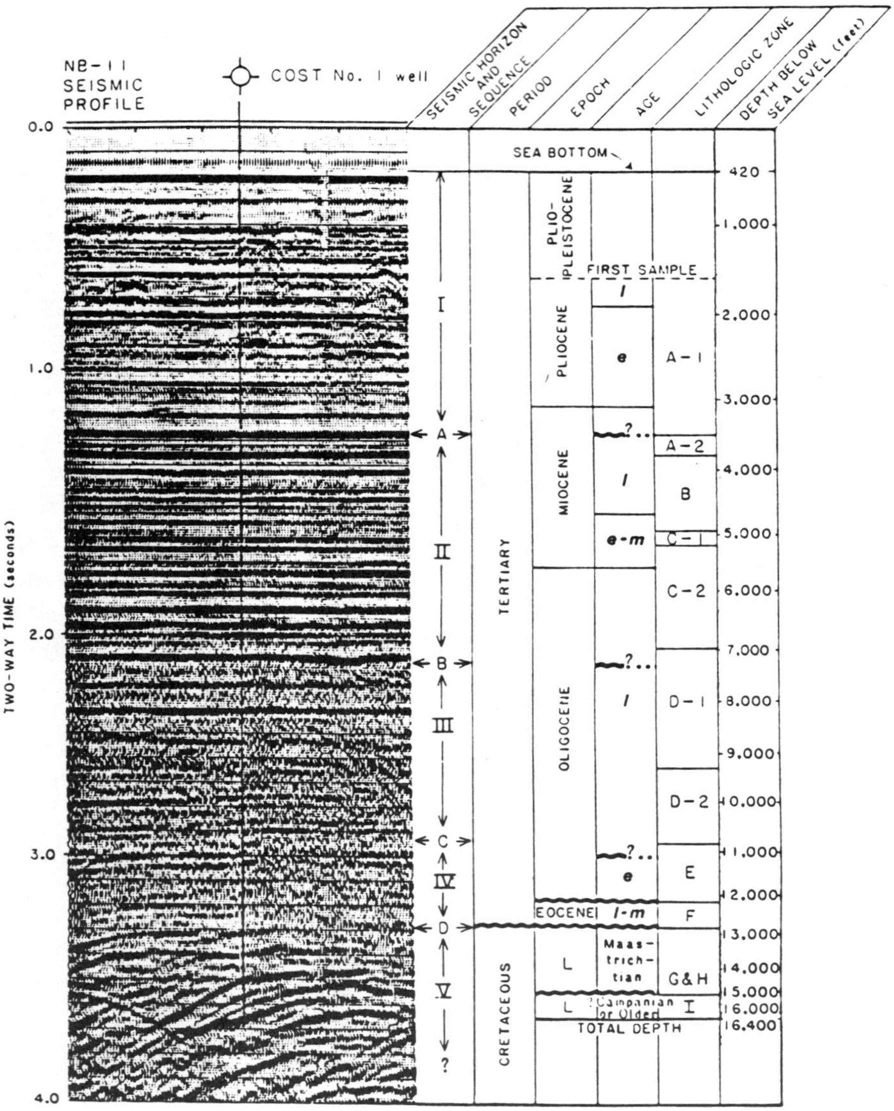

FIGURE 7.2 Seismic profile across the region containing the Navarin basin COST No. 1 well and corresponding geologic column from the well.

consists of poorly sorted, silty, sandy mudstone and diatomaceous ooze. Zone B consists of bioturbated, muddy, very fine and fine-grained sandstone interbedded with sandy mudstone. Zone C is composed of fine-grained and very fine-grained muddy sandstone and siltstone interbedded with mudstone and claystone. Zone D is characterized by sandy mudstone, fine-grained muddy sandstone, and claystone with rare lenses of siltstone and sandy carbonate. Zone E consists of poorly sorted gray claystone, mudstone, and sandy mudstone with abundant detrital clay matrix. Zone F consists of dark gray calcareous claystone and sandy mudstone. Zones G/H are characterized by siltstone, very fine-grained sandstone, mudstone, claystone, and coal and are intruded by diabase and basalt sills. Zone I is composed of claystone, siltstone, tuff, and mudstone deposited in a marine environment.

There is a major diachronous angular unconformity between late Cretaceous and younger Tertiary sedimentary rocks. The vitrinite reflectance data are abnormally high in the late Cretaceous section below the unconformity because of the thermal intrusion, which makes it difficult to simulate the thermal history of these sediments. Accordingly, the inverse model is applied only to the sequence above the unconformity, that is, the Cenozoic rocks. Table 7.1 lists the basic input data of the Navarin COST No. 1 well. The Cenozoic sequence is split into 13 modeling layers from bottom to top, based on lithology, paleontology, biostratigraphy, and deposition environments. Layers 2 and 4 represent unconformities and are treated as hiatuses, marking eustatic events (Turner *et al.,* 1984). The temperature at the sediment–water interface and the bottom-hole temperature are based on the temperature data of Fig. 7.3. The vitrinite reflectance data with depth are shown in Fig. 7.4. The kerogen content data are generalized from the geochemical data in Turner *et al.* (1984).

A. Comparison of the Modeled Results with Real Data

Based on the data available, the following comparisons are made of the model results with the observed data (Cao and Lerche, 1988).

1. Thickness

Table 7.2 gives the formation thickness changes of each modeled layer during compaction. The table shows that the thickness of each layer decreases after its deposition time and that the compaction rate is different at different compaction stages. At the early (shallow depth) stages of compaction, the sediment thickness decreases faster than at the later (deeper depth) stages of compaction. For example the compaction

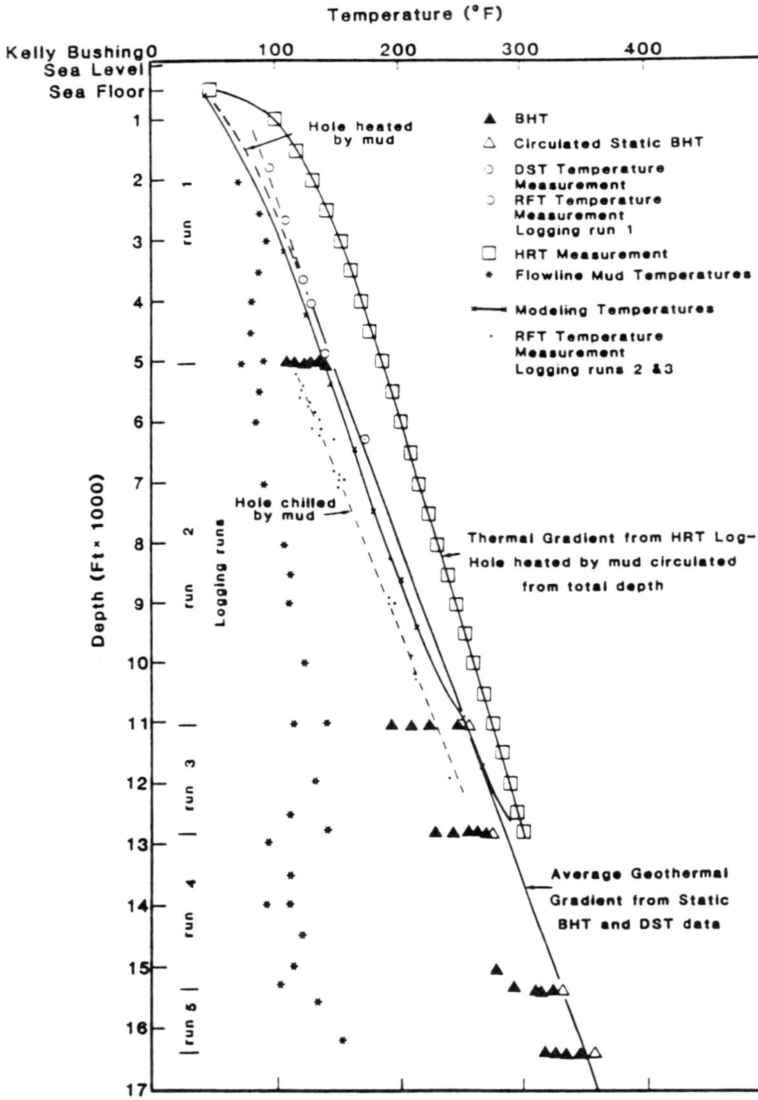

FIGURE 7.3 Several measured present-day temperature profiles with depth. In order to obtain a present-day heat flux, we used the modeling temperature–depth curve shown. This curve was obtained to satisfy the best average geothermal gradient.

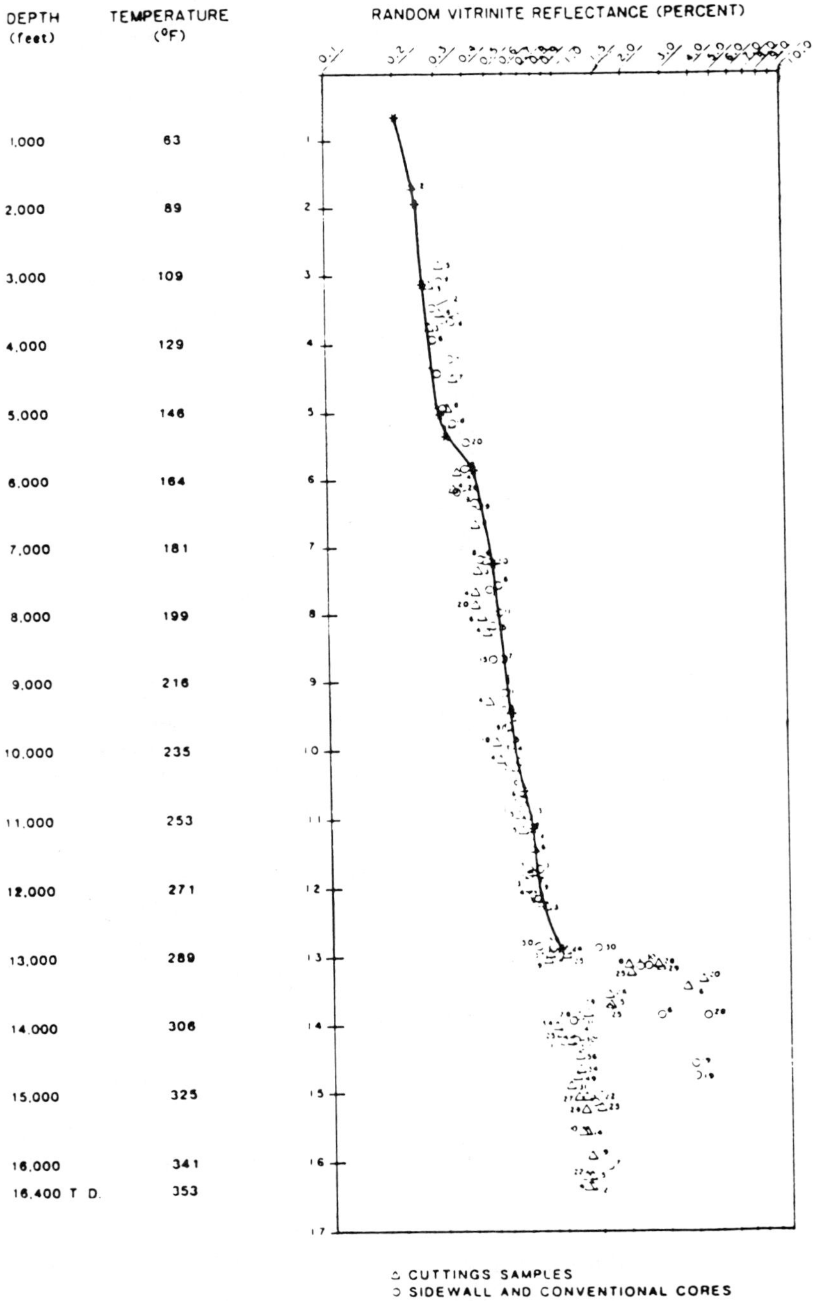

FIGURE 7.4 Plot of vitrinite reflectance observations with depth (symbols). Superimposed is a solid curve representing the predicted vitrinite reflectance obtained by the thermal inversion procedure.

Table 7.1 Basic Input Data for Navarin Basin COST No. 1 Well

Layer	Depth at base of formation (ft)[a]	Age at base (MYBP)[b]	Lithology[c]	Paleowater depth (ft)[c]	Kerogen content[d]	
					Type II	Type III
13	1,920.0	3.4	Mudstone	432	—	—
12	3,180.0	5.3	Sandy mudstone	300	—	—
11	3,860.0	7.9	Fine-grained muddy sandstone	600	—	—
10	5,010.0	18.6	Fine-grained muddy sandstone	600	0.4	0.6
9	5,360.0	21.2	Fine-grained muddy sandstone	600	0.4	0.6
8	5,704.0	23.7	Fine-grained muddy sandstone	600	0.4	0.6
7	7,130.0	24.8	Fine-grained muddy sandstone	300	0.4	0.6
6	9,450.0	26.7	Sandy mudstone	600	0.4	0.6
5	11,100.0	28.0	Mudstone	1,500	0.4	0.6
4	11,100.0	32.8	Hiatus	0	0.0	0.0
3	12,280.0	34.0	Mudstone	1,500	0.5	0.5
2	12,280.0	38.0	Hiatus	0	0.0	0.0
1	12,780.0	45.0	Mudstone	600	0.7	0.3

[a] Depths are measured from the Kelly Bushing, which was 85 ft above mean sea level.
[b] Age is based on the Geological Society of America (GSA) 1983 geologic time scale.
[c] Lithology and paleowater depth and kerogen content are based on OCS Report MMS 84-0031 (Turner *et al.*, 1984).
[d] No type I kerogen recorded.

rate of layer 1 (shale) decreases from 4.25 m/Myr (13.9 ft/Myr) at the end of layer 4 deposition to 2.71 m/Myr (8.9 ft/Myr) at the end of layer 10 deposition. The total modeled thickness of the whole sequence at the present day is 3729.0 m (11,760.4 ft) and the true thickness is 3737.7 m (11,763.4 ft), with a relative error of about ±1% for individual layers. Table 7.2 shows similar predictions of layer thicknesses for all subsequently deposited layers.

2. *Porosity*

Figure 7.5 shows present-day porosity versus depth for the Navarin COST No. 1 well. The solid line is the porosity curve obtained from the dynamic model and the points are measured porosity values for 114 samples from sidewall and conventional cores. No measured porosity

Table 7.2 Thickness Change in Meters of Each Modeled Layer with Time (Myr).[a]

Time (Myr)	Layer[b] 1	3	5	6	7	8	9	10	11	12	13	Lithology
7	230.0											Shale
11	229.4											Unconformity, depositional hiatus
13	196.8	477.2										Shale
17	187.5	436.8										Unconformity, depositional hiatus
18.3	181.0	401.1	691.5									Shale
20.2	171.3	385.0	558.1	800.1								Sandy, mud
21.3	164.4	373.4	530.1	734.1	553.5							Sandy
23.8	161.3	368.3	528.2	729.8	514.6	157.1						Sandy
26.4	159.1	364.8	524.6	726.8	508.7	123.3	154.0					Sandy
37.1	153.3	357.0	515.5	717.5	487.1	118.3	115.8	426.8				Sandy
39.7	151.4	355.9	513.6	712.4	476.0	115.5	113.1	380.5	261.5			Sandy
41.6	148.0	355.1	511.3	705.0	452.9	109.6	107.3	361.8	211.3	440.5		Gray shale
45.00	144.6	353.4	508.7	701.8	442.2	107.0	104.8	353.6	206.7	375.0	431.2	Mudstone
Measured	152.4	359.7	502.9	707.1	434.6	104.8	106.7	350.5	207.3	384.1	427.6	

[a] Total thickness (present day): 3729 m predicted, 3737.7 m measured.
[b] Layers 2 and 4 are depositional hiatuses characterized by unconformities.

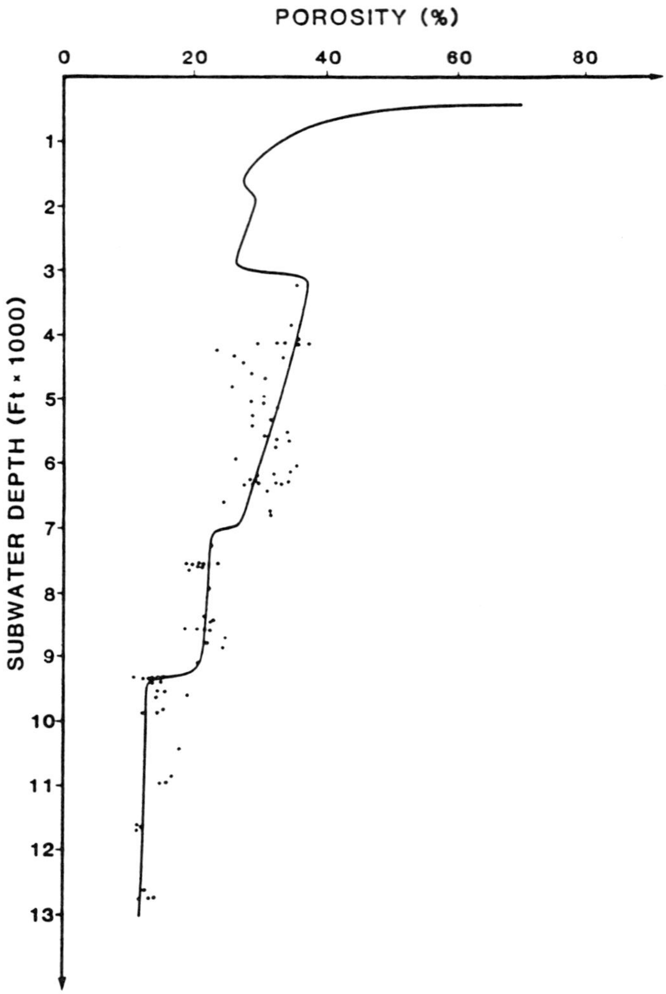

FIGURE 7.5 Porosity measurements with depth in COST No. 1 (dots) and the predicted porosity variation (solid line) obtained by running the fluid flow/compaction code.

data are available in the interval less than 914 m (3000 ft) deep. It is apparent that the modeled porosity fits the measurements rather well from 914 m (3000 ft) to 3962 m (13,000 ft). Table 7.3 gives a comparison of modeled porosity and measured porosity for each layer, showing a maximum fractional error of 17% and an average fractional error of order 10%.

Table 7.3 Comparison of Modeled Porosity with Measured Porosity

	Porosity (%) for layer										
	1, shale	3, shale	5, shale	6, sandy shale	7, muddy sand	8, muddy sand	9, muddy sand	10, muddy sand	11, muddy sand	12, sandy mud	13, mud
Modeled average porosity	12.1	12.4	12.7	21.9	29.2	32.0	33.1	35.1	37.1	28.2	36.5
Measured average porosity	12.9	11.5	15.2	18.7	37.0	31.5	31.5	31.0	35.1	—	—

3. Formation Pressure

One of the major differences between an isostatic model and a dynamic model is that the dynamic model can simulate the evolution of formation pressure. It is important to examine the degree to which the modeled formation pressure with depth is in accord with the measured formation pressure. The existence of abnormally high pore pressure in the Navarin COST No. 1 well has been shown by wireline logs, formation drill stem tests, and the drilling exponent method (Turner *et al.*, 1984). The wireline log and formation test data clearly identify a major zone of abnormal formation pore pressure in the interval below 2874 m (9430 ft) in the Navarin Basin COST No. 1 well. A shallow zone of excess pore pressure between 762 m (2500 ft) and 1170 m (3840 ft) also appears to be present.

Figure 7.6 shows the pressure profile for the Navarin basin COST No. 1 well. The thick dashed line is the modeled formation pressure curve and shows two abnormal pore pressure zones. The deep zone starts at a depth of about 2804 m (9200 ft) and extends to the intrusive unconformity. A reservoir formation test (RFT) at 3098 m (10,163 ft) (subsea depth) within the zone yielded 362 atm (5320 psi) on pressure buildup—that is, 49 atm (726 psi) greater than the anticipated 313 atm (4594 psi) if the well had been normally pressured. The modeled pressure at the same depth is 358 atm (5256 psi), 45 atm (662 psi) greater than anticipated for a normally pressured section. The shallow abnormal pressure zone is predicted by the dynamic model to occur in the interval of 305 m (1000 ft) to 1158 m (3800 ft)—as seen in Fig. 7.6. (Details of the effective stress law used are provided in the Appendix to Section I of this chapter, as is the pragmatic way of determining the one unknown constant in the permeability–void ratio law and a similar constant in the effective stress–void ratio law).

4. Permeability Variations with Depth

Figure 7.7 shows the logarithm of measured formational permeability variations (in millidarcies) with depth (asterisks). Superimposed on the figure is the permeability variation predicted with the fluid flow/compaction model, which allows for dynamical evolution (see the Appendix to this section for details of the permeability law used). To be noted from Fig. 7.7 is the excellent agreement of the predicted variation and the measured values—in a general trend sense. In particular, the agreement between 914 m (3000 ft) and 3048 m (10,000 ft) is extremely satisfactory. Below about 3048 m (10,000 ft) the dominant lithology is shale with thin interbedded sand layers. The measurements of permeabil-

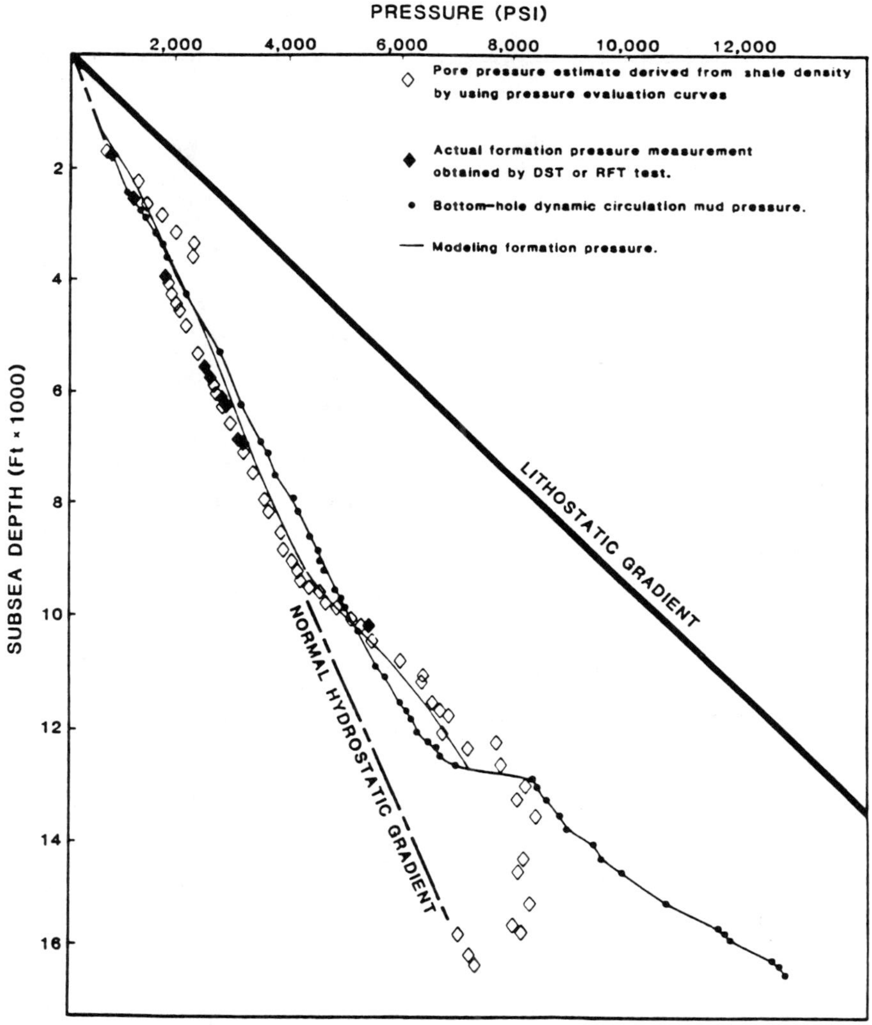

FIGURE 7.6 Pore pressure measurements with depth. The three different symbols represent different methods of estimating pore pressure, as explained in Turner *et al.* (1984). The light solid curve is the predicted variation of pore pressure with depth obtained by running the fluid flow/compaction code.

ity below 3048 m (10,000 ft) are, of course, made on the sand layers, as potential hydrocarbon reservoir sites. From the observed high overpressure we are fairly certain that the *bulk* permeability is quite small below 3048 m (10,000 ft) since overpressuring continues to increase. Hence the

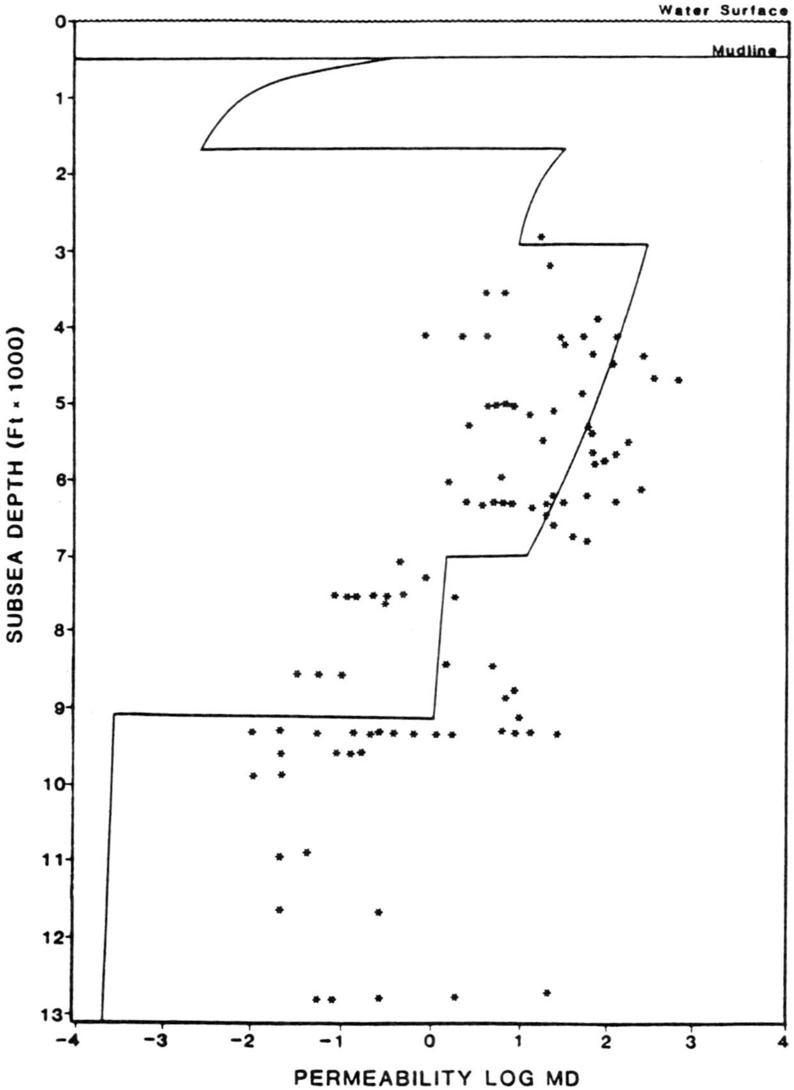

FIGURE 7.7 Permeability variations with depth. Asterisks denote measured values. Between about 2000 and 9000 ft, the dominant lithology is sand with shale stringers. Above about 2000 ft and below about 9000 ft, the dominant lithology is shale with sand stringers. Since the permeability measurements are made in the sand stringers below about 9000 ft, they misrepresent by a large factor the bulk permeability of the dominant shale unit. The solid line is the predicted variation of permeability with depth obtained by running the fluid flow/compaction code.

thin sand permeability measurements can be misrepresenting the overall variation of bulk permeability.

Above 914 m (3000 ft) there are, apparently, no permeability measurements, but the shallow zone of overpressuring observed, and also predicted by the fluid flow/compaction model, indicates that the permeability must be fairly low in order to maintain the shallow overpressure. The predicted variation of permeability in the shallow [≤610 m (2000 ft)] zone is, therefore, consistent with the overpressure observations.

When it is borne in mind that the individual formation thicknesses with depth, porosity variations with depth, fluid overpressuring with depth, and permeability variations with depth are all correctly obtained (with only two "free" numbers adjusted to minimize discordances between observations and predictions, as outlined in the Appendix), it seems that fluid flow/compaction modeling is a ruggedly stable tool capable of being used in a predictive capacity with relative accuracy in predicted bulk properties.

5. Temperature

Figure 7.3 shows the temperature profile for the well. The modeled temperature variation with depth fits the measured temperature data well. The modeled temperature curve shows two zones of high thermal gradient, consistent with the abnormally high-pressure zones (shales) and the one normal-pressure zone (sandstone). Within the shallow overpressure zone [above 1158 m (3800 ft)] the average predicted thermal gradient is 4.0°C/100 m (2.2°F/100 ft) [cf. 4.6°C/100 m (2.5°F/100 ft) in Turner *et al.*, 1984]; within the normal pressure-zone [1158–2804 m (3800–9200 ft)] the average predicted thermal gradient is 3.3°C/100 m (1.79°F/100 ft); and within the deep overpressure zone the average predicted thermal gradient is 4.7°C/100 m (2.58°F/100 ft). In Turner *et al.* (1984) an average thermal gradient of 3.2°C/100 m (1.78°F/100 ft) from 1158 m (3800 ft) to 4999 m (16,400 ft) was estimated without considering the thermal gradient change in the deep overpressured zone. The reason for the higher thermal gradient in the overpressured zones is that the thermal conductivity of overpressured sedimentary rocks is lower than that of normally pressured sedimentary rocks, because the overpressured zone has an higher fluid content and water has a lower average thermal conductivity than fully compacted matrix. Hence the temperature gradient increases to balance the heat flux through the sediments at each instant of time.

6. Vitrinite Reflectance

Figure 7.4 shows the vitrinite reflectance profile of the well. The dark line is the predicted vitrinite reflectance with depth based on the inverse methods, which provide a variable heat flux with time, stopping at the unconformity mentioned earlier. The curve in Fig. 7.4 generally reflects the change of measured vitrinite reflectance with depth above the unconformity. To predict the abnormally high vitrinite reflectance below the unconformity an intrusion thermal model is needed (Wang *et al.*, 1989).

B. Geohistory, Thermal History, and Hydrocarbon Generation History of the Navarin Basin COST No. 1 Well

1. Geohistory

Figure 7.8 shows the geohistory of the Cenozoic sequence in the Navarin Basin COST No. 1 well. Since the deposition of dark gray calcareous claystone and sandy mudstone in the Eocene, the basin has undergone two major uplifts: one between the Eocene and the early Oligocene, and the other at the early Oligocene (Turner *et al.*, 1984). During these two uplifts, coarse-grained sediments were deposited in shallow water. After the early Oligocene uplift, the basin subsided quickly during the rest of the Oligocene; this was followed by a slow subsidence with a nearly constant subsidence rate during the Miocene, when the depocenter was depositing dominantly regressive sandstone, which forms the most promising reservoir rock (Steffy *et al.*, 1985). From Pliocene to present, the basin has undergone subsidence with fine-grained material deposition. Figure 7.9 shows the basement subsidence with time, assuming a basement depth of 9144 m (30,000 ft) (Steffy *et al.*, 1985). The dark line in Fig. 7.9 is the total basement subsidence; the dotted line is the basement subsidence caused by tectonism. It can be seen that the tectonic basement subsidence is dominant from Eocene to Oligocene, and the sediment loading effect played a more dominant role thereafter in influencing the basement subsidence. Figure 7.10 gives the rates of total basement subsidence, tectonic basement subsidence, and sedimentary deposition with time. The curves of Fig. 7.10 show a maximum rate during the Oligocene and a nearly constant rate during the Miocene.

In summary, examining the whole history of fluid flow and compaction in the COST No. 1 well, we find that the Eocene shale (zone F) and early Oligocene shale (zone E) were overpressured soon after their deposition and these zones are predicted to have undergone fracture formation

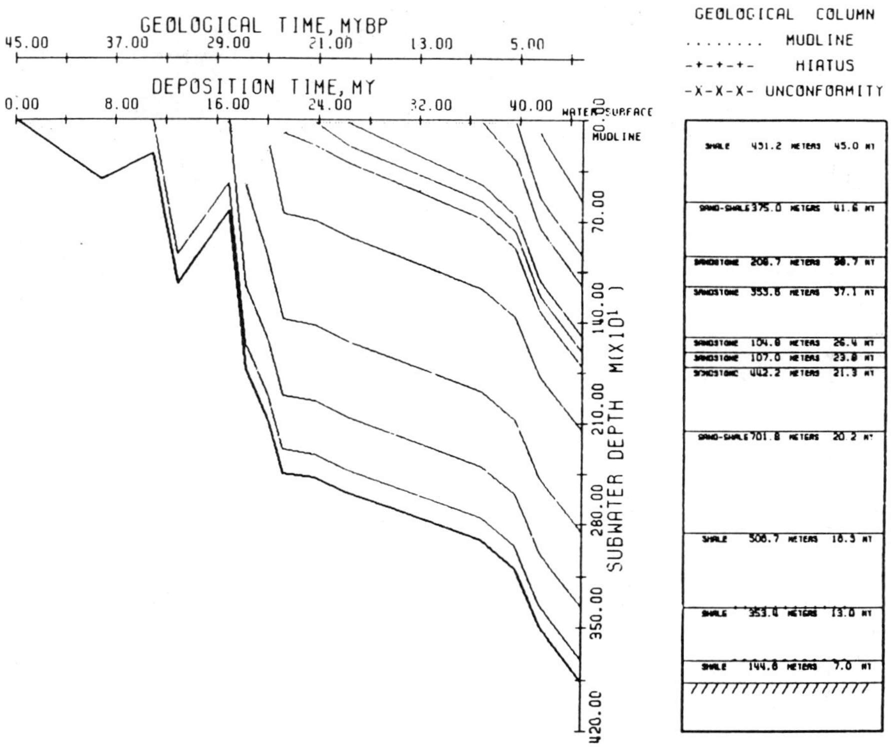

FIGURE 7.8 Reconstructed geohistory of COST No. 1 well, allowing for overpressuring with time. Note the rapid subsidence event at about 28 Ma, relatively slow subsidence thereafter, and two major unconformities prior to the rapid subsidence.

during the late Oligocene. Indeed, Fig. 7.11 shows the profiles of porosity, permeability, pressure, and fluid flow rate with depth at the end of layer 6 deposition (24.84 MYBP). There is then a fracture zone at around 1400 m (4593.4 ft) with a jump in permeability and fluid flow rate. The jump persists until the fluid pressure drops below a critical value (set at about 0.85 of lithologic overburden) at which the fractures are taken to shut.

2. Thermal History

Figure 7.12 shows the heat flux change with time, reflecting a warming trend from past to present with a present heat flux of 1.7 HFU. Based on the modeled heat flux change with time obtained from the fluid flow and burial history, in agreement with downhole formation thicknesses, porosity, permeability, fluid pressure, and vitrinite reflectance measure-

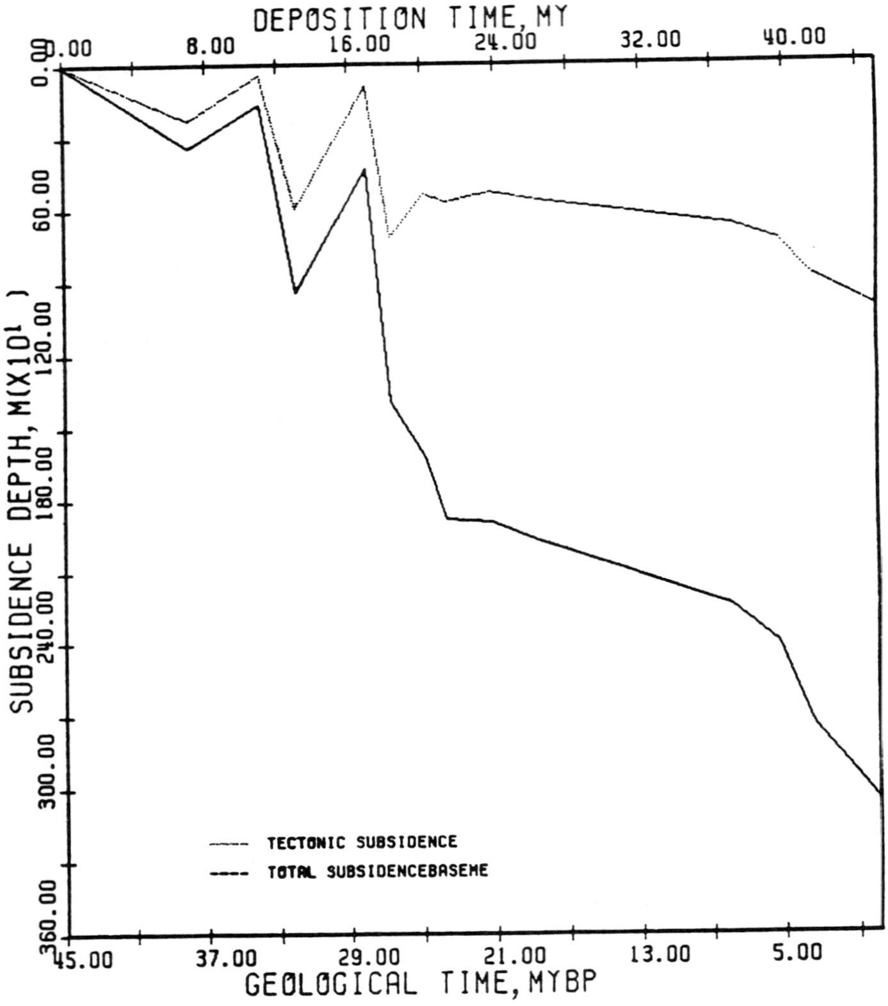

FIGURE 7.9 Curves of basement subsidence with time. The solid curve represents the total basement subsidence, including the effects of sediment loading; the dotted curve represents the tectonic subsidence when the isostatic compensated subsidence due to sediment load is removed in the manner of Steckler and Watts (1978).

ments, we can reconstruct the thermal temperature history (Fig. 7.13), which shows a variable thermal gradient both in the past and with depth. Based on the temperature history in Fig. 7.13, Figs. 7.14 and 7.15 show the thermal maturity evolution of the Cenozoic sequence, using isoreflectance lines as a maturity indicator (Fig. 7.14) and Waples' TTI (Waples,

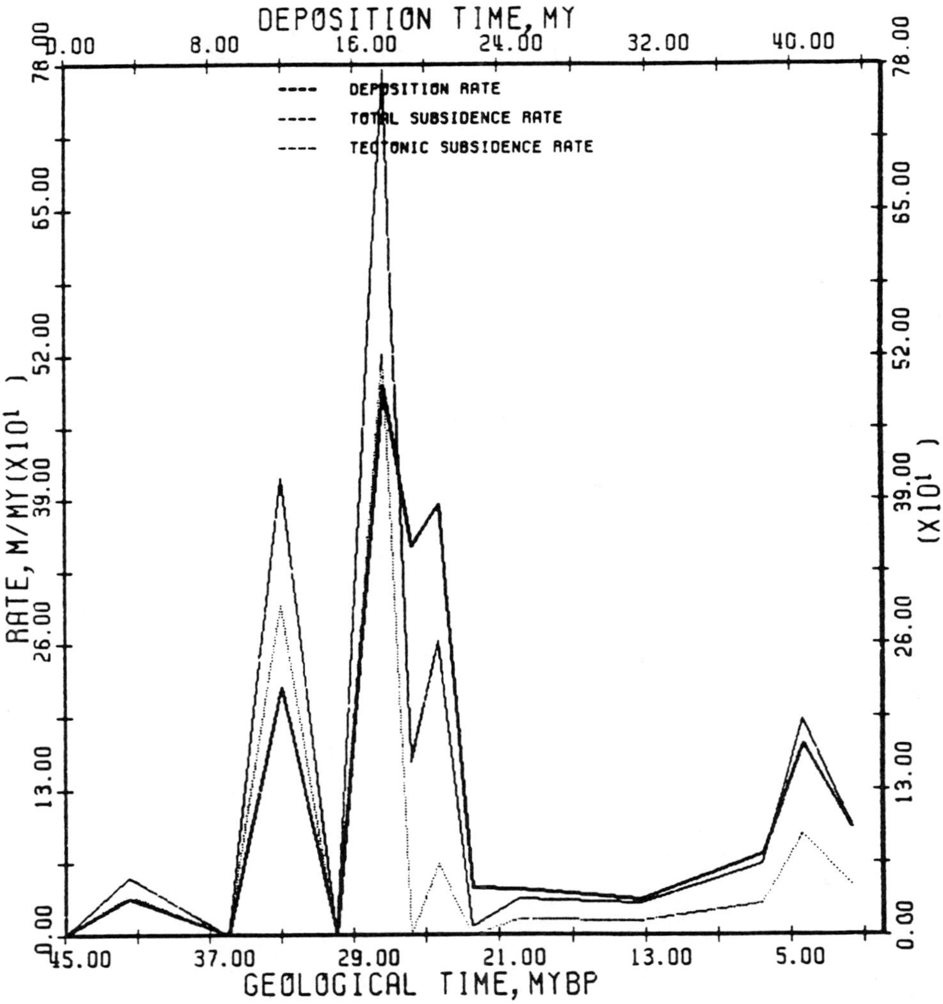

FIGURE 7.10 The three curves show the rate of subsidence including sediment load (light dashed line); the rate of basement subsidence with the isostatic compensation due to sediment load removed (heavy dashed line), and the sedimentary deposition rate (light dotted line).

1980) (Fig. 7.15). The isovitrinite reflectance lines, plotted in Fig. 7.14, show the thermal maturity of each layer. If a vitrinite reflectance value of 0.6 is chosen as defining the onset of the "oil window" (Waples, 1980), then only the Eocene and early Oligocene shales (layers 1, 3, and 5) are thermally mature and within the oil window; younger sediments are immature. Fig. 7.15 indicates that only layer 1 and layer 3 are mature and

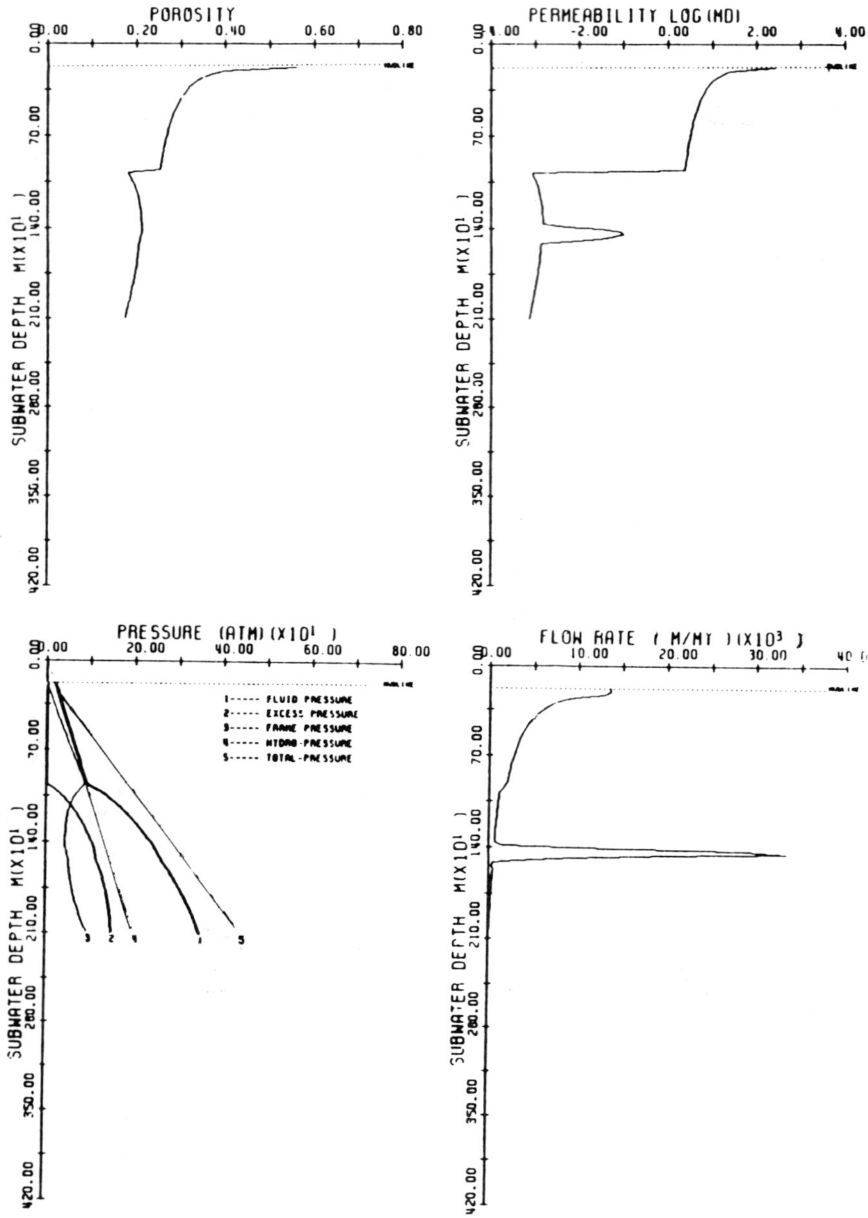

FIGURE 7.11 Profiles of predicted variations with depth of porosity, permeability, pressures, and fluid flow rate at 24.8 MYBP. Note the pressure of a relative spike in permeability at about 1400 m depth, representing the fracturing of sediments at that depth at that time. The fluid flow also shows a spike at the same depth, corresponding to a total fluid pressure in excess of 0.85 of the lithologic overburden.

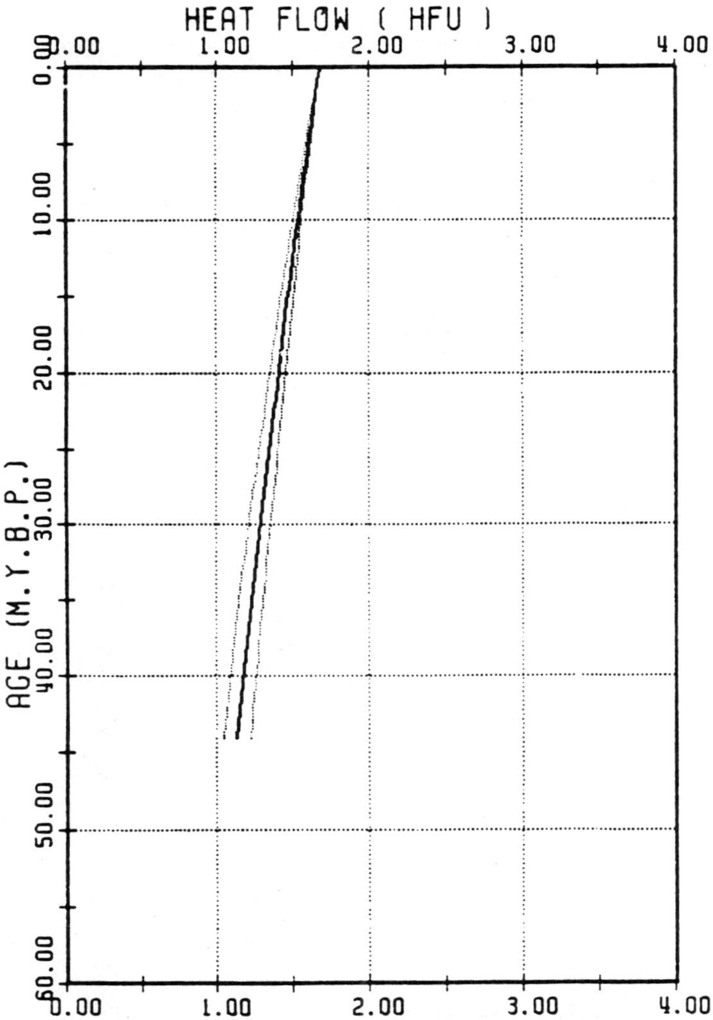

FIGURE 7.12 Variation of heat flux with time obtained by thermal inversion of the vitrinite reflectance values recorded in the well, along the burial paths of the sediments. The heat flux variation obtained is the best one-parameter variation capable of providing minimum discordance with the observations and represents a slow warming trend in the region by about 0.7 HFU over the last 45 Ma.

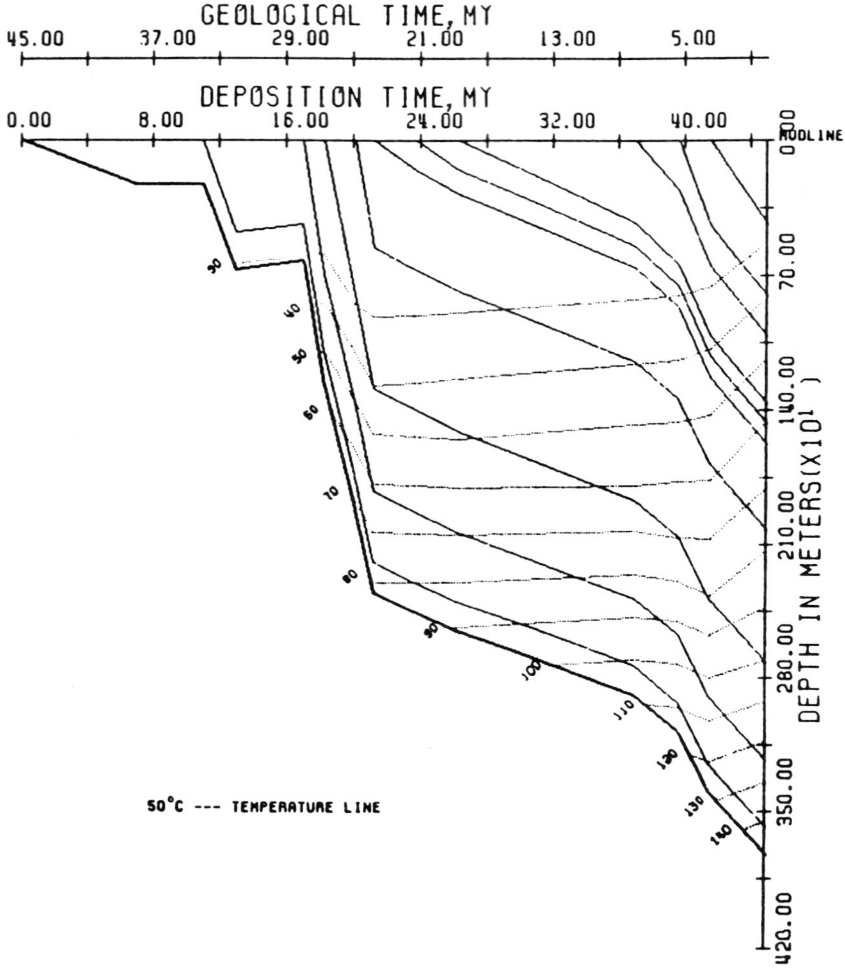

FIGURE 7.13 Superposition of burial history of the COST No. 1 well and isotemperature lines obtained by using the best paleoheat flux variation from the thermal inversion. Note that the 100°C isotherm, occasionally considered a maturation indicator, occurs at about 2800 m and starts at about 15 Ma, indicating kerogen maturity at the present day only for formations deeper than 2800 m which were deposited earlier than 29 MYBP.

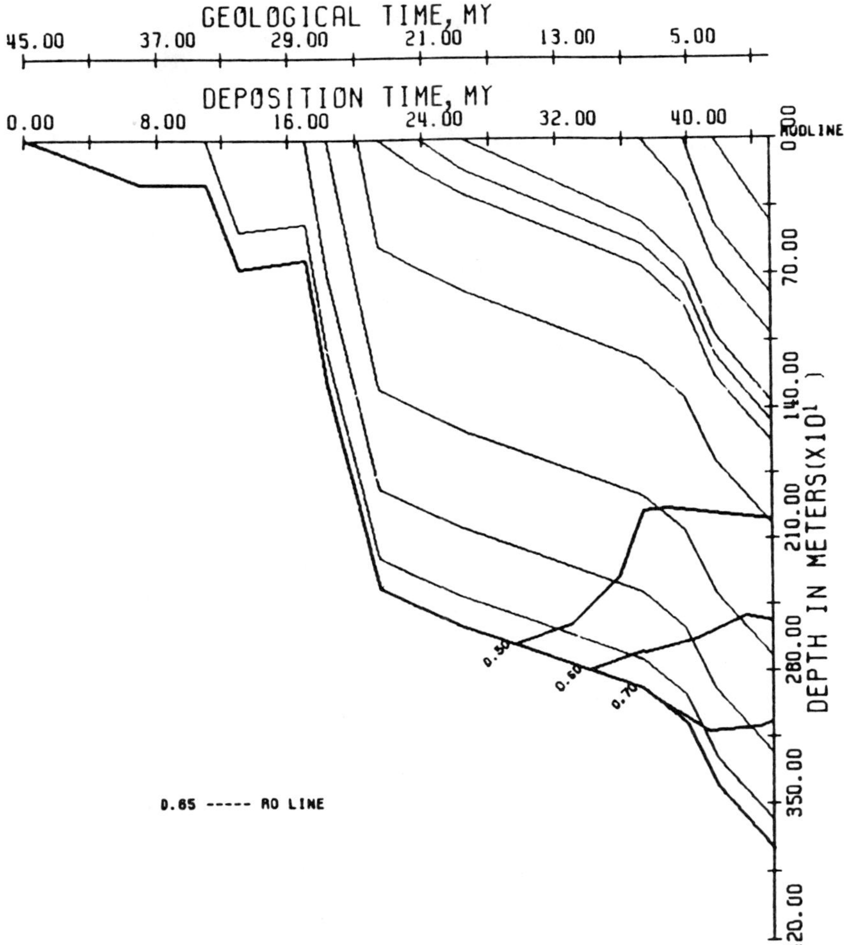

FIGURE 7.14 Superposition of burial history of the COST No. 1 well and isoreflectance lines obtained by using the best paleoheat flux variation from the thermal inversion. Note that the $R_0 = 0.6$ isoreflectance line (occasionally considered indicative of formations entering the oil window), varies from 2800 m at about 20 MYBP to about 2500 m at the present day, suggesting that hydrocarbon maturity has occurred only for formations deeper than about 2500 m and deposited earlier than about 27 MYBP.

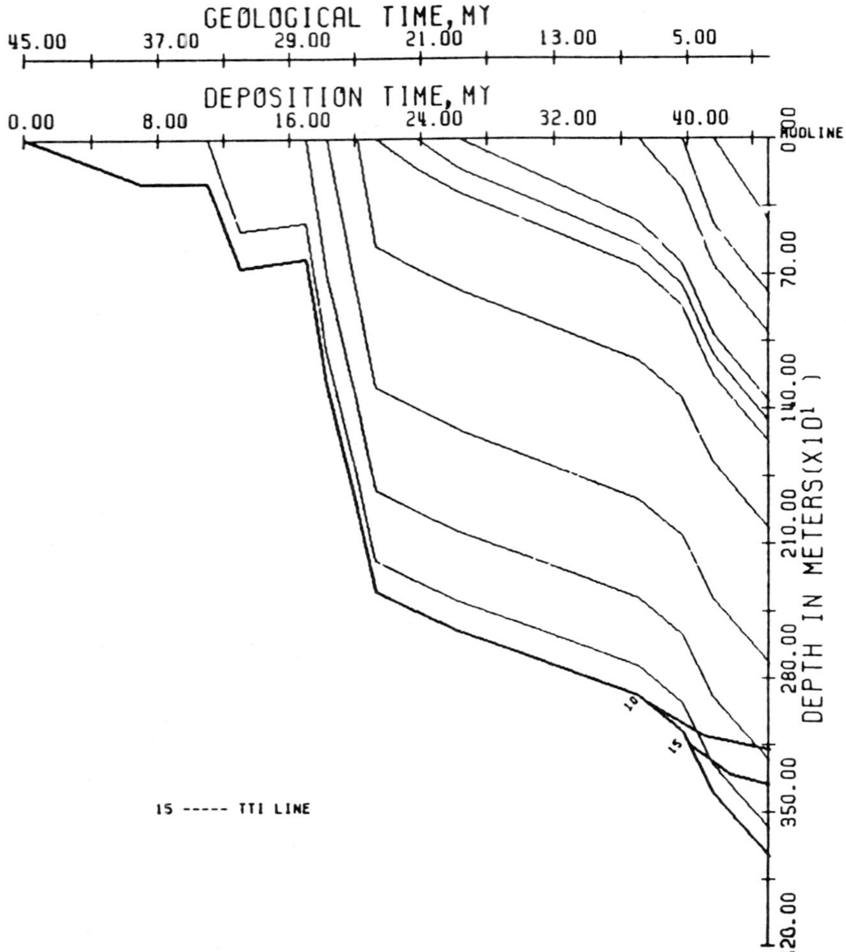

FIGURE 7.15 As for Fig. 7.14 but with iso-TTI contours. A TTI of 15 is often considered to represent the start of oil generation and in this case would suggest maturation only for formations deeper than about 3200 m deposited earlier than about 30 MYBP.

in the oil window if a TTI value of 10 is chosen as the threshold (Waples, 1980). Also, the depth and time at which each layer entered the oil window can be read from Figs. 7.14 and 7.15.

3. Hydrocarbon Generation History

Figure 7.16 shows the hydrocarbon generation history of the Cenozoic sequence in the Navarin Basin COST No. 1 well based on the temperature

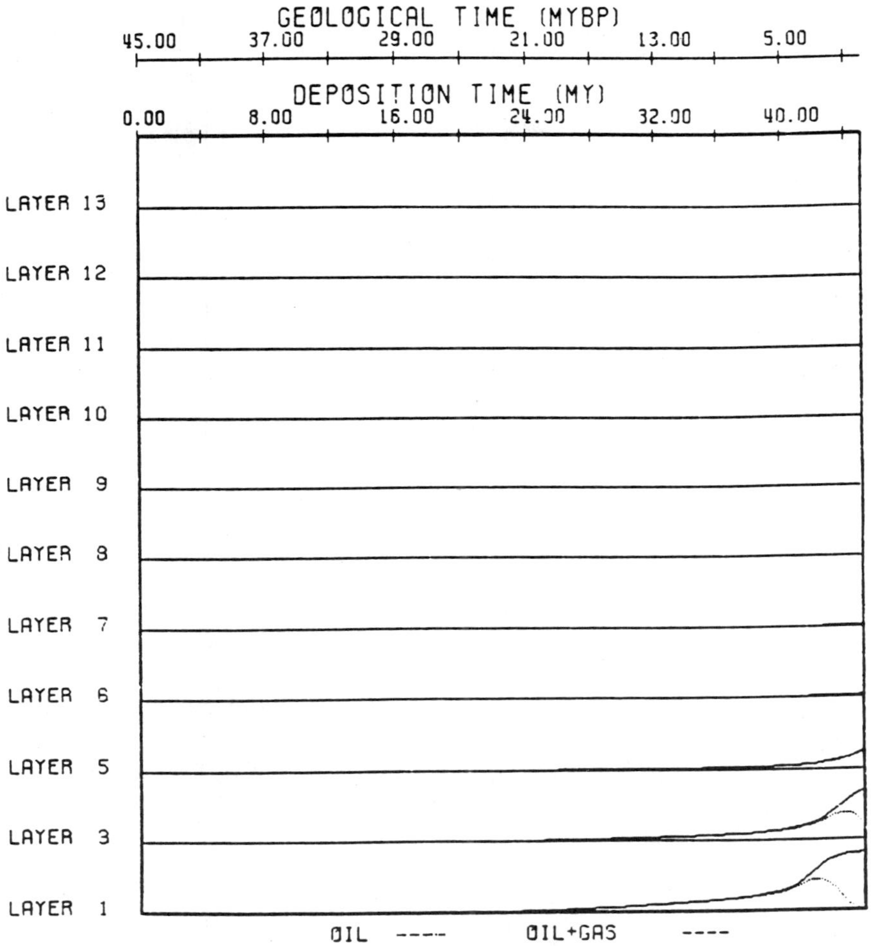

FIGURE 7.16 Hydrocarbon generation history of various layers using a Tissot-type model supplemented by the fluid flow/compaction burial history and the variation of heat flux with time obtained by thermal inversion of the vitrinite reflectance data. Note that only the deepest layers are suggested to be in the oil window with generation having started around 10–5 MYBP.

history and burial history created in the model. The figure indicates that the shales of Eocene and early Oligocene are good source rocks in terms of thermal maturity (layers 1 and 3), layer 5 has begun to generate hydrocarbons, and layer 6 and layers above are immature. Figures 7.17 and 7.18 provide more details concerning hydrocarbon generation for layers 1 and 3. The figures shown are based on the hydrocarbon

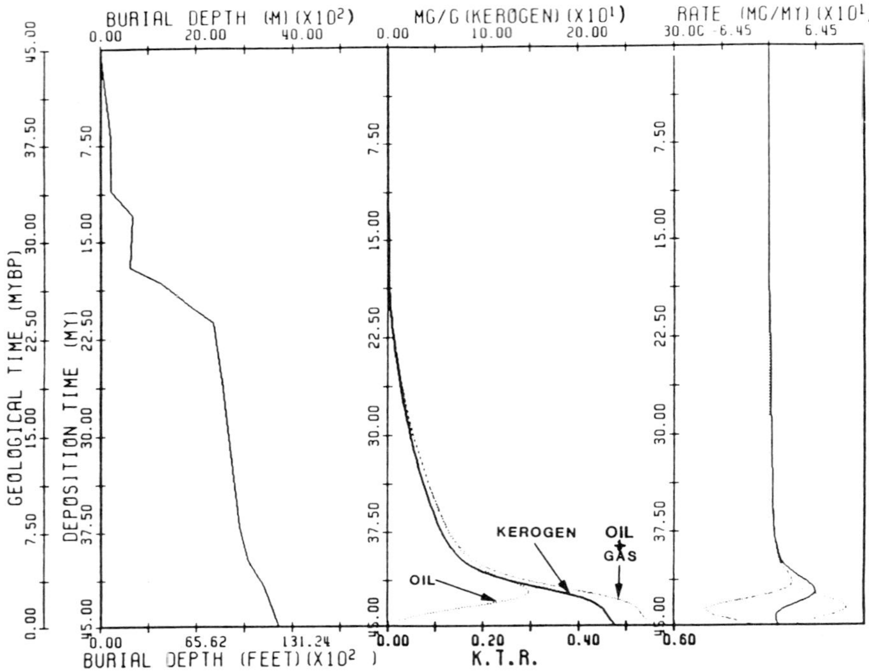

FIGURE 7.17 Detailed hydrocarbon generation history of the Eocene shale (layer 1) in the COST No. 1 well. Shown are the burial depth of the layer with time, the cumulative oil and gas production with time, and the rate of oil and gas production. Note that only in the last 8 Ma or so has the cumulative amount of hydrocarbons exceeded 0.5 mg/g kerogen.

generation model presented by Tissot and Welte (1978). The modified hydrocarbon generation model given in Chapter 5 did not provide a significant discordance with the Tissot–Welte results (and therefore is not shown). The burial histories of layers 1 and 3 are shown in the left-hand plots of Figs. 7.17 and 7.18 and give the changes in the depth of the base of layers 1 and 3 with time. The total amounts of hydrocarbons generated for layers 1 and 3 with time are shown in the middle plots of Figs. 7.17 and 7.18, as is the transformation ratio of kerogen in layers 1 and 3. The right-hand plots give the oil generation, gas generation, and kerogen transformation ratio with time. From Fig. 7.17 we see that the Eocene shale (layer 1) arrived at the oil window in the early Pliocene (7 MYBP, 2000 m), reached peak oil generation in the late Pliocene (3 MYBP, 3500 m), and probably reached peak gas generation in the very late Pliocene (1.4 MYBP). The total hydrocarbon production (oil and gas)

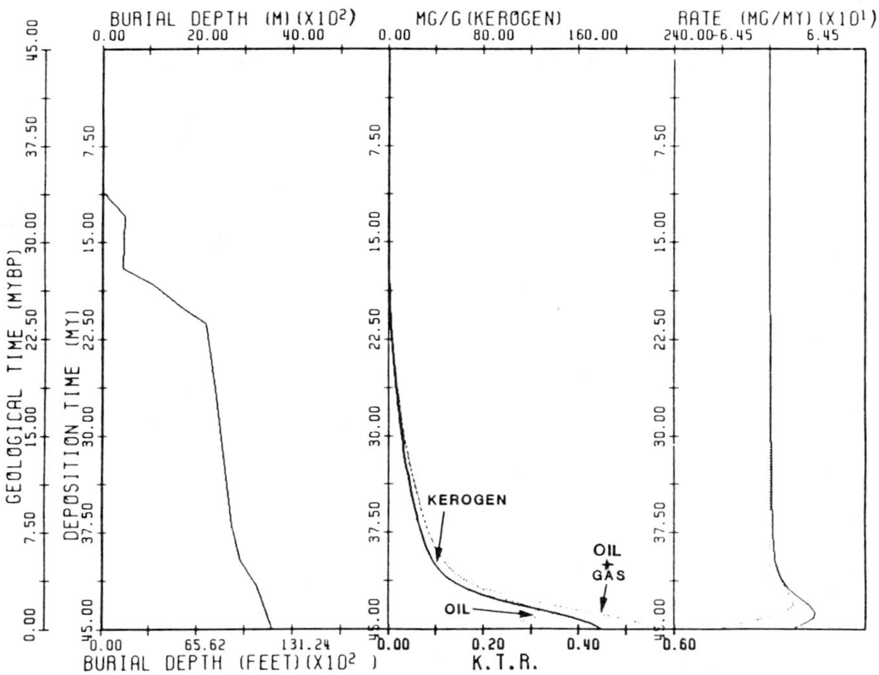

FIGURE 7.18 As for Fig. 7.17, but for the early Oligocene shale (layer 3). Note that significant cumulative production has occurred only in the last 5 Ma or so.

from the Eocene shale is about 275 mg/g kerogen, with a kerogen transformation ratio of 0.48.

The early Oligocene shale (layer 3) (Fig. 7.18), has a hydrocarbon generation history similar to that of the Eocene shale, entering the oil window around middle Pliocene (4.3 MYBP, 3000 m) and having peak oil generation at about the end of the Pliocene (1.4 MYBP, 3500 m). The Oligocene shale is still in its peak gas generation stage with a total hydrocarbon production (oil and gas) of 154 mg/g of kerogen, corresponding to a kerogen transformation ratio of 0.45.

Layer 5 (late Oligocene mudstone) is just entering the oil window, with a total hydrocarbon generation of 84 mg/g of kerogen (78 mg oil and 6 mg gas) and a kerogen transformation ratio of 0.18, but probably cannot be an effective source rock. The simulation does not consider biogenic hydrocarbon formation from kerogen.

C. Discussion

From the combined geohistory, thermal history, and hydrocarbon generation history of the Navarin COST No. 1 well simulated by the model consistent with the observed downhole data, and including other geological and geophysical data collected in the basin, the following considerations are obtained and may be of use for further hydrocarbon exploration in the basin.

1. Source Rock

The Eocene gray to dark gray mudstone and shale (zone F) constitute the best source rock in the Navarin basin, with TOC ranging from 1 to 2%. The Eocene shale reached thermal maturity during the late Miocene (Fig. 7.17) and arrived at peak oil generation at the late Pliocene with a total amount of generated hydrocarbons of about 275 mg/g kerogen. The early Oligocene shale (zone E) is the next best source rock, with TOC around 1%; it matured in the late Miocene and arrived at peak oil generation at the end of the Pliocene, with a total amount of hydrocarbons generated of about 154 mg/g kerogen. The late Oligocene mudstone (layer 3) has just reached maturity but probably cannot be considered an effective source rock in the Navarin COST No. 1 well. Other, younger, fine-grained sediments are unlikely to be good source rocks because they are thermally immature and have a low TOC (less than 0.5%).

There are three subbasins in the regional-scale Navarin basin, and the COST No. 1 well is located on a structure bench, indicating that the Eocene and early Oligocene (layers 3 and 5) will more likely be generating significant amounts of hydrocarbons in the subbasins. The Cretaceous rocks below the major unconformity are thermally mature enough to be good source rocks.

2. Reservoir Rock

Miocene sandstones are the most promising reservoir rocks because of their high porosity and permeability. There probably exist some potential reservoir rocks that were deposited during the two major uplifts in the Tertiary (one between the Eocene and early Oligocene and one in the early Oligocene), because these rocks are closest to the major Eocene and early Oligocene source rocks. Also, the late Cretaceous rocks directly below the major unconformity may be potential reservoir rocks reflecting a variety of depositional environments, because (1) they are coarse-grained rocks deposited in streams and distributary channels and possibly are strongly influenced by the deltaic distribution system (Turner *et al.*,

1984); (2) the erosion during the major unconformity period might lead to both coarse-grained sediments and secondary and fracture porosity, and (3) the Cretaceous rocks are close to the major source rock above the unconformity, which may increase their chances of receiving downward-migrating hydrocarbons generated from the source rock.

3. Exploration Targets

Based on the geohistory, thermal history, and generation history from the model and other geological and geophysical data, exploration for hydrocarbons in the Navarin basin should most likely be targeted at (1) Eocene and Oligocene strata and (2) late Cretaceous strata right below the late Cretaceous–early Tertiary unconformity.

a. Eocene and Oligocene Strata. Eocene and early Oligocene organic-rich mudstones and shales (zones E and F) are thermally mature and are able to generate significant amounts of hydrocarbons. Late Oligocene shelf sands are the best reservoir rocks encountered in the COST No. 1 well. Oligocene and Eocene turbidite deposition may be present along subbasin margins and as detrital aprons around basement highs (Steffy *et al.*, 1985). Also, during the two major uplifts in the Eocene and Oligocene, some reservoir rock types and favorable facies changes were probably formed in the basin. In addition to these possible stratigraphic traps, seismic surveys indicate the existence of anticlines and growth faults along the flanks of the subbasins (Marlow, 1979), which means that potential structural traps and growth fault-related traps may also exist.

b. Late Cretaceous Strata Directly Below the Unconformity. Eocene and early Oligocene source rocks can be not only the major hydrocarbon sources for late Cretaceous reservoir rocks right below the unconformity but also effective seals for these reservoir rocks. The late Cretaceous–early Tertiary unconformity may act as a migration path for hydrocarbon generated from the Eocene and early Oligocene source rocks. Various stratigraphic traps, in the depth range of the late Cretaceous–early Tertiary unconformity, should then be considered as potential candidates for hydrocarbon accumulation.

This study leads to the following conclusions:

1. The dynamic model is a useful tool for hydrocarbon explorationists for quantitative studies of geohistory, thermal history, hydrocarbon generation and migration history in a sedimentary

basin. A one-dimensional fluid flow model is an appropriate first vehicle to use in a frontier area where only one or a few wells have been drilled, provided that the escape of fluid from the sedimentary fill is indeed predominantly vertical.

2. The major source rocks in the Navarin basin are likely to be Eocene and early Oligocene organic-rich mudstone and shales, which are mature enough to generate significant amounts of hydrocarbons.

3. The most prospective exploration plays in the Navarin basin are the Eocene and Oligocene strata and, perhaps, the late Cretaceous strata right below the late Cretaceous–early Tertiary unconformity, although the effect of the igneous intrusive on the thermal maturity of sediments in the vicinity of the intrusive has yet to be investigated (Wang *et al.*, 1989).

4. The two major source rocks have been overpressured since Oligocene time, providing a mechanism for the expulsion of generated hydrocarbons from the source rocks into surrounding reservoir rocks.

D. Appendix to Section I: Permeability and Frame Pressure Laws and Method of Solution

The choice of dependence of permeability and frame pressure on void ratio for various lithologies has had a long and chequered history (Ungerer *et al.*, 1984). Proponents can be found for determining the behavior *ab initio,* while others note that the complexity of physical processes over geologic time is so variable that quasi-empirical relationships are to be preferred. We have sided with this second viewpoint and, on the basis of measurements from several hundred wells (Glezen, 1983), we have used the following relationships in our calculations (cgs units):

1. Frame pressure P_f is a function of void ratio e:

$$\text{Shale:} \quad P_f = P_{f*}(e/e_*)^{-A} \tag{A1}$$

with the surface value $e_* = 3$ (corresponding to 75% porosity) and the scaling pressure $P_{f*} = 6.9 \times 10^5$ dyn cm^{-2}. The parameter A has a value of around 3.5 for an average of several dozen Gulf Coast wells.

$$\text{Sand:} \qquad e = e_* - C(P_f - P_{f*}) \tag{A2}$$

where $C \cong 7 \times 10^{-10}$, $e_* \cong 0.6$, $P_{f*} \cong 1.0$ dyn cm^{-2}.

2. Permeability k is a function of void ratio:

$$\text{Shale:} \qquad k = k_*(e/e_*)^B, \quad \text{with } k_* \cong 10^{-12}\ \text{cm}^{-2} \tag{A3}$$

$$\text{Sand:} \qquad k = k_*(e/_*)^B, \quad \text{with } k_* \cong 3 \times 10^{-7}\ \text{cm}^{-2} \tag{A4}$$

and $B \sim 4.0$ for shale and $B \sim 6.5$ for sand.

Note that the shale permeability is about five orders of magnitude smaller than the sand permeability, so that the flow and the overpressuring are dominantly controlled by the shale layers.

The two parameters A and B for shale are at our disposal, since changes in those parameters markedly influence the fluid pressure gradient and permeability.

Since the excess fluid pressure gradient is limited to being less than about 0.5 psi/ft, the dominant control on the dynamical evolution is through the exponent B of the shale permeability. Both A and B appear to be about 4.0, but there is sufficient variation in their values from well to well that we have treated them as empirical constants to be determined. This treatment then leads to the method of solution. First, "freeze" the parameter A, since fluid pressure gradients do not vary by orders of magnitude. Then run the calculation of fluid flow/compaction for a fixed value of B. Check the degree to which the predicted formation thicknesses agree with observed values at the present day. Adjust B until a minimum discordance is obtained. (Experience with many wells has shown that with an initial B set at 4.0, the final B will be between about 2.5 to 5.0, and convergence is obtained after about three iterations.) Then freeze B at its best value, and let A be allowed to vary until either the predicted porosity with depth or the predicted fluid pressure with depth (but not both) comes into minimum discord with the observed values. (Again, experience has shown that an initial A set at 4.0 yields a final A with range of 3.0 to 5.0 after about three iterations.)

The sequence is repeated: A is once more frozen and B adjusted. This second sequence of operations normally adjusts B by about 10%. The refreezing of the "new" B and subsequent redetermination of A normally varies A by about 10%. Further repetitions of the sequence vary A and B hardly at all.

The permeability and frame pressure variations with void ratio are then determined empirical relations. The dynamic fluid flow/compaction code then determines the evolution of fluid flow, formation thickness, porosity with depth, fluid pressure with depth, and permeability with depth with no other "free" parameters being varied. The results are as shown in the text.

II. Iwafune-Oki Area, Japan

As we showed in Chapter 6, two-dimensional computer models have been developed in which the time-dependent and (spatially) two-dimensional partial differential equations are solved to simulate fluid flow/compaction and heat transfer processes. The models also simulate the petroleum generation process by using mathematical models of the Tissot–Espitalie (1975) and Tissot–Welte (1978) varieties. For migration, separate oil and gas phases are invoked as main modes of migration and three forces (hydrodynamic, capillary, and buoyancy) are considered to be exerted at each spatial location. Various kinds of graphic outputs are available from each such model, including two-dimensional views of basin development through time with fluid migration patterns. Here we examine the application of such a model to a particular area, using simplified geological conditions, to illustrate the sort of patterns of behavior that are quantifiable for the generation, migration, and accumulation of hydrocarbons in relation to evolving dynamical and thermal behaviors in a sedimentary basin.

A. Geologic Setting

1. Regional Geology

The Iwafune-Oki oil field is a recent Japanese discovery; it is located in the Japan Sea, in offshore Niigata, the main petroleum province of Japan (Fig. 7.19). Oil and gas were discovered in 1983. The tectonic setting is part of a back arc basin characterized by a fairly high geothermal gradient (4.2°C/100 m). Thick clastics (6000 m) were deposited over a short geologic time interval (since the middle Miocene) and were influenced by strong volcanic activity. The general stratigraphy of the area is shown in Fig. 7.20 (Nakayama, 1986).

The tectonic event associated with the "Green Tuff" triggered sedimentation over the area some time in the Miocene (20 Ma). Volcanic lava and its related tuff (the Green Tuff Member in the Nanatani Formation)

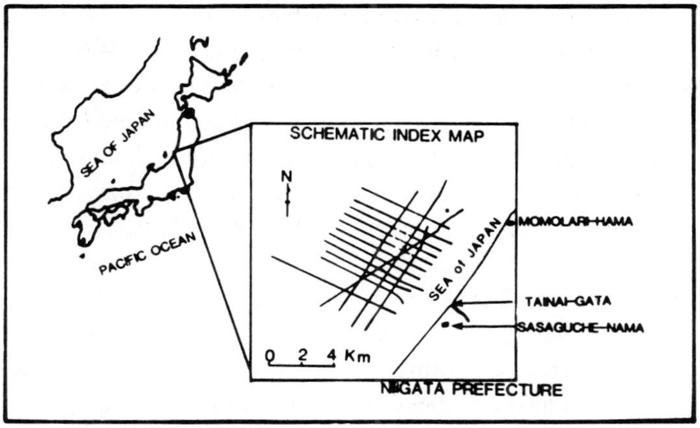

FIGURE 7.19 Location of Iwafune-Oki area, Japan. Lines show locations of the seismic survey and dots show locations of the wells.

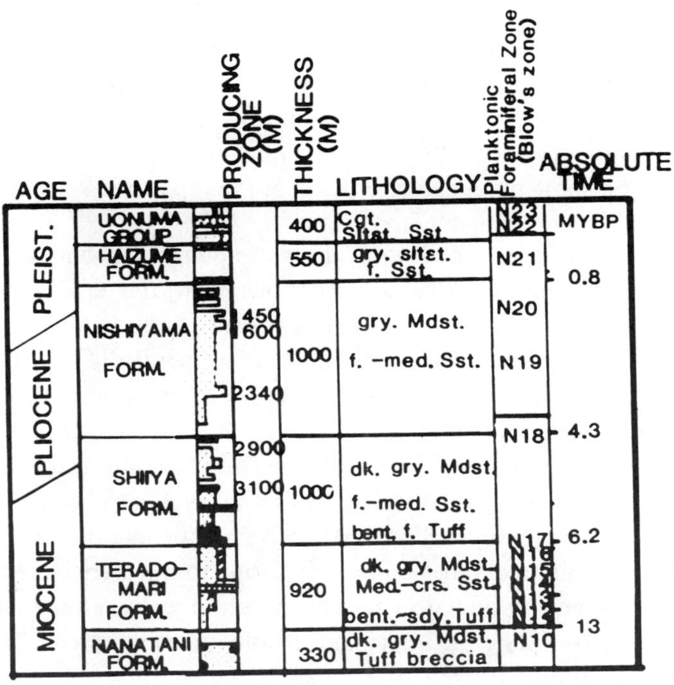

FIGURE 7.20 Regional stratigraphy of the Niigata basin (from Nakayama, 1986).

cover metamorphic rocks and granites which range in age from Paleozoic to Mesozoic. When considering the petroleum geology of the area, these older rock units form the basement. After a few megayears of active volcanic movement, areal subsidence was accompanied by the slow sedimentation event associated with the Nanatani mudstone member. This mudstone eventually covered the entire area and can be considered in part to be lateral facies equivalent to the Green Tuff. Subsequently, as the depositional setting deepened, sedimentation is represented by the "black shale" of the Teradomari Formation. At 4.3 MYBP, the area shallowed somewhat, but relatively deep conditions persisted with the deposition of turbidites representing the Shiiya Formation. Thin tuffaceous sandstone layers become more frequent in the upper portions of this unit. The interpretation of these alternations of shales and sandstones as turbidite deposits was made by Katahira (1974). These sediments probably accumulated at the edge of the continental shelf of this time. The subsequent Nishiyama stage is also characterized by the shale/sandstone sequence. The sedimentation rate increased and became extremely high (more than 1000 m/Ma). After 0.8 MYBP, the depositional setting of the area suddenly shallowed and the coarser sediments of the Haizume Formation or the Uonuma Group were deposited. Subsidence was rapid, so that very thick sediment piles were formed. This sedimentation lasted until very recently, when the final uplift of the area took place. There may be some evidence of erosion at the top of the sequence but its effects seem to be negligible.

The thin sandstones which form the petroleum reservoirs are found in the Shiiya and the Nishiyama formations near the culmination of a plunging anticlinal structure. The wells drilled at the culmination of the anticline (about 3 km north of the present oil field) did not detect any sandstone layers, so the trap is thought to be stratigraphic. The extension of the sand bodies may be limited.

Another possibility is that the oil/gas accumulations occur in the lowest member Green Tuff, because this unit is one of the highly potential reservoirs in this province. Since there is very little likelihood of a source rock for petroleum available under this formation, it is believed that the oil/gas must have migrated downward stratigraphically. Understanding of the paleohydrogeology is important if the possibilities of petroleum accumulation in the Green Tuff are to be predicted.

2. *Data Available for Input*

a. Cross Section. A number of seismic records are available for the area. Among them, line A–A' was selected to provide a two-dimensional

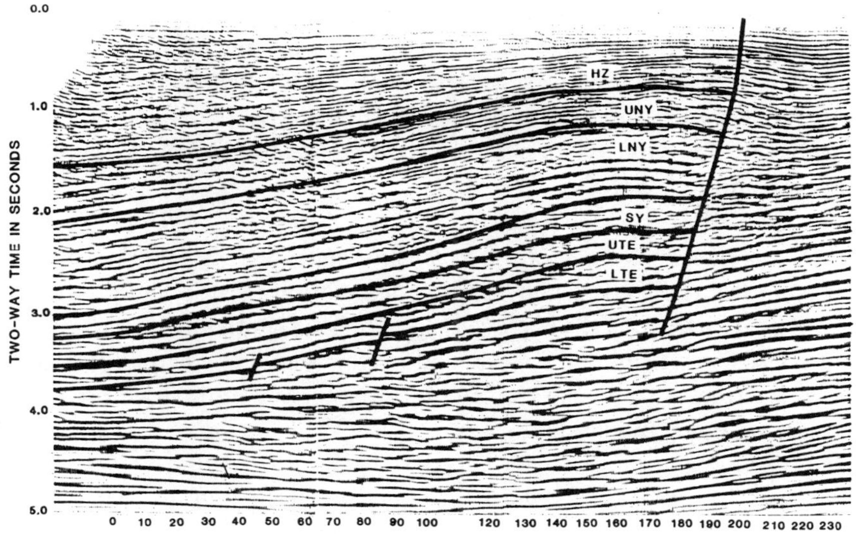

FIGURE 7.21 Seismic section *A–A′*.

view of the area (Fig. 7.21). Note that the anticlinal structure seen in this section does not represent the actual culmination of this feature because the structure itself is dipping toward the south. At the highest point on the section, well X was drilled. Knowing the seismic interval velocity at the point, the depths of formation boundaries from well data were converted to two-way travel times so that they could be correlated with the seismic events on the section. Then such events were extrapolated using regional geological ideas to make up the cross section (Fig. 7.22). The fault at the

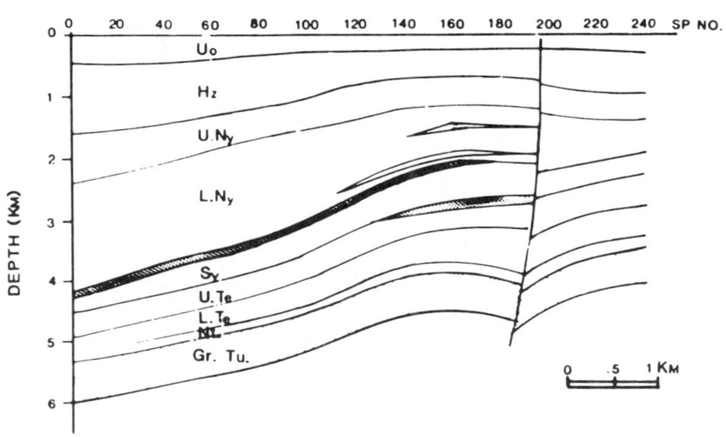

FIGURE 7.22 Cross section for the model.

east side of the structure does not appear in the model application because the model did not then have the capability to handle fault events (see, however, Wei and Lerche, 1989). However, the fault can be simulated by artificial lithology inputs. In this section, the less permeable shale is assigned at the fault location to simulate the sealing effect of this fault through time.

b. Temperature. Temperature measurements are available from both drill stem test (DST) data and borehole temperature (BHT) at the logging time. The data from DST (Table 7.4) are much more reliable than the BHT data because measurements were made under equilibrium conditions and are considered to represent the true formation temperatures without any corrections being made (Jones, 1975).

c. Porosity. Porosities were calculated from the bulk density values of the formation density logs (FDC), assuming 2.65 g/cm^3 for the matrix density and 1.00 g/cm^3 for the water density (Table 7.5). Since no data are available for the cementation/dissolution process, all the pores are assumed to be primary pore space and unaffected by cementation.

d. Pressure. Pressure values are also available from the DST data. Shut-in pressure is considered as the true formation pressure for the present application.

e. Vitrinite Reflectance. Vitrinite reflectance values were measured for 20 samples from cuttings and cores (Table 7.6). Although the data are scattered, the better values are associated with the higher number of measurements for each sample. Data measured with less than 80 counts were not used for input to the model.

Table 7.4 Temperature Data from the Drill Stem Test (DST)

Test number	Depth (m)	Temperature (°C)	Type of fluid
5	1484–1499	77.8	Gas
4	1962.5–1976	100.0	Oil/gas
3	1986–1999	97.2	Oil/gas
2	2113.5–2143	105.6	Oil/gas
1	2156–2183	107.5	Oil/gas

Table 7.5 Porosity Calculated from the Formation Density Log (FDC)[a]

Depth (m)	Bulk density from FDC (g/cm^3)	Calculated porosity (%)
650.0	2.07	35.2
750.0	2.03	37.6
850.0	2.06	35.8
950.0	2.08	34.5
1000.0	2.11	32.7
1100.0	2.07	35.2
1200.0	2.17	29.1
1300.0	2.20	27.3
1380.0	2.22	26.1
1550.0	2.24	24.8
1650.0	2.27	23.0
1750.0	2.28	22.4
1850.0	2.32	20.0
2050.0	2.38	16.4
2250.0	2.43	13.3
2350.0	2.44	12.7
2450.0	2.45	12.1
2550.0	2.44	12.7
2650.0	2.46	11.5
2840.0	2.48	10.3

[a] Matrix density, 2.65 g/cm^3; water density, 1.00 g/cm^3.

f. Kerogen Type. Visual kerogen analysis was also done. The data are summarized according to formations (Table 7.7). The kerogen type of other formations not listed in Table 7.7 is assumed to be the same as the nearest known type in the table.

g. Total Organic Carbon. TOC data are available from Rock Eval analysis and are listed in Table 7.8; they, too, are averaged according to formation and are used in the same way as the kerogen type data.

B. Calibration of Parameters

Since not all the input data for the system are available, some input parameters have to be estimated through a process called calibration. The

Table 7.6 Vitrinite Reflectance Data

Depth (m)	Number of measurements	Vitrinite Reflectance (%)
1122.0	100	0.33
1240.0	48	0.31
1297.0	85	0.33
1397.0	100	0.40
1515.0	100	0.39
1536.0	100	0.38
1598.0	100	0.38
1700.0	100	0.38
1747.0	68	0.37
1800.0	100	0.41
1907.5	100	0.41
1988.7	100	0.49
2003.0	13	0.35
2085.0	100	0.44
2173.0	100	0.42
2222.5	38	0.43
3216.0	1	0.40
2421.5	100	0.50
2514.5	70	0.40
2593.5	43	0.48

main idea behind the calibration procedure, as in the one-dimensional case, is to provide results consistent with the observed values at least for the present-day situation. A one-dimensional situation at the well location is set up to find out suitable input parameter values at this location. Here the postulate is that the parameter values derived from the one-dimensional modeling are still effective for the two-dimensional case. The difference will be discussed later when the values are applied to the two-dimensional case.

1. Compaction Constants

When the model was run with the default compaction constants, the trend of the calculated porosity values with depth did not match that of the observed values, implying that we need to calibrate the slope variables on both shale and sandstone frame pressure versus void ratio equations. Depth can be converted to a value of $P_f - P_{f^*}$, assuming that the excess

Table 7.7 Data from Visual Kerogen Analysis.

		Percent by volume		
Formation	Depth	Type I, amorphous	Type II, herbaceous	Type III, woody/coaly
Lower Nishiyama (3)	1122.0	5	49	46
	1204.0	7	40	53
	1297.0	8	46	46
	1395.0	19 (10)[a]	35 (42)	46 (48)
Lower Nishiyama (2)	1512.0	17	41	42
	1535.0	6	41	53
	1597.0	15	40	45
	1680.0	21	44	35
	1698.0	27	40	33
	1713.0	0	25	75
	1747.0	32	26	42
	1798.0	34 (19)	33 (36)	33 (45)
Sandstone member (3)	1902.5	15	43	42
	1972.0	19	46	35
	1988.8	11	46	43
	2003.3	51 (24)	34 (42)	15 (34)
Lower Nishiyama (1)	2081.5	52	29	19
Sandstone member (2)	2173.0	23	51	26
Shiiya	2219.5	52	31	17
	2316.0	39	47	14
	2424.5	32	48	20
	2514.5	58	31	11
	2595.5	52	35	13
	2677.0	52 (48)	31 (37)	17 (15)
Upper Teradomari	2815.0	39	39	22

[a] Numbers in parentheses are average values within the formation.

pressure and the generation pressure are negligible. Figure 7.23 shows $P_f - P_{f*}$ versus void ratio for observed values. Starting with the original equation

$$\frac{e}{e_*} = \left[\frac{(P_f - P_{f*}) + P_{f*}}{P_{f*}}\right]^a \tag{7.1}$$

with the assumption $P_f - P_{f*} >> P_{f*}$, we get

Table 7.8 Total Organic Carbon Data from Rock Eval Analysis

Formation	Depth (m)	TOC (%)
Upper Nishiyama	815.5	0.76
	817.5	0.46
	977.0	0.49 (0.57)
Lower Nishiyama (3)	1126.0	0.58
	1203.0	0.56
	1290.0	0.33
	1295.0	0.50
	1393.0	0.56 (0.51)
Lower Nishiyama (2)	1495.5	0.38
	1510.0	0.51
	1543.0	0.45
	1601.0	0.39
	1690.0	0.77
	1702.0	0.84
	1752.5	0.55
	1802.0	0.63
	1893.0	0.42 (0.55)
Lower Nishiyama (1)	1943.0	0.63
	1975.0	0.63
	1994.7	0.72
	2000.3	1.13
	2003.3	0.89
	2017.5	0.58
	2074.0	0.84 (0.77)
Shiiya	2125.0	0.90
	2146.0	0.53
	2208.0	0.77
	2312.0	0.83
	2418.0	0.59
	2499.5	0.74
	2602.0	1.12 (0.78)
Upper Teradomari	2698.0	0.73
	2817.0	0.52 (0.63)

[a] Numbers in parentheses are average values within the formation.

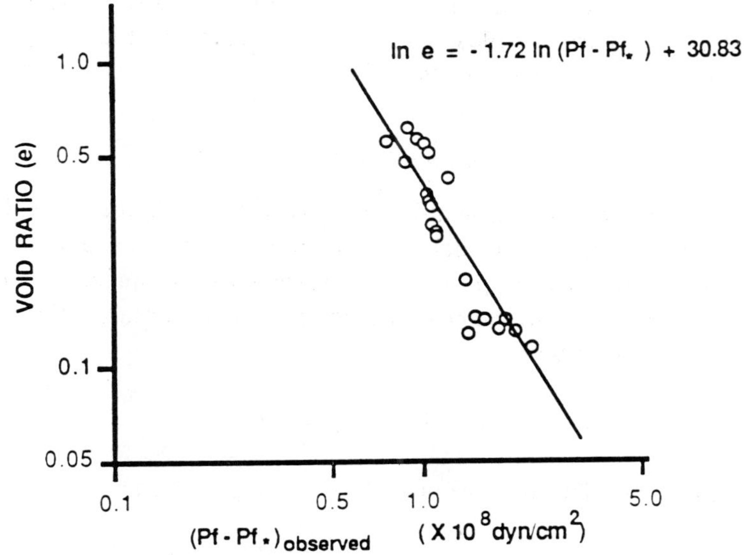

FIGURE 7.23 Graph of void ratio versus estimated frame pressure. Data are from Table 7.2.

$$\ln(e) = a \ln(P_f - P_{f*}) + [\ln(e_*) - a \ln(P_{f*})] \tag{7.2}$$

From Fig. 7.23 we can estimate the rough values

$$a = -1.72, \quad \ln(e_{*sh}) - a \ln(P_{f*}) = 30.83$$

If we set $e_{*sh} = 3.0$ as the default value,

$$P_{f*} = 0.3215 \times 10^8 \tag{7.3}$$

For sandstone, setting $e_{*ss} = 0.612$ and P_{f*} as above, with a void ratio of 0.30 at $P_f = 1.674 \times 10^8$, we solve to obtain the rough estimate

$$C = 2.317 \times 10^{-9} \tag{7.4}$$

We now have first passes for all the variables (e_{*sh}, e_{*ss}, a, C, P_{f*}) in the system. However, the excess porosities obtained by using the first-pass variables indicate that the trend of calculated values is similar to that of observed values but with some offset. So, once again, e_{*sh} is reevaluated to get the best fit to the observations because shale is the dominant lithology of the section. Trial-and-error tests suggest that $e_{*sh} = 1.75$ gives the smallest mean square residuals (Fig. 7.24). The distributions of porosities calculated in this way and observed values are shown in Fig. 7.25.

2. Permeability Constants

Direct measurements of permeability and porosity (void ratio) provide the values of b_{sh} and b_{ss}. When such measurements are not available, b_{sh} and b_{ss} can be estimated from pressure measurements. A pair of b_{sh} and b_{ss} values was tested to find the best fit to the resultant pressures (Fig. 7.26). The selected values were

$$b_{sh} = 3.4; \quad b_{ss} = 5.76$$

The pressure distributions are also shown in Figure 7.27.

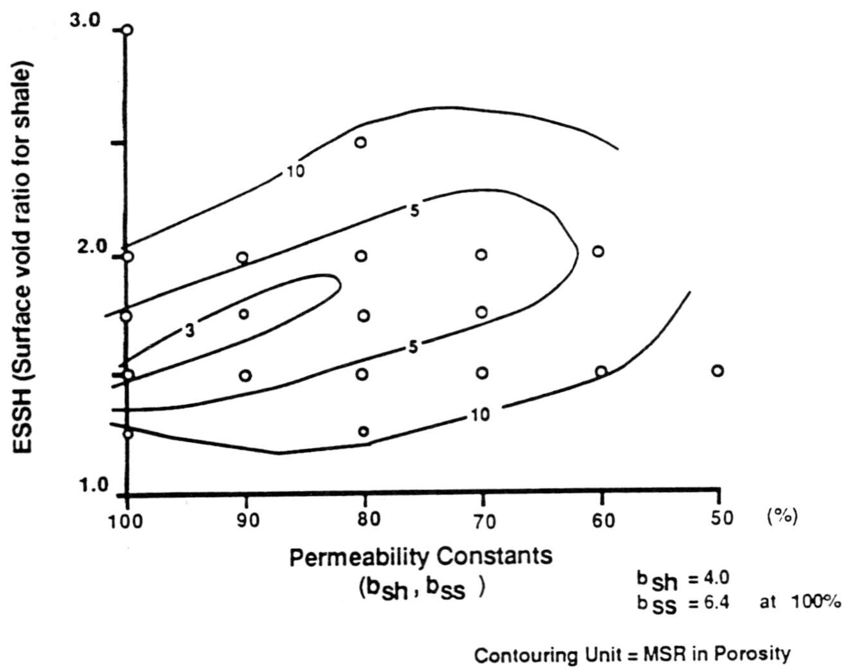

FIGURE 7.24 Mean square residuals of porosity between observed and calculated values (contour in units of percent). The best fit is obtained when e_* is 1.75 and permeability constants are 90%.

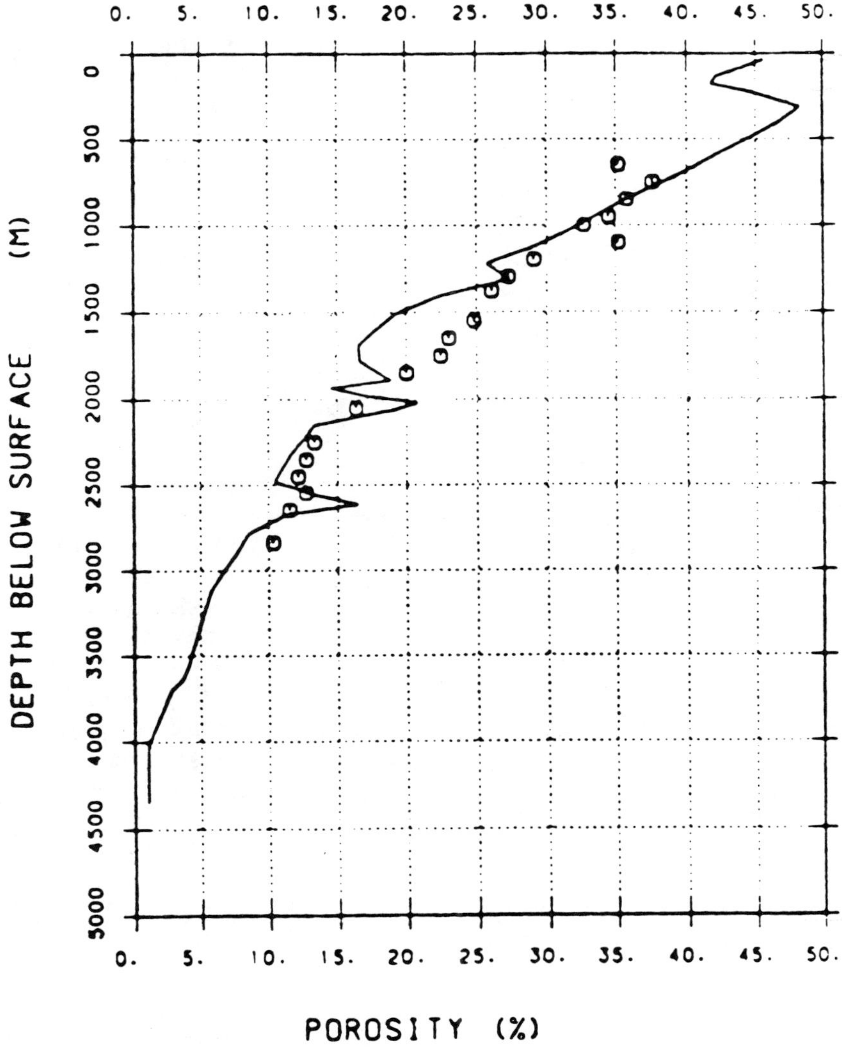

FIGURE 7.25 Porosity distributions calculated (solid line) and observed (symbol).

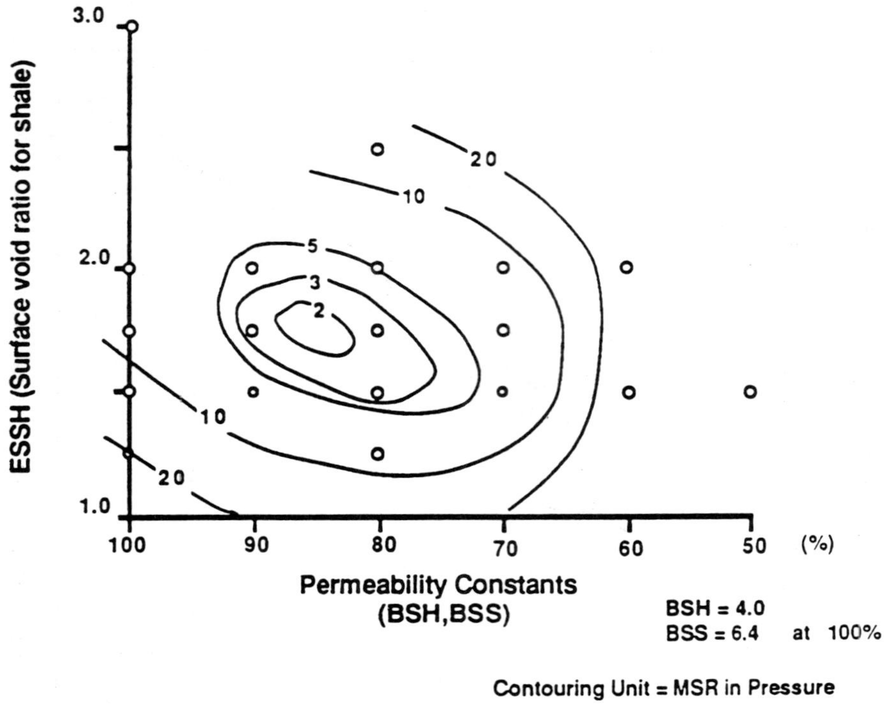

FIGURE 7.26 Mean square residuals of pressure between observed and calculated values (contour in units of kilograms per square centimeter). The best fit is obtained when e_* is 1.75 and permeability constants are 80%.

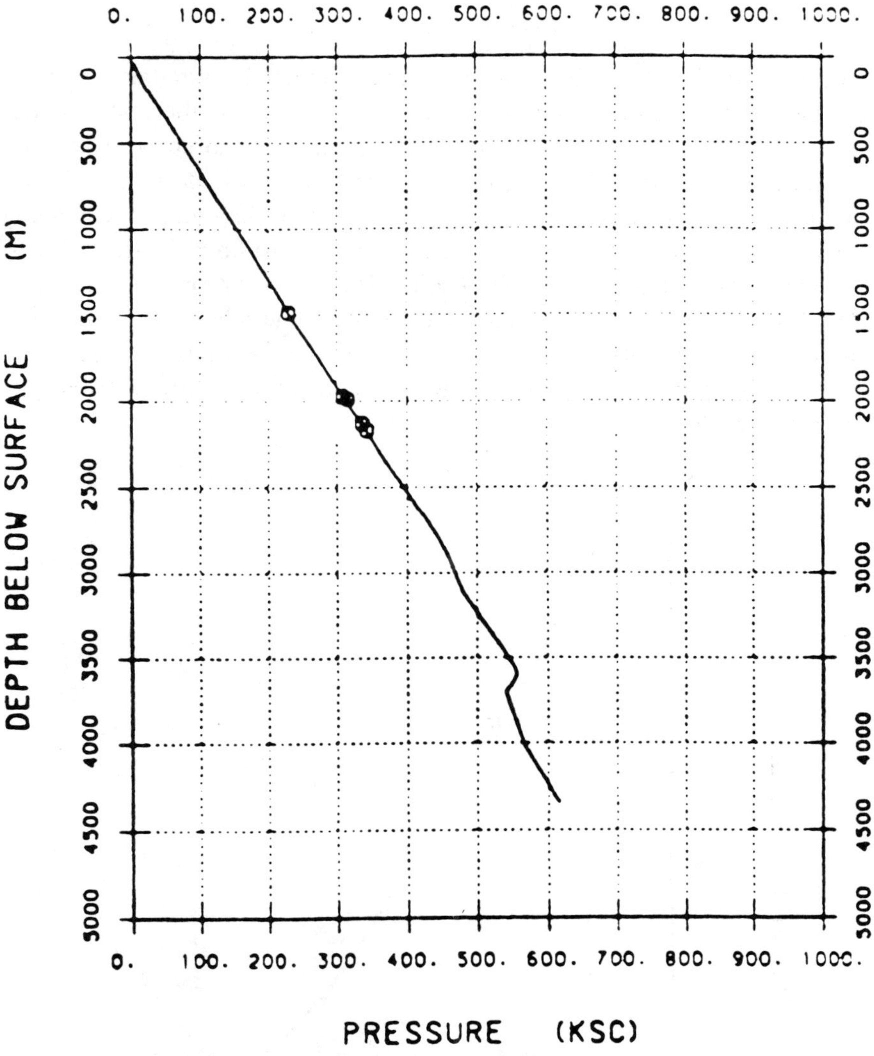

FIGURE 7.27 Pressure distributions calculated (solid line) and observed (symbol). KSC, kilograms per square centimeter.

3. Present-Day Heat Flow

The present-day heat flow can be back-calculated from the observed temperatures, although nonlinear relationships of the thermal conductivity with depth make it difficult to back-calculate directly. Since the temperature is one of the output variables from the system, a trial-and-error method was used to obtain present-day heat flow values. Changing the heat flow values within a dynamic range, we can find the best value to fit the observed temperature distribution (Fig. 7.28). The present-day heat flow (Q_0) of 1.50 HFU is a quite reasonable value when compared with direct measurements of heat flow over the area characterized by a back-arc tectonic setting. Figure 7.29 shows the temperature distributions from the model.

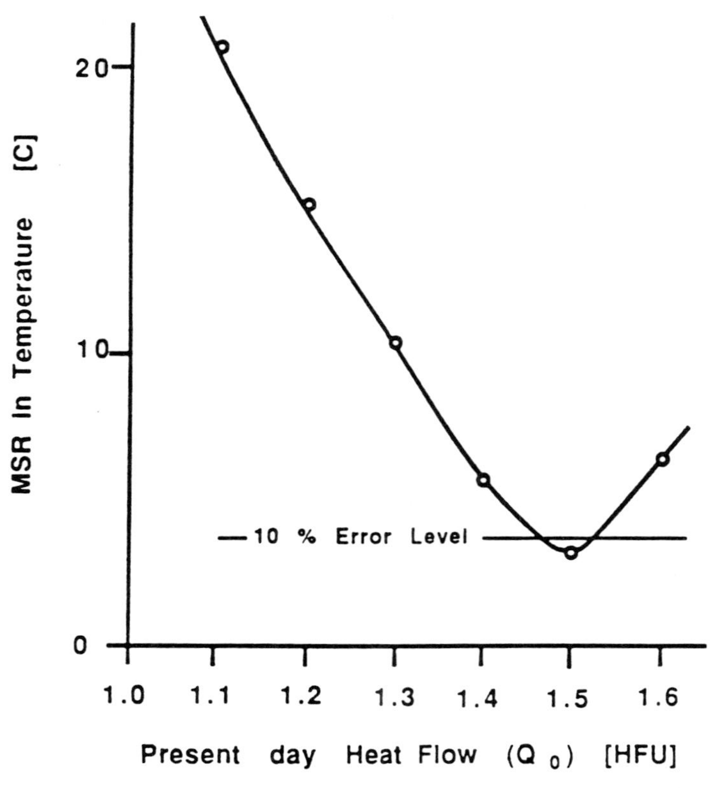

FIGURE 7.28 Mean square residuals of temperature between observed and calculated values as a function of present-day heat flow (Q_0).

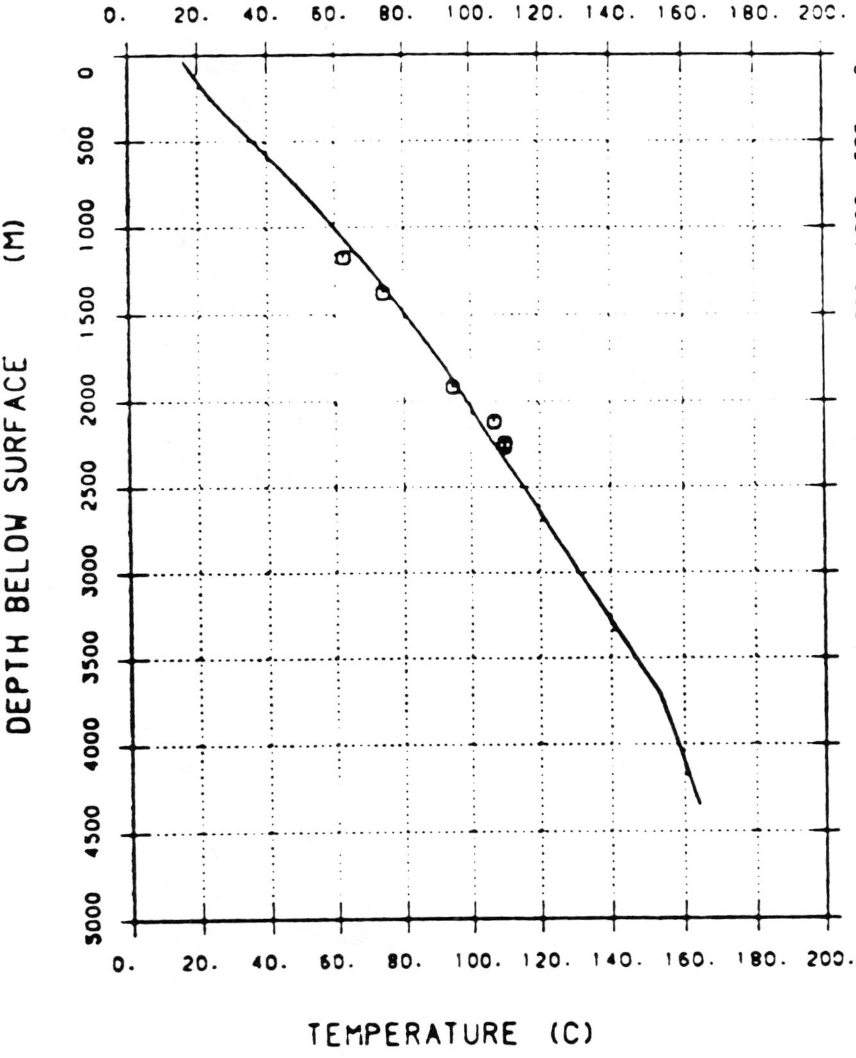

FIGURE 7.29 Temperature distributions calculated (solid line) and observed (symbol).

4. Paleoheat Flow

The paleoheat flow is not a directly measurable parameter but must have influenced the values of any maturation index, vitrinite reflectance in this application. A trial-and-error method is again used to determine the paleoheat flow value by the technique of Lerche *et al.* (1984). The value of the fractional heat flux coefficient β is varied and the mean square residuals between the calculated and the observed temperatures are calculated (Fig. 7.30). The value $\beta = -0.02\ (\mathrm{Ma})^{-1}$ gives the best fit to the observed data, but the error bars of uncertainty are large. This uncertainty suggests that the β value is not sensitive to the vitrinite reflectance values because we have data only at the shallow part of the section (less than 6 MYBP). The heat flow probably did not change significantly during the last 20 Ma, a rather short interval of geologic time (Ogura, 1981). The fact that β is negative implies a warming trend from past to present, which seems to have occurred only recently when the main maturation took

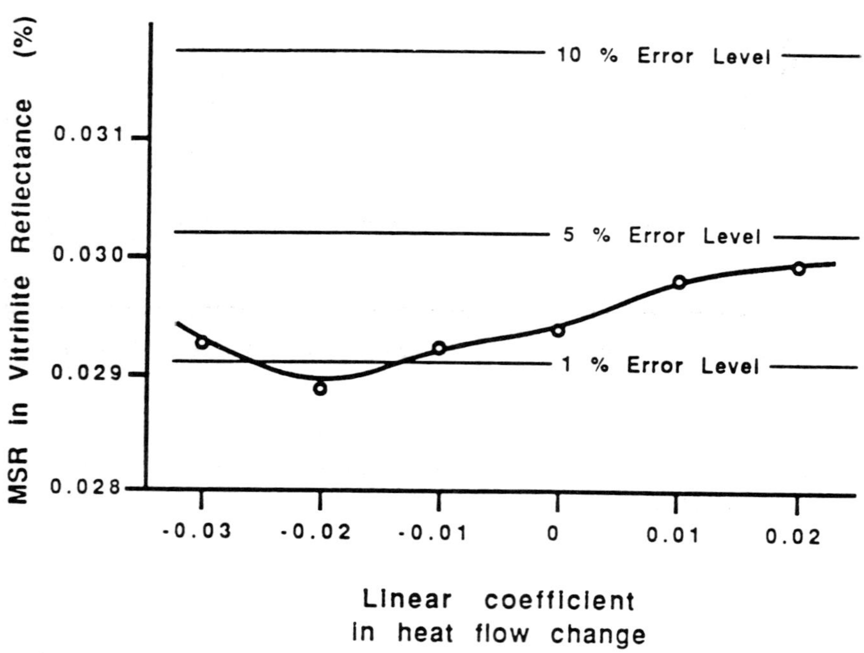

FIGURE 7.30 Mean square residuals of vitrinite reflectance between observed and calculated values as a function of linear change in heat flow coefficient β, in (million years)$^{-1}$.

place. Heat flow prior to 6 MYBP did not influence the maturation history because the total burial involved was limited. The vitrinite reflectance distribution from the model is shown in Fig. 7.31.

5. Blind Test

To test the appropriateness of the calibrated parameters, a blind test was performed. Well Y is near the basin center but southwest (about 30 km) of the structure. Well Y is not related structurally to the Iwafune-Oki area; however, the sedimentation pattern has the same characteristics. Treating well Y as being in an equivalent setting at the basin center westward of well X, the model was run with the parameter values selected as above, and again the mean square residuals were calculated for the four

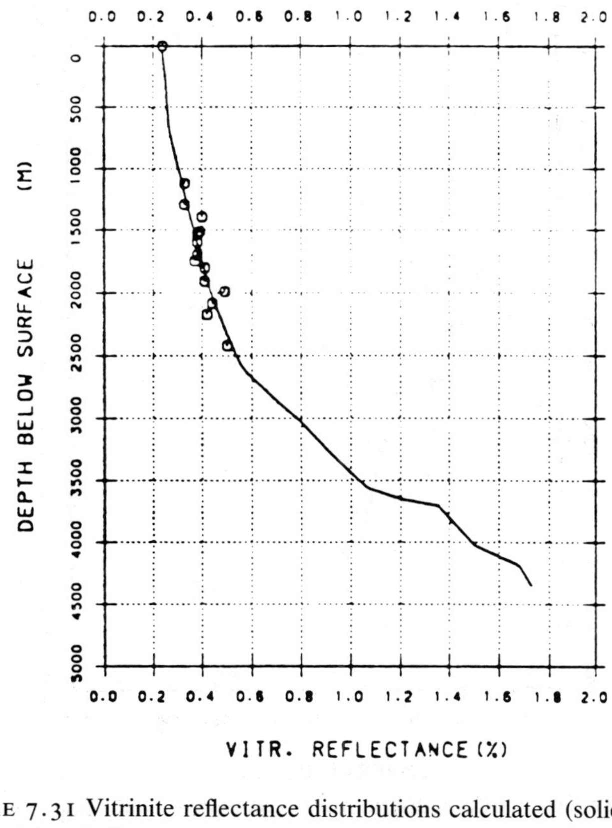

FIGURE 7.31 Vitrinite reflectance distributions calculated (solid line) and observed (symbol).

variables. At first, we suppose that only lithologic information is given. The error in the estimate of temperature distribution (Fig. 7.32) is so large that it would give an extremely erroneous maturation history. Then some of the temperature measurements are given to adjust the present-day heat flow (Q_0). The value of Q_0 at well Y is 1.15 HFU, compared with 1.50 HFU at the position of well X in the Iwafune-Oki area. Using this value with the previously found parameter set, we obtain better predictions of porosity, pressure, and temperature (Figs. 7.33, 7.34, and 7.35). For the vitrinite reflectance, the prediction does not fit values for the deeper parts of the section (Fig. 7.36), probably because the measurements from the

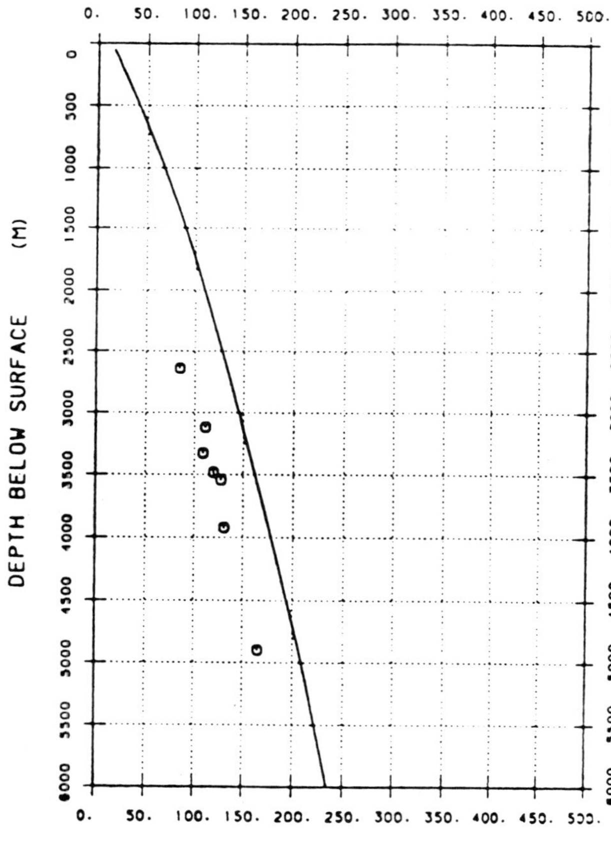

FIGURE 7.32 Temperature distributions calculated (solid line) and observed (symbol) at well Y, results of the blind test. Q_0 = 1.5 HFU was used.

Iwafune-Oki field were limited to the shallower depths and were insensitive to paleoheat flow (Fig. 7.30). The implications are that either the linear coefficient β was much lower near well Y or the calculated conversion factor was not accurate enough to estimate the vitrinite reflectance values in the deeper section. This sort of uncertainty should be taken into account when the outputs from models are interpreted since it affects the sensitivity of the results. If we did the blind test in the converse mode (i.e., knowing the well data basinward, predict the parameters in the structural high region), the error bar would be much smaller. Information from the deeper parts of a basin is thus very important.

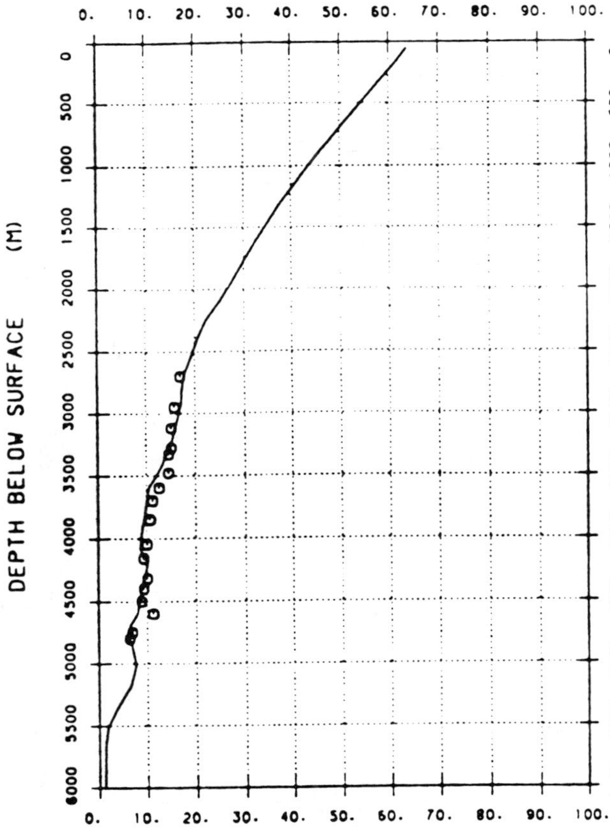

FIGURE 7.33 Porosity distributions calculated (solid line) and observed (symbol) at well Y, results of the blind test.

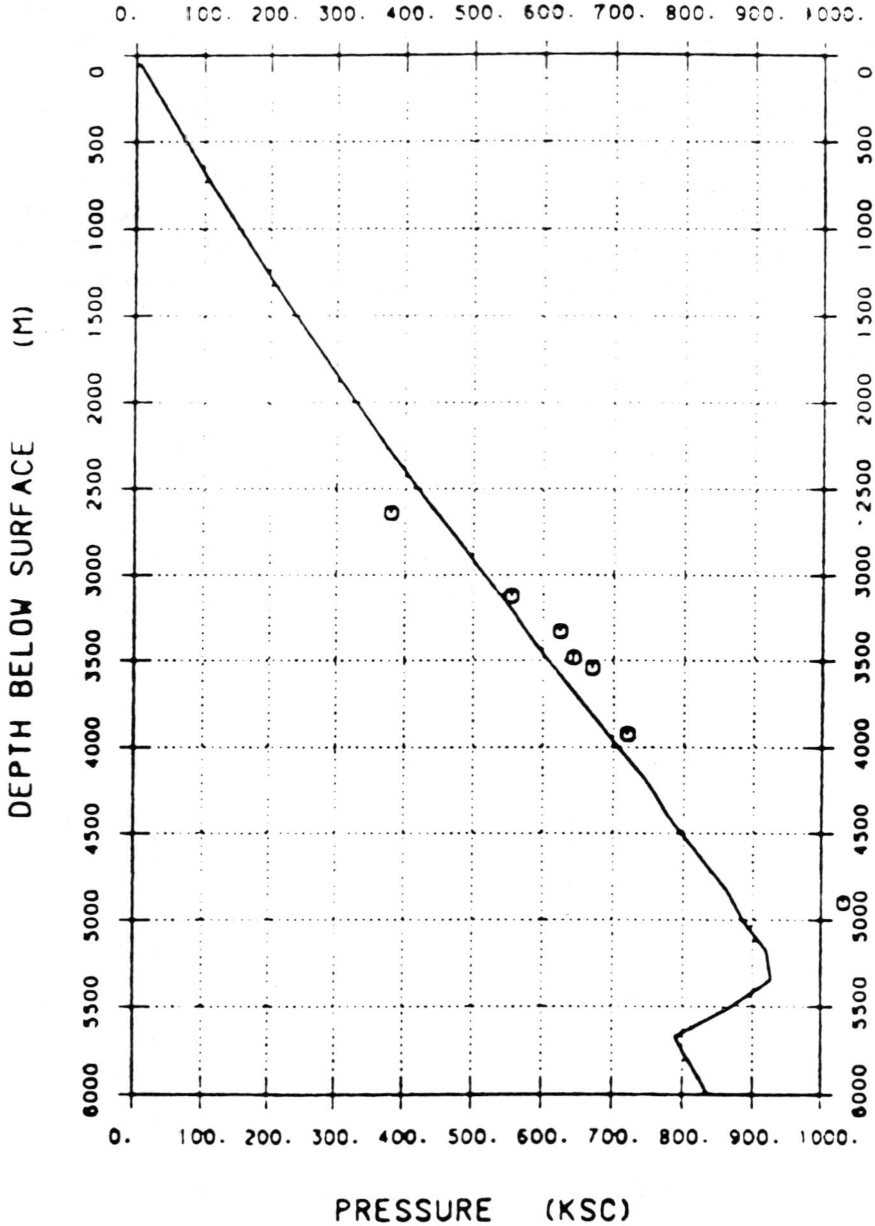

FIGURE 7.34 Pressure distributions calculated (solid line) and observed (symbol) at well Y, results of the blind test.

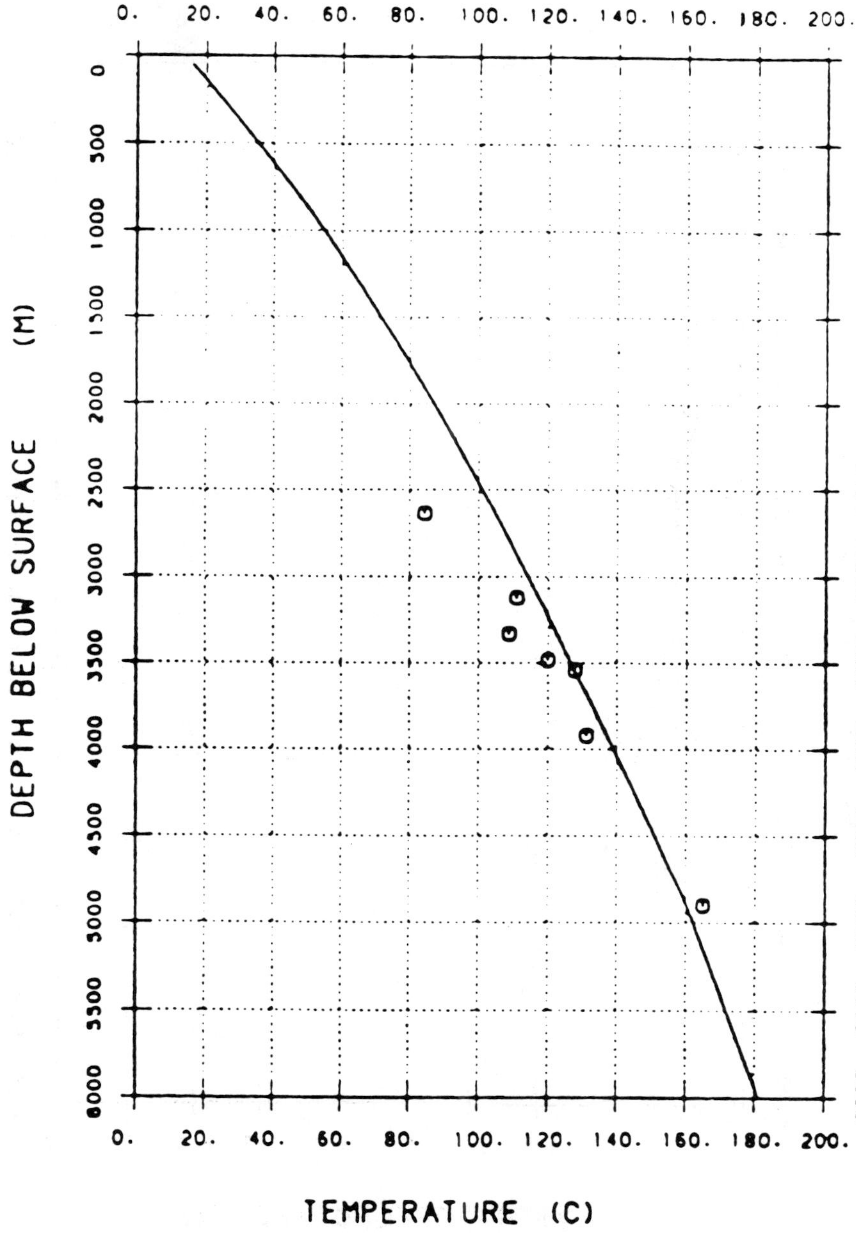

FIGURE 7.35 Temperature distributions calculated (solid line) and observed (symbol) at well Y, results of the blind test.

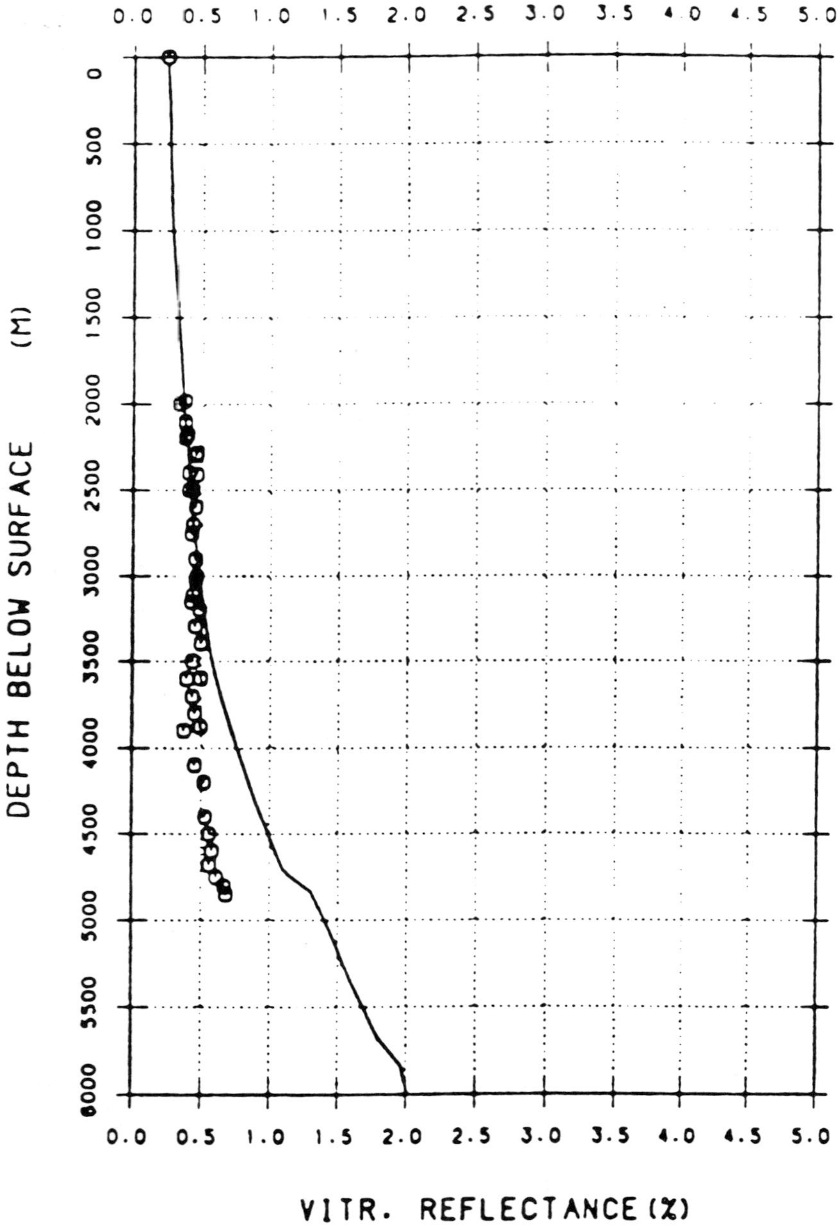

FIGURE 7.36 Vitrinite reflectance distributions calculated (solid line) and observed (symbol) at well Y, results of the blind test. The values below 3500 m do not fit well.

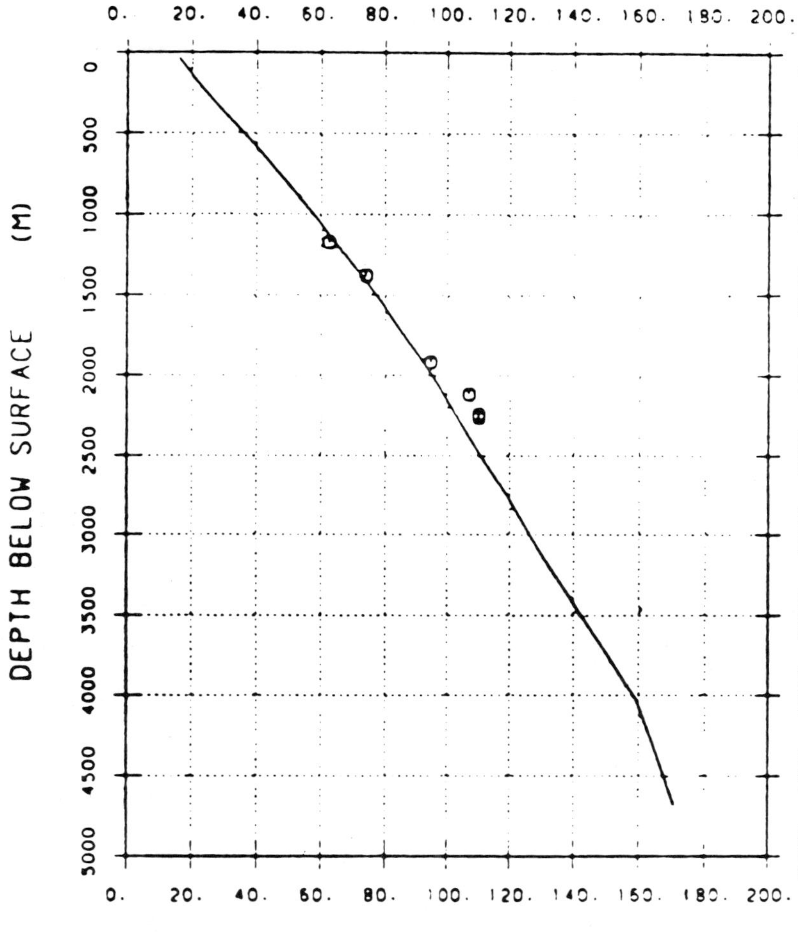

FIGURE 7.37 Temperature distributions calculated (solid line) and observed (symbol) after two-dimensional (2-D) modeling. Calculated values are slightly lower than in the one-dimensional case (Fig. 7.29).

C. Basin Analysis Using the Outputs

After the previously defined regional parameters were modified by the information from well Y in the deeper part of the basin (Table 7.6), the model was run for the entire section to characterize the manner of petroleum generation and migration in this basin.

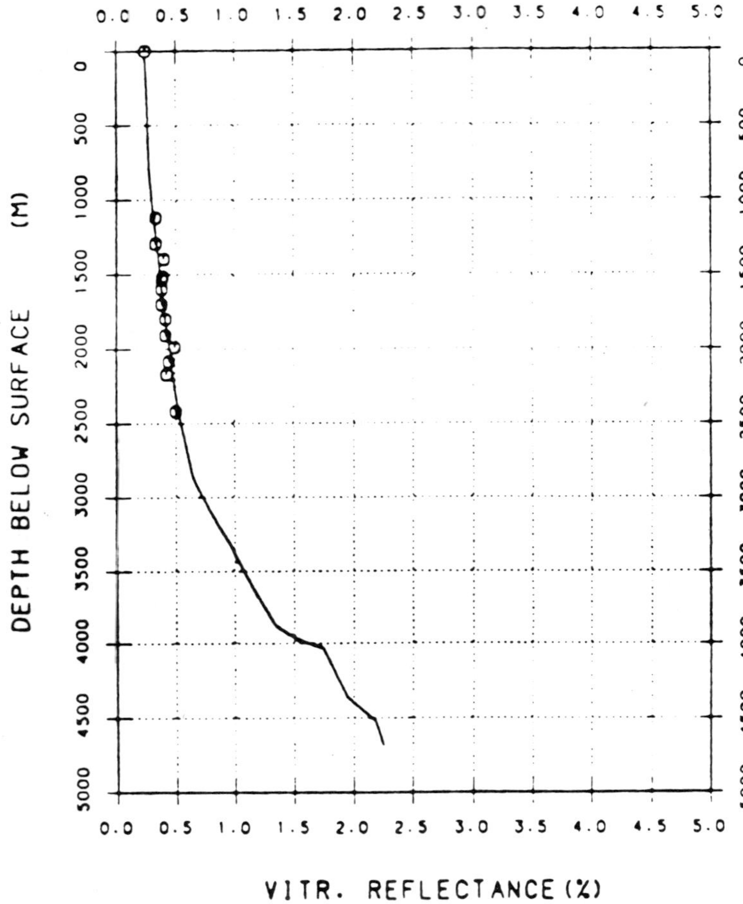

FIGURE 7.38 Vitrinite reflectance distributions calculated (solid line) and observed (symbol) after 2-D modeling. A two-dimensional effect is not visible.

1. Point Data Analysis

a. Two-Dimensional Effects. The calibration of the input parameters was done by using a one-dimensional geologic setting, whereas the actual process of simulation used a two-dimensional situation. The difference between the temperature and the vitrinite reflectance predictions (Figs. 7.37 and 7.38) for the two models is essentially negligible. In the case of the pressure prediction, the values are slightly lowered (Fig. 7.39) although higher values were expected because fluids should concentrate and increase the pressure within the structural high. As a result of

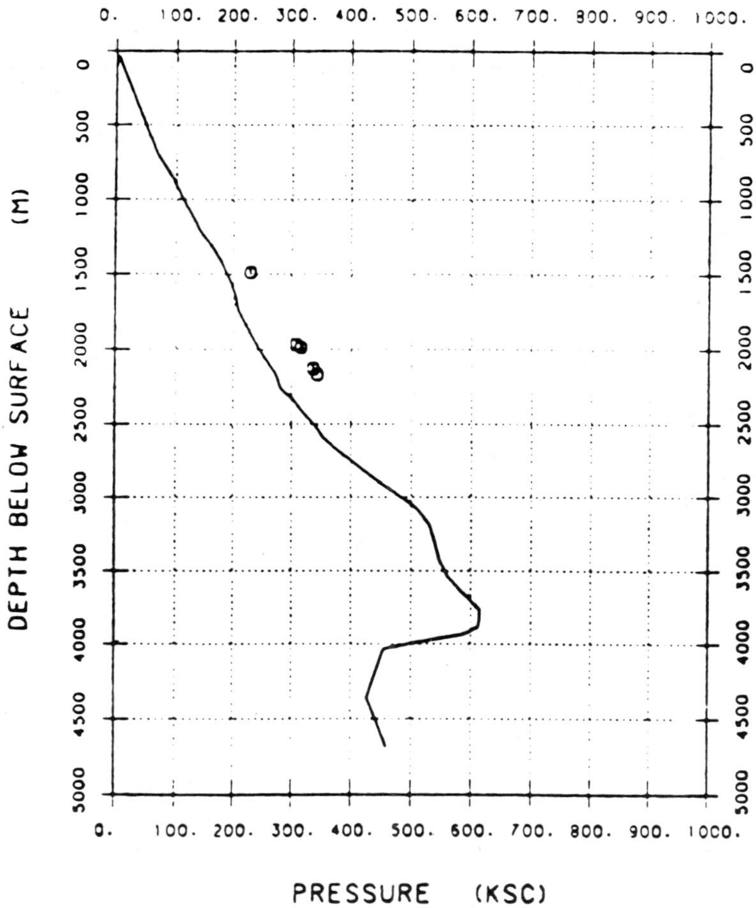

FIGURE 7.39 Pressure distributions calculated (solid line) and observed (symbol) after 2-D modeling. Calculated values are significantly lower because of the lateral movement of the fluid.

pressure reduction, the porosity values are also lowered (Fig. 7.40). This phenomenon may be accounted for by the regional pressure distribution and the structural development. Well X was not always located at the culmination of structure; therefore the fluid is escaping laterally. Such lateral migration does not take place in the one-dimensional case. A calibration procedure is again necessary to adjust for the two-dimensional effect. The values of b_{sh} and b_{ss} then increased slightly ($b_{sh} = 5.0$, $b_{ss} = 8.0$) and e_{*sh} changed to 2.50 to obtain a best fit for the two-dimensional section.

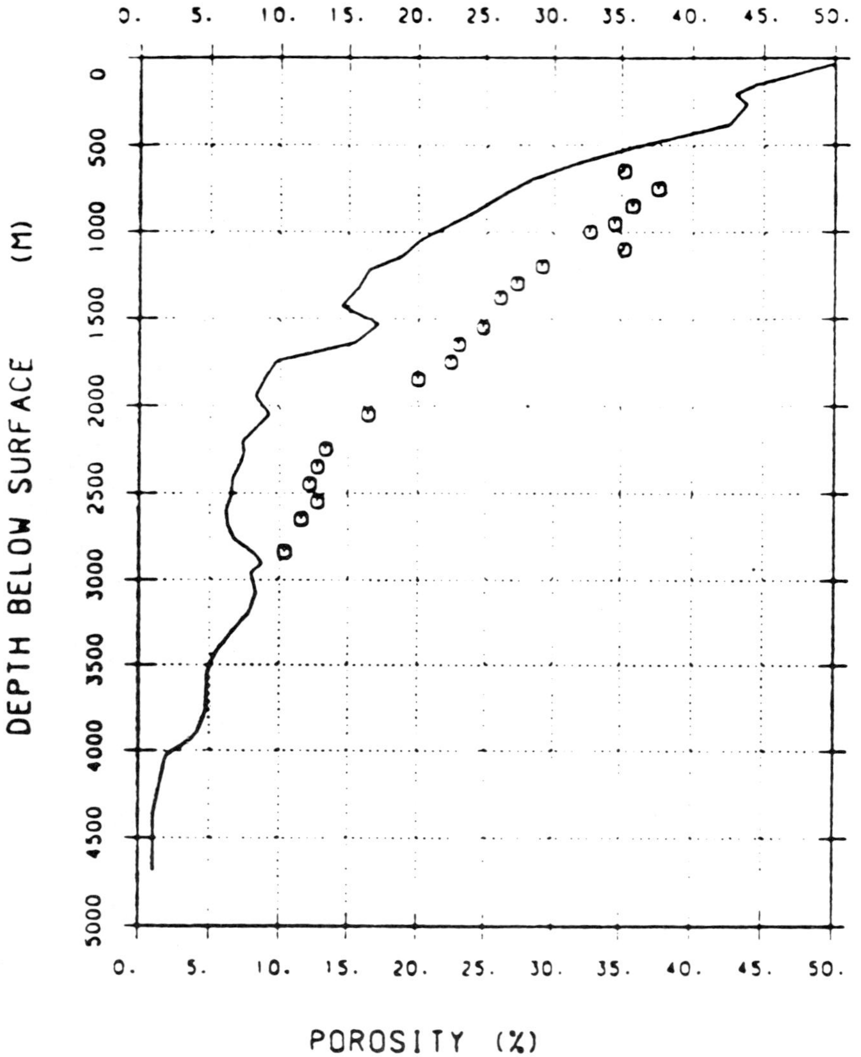

FIGURE 7.40 Porosity distributions calculated (solid line) and observed (symbol) after 2-D modeling. Calculated values are significantly lower because of the reduction of pressure.

b. Basin Center Position. The same kinds of point data outputs are available for the basin center position. The trends of porosity, pressure, and temperature with depth are almost the same regardless of the locations (Figs. 7.41, 7.42, and 7.43). However, if we compare the samples from the same depths (shallower than 4500 m), the vitrinite

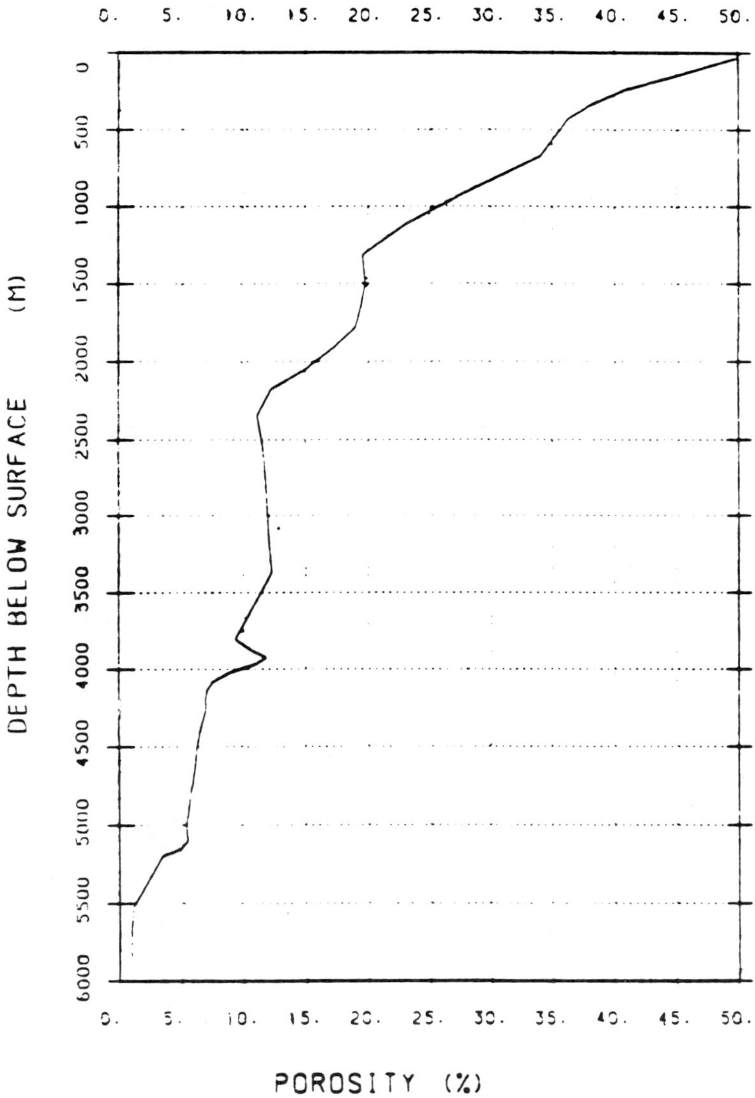

FIGURE 7.41 Predicted porosity distribution at a location in the basin center within 2-D model.

reflectance values are lower at the basin center (Fig. 7.44). This is caused by the difference of time involved in the maturity of vitrinite reflectance, as a result of which the same maturation level occurs deeper in the basin center. The pressure trend with depth at the basin center is the same as that at the well location—probably because of the small distance between

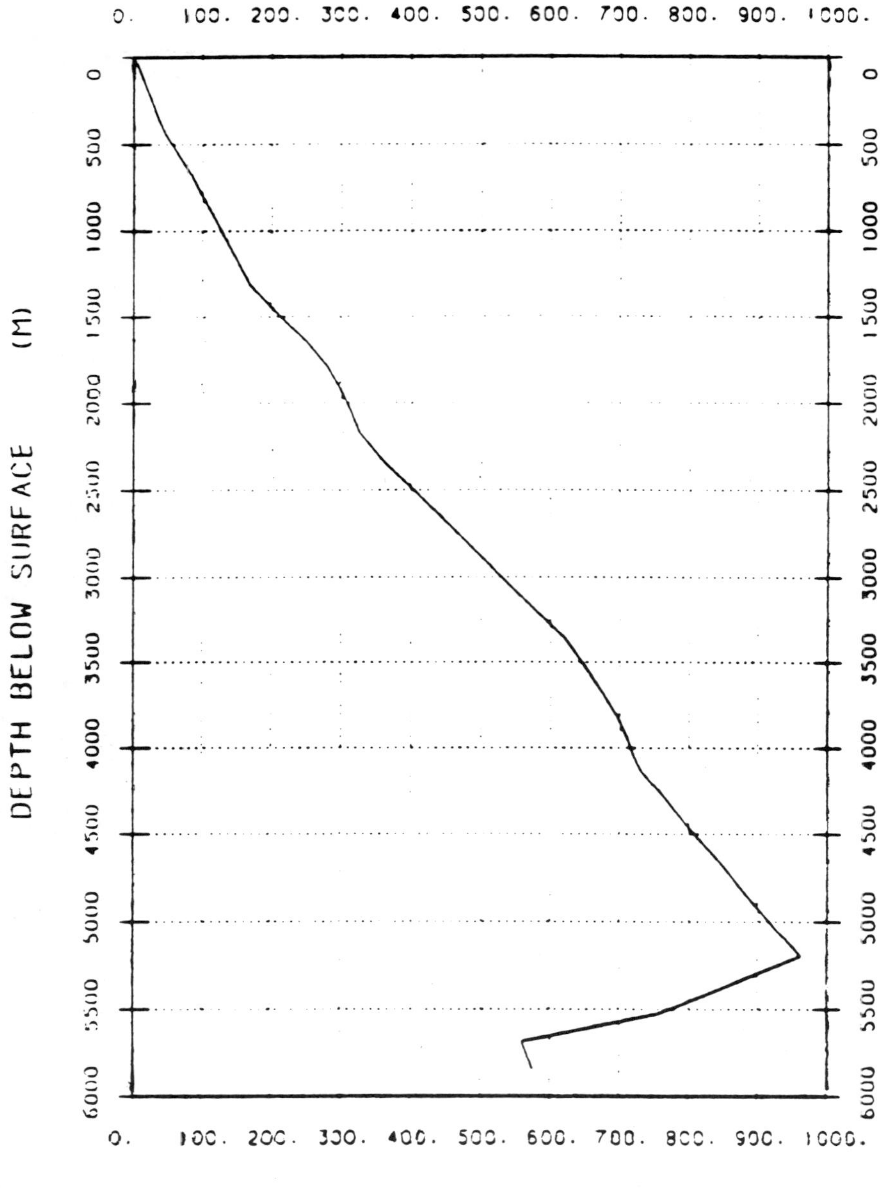

FIGURE 7.42 Predicted pressure distribution at a point in the basin center within the 2-D model.

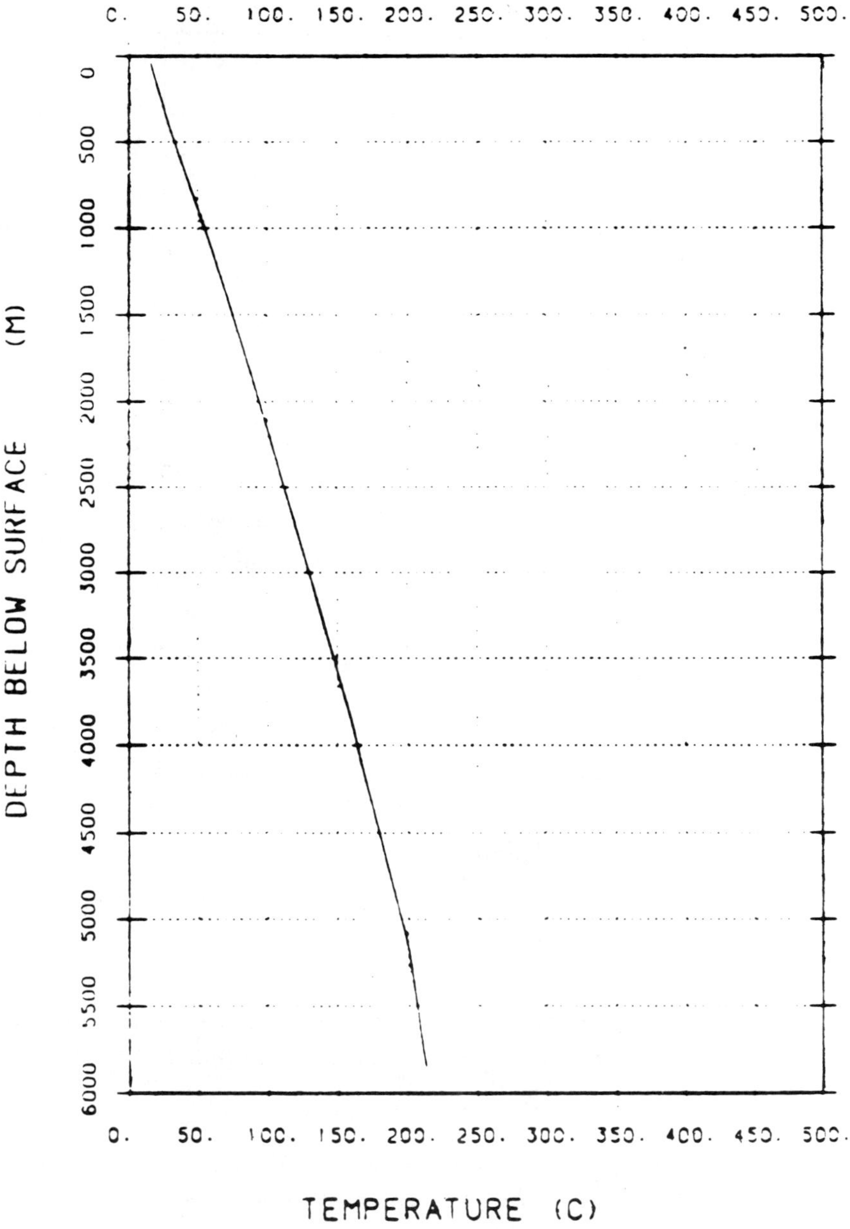

FIGURE 7.43 Predicted temperature distribution at a point in the basin center within the 2-D model.

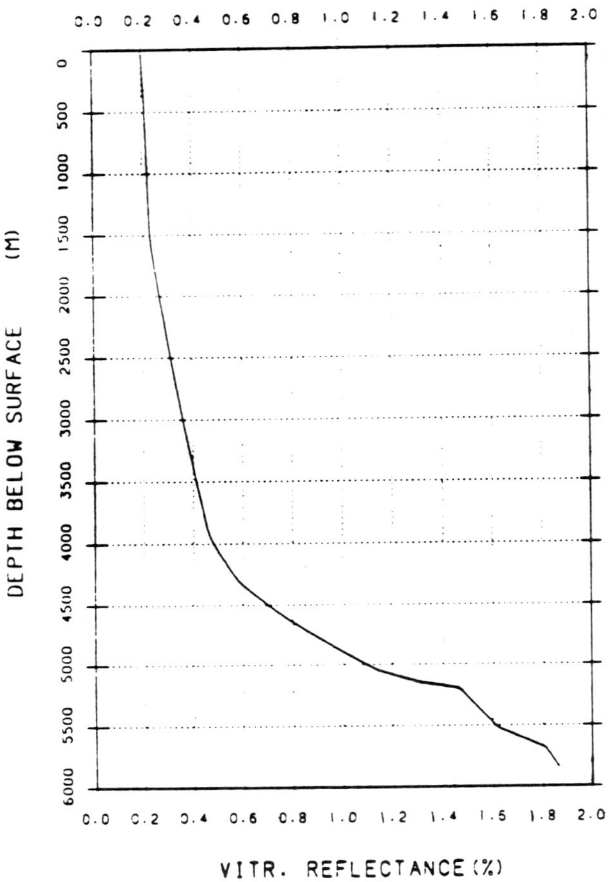

FIGURE 7.44 Predicted vitrinite reflectance distribution at a point in the basin center within the 2-D model.

them. In this application the total width of the section is only 6 km; therefore any lateral pressure difference equilibrates rather quickly.

2. *Geohistory Analysis*

Geohistory analyses are also available at the two points: one at well location X and the other at the basin center, well location Y.

a. Temperature Development (Fig. 7.45a and b). The isotemperature lines in the past change slightly in terms of depth since the paleoheat flow determined from inversion of vitrinite reflectance data is lower than the present heat flux. Through these outputs from the model we may examine the temperature history of particular formations.

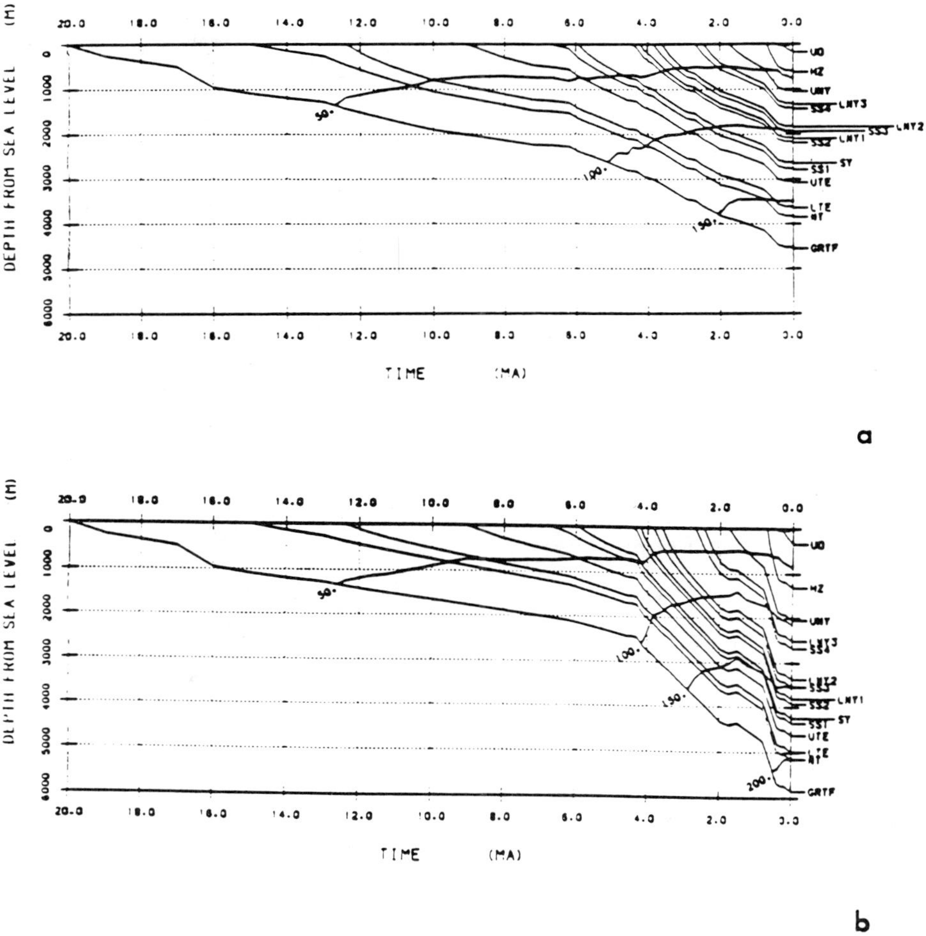

FIGURE 7.45 Geohistory diagram with temperature development (a) at the well location and (b) at the basin center within the 2-D model.

b. Maturation Development (Figs. 7.46a and b and 7.47a and b). Waples' TTI value at 3200 m depth is only 15 (Fig. 7.46), whereas the vitrinite reflectance value at the same depth is about 0.48 (Fig. 7.47). In addition, the characters of maturation development based on vitrinite reflectance and on TTI are manifestly different on the geohistory diagrams. This is a typical example in which the Waples TTI value does not calibrate correctly against observed data because the extremely rapid sedimentation (rapidly increasing temperature) causes a different maturation history than that predicted by Waples' TTI. Maturation reliability is retained by the predicted vitrinite values, since they are in accord with the observed

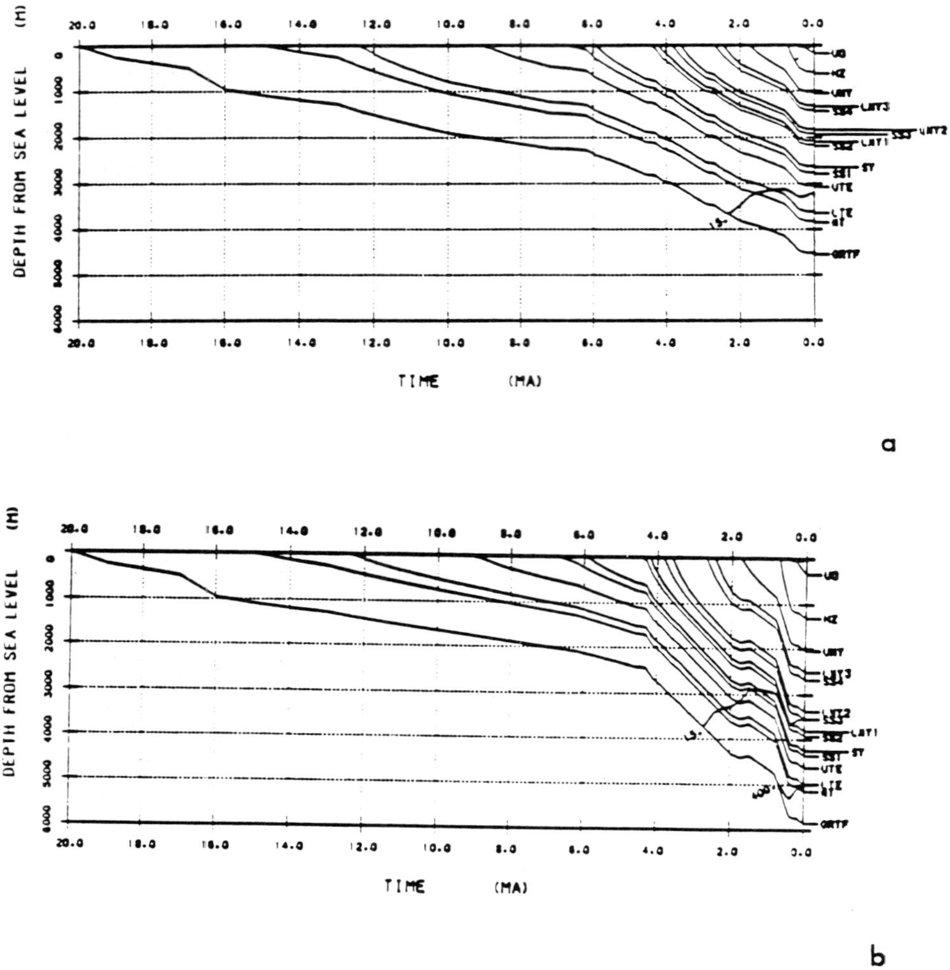

FIGURE 7.46 Geohistory diagram with Waples' TTI development (a) at the well location and (b) at the basin center within the 2-D model.

data. But, in either case, we observe that the maturation history is quite different at well X and at the basin center after 4.3 MYBP.

c. Basement Subsidence (Fig. 7.48a and b). Subtracting the sediment load effect from the total subsidence, we obtain the tectonic subsidence through time. At both at well X and the basin center, the rate of tectonic subsidence changed at 6.0 MYBP. However, it should be noted that the paleobathymetry was assumed to be zero in this application since reliable data were not available.

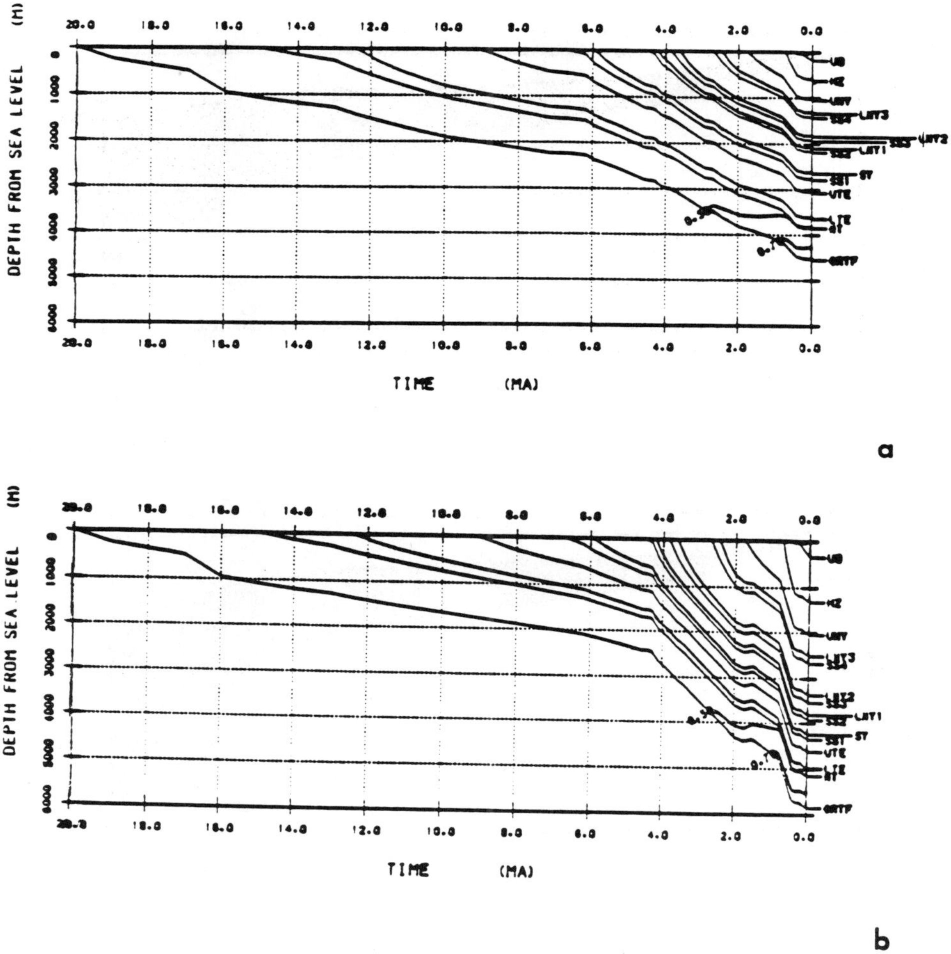

FIGURE 7.47 Geohistory diagram with vitrinite reflectance development (a) at the well location and (b) at the basin center within the 2-D model.

d. Excess Pressure (Fig. 7.49a and b). Excess pressure mainly developed after 4 MYBP at both well X and the basin center. However, the magnitude is greater at the basin center than at well X. This difference of excess pressure drives the lateral, eastward-trending fluid flow with time.

e. Oil/Gas Generation (Figs. 7.50a and b and 7.51a and b). The rate of oil/gas generation is expressed in terms of weight (milligrams) that a unit weight of rock can produce during 1-Ma interval. At well X, oil generation (0.50 mg/g rock) occurred only for the Nanatani through the upper

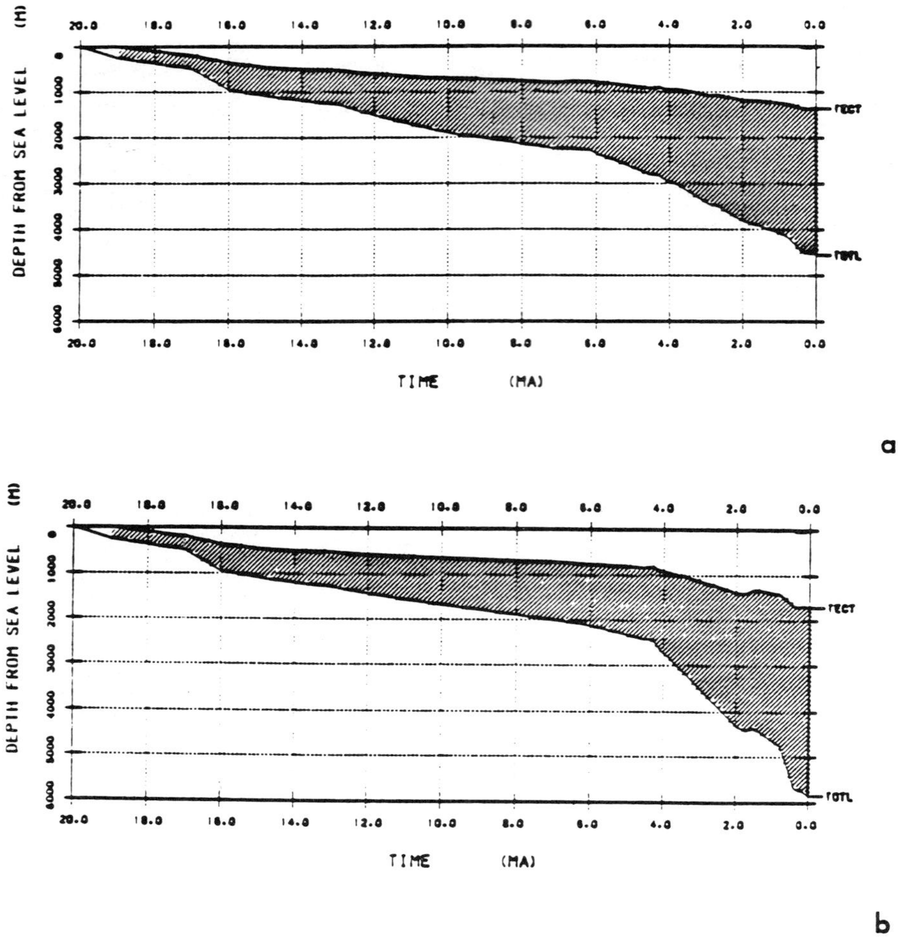

FIGURE 7.48 Total subsidence and tectonic subsidence as a function of time (a) at the well location and (b) at the basin center within the 2-D model.

Tertiary formations after 3.0 MYBP (Fig. 7.50a), whereas in the basin center it occurred for the wider range of formations from the Nanatani to the lower Nishiyama (Fig. 7.50b). It should be noted that the time of generation at the basin center is the same as that at well X. Gas generation (>0.50 mg/g rock) occurred in the same manner as oil generation. The time of generation for both locations is after 3.0 MYBP. However, the formations where gas was generated cover a wider portion of the basin center (Fig. 7.51a and b).

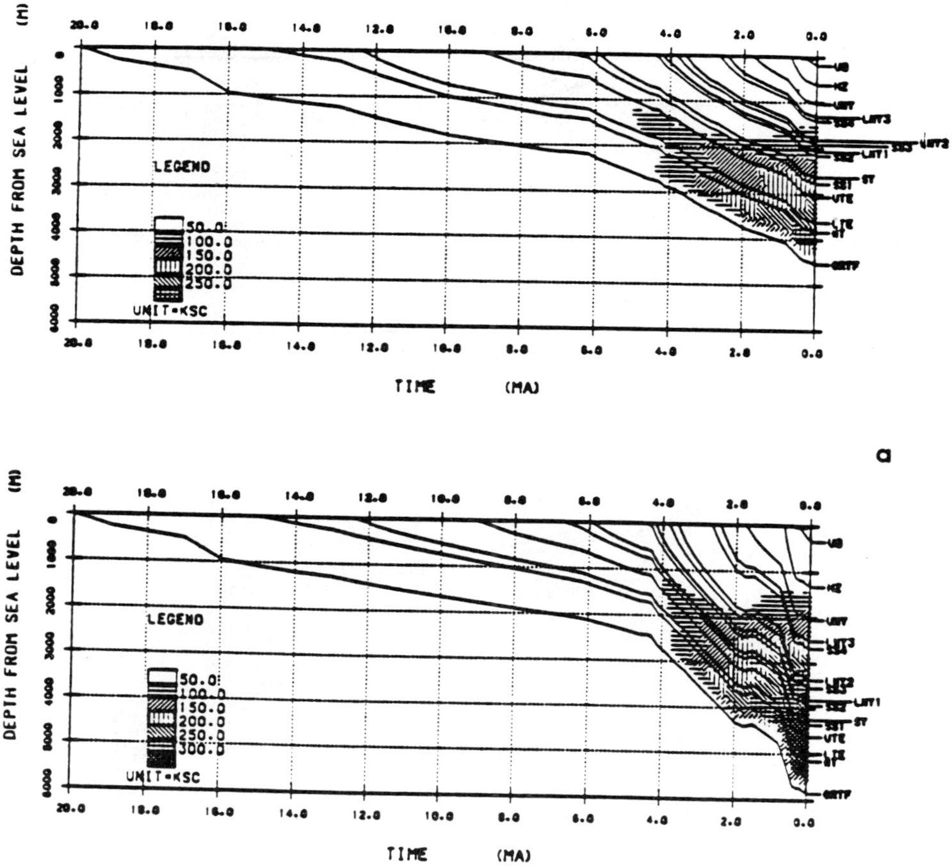

a

b

FIGURE 7.49 Geohistory diagram with excess pressure development (a) at the well location and (b) at the basin center within the 2-D model.

f. Oil/Gas Saturation (Figs. 7.52a and b and 7.53a and b). Oil/gas saturation represents a relative amount of the percentage of the pore space. Oil saturation started to increase in the SS2 and SS3 layers after 3.0 MYBP, coinciding with the time of generation in the Nanatani (NT) and Lower Teradomari (LTD) formations at the same location. However, oil trapped in the SS2 and SS3 reservoirs does not come from the lower source rocks of this area. As the two-dimensional view will reveal later, the oil comes from basinward locations. This behavior emphasizes the

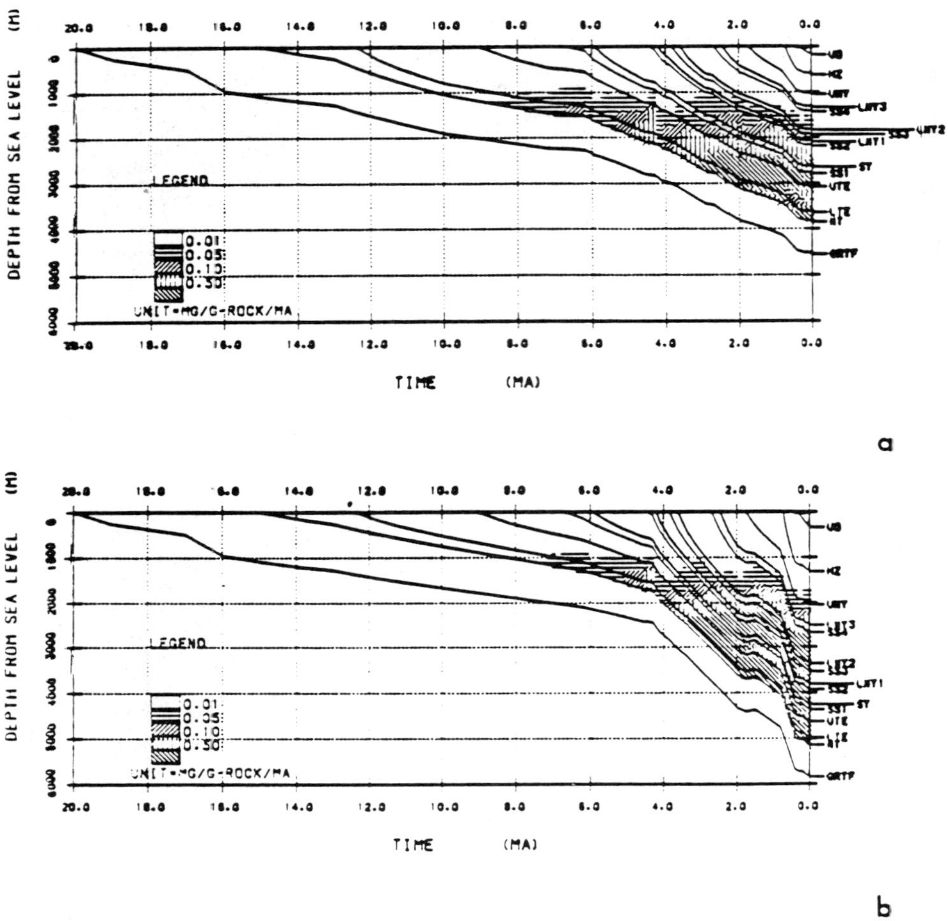

FIGURE 7.50 Geohistory diagram with oil generation rate (a) at the well location and (b) at the basin center within the 2-D model.

points that hydrocarbon generation and accumulation are totally different phenomena which involve the consideration of two-dimensional or even three-dimensional effects. Note that oil begins to accumulate in the reservoirs at well X while no generation takes place around them. The same pattern can be recognized on the gas saturation diagram (Fig. 7.53a and b).

g. Oil/Gas Accumulation (Figs. 7.54a and b and 7.55a and b). Because the value of oil/gas saturation is a relative percentage or fraction, the *absolute* amount (in terms of the volume at the surface in stock tank

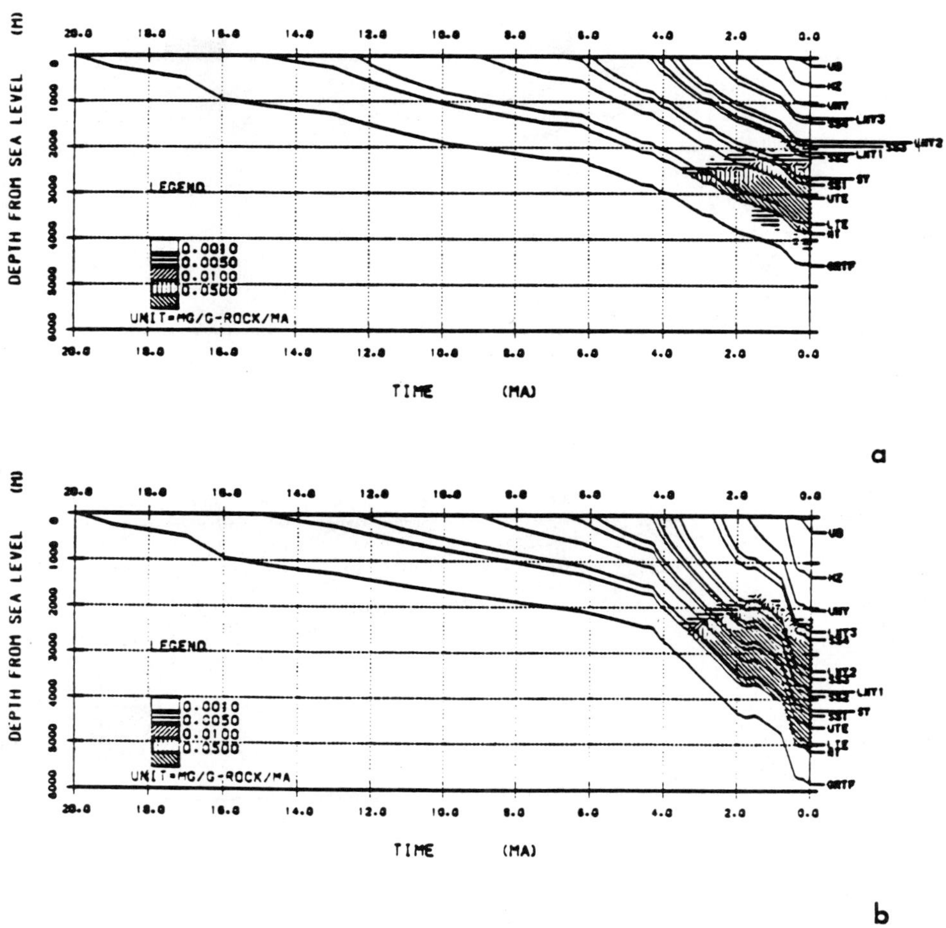

FIGURE 7.51 Geohistory diagram with gas generation rate (a) at the well location and (b) at the basin center within the 2-D model.

barrels that a unit volume of rock contains at an instant of time) is also mapped on the geohistory diagram and illustrates the scattered distribution of over the area, whereas the gas is concentrated in the reservoir. The implication is that the migration of oil is less efficient than that of gas per unit time, predominantly because of the lower buoyancy of oil.

h. Timing Chart Analysis. The pictorial representation of information on a geohistory diagram is rather qualitative. In this section, we introduce a more effective way to express the timing of petroleum generation and accumulation.

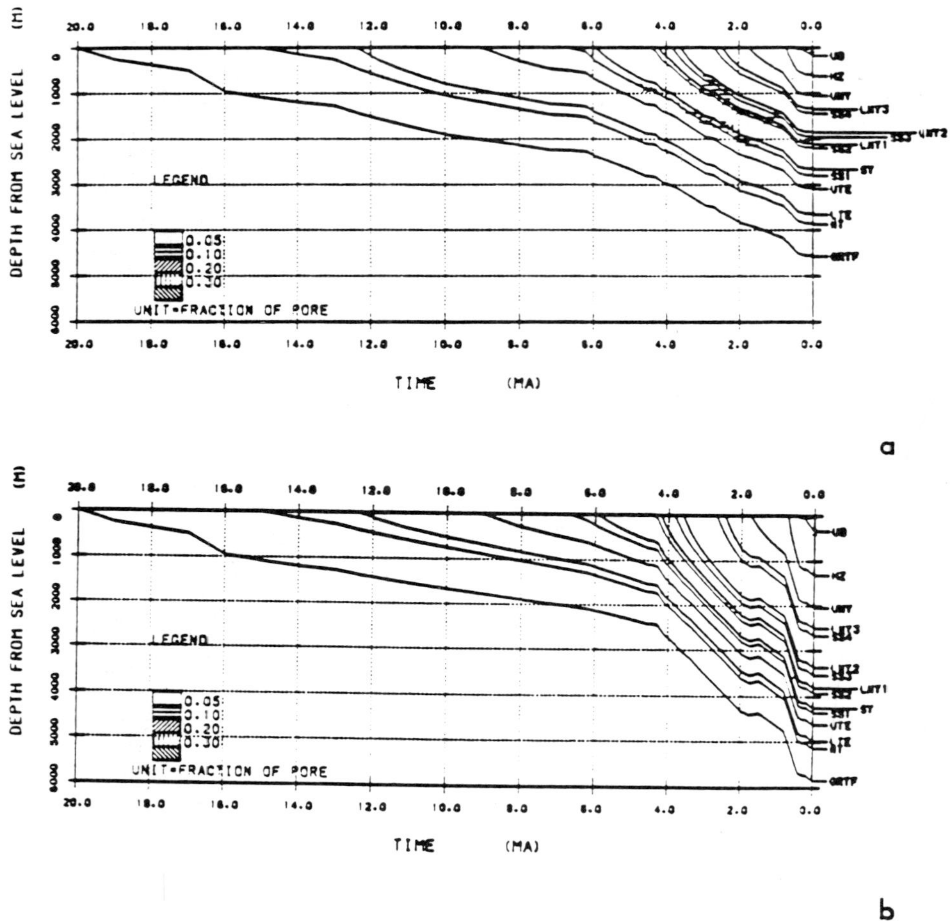

FIGURE 7.52 Geohistory diagram with oil saturation development (a) at the well location and (b) at the basin center within the 2-D model.

Parameter values for a particular geographic location, depth, and time are averaged or summed according to the formation or unit. Then the values are normalized using the maximum value to represent a *relative* magnitude for that formation. These values are plotted on the time axis for each formation. Although we can choose any output parameter to be represented in this way, usually those from the oil/gas generation rate (Figs. 7.56 and 7.57), oil/gas saturation (Figs. 7.58 and 7.59), and oil/gas accumulation (Figs. 7.60 and 7.61) are presented. These example figures for well X clearly show the timing of filling the reservoir. The oil/gas

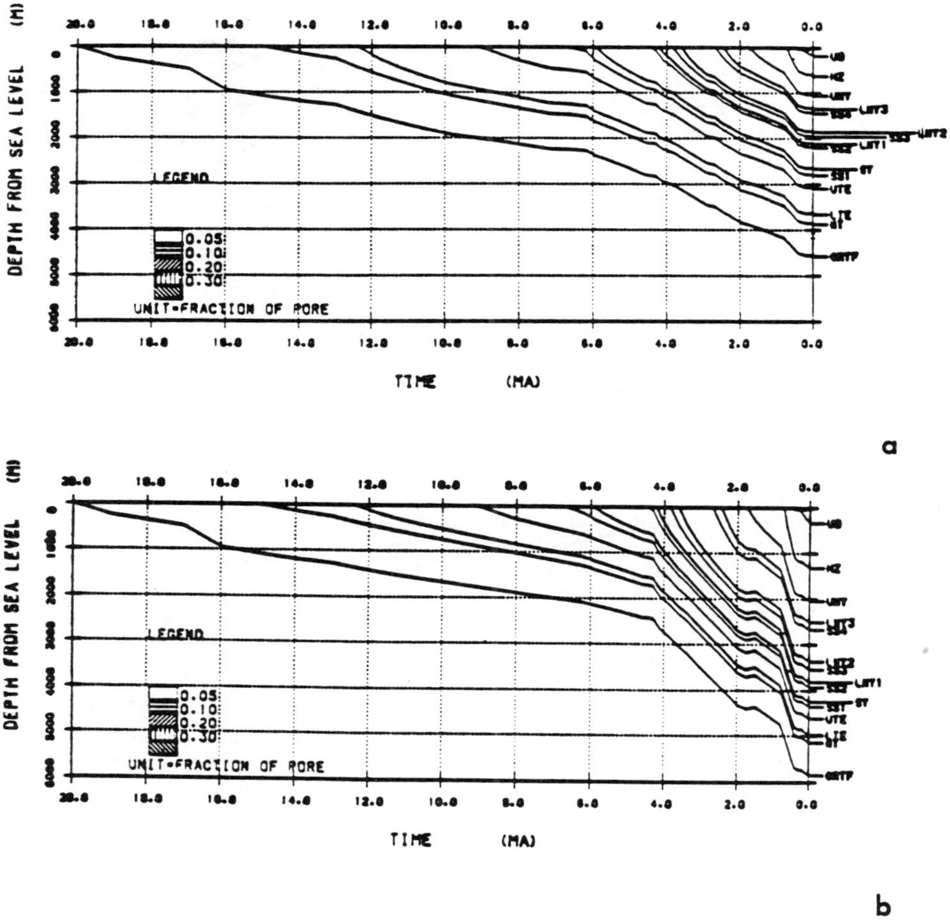

FIGURE 7.53 Geohistory diagram with gas saturation development (a) at the well location and (b) at the basin center within the 2-D model.

generation rate of the NT, LTE, and UTE formations is active only after 4–2 MYBP (Fig. 7.60). Some gas generation, seen in layer SS, occurred in recent years because the previously migrated oil was able to produce some gas.

D. Two-Dimensional Analysis

Two-dimensional graphic outouts are available for any geologic data at any time in the past. In this example, we analyze the section at six selected stages (6.2, 4.3, 2.0, 0.8, and 0.4 MYBP and present) during the basin development.

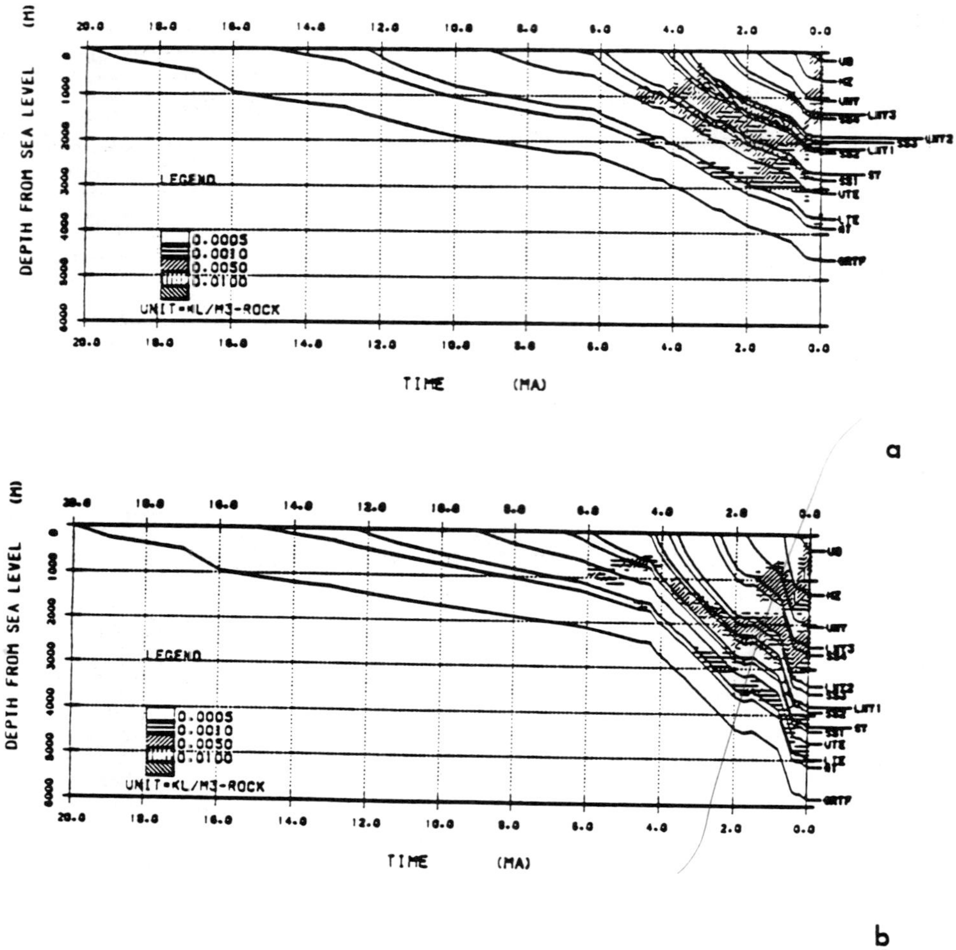

FIGURE 7.54 Geohistory diagram with oil accumulation development (a) at the well location and (b) at the basin center within the 2-D model.

1. Excess Pressure Distribution (Fig. 7.62a–f)

In general, the sequence of two-dimensional graphic outputs shows a very clear paleostructural development over the area. From Fig. 7.62a and b we deduce that the basin depocenter was near the well location until Shiiya age (4.3 MYBP). During the next age, the Nishiyama, the basin center shifted westward, producing the structural highs of the eastern part including the Iwafune-Oki structure at point 9 of the cross section.

Excess pressure developed very rapidly during the early Nishiyama stage (Fig. 7.62d). Since then the core of the excess pressure has been

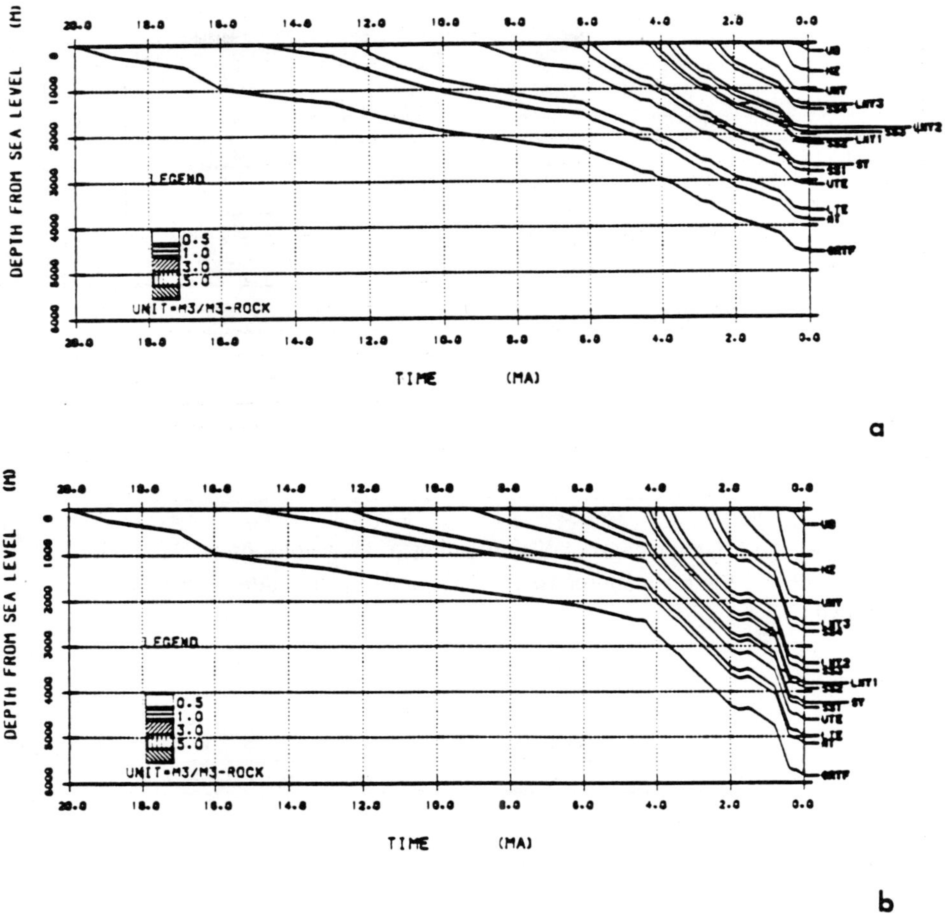

FIGURE 7.55 Geohistory diagram with gas accumulation development (a) at the well location and (b) at the basin center within the 2-D model.

located in the deepest part of the western section. The excess pressure is thought to be the main cause of downward migration from the Nanatani mudstone member to the Green Tuff member. However, the downward pressure gradient developed at the boundary only before 4.3 MYBP.

2. *Temperature Distribution (Fig. 7.63a–f)*

Since no lateral change of basement heat flux was taken into account across the section (6 km), the temperature distribution has a horizontal character through time. However, the excess pressure maintains a

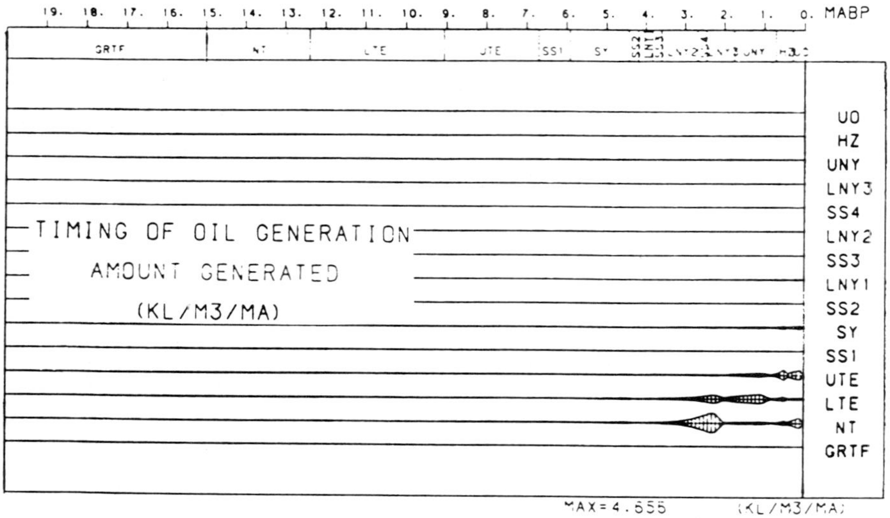

FIGURE 7.56 Timing chart of oil generation at the well location. NT, LTE, and UTE are the main source rocks. The time of generation is also clearly shown.

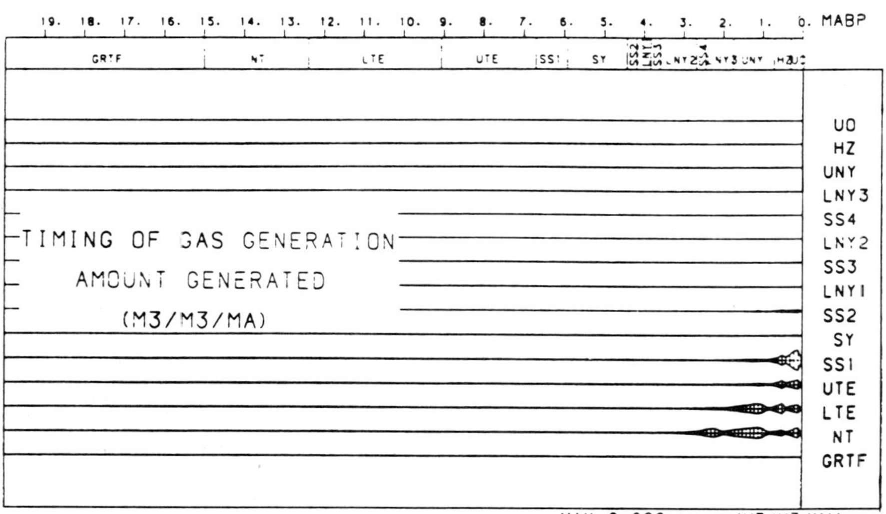

FIGURE 7.57 Timing chart of gas generation at the well location. The timing of generation is slightly shifted toward recent compared with oil generation (Fig. 7.56). Note that the generation in layer SS1 should come from the oil once trapped there.

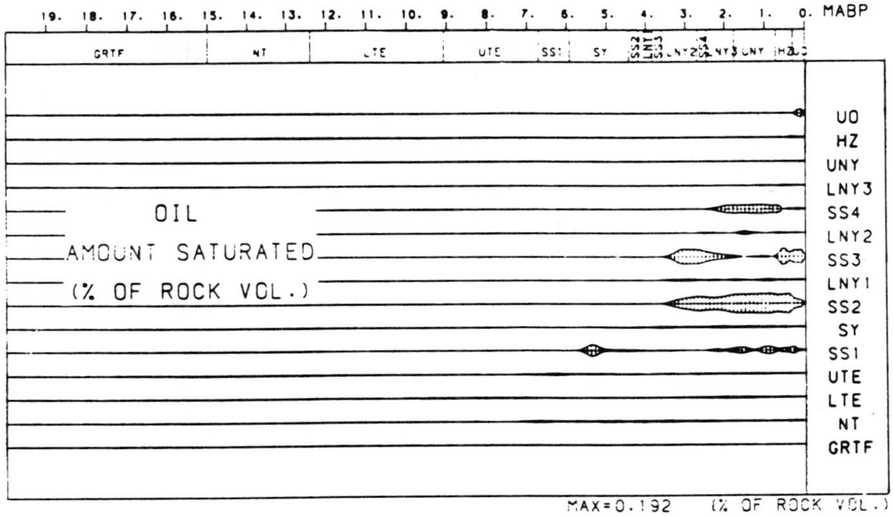

FIGURE 7.58 Timing chart of oil saturation at the well location. Most of the oil here is considered to have migrated from the basin.

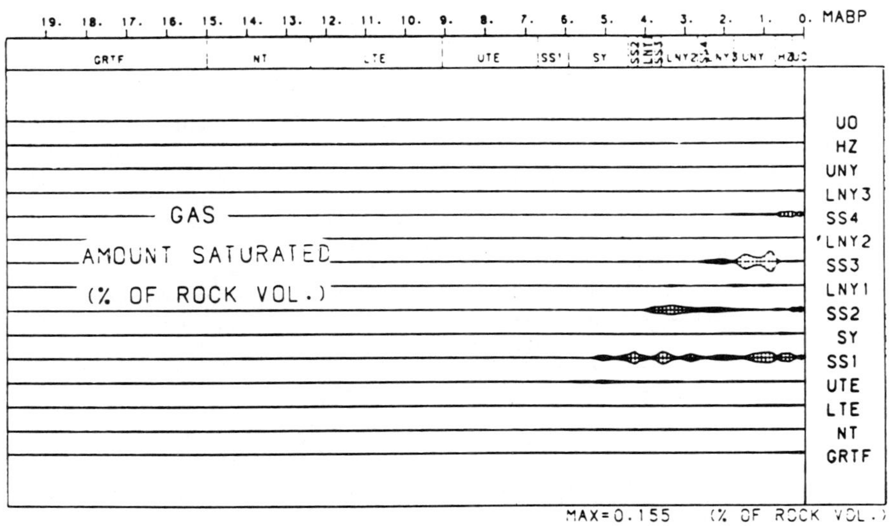

FIGURE 7.59 Timing chart of gas saturation at the well location.

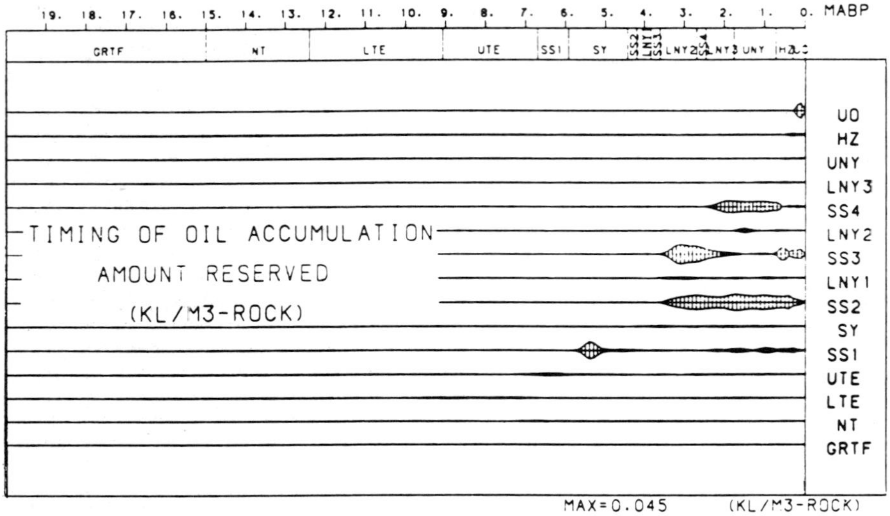

FIGURE 7.60 Timing chart of oil accumulation at the well location. The unit is the volume of stock tank at the surface that a unit volume or rock can produce.

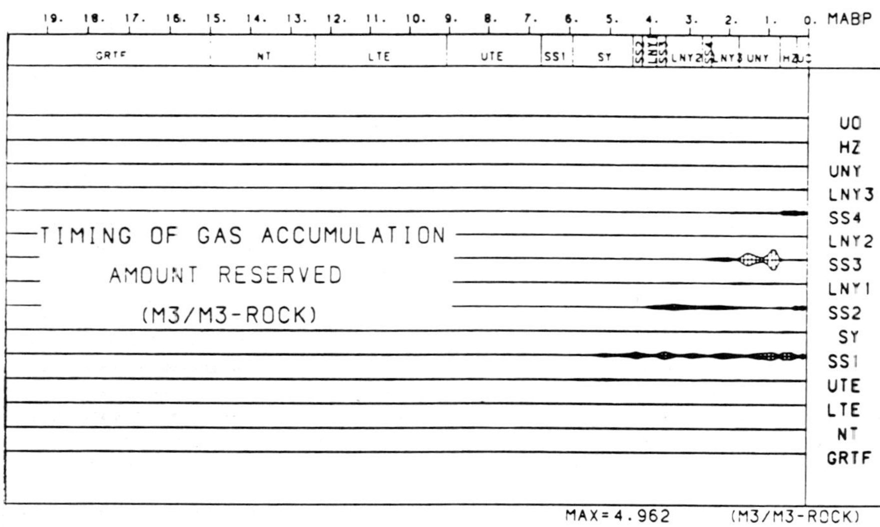

FIGURE 7.61 Timing chart of gas accumulation at the well location. The unit is the same as for oil but in cubic meters.

high-porosity zone which decreases the thermal conductivity of the rock. Because heat tends to flow along the least resistive paths, it tends to bypass less conductive zones. This effect is not large but is visible as isotemperature lines are slightly tilted toward the basin center in the present-day section (Fig. 7.63f).

3. Oil Generation and Migration (Fig. 7.64a–f)

The amount of oil accumulated is shown in units of milligrams per gram of rock. In Fig. 7.64c, for example, the Lower Teradomari formation (LTE) and the Nanatani mudstone member (NT) generated oil such that 1.0 mg was produced per gram of rock during the interval 4.3–2.0 MYBP. The movement of oil is also plotted, yielding the approximate migration pathway. The main oil generation occurred below 2500–3500 m in depth (the zone which produced more than 0.1 mg/g rock) from 2.0 MYBP to the present (Fig. 7.64a–f).

4. Gas Generation and Migration (Fig. 7.65a–f)

The amount of gas generated is shown in the same manner as for oil. The main gas generation (the zone with more than 0.1 mg/g rock) occurred from 2.0 MYBP onward at depths from 3000 m (at 2.0 and 8.0 MYBP) to 4000 m (at 0.4 MYBP and the present). Some reservoirs with no TOC content (SS1-4 and GRTF) can still generate a significant amount of gas (more than 0.5 mg/g rock) because oil had already migrated into those reservoirs and later cracked to gas. The migration pathways are generally similar to those of oil but are more vertically inclined because of the extra buoyancy effect of the gas.

5. Oil Saturation (Fig. 7.66a–f)

The development of oil saturation at each spatial location is shown with the same migration arrows used previously. At the Iwafune-Oki area, the oil started to be trapped in layer SS1 from 4.3 MYBP onward, whereas for layers SS2–4 this trapping occurred from 2.0 MYBP. Also note that some oil was concentrated in the lowest layer of Green Tuff but has diminished recently (Fig. 7.66f), matching the cracking stage of oil to gas.

6. Gas Saturation (Fig. 7.67a–f)

As with oil saturation, we can also trace the development of gas saturation. Figure 7.67a–f indicates that layers SS2, SS1, and Green Tuff (GRTF) are relatively good reservoirs.

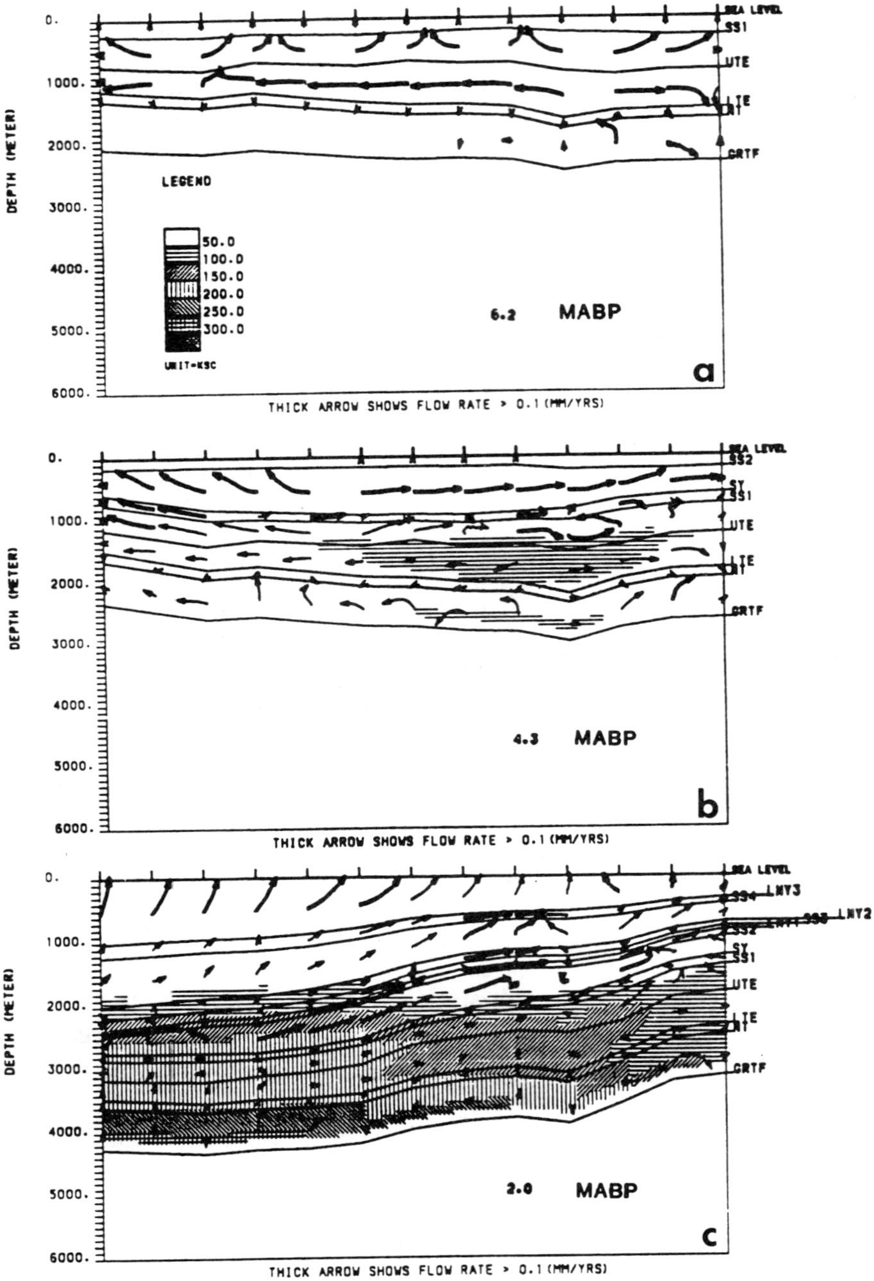

FIGURE 7.62 Development of excess pressure through time.

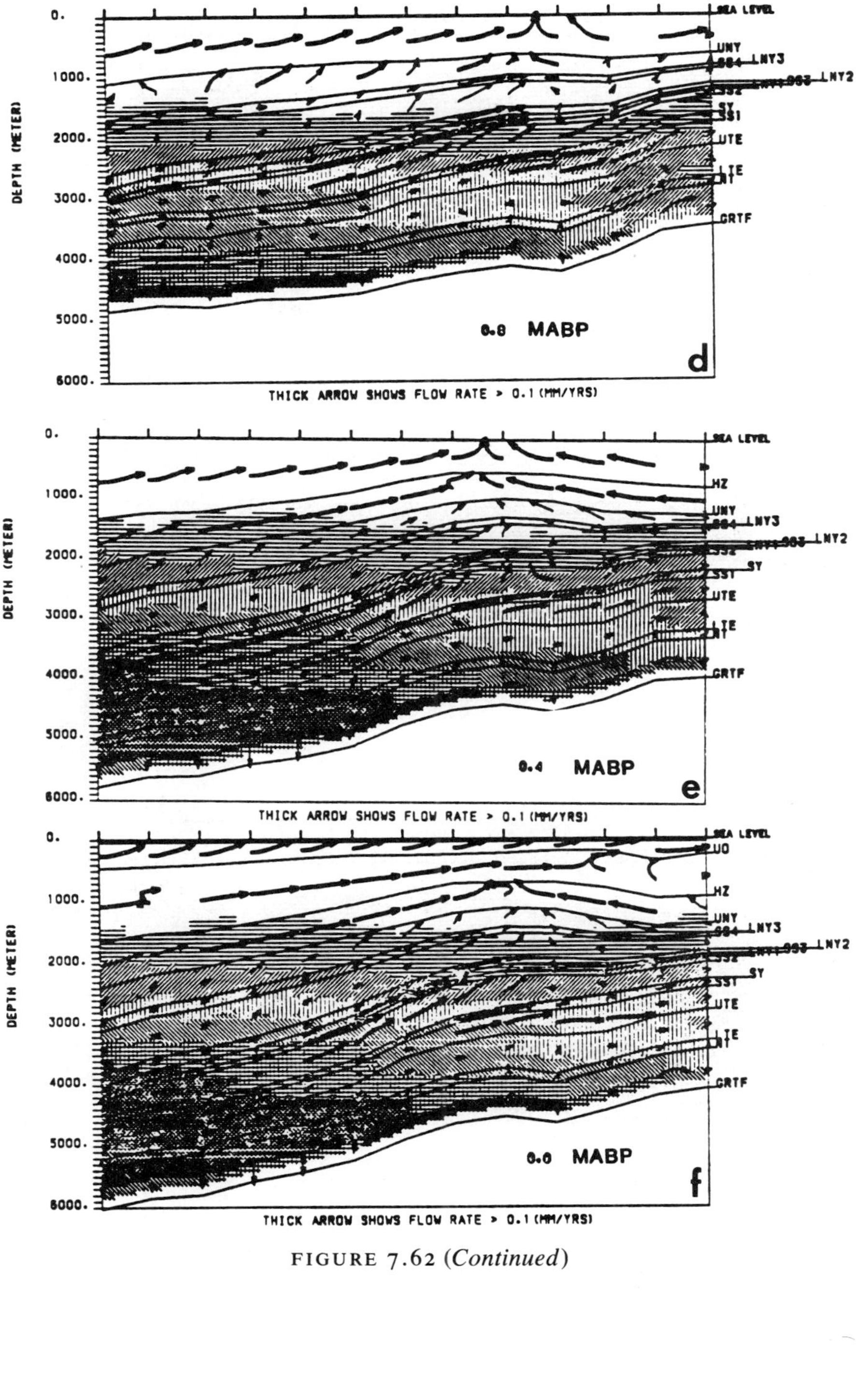

FIGURE 7.62 (*Continued*)

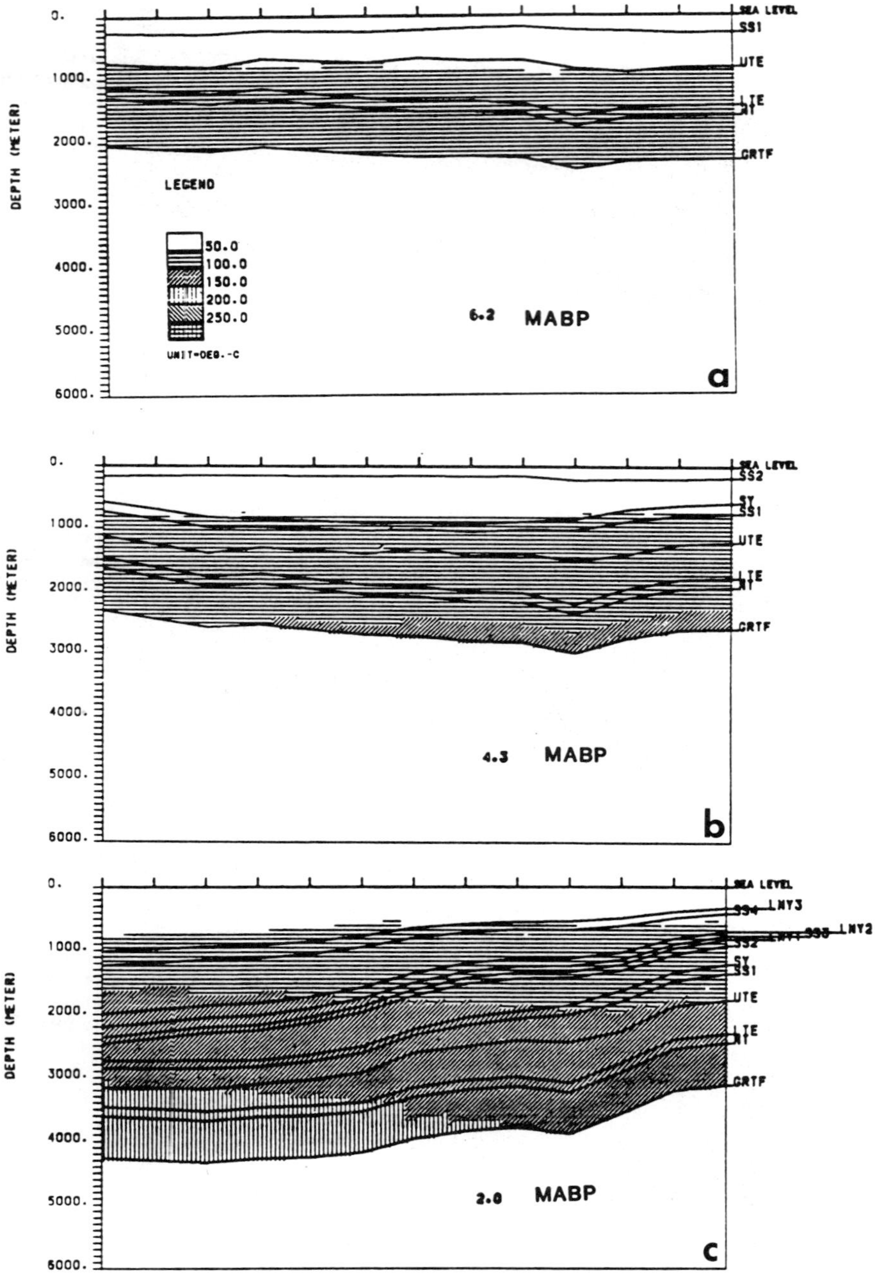

FIGURE 7.63 Development of temperature through time.

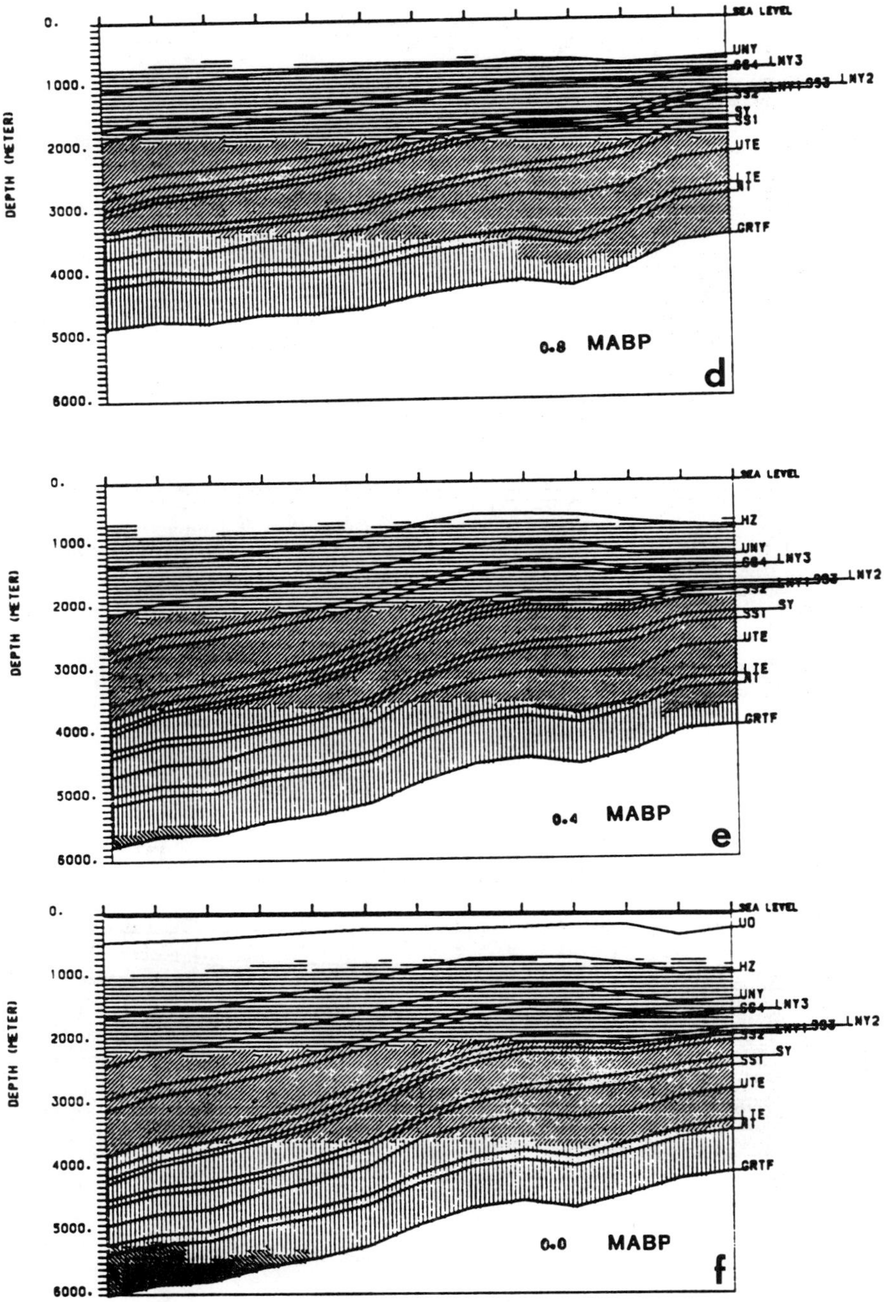

FIGURE 7.63 (*Continued*)

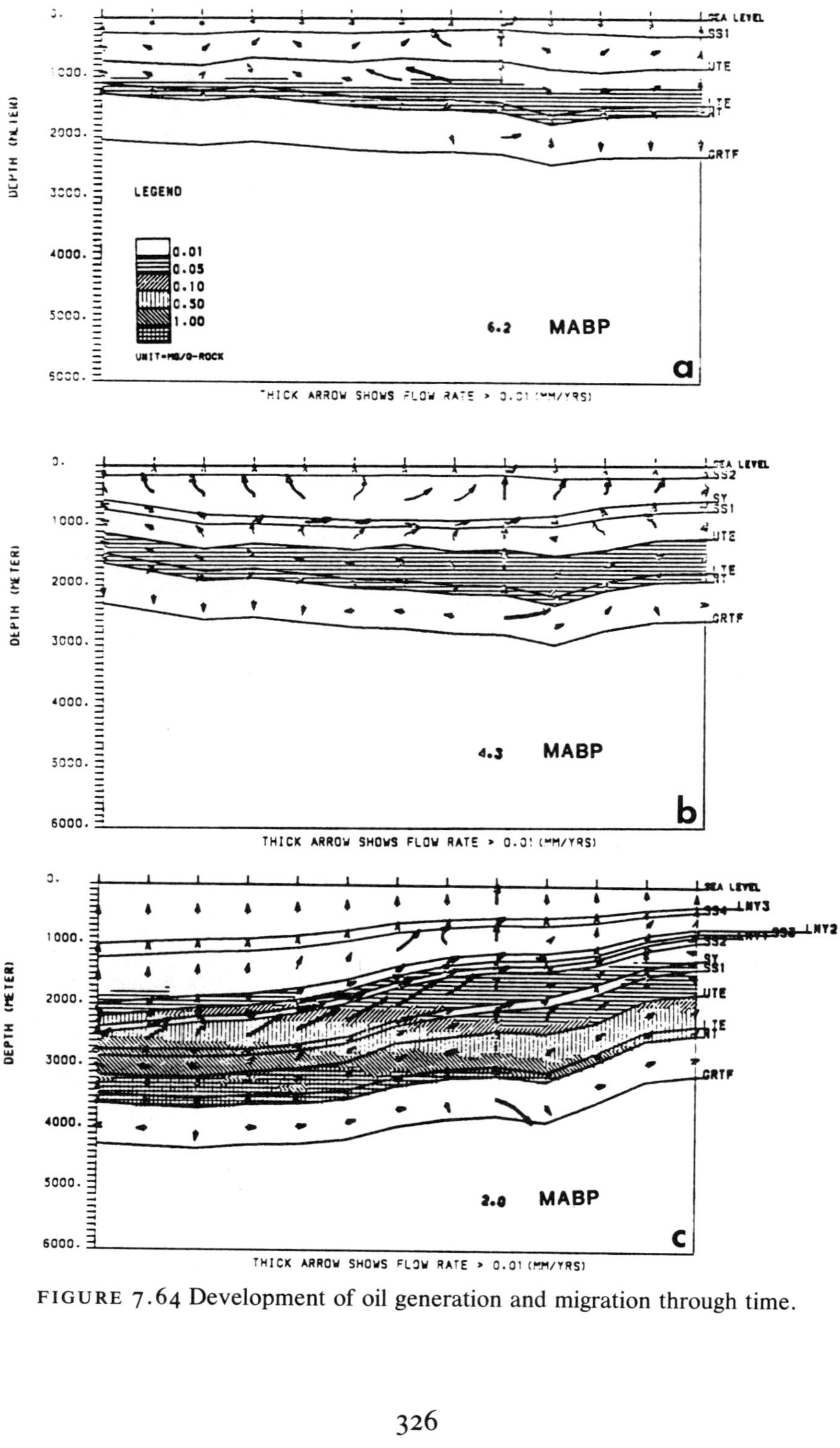

FIGURE 7.64 Development of oil generation and migration through time.

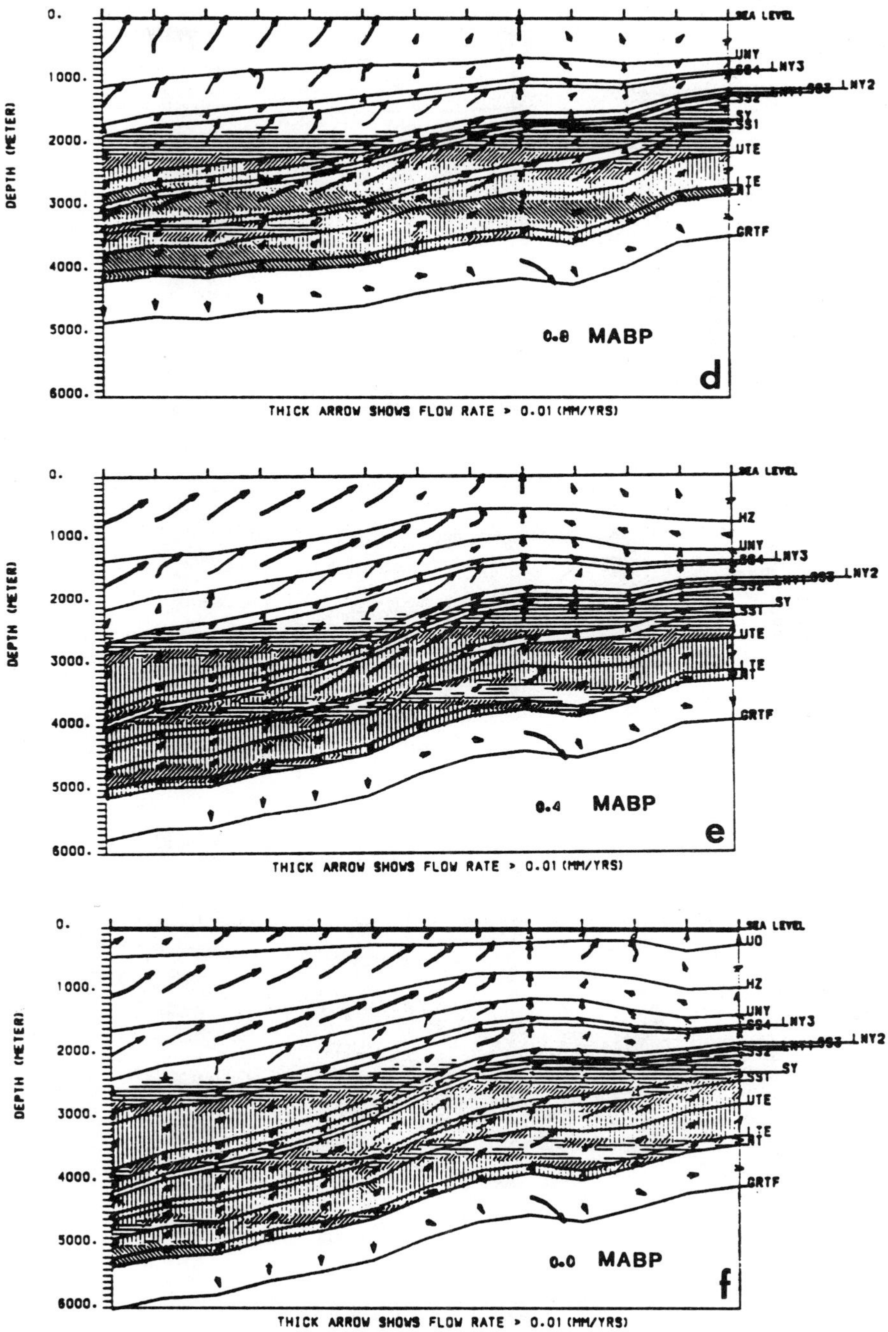

FIGURE 7.64 (*Continued*)

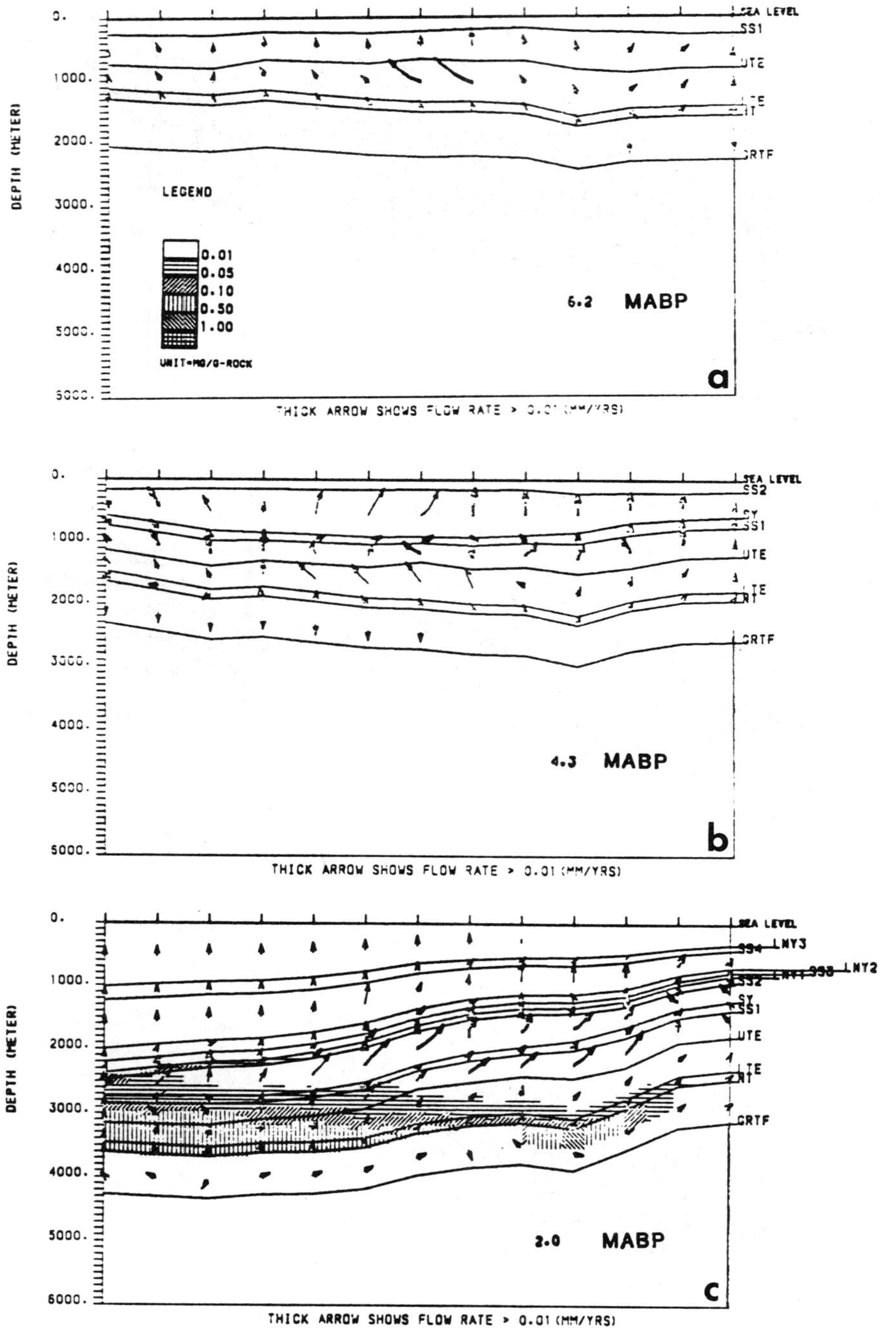

FIGURE 7.65 Development of gas generation and migration through time.

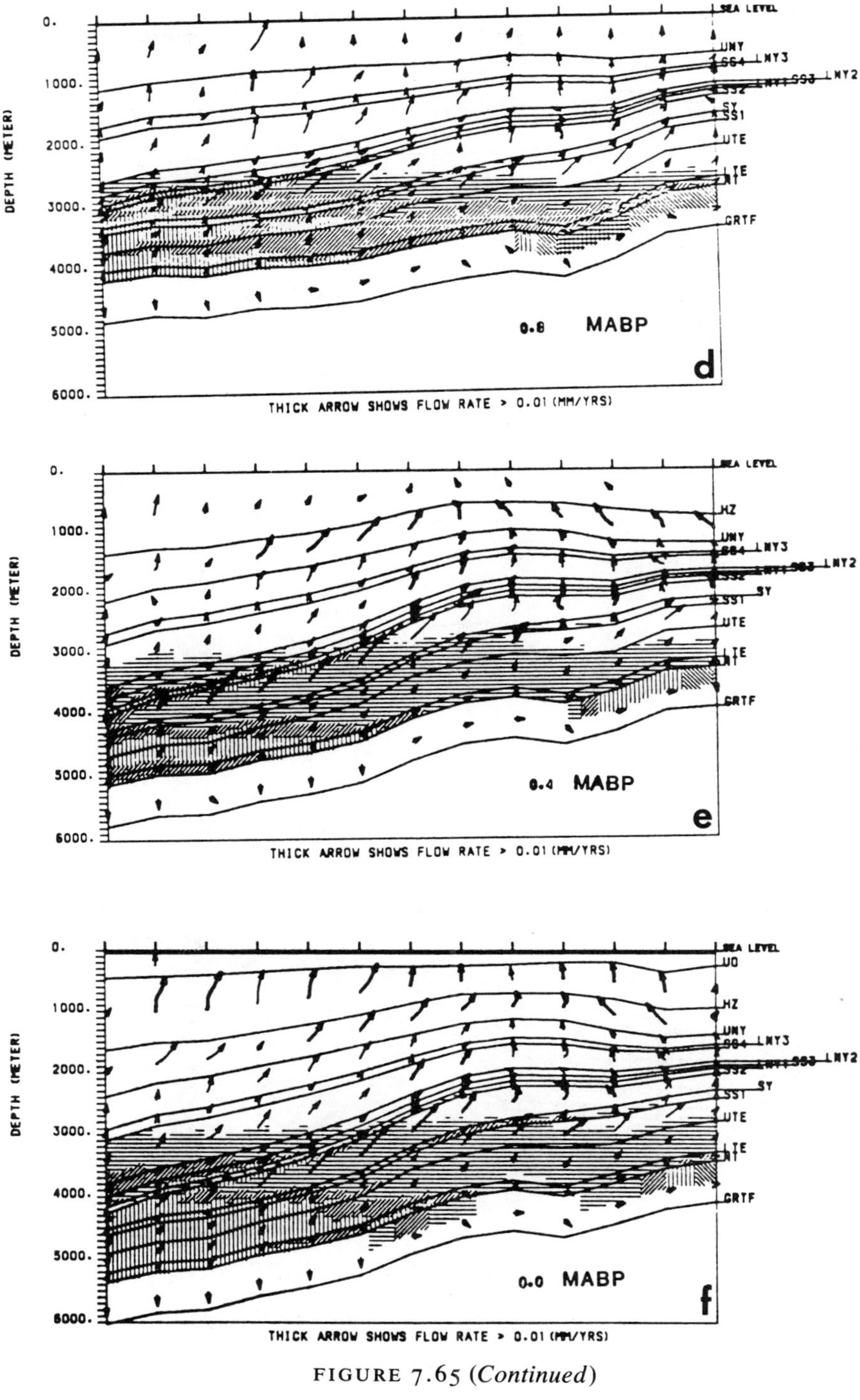

FIGURE 7.65 (*Continued*)

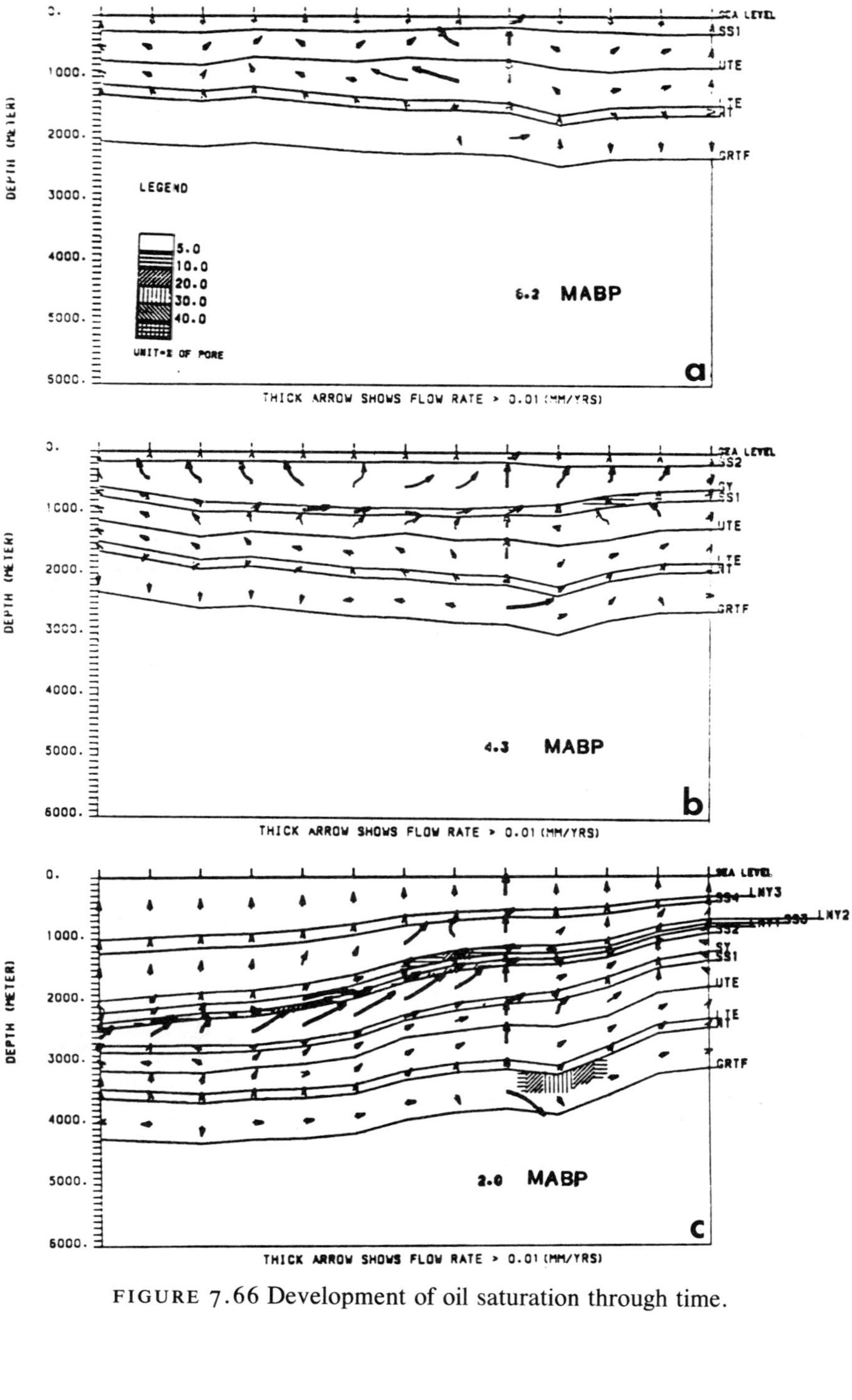

FIGURE 7.66 Development of oil saturation through time.

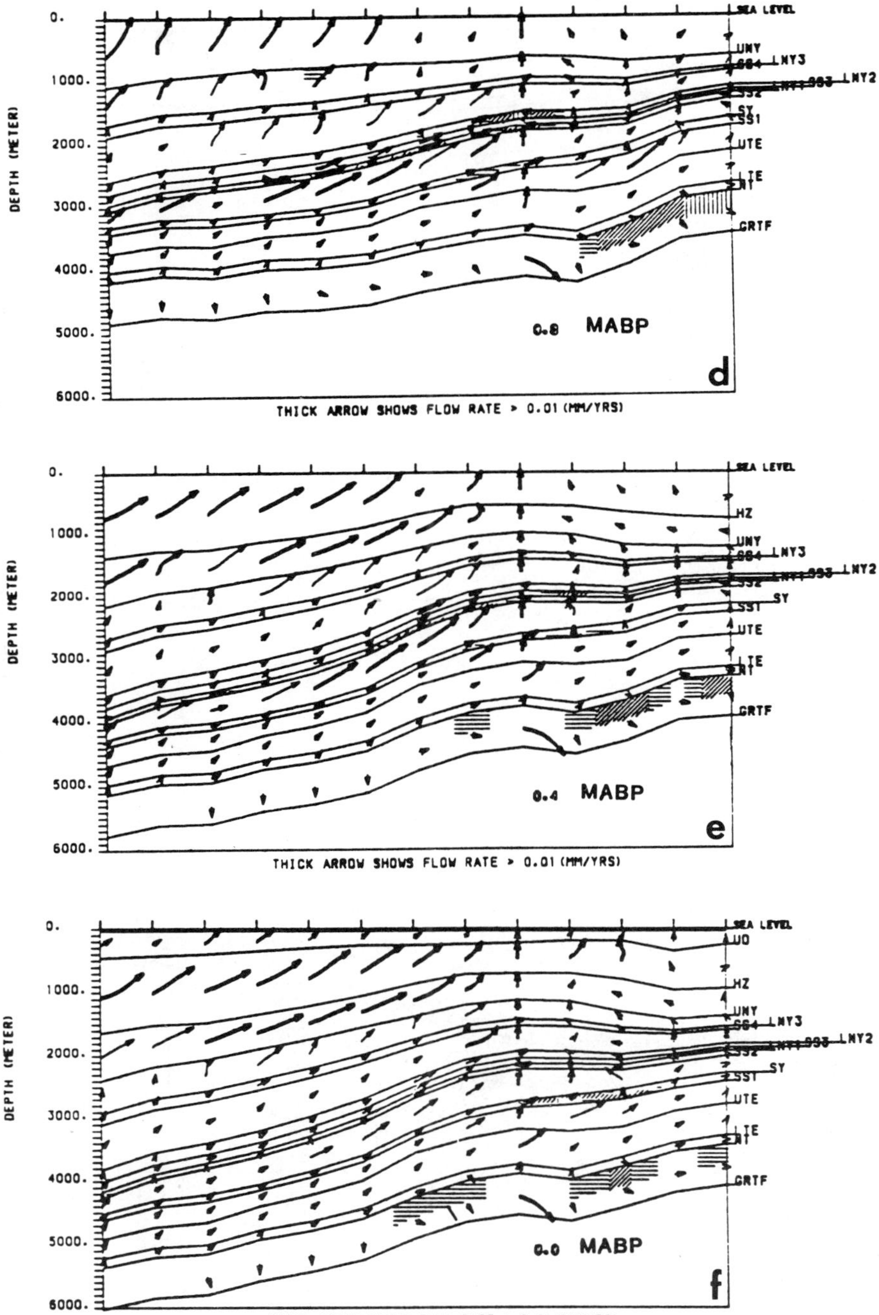

FIGURE 7.66 (*Continued*)

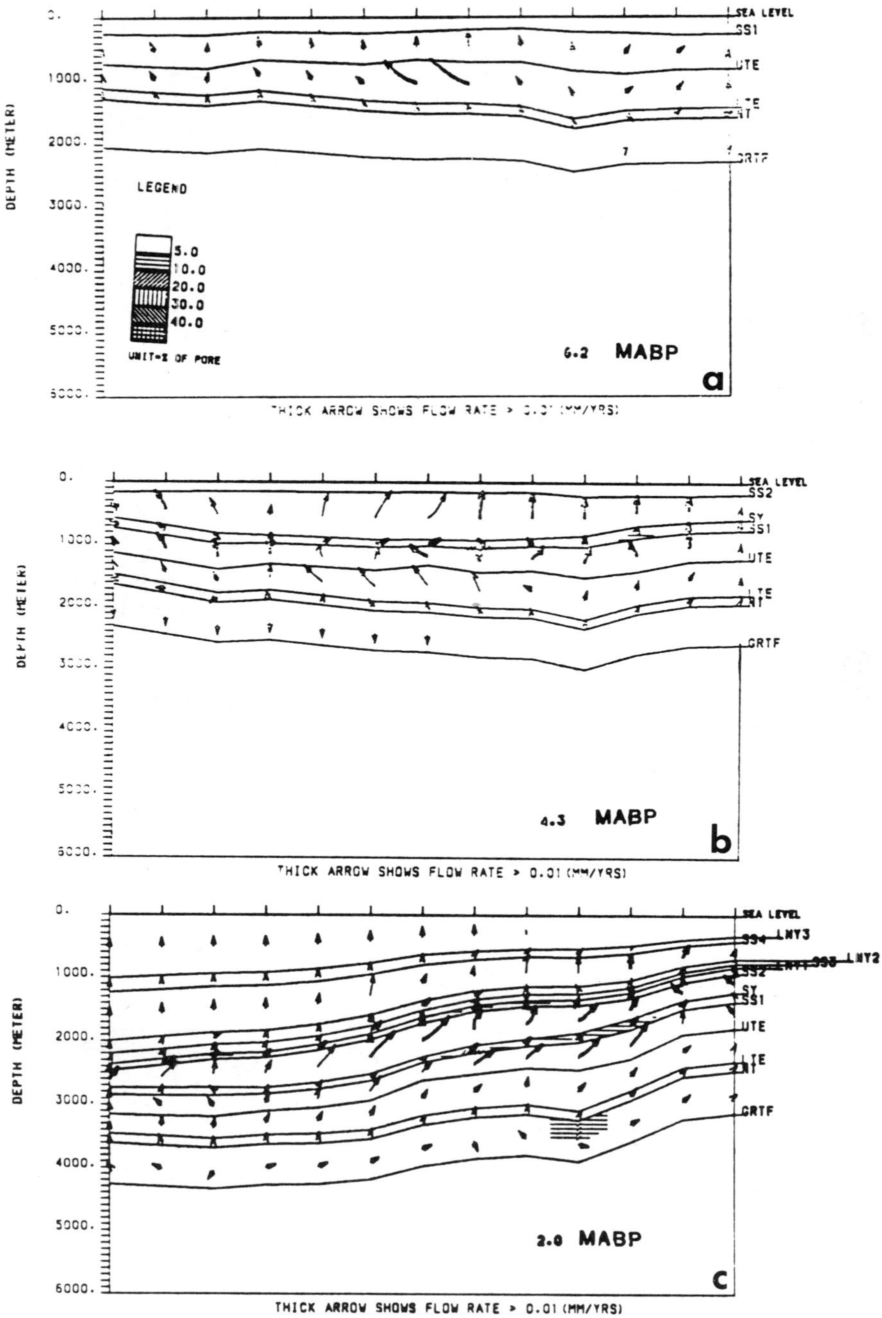

FIGURE 7.67 Development of gas saturation through time.

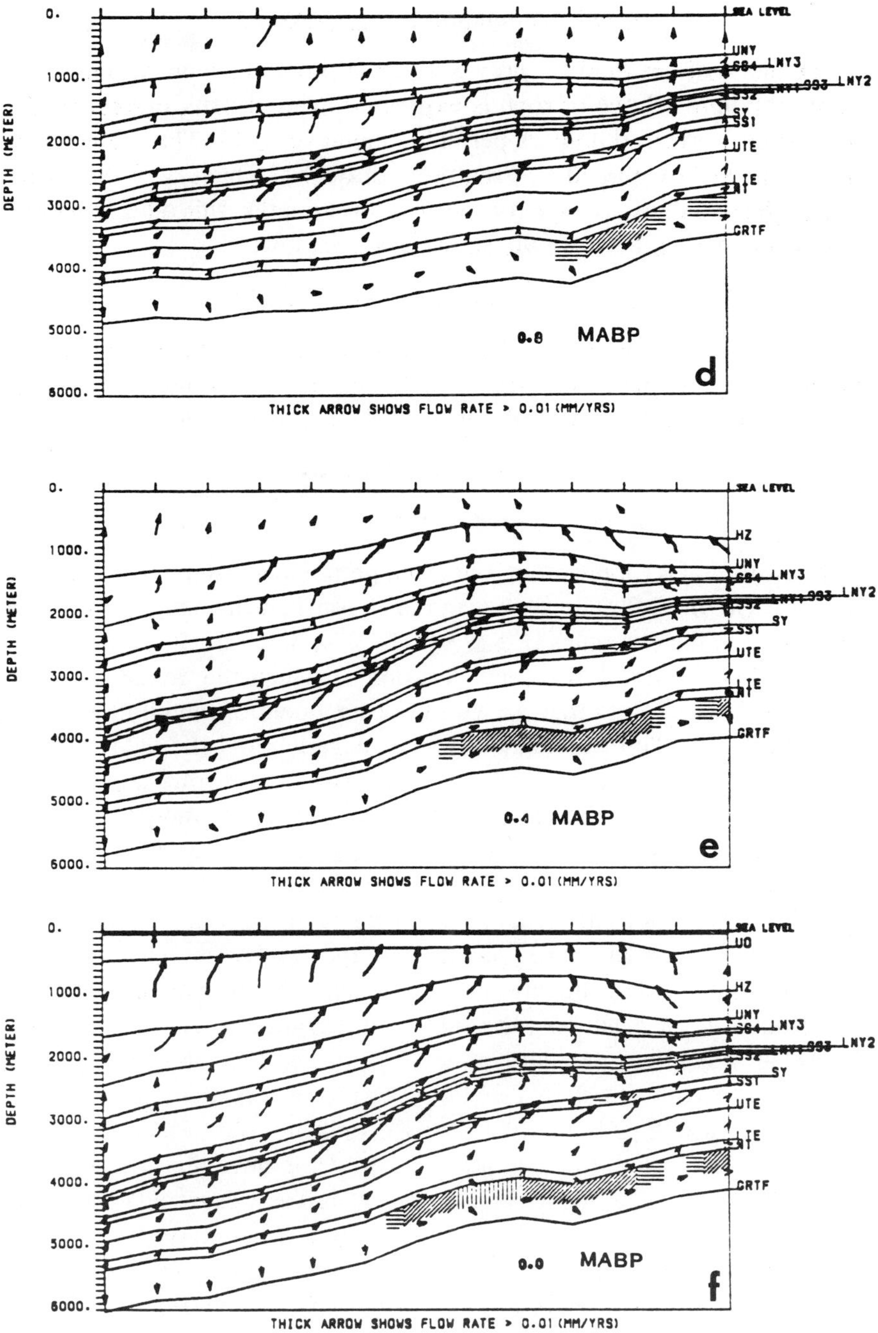

FIGURE 7.67 (*Continued*)

7. Oil Accumulation (Fig. 7.68a–f)

Using the same system as previously, the amount of oil accumulated within the unit volume of rock is expressed in stock tank barrels as if it were at the surface (kiloliters per cubic meter of rock). The sequence of figures shows how the oil was scattered through the history of the basin. Although the values are high for the shallower depths, this oil is immobile and uneconomical because the saturation is so low.

8. Gas Accumulation (Fig. 7.69a–f)

Gas accumulations are mapped in the same manner as oil in units of cubic meters per cubic meter of rock. The figures show that the gas is not as scattered as the oil because of the greater buoyancy of the gas.

E. Sensitivity Analysis

1. Capillary Force as a Mode of Migration

In the Iwafune-Oki application, oil/gas migration occurred into the lowest member, Green Tuff. The arrows in the Green Tuff member are directed downward in the examples (Fig. 7.64a–f). The flow rate at each location was checked to detect the exact location at which the downward flow occurred. It was found that the vertical component of the oil flow at the Nanatani/Green Tuff boundary becomes negative and the values inside the Green Tuff are almost negligible. Therefore, the arrows are directed downward for almost the entire geologic history of the model, although the *hydrodynamic* pressure gradient does not decrease downward (Fig. 7.62a–f).

Now, as well as hydrodynamic pressure, both capillary pressure and the buoyancy pressure are included in the dynamical model. Since buoyancy cannot be a downward force, capillary pressure must play a significant role in producing the downward migration. A test was conducted in which the capillary effect was turned off for the migration process. The result is shown in Fig. 7.70a–c, where the amount of oil generation is mapped and the arrows indicate the direction of oil migration. Note that the migration arrows are now directed upward except in Fig. 7.70a. Since the capillary force is always exerted from small pore throats to large pore throats, its magnitude overcomes both the regional hydrodynamics and buoyancy in the area close to the formation boundary. Thus the high-porosity (large-pore) part of the Green Tuff must lie directly adjacent to the Nanatani mudstone member, as was assumed in this application. In fact, the Green Tuff has a wide variety of facies, so the detection of such a contact with the Nanatani mudstone is a key to the exploration for new reserves and reservoirs in the Green Tuff.

2. Sandstone Geometry

The sand geometry of the area is unknown. In this application the sandstone geometry is interpreted from the seismic record and is used as a known factor in the model (Fig. 7.71), so that the sand layers SS1, SS3 and SS4 are rather restricted beyond the field, whereas layer SS2 is assumed to be a widespread blanket sand. The dynamical model was tested taking *all* the sandstones to be widespread blanket layers. The results (Figs. 7.71a–c, 7.72a–c, and 7.73a–c) show a tremendous change of flow pattern for all oil, gas, and water migration (compare with Figs. 7.66, 7.67, and 7.62). As a result, the oil/gas saturation increased in the reservoirs, while excess pressure in the deeper parts of the basin decreased slightly (Figs. 7.71c, 7.72c, and 7.73c). The sandstone geometry is, therefore, a very important factor, not only for evaluating reservoir potential but also for understanding regional fluid migration patterns. This point was also made earlier in connection with the effective bulk permeability of volumes in which the sand/shale units remain discrete but intermingled.

F. Discussion

From the application to the Iwafune-Oki field, Japan, the following general points can be made:

1. A two-dimensional setting is required to derive suitable calibrated values for inputs. A one-dimensional setting at a set of well locations is a convenient way to get approximate initial values; however, those values should then be checked against the more correct two-dimensional setting. In particular, permeability parameters must be carefully controlled since they determine the fluid flow amount and time scale and thus the formation thicknesses, porosity, and fluid pressure with depth at the present day at each location in the basin.
2. The "blind" test shows that some form of temperature (or heat flow) control is necessary when we extrapolate parameter values from shallow to deep parts of a basin. But the error in prediction will be much smaller when we estimate shallower-depth behavior using data from deeper sections. The data from deeper wells have a significant effect on the evaluation of the basin.

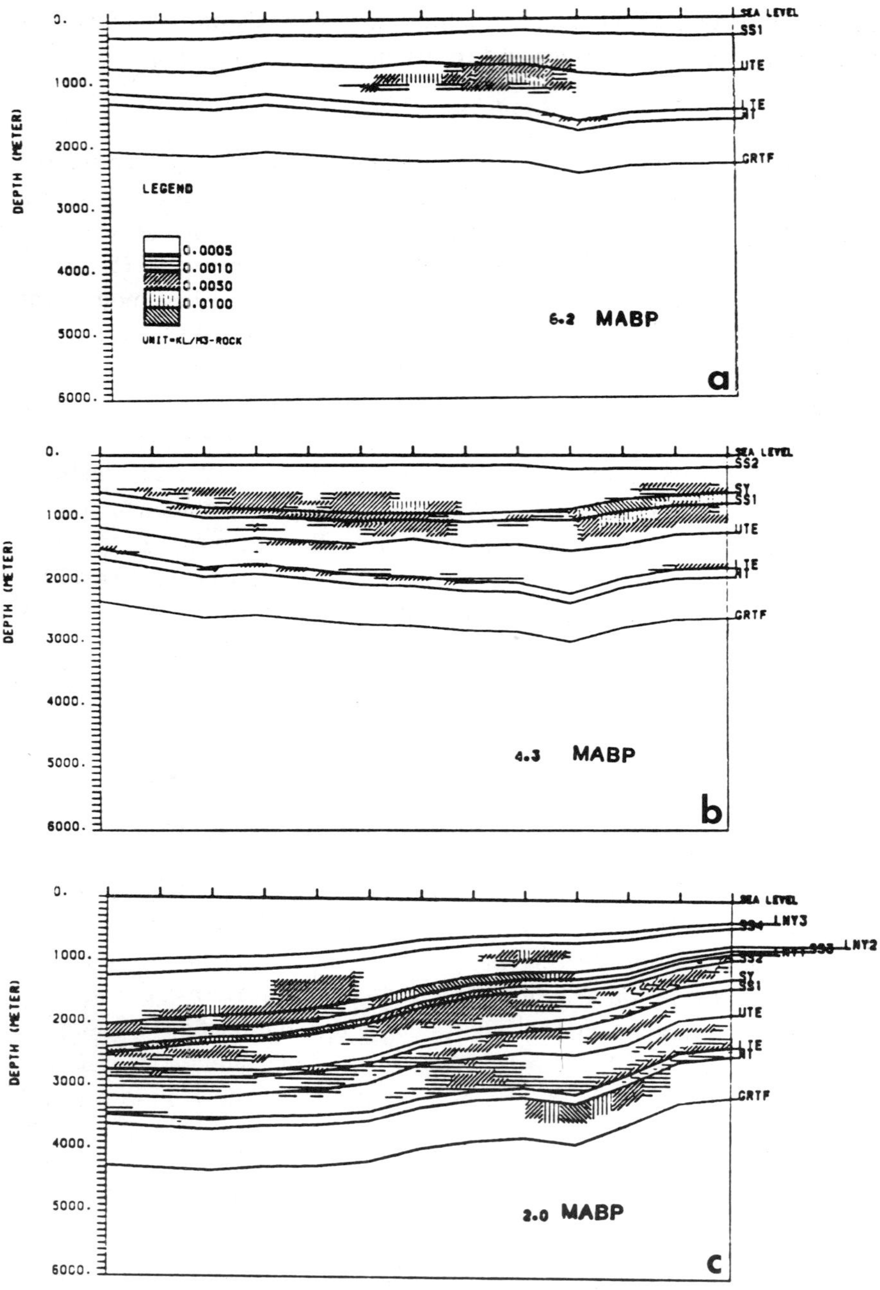

FIGURE 7.68 Development of oil accumulation through time.

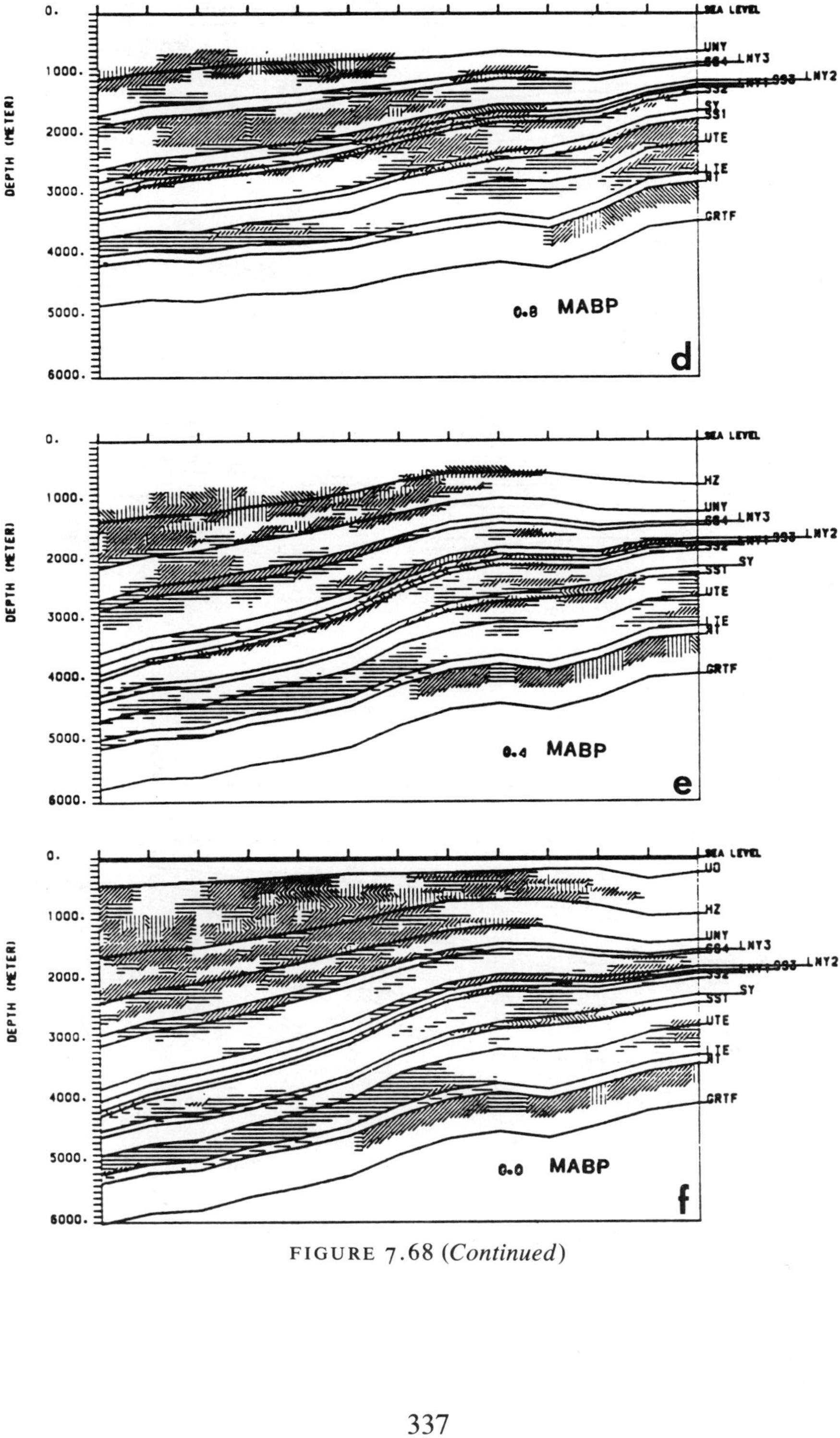

FIGURE 7.68 (*Continued*)

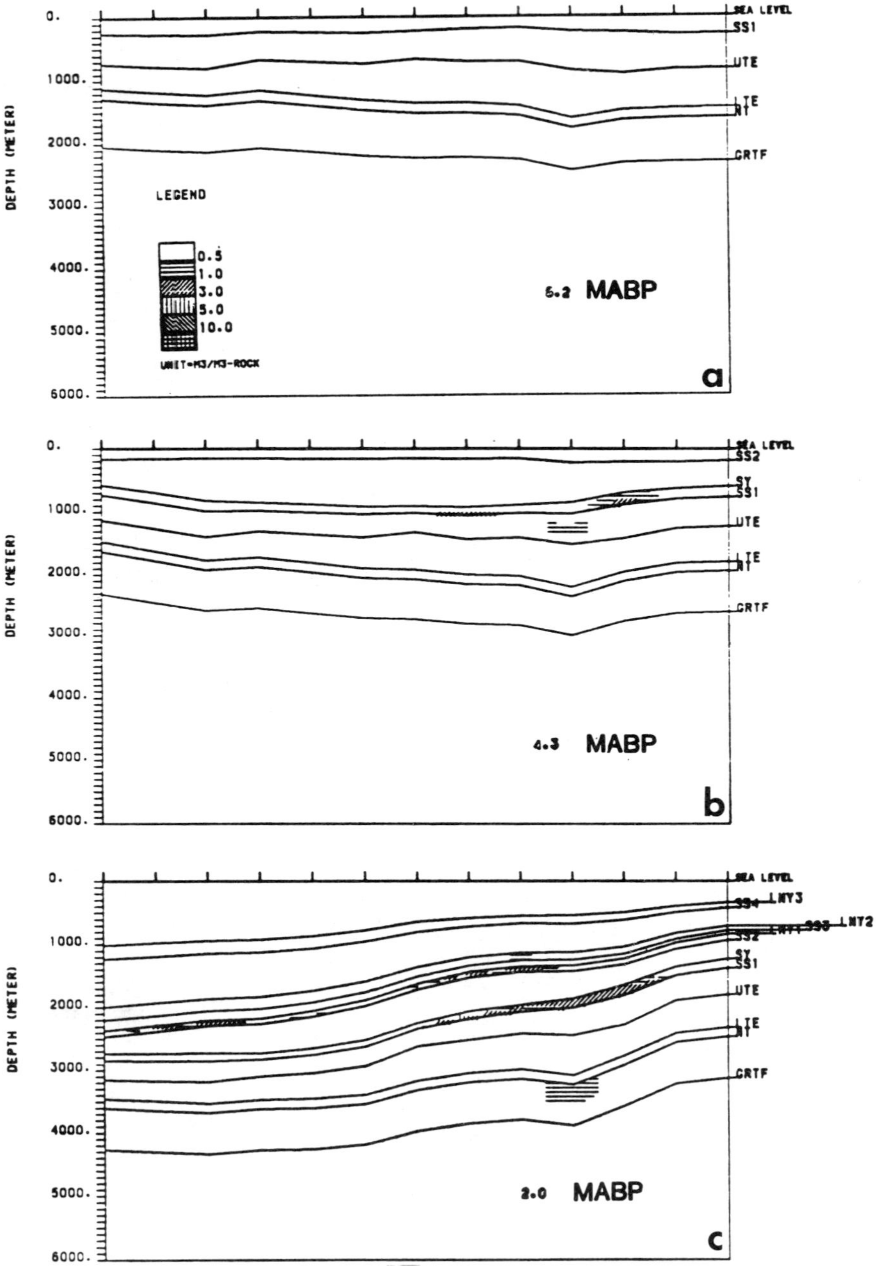

FIGURE 7.69 Development of gas accumulation through time.

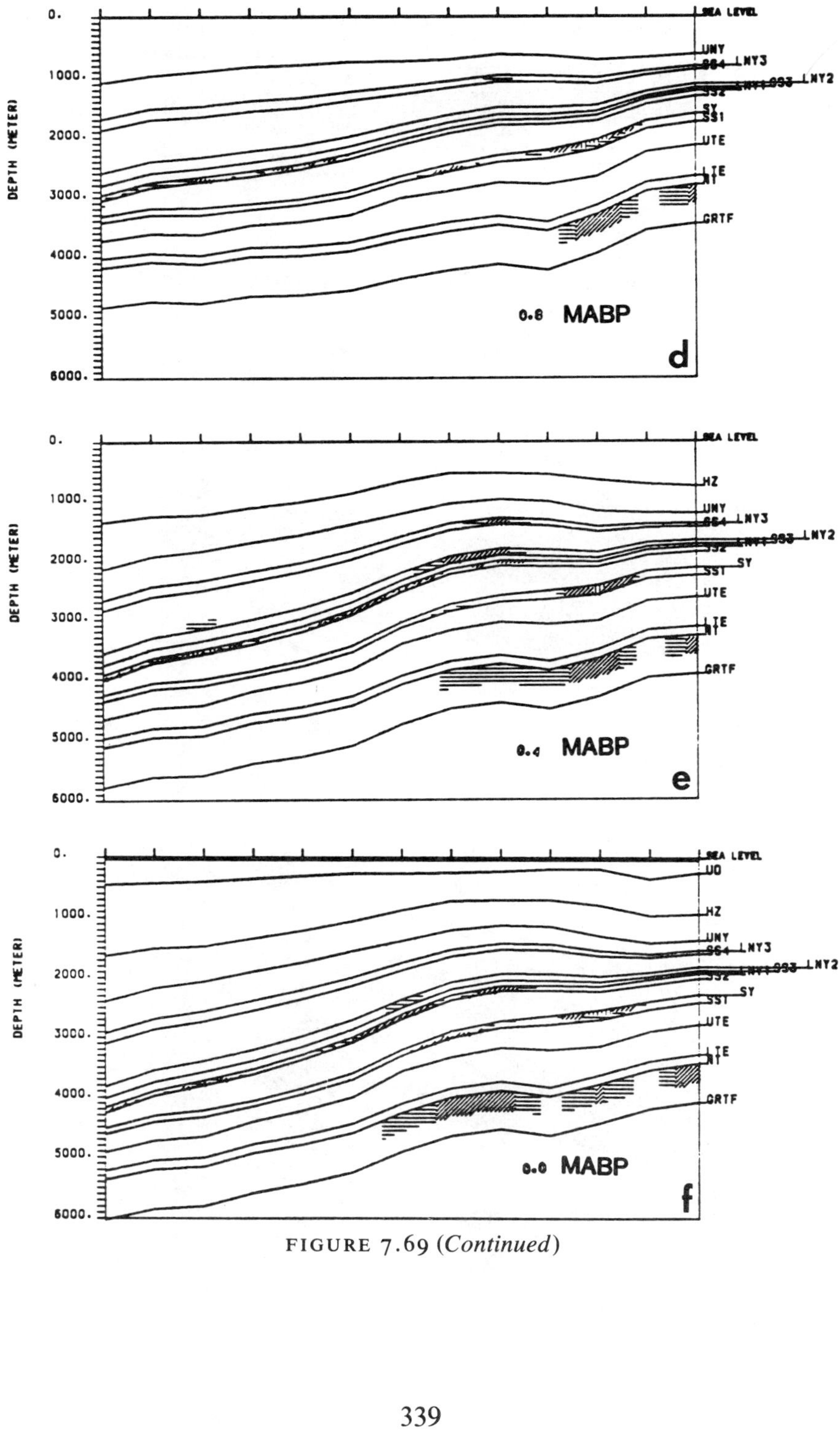

FIGURE 7.69 (*Continued*)

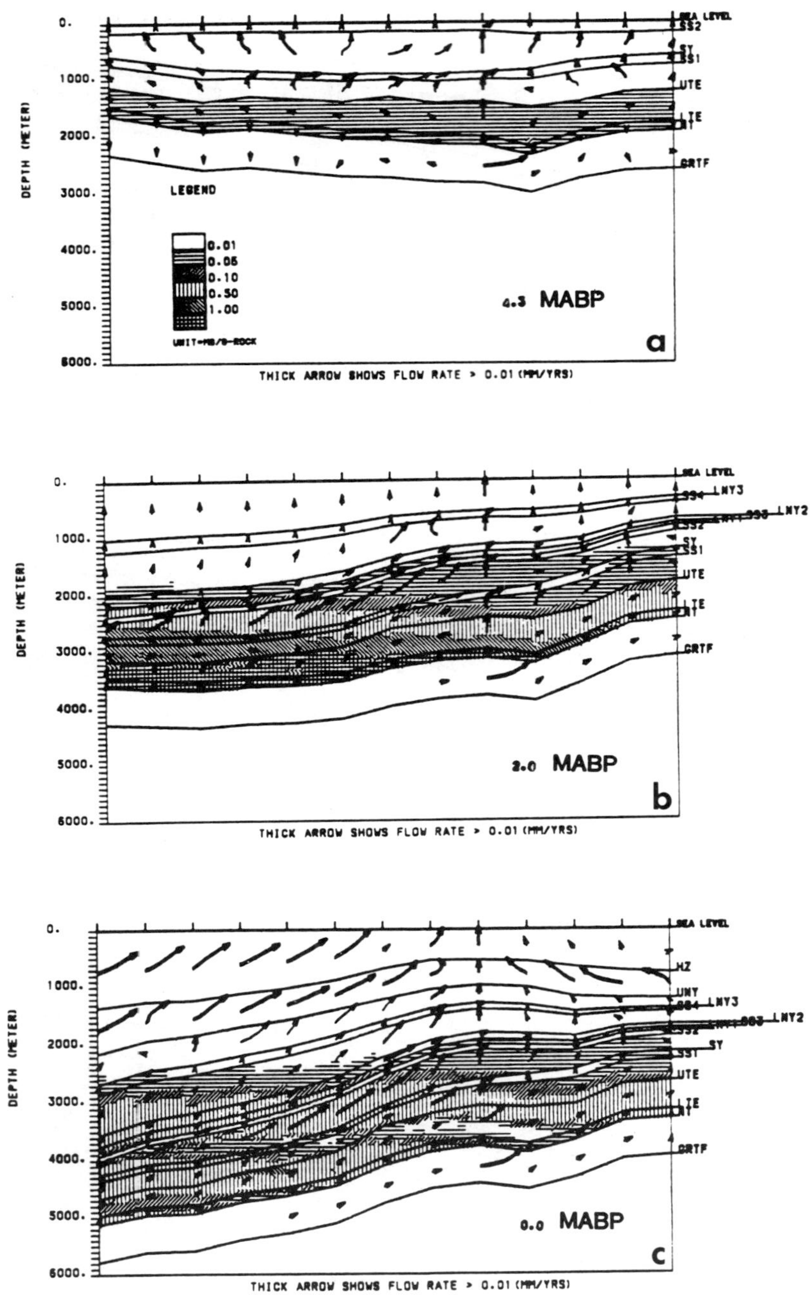

FIGURE 7.70 Development of oil generation and migration without capillary effect.

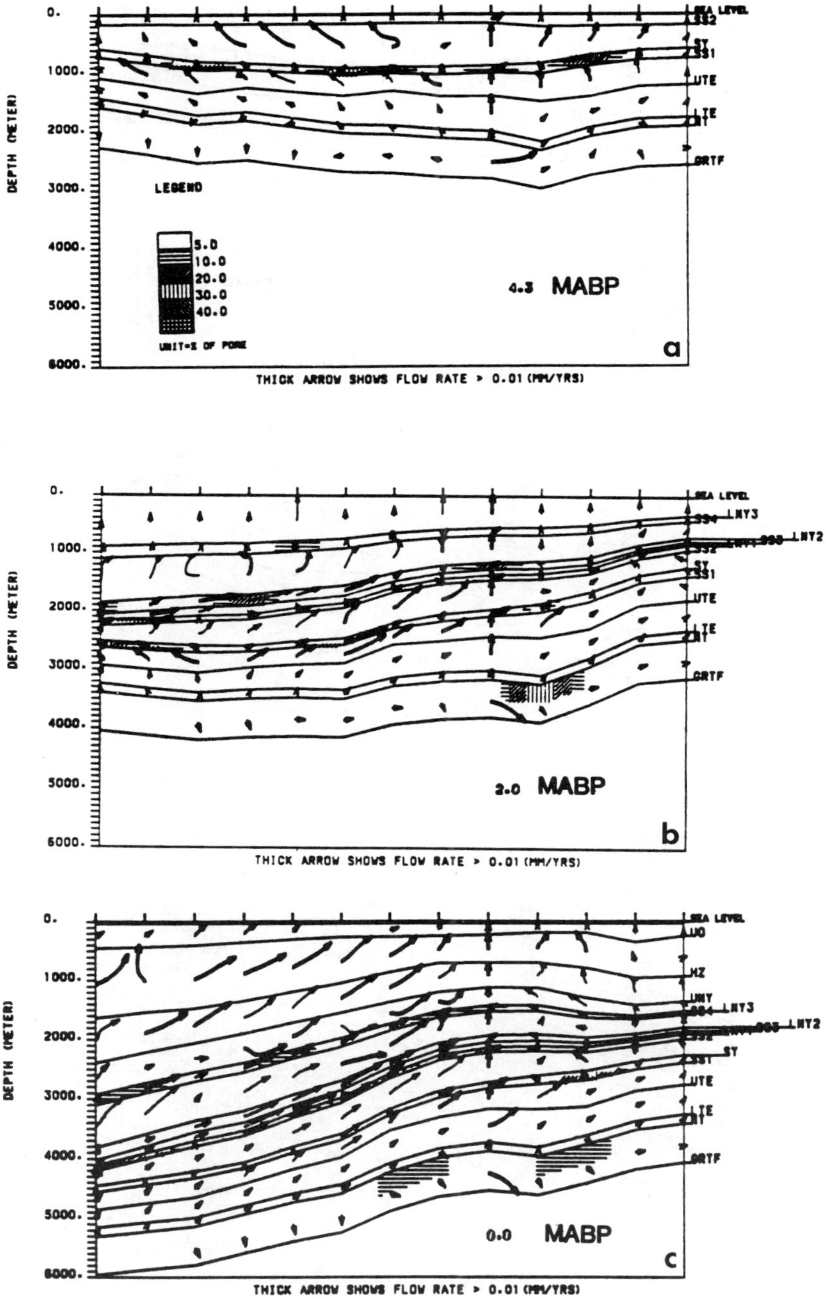

FIGURE 7.71 Result of sandstone geometry test showing development of oil saturation and migration pattern. All sandstones are extended toward the basin center.

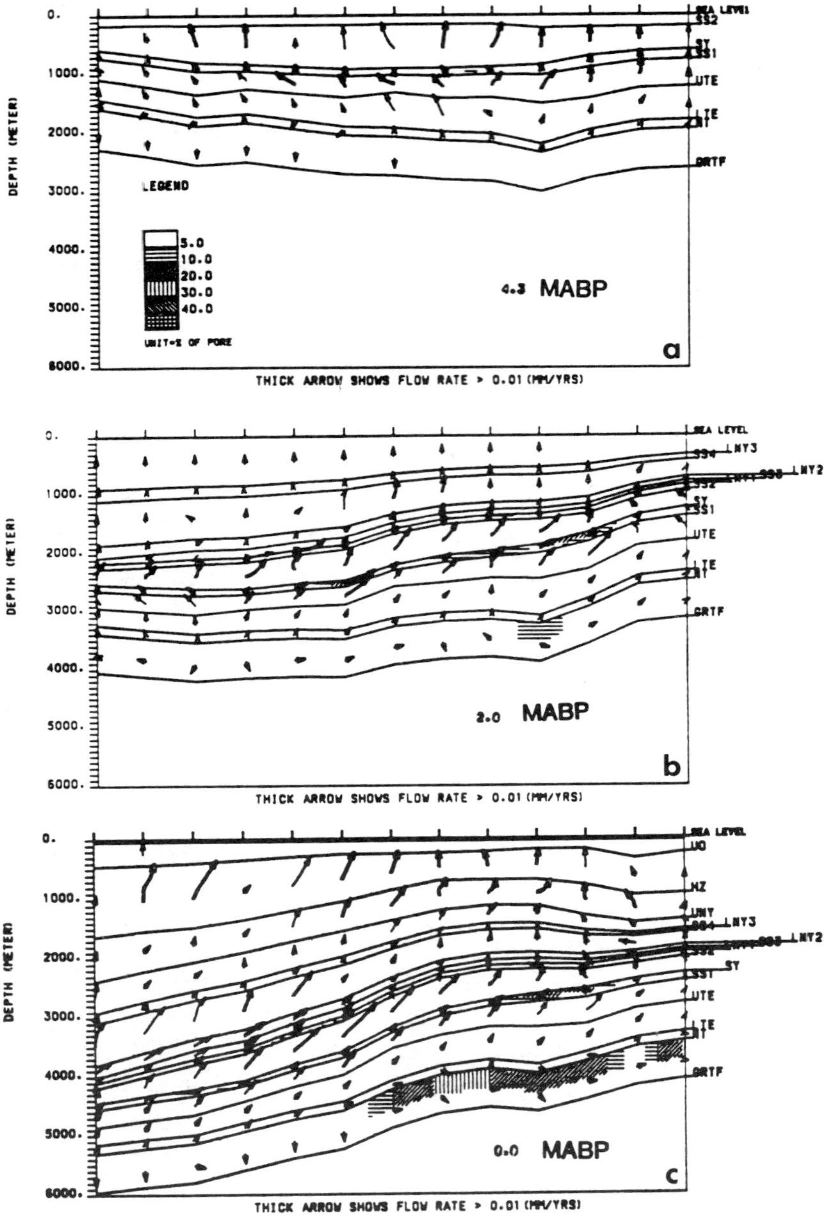

FIGURE 7.72 Result of sandstone geometry test showing development of gas saturation and migration pattern. All sandstones are extended toward the basin center.

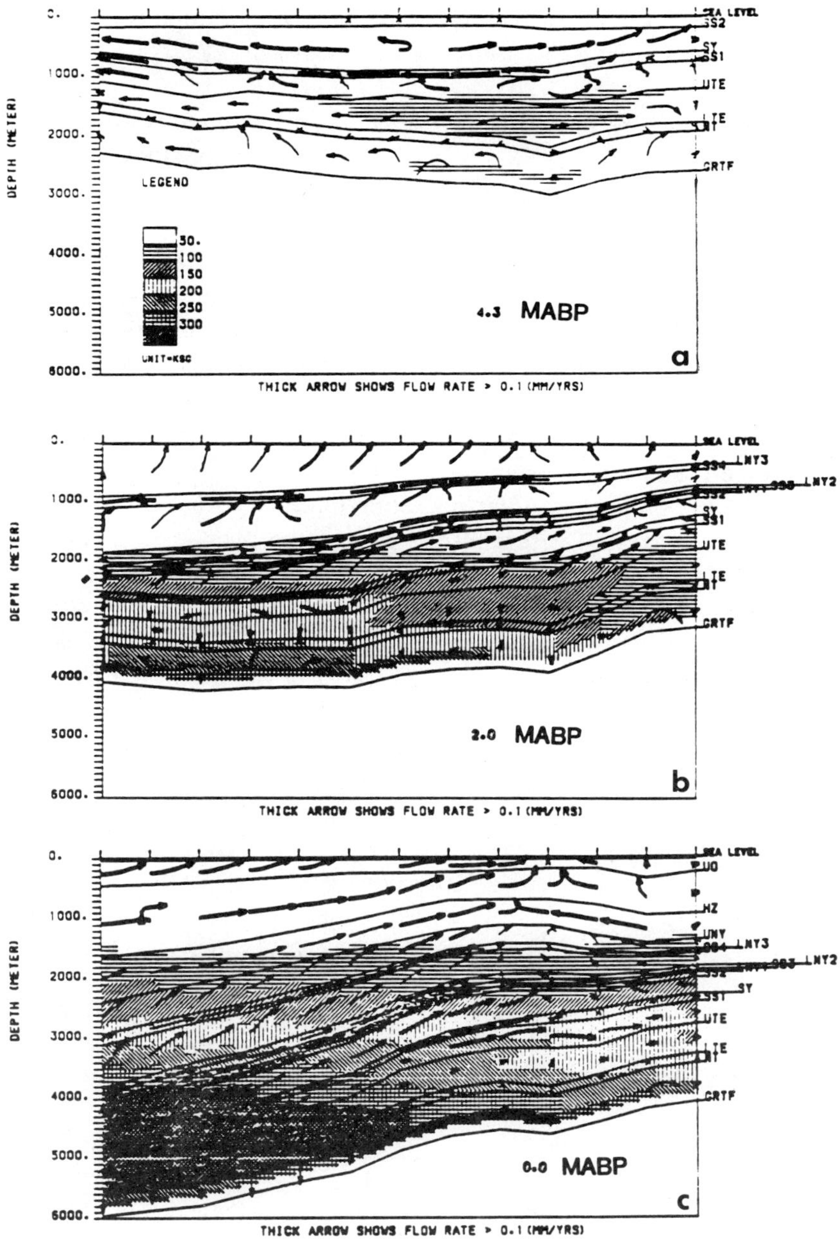

FIGURE 7.73 Result of sandstone geometry test showing development of excess pressure and water migration pattern. All sandstones are extended toward the basin center.

In addition, for the petroleum geology of the Iwafune-Oki field in particular, we can make the following observations:

1. The structure of the Iwafune-Oki became prominent after 4.3 MYBP. Before that, the area had subsided almost equally with the Iwafune-Oki area near a depocenter.

2. The depth of the main hydrocarbon generation is below 2500 m for oil and below 3000–4000 m for gas, with a slight dependence on geologic age.

3. The generation was active from 2.0 MYBP through to the present for both oil and gas. There is no great difference in timing in generation in the structural highs or in the basin center.

4. The accumulation of oil/gas in the reservoirs has been continuous since the onset of generation.

5. The possibility of oil/gas migration into the lowest Green Tuff member is suggested if the porous part of the member has direct contact with the overlying Nanatani mudstone. Capillary force plays a dominant role in any such downward migration. Downward migration appears to have occurred throughout most of the geologic history of the section being modeled. Most of the gas in the Green Tuff was formed from oil at the reservoir site.

6. The sandstone geometry can change the flow pattern over the area and is one of the most important parameters to be assessed as accurately as possible in the evaluation of this basin, or, indeed, of any other basin.

8

Paleoheat Flux, Activation Energy, Stratigraphic Age, and Unconformity Determinations from Thermal Indicator Inversions

I. Paleoheat Flux and Activation Energy

This part of the chapter describes paleoheat flux determinations by inversion from independent thermal indicators, vitrinite reflectance and sterane and hopane isomer data, for a single well, well X, from the northern North Sea. When compared and contrasted, these inversions give the same paleoheat flux variation with time to within error assessment of the data quality and the inversion schemes. This consistency provides firm corroborative support for the basic physics and chemistry used to describe the three indicators: if a significant error were being made, no such consistency would be expected, contrary to determination. In addition, by exhaustive trial-and-error searches to determine a minimum residual mismatch between theory and observations, we obtain estimates of 70 ± 40 kJ/mol and 60 ± 40 kJ/mol for the activation energies of sterane isomerization and hopane isomerization, respectively. A neighboring well, Y, within 10 km of well X had sufficient hopane isomer data for inversion to be performed. It, too, yielded an activation energy of 60 ± 40 kJ/mol, confirming the result obtained for well X and providing an assessment of the accuracy of this method of determining chemical parameters. Well Z, within 50 km, also had enough sterane isomer data for an inversion to be performed. It yielded an activation energy of 65 ± 35 kJ/mol, comparable to that from well X. The consistency of the three independently determined paleoheat flux variations with time, and of the chemical activation energies, strongly implies that the inversion of different downhole thermal indicator data is a valid technique that yields consistent behavior. This is important in the assessment of the thermal history of source rocks; it forms the basis for a systematic method of assessing basinal thermal maturity, independent of model behaviors assumed for paleoheat flux without supporting evidence.

The use of a single thermal indicator, such as vitrinite reflectance, to determine paleoheat flow is subject to possible systematic and statistical bias. It also leaves open the charges that (1) the answers are inconsistent with proposed geologic models and (2) the basic assumptions of the method rely too heavily on a complex of interlocking physics and chemistry, whereas one may intuitively "feel" that the behavior of the thermal indicator should be "simpler" (in some undefined sense) than proposed. Clearly, without other, independent, thermal indicators to provide paleoheat flux estimates, it is difficult to refute such potentially serious charges.

However, with more than one thermal indicator in a single well, we can use the indicators as cross-checks on each other to determine the systematic and statistical biases recorded by one thermal indicator in relation to the others, when reconstructing, by inversion, the paleoheat flux variations with time. Here we consider information obtainable from three thermal indicator data sets in a single well: vitrinite reflectance with depth, sterane isomeric ratio with depth, and hopane isomeric ratio with depth.

We allow for a paleoheat flux varying linearly with time: $Q(t) = Q_0(1 + \beta t)$, where β is the rate of change of heat flow.

The use of a linear heat flux with time is based on much practical experience with many wells where we attempted to determine $Q(t)$ beyond the linear term (Lerche *et al.*, 1984). By and large, we found that the data quality is not sufficient in general to permit extraction by inversion of any terms beyond the linear one with any degree of confidence. This does *not* imply that we consider the actual heat flux to vary linearly with time; rather it implies that the data are sufficiently "scruffy" that one cannot do any better than as given by a linear heat flux. A drawback of this situation is that if β is negative, the inferred linear heat flux will become zero at some point in the past, implying that the heat flux probably was indeed cooler, but for low heat fluxes the poor-quality data lose all capability of resolution.

One would also hope for regional consistency when looking at β values basinwide. This point was examined in Chapter 2 for 20 wells in the North

Sea, and wells grouped closely together did have similar β's. Also, when the heat flow at the present day and the heat flow at 100 Ma were compared, the contoured heat flow pattern did not become more complex or produce "bull's-eyes." In fact, the basinal heat flux inferred from inversion was consistent both regionally and temporally.

A. Determination of the Activation Energies for Steranes and Hopanes

In Chapter 3 we saw that the mismatch between observations of the sterane or hopane isomeric ratio with depth and predictions from the Arrhenius formulation of isomeric redistribution depends on two quantities: (1) the heat flow with time, parametrized through β, the coefficient of the assumed linear heat flow variation, and (2) the activation energy E. It is also clear that both E and β enter the formula for determining MSR, the goodness of fit to the observed data, in a nonlinear manner. Further, the behavior is not proportionate; that is, doubling E and β will not lead to the same functional behavior. Thus E and β enter in nonlinear, but different, manners. We can use these facts to estimate the best activation energy consistent with the well data as follows. By varying *both* E and β it is possible to do an exhaustive trial-and-error search to determine where the minimum MSR occurs in the parameter space of E and β.

Figure 8.1 shows contour plots of MSR in an E–β space for well X; Fig. 8.1a is for the sterane isomeric ratio data and Fig. 8.1b for the hopane isomer data.

We expect that the respective activation energies are constants basinwide, since they should represent chemical parameters and not geologic variables. From Fig. 8.1 we see that the best values of activation energy (i.e., those with the least mean square residual fit) are currently consistent with the downhole isomer data when $E = 70 \pm 40$ kJ/mol for the sterane isomerization and $E = 60 \pm 40$ kJ/mol for the hopane isomerization. The large uncertainties in these values are closely tied to the sparseness of the available downhole sterane and hopane isomer data for this well, but data from other wells confirm these values (see later in this chapter).

B. Comparison of Paleoheat Flux Determinations

In well X in the North Sea, near the Viking Graben, we have available well data extending back nearly 200 Myr. These data consist of (1) lithostratigraphy with depth, (2) bottom-hole temperatures, (3) paleontological age dates for each formation, (4) formation thicknesses, (5) vitrinite reflectance measurements with depth, (6) measurements of

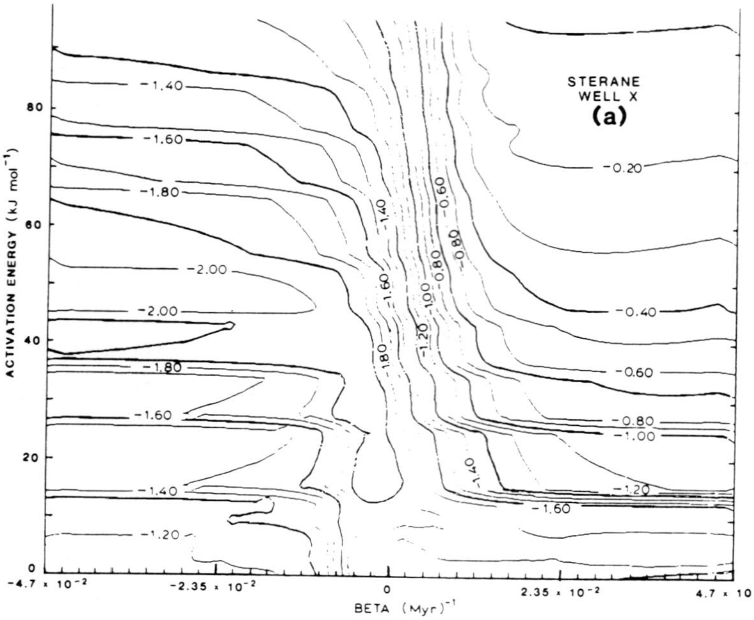
STERANE
WELL X
(a)
-1.40
-1.60
-1.80
-2.00
-2.00
-1.80
-1.60
-1.40
-1.20
-1.40
-1.60
-1.80
-0.60
-0.80
-0.80
-1.00
-1.20
-1.40
-1.60
-0.20
-0.40
-0.60
-0.80
-1.00
-1.20
80
60
40
20
0
ACTIVATION ENERGY (kJ mol^{-1})
-4.7 x 10^{-2}
-2.35 x 10^{-2}
0
2.35 x 10^{-2}
4.7 x 10
BETA (Myr)$^{-1}$

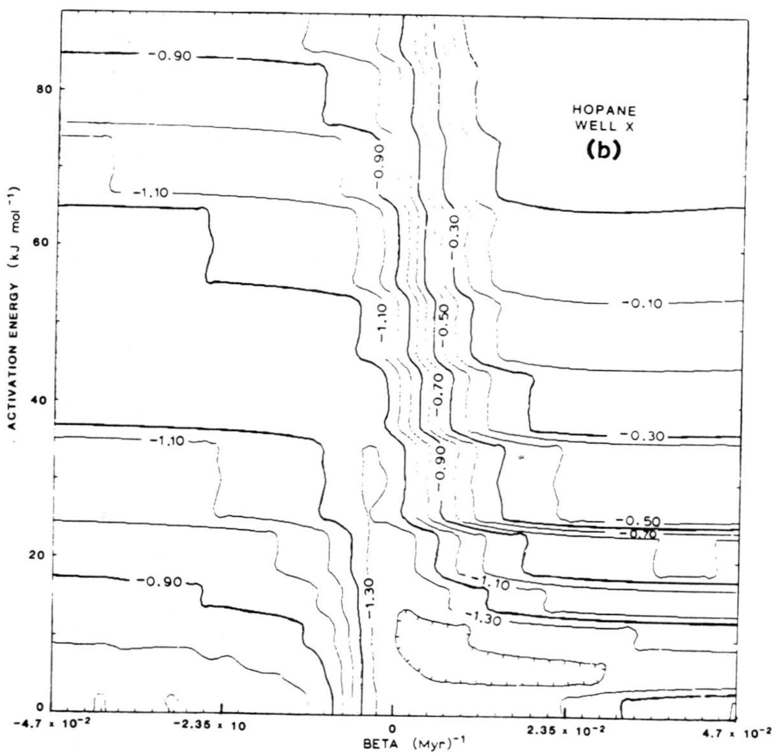
HOPANE
WELL X
(b)
-0.90
-1.10
-1.10
-0.90
-0.90
-1.10
-0.70
-0.90
-1.30
-0.30
-0.50
-0.10
-0.30
-0.50
-0.70
-1.10
-1.30
80
60
40
20
0
ACTIVATION ENERGY (kJ mol^{-1})
-4.7 x 10^{-2}
-2.35 x 10
0
2.35 x 10^{-2}
4.7 x 10^{-2}
BETA (Myr)$^{-1}$

the sterane isomeric ratio with depth, (7) measurements of the hopane isomeric ratio with depth, and (8) paleobathymetric data from which a variation of paleowater depth with time is determined.

We use these data, in conjunction with a burial history code, to "backstrip" the sediments with time and also to invert the thermal indicator data to obtain paleoheat flux evolution with time.

1. Vitrinite Reflectance Method

Based on the vitrinite data shown in Fig. 8.2, we have followed the inversion procedure presented in Chapter 3. The best linear heat flow with time obtained by this scheme is shown as the central solid line in Fig. 8.3. The enveloping solid lines represent a quantitative assessment of the uncertainty in the paleoheat flow determination; they are determined by finding the minimum MSR between predicted and measured vitrinite reflectance values (which yields the best β) and then adding 10% to this minimum value. The corresponding β's are then recorded for $MSR = 1.1 \times (MSR)_{min}$. A plot of MSR against β is given in Fig. 8.4. Note that the minimum MSR = −2.45.

In Fig. 8.2 we also show the theoretical vitrinite reflectance curves—the central solid line corresponding to the best linear heat flux (best β) and the enveloping solid lines corresponding to the uncertainty assessment curves of paleoheat flux variations shown in Fig. 8.3.

In addition, unlike the case of steranes and hopanes, we have very little uncertainty in the chemical parameters, T_D and T_c, entering the vitrinite reflectance inversion method, for they have been determined by exhaustive trial-and-error searches on over 300 wells taken from many different basins around the world.

Reading off from Fig. 8.3 we can write

$$\beta_{vr} = 3 \times (1 \pm 1) \times 10^{-3}\ (\text{Myr})^{-1}$$

as the best value of β (fractional rate of change of heat flow) using the vitrinite reflectance method. Note that the heat flow varied only slightly in

---

FIGURE 8.1 (a) Contours of constant MSR between predicted sterane isomer values and measured sterane isomer values for variations in activation energy (*E*) and β for well X. The ordinate is activation energy in kilojoules per mole; the abscissa is β in (million years)$^{-1}$. (b) As for (a) but for hopane isomer data from well X.

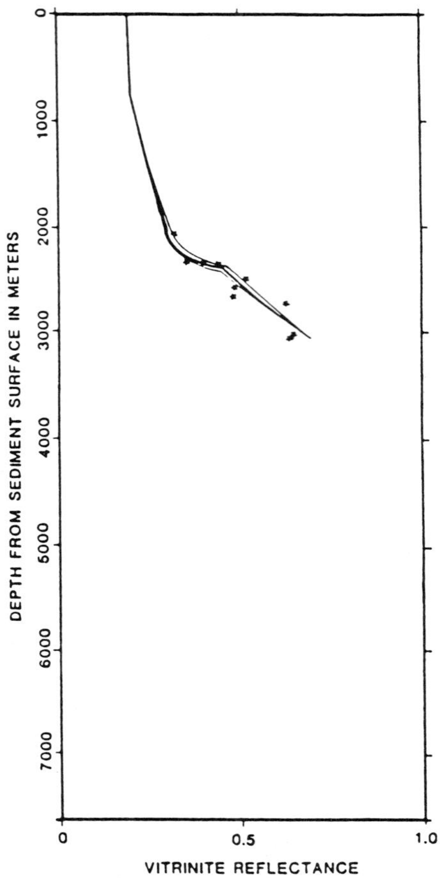

FIGURE 8.2 Vitrinite reflectance data with depth for well X. The central solid curve is for the predicted reflectance values with the best fit to the measured reflectance data for a variety of linear heat flows, using the paleoheat flux inversion method given in the text. The two enveloping solid curves are the error assessment curves. The asterisks are measured vitrinite values.

the past for this well. Even in the Permian, the best heat flow was only 1.5 times the present-day value.

2. Hopane Isomeric Method

Because of the sparseness of the hopane data, the uncertainties in paleoheat flux determination are quite large due to (1) an inability to be precise about the best activation energy and (2) the sparse and clustered

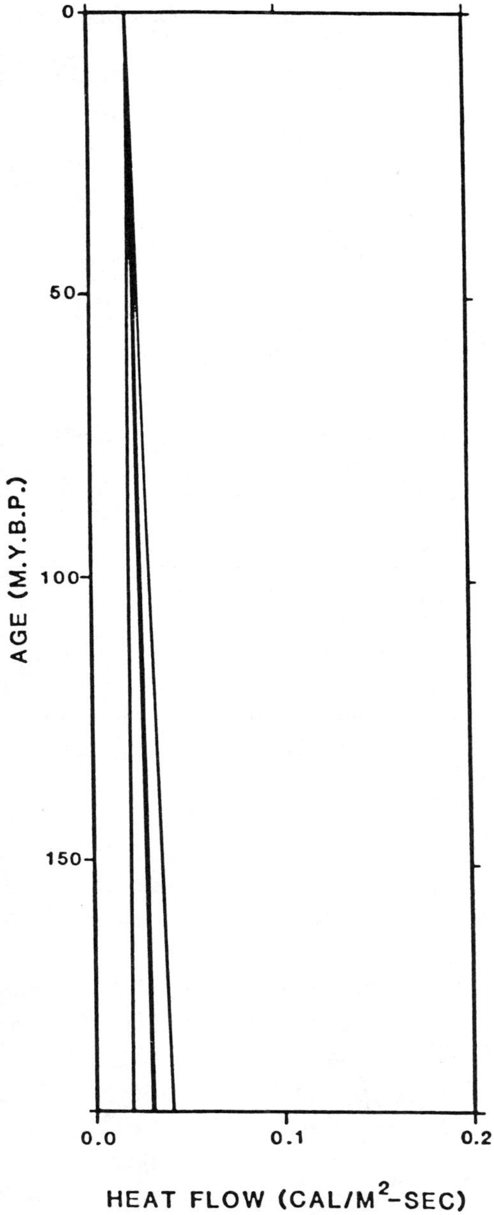

FIGURE 8.3 Linear paleoheat flow with time for well X determined from the match between vitrinite reflectance data and the values predicted using the inversion method given in the text. The ordinate is time (million years before present), the abscissa is heat flow in heat flow units (1 HFU = 40 mW/m^2). The central solid curve is the heat flow value that produce the predicted vitrinite values with the best fit to the measured data; the enveloping solid curves are error assessment curves determined as described in the text.

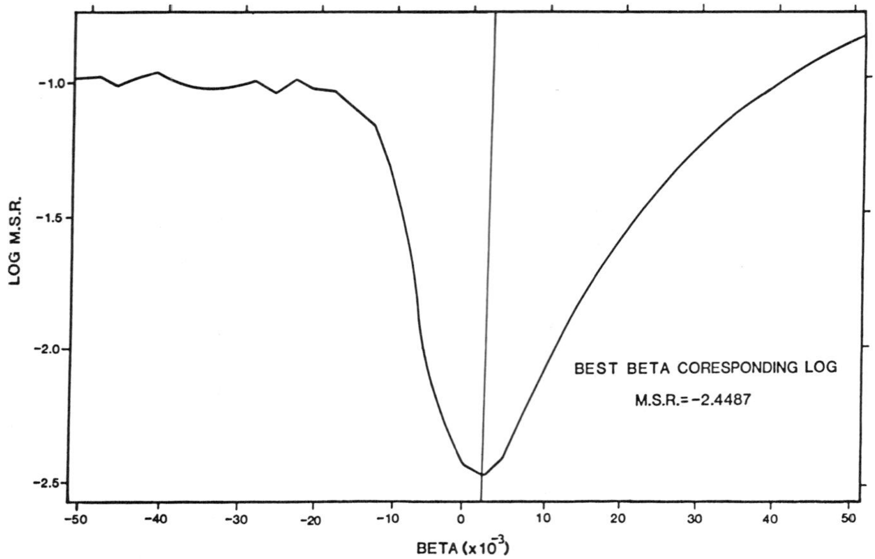

FIGURE 8.4 Plot of MSR (between predicted vitrinite reflectance values and measured vitrinite values) versus β used for determining the best linear paleoheat flow using vitrinite reflectance data from well X. The best fit occurs for the β which gives the minimum MSR between predicted and measured vitrinite reflectance values.

sampling of the data, which provides basically only two groups of points, so that the thermal history is not being recorded uniformly throughout the well.

This sparseness of data, the clustering problem, and the data scatter within the clusters make themselves felt in the MSR versus β plot (Fig. 8.5), where we see that the minimum MSR is only about -1.3—more than a full order of magnitude poorer in fit than given by the vitrinite reflectance inversion.

The corresponding paleoheat flux variation with time is shown in Fig. 8.6, again with the central line representing the best fit and the two enveloping lines being determined from MSR values 10% larger than the minimum. The corresponding theoretical hopane isomer curves are shown on Fig. 8.7 superimposed on the hopane data.

Inspection of Fig. 8.6 and/or Fig. 8.7 shows that the uncertainty is quite large, as anticipated. The corresponding β can be written

$$\beta_H = 8 \times (1^{+2}_{-1}) \times 10^{-3}\ (\text{Myr})^{-1}$$

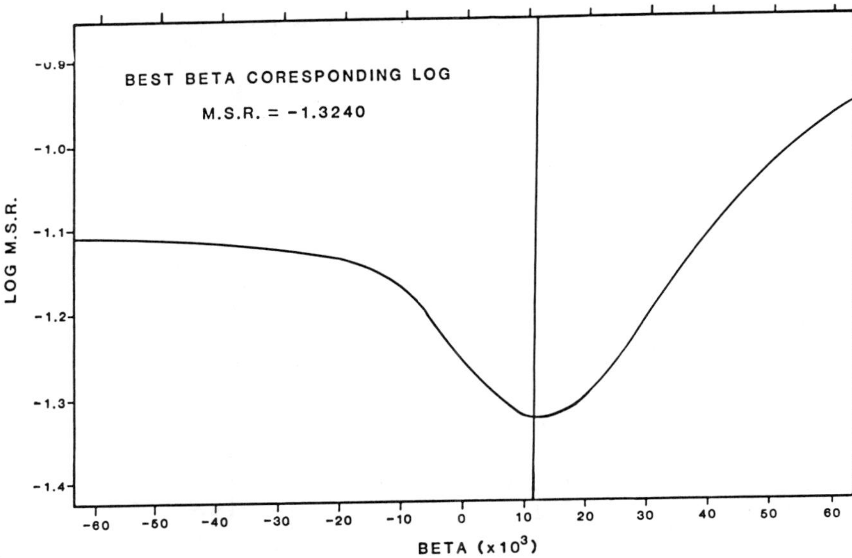

FIGURE 8.5 Plot of MSR (between predicted hopane isomer values and measured hopane isomer values) versus β for determining the best linear paleoheat flux using hopane isomer data from well X. The best fit between predicted and measured values is at the minimum MSR.

so that the range of uncertainty in β more than encompasses the better vitrinite reflectance determination of paleoheat flux.

The hopane isomeric method also predicts a higher heat flow in the past, but until more and better data are available we should treat this only as qualitative confirmation of the more precisely determined paleoheat flux variations from the vitrinite reflectance data. Nevertheless, it does highlight the point that with an increase in hopane isomer measurements better evaluations could be made.

3. Sterane Isomer Method

The sterane isomer record with depth has also been inverted following the procedure laid down in Chapter 3. Here we have used an activation energy of $E_A = 70$ kJ/mol, although the uncertainty of ± 40 kJ/mol in this number informs us that once again the sterane isomeric data are not sufficient to determine accurately the paleoheat flux variations with time. This fact is brought to light explicitly in Fig. 8.8, where the best MSR is about -2.05, nearly a factor of 3 poorer fit to the sterane isomer data than the MSR of -2.45 for the vitrinite reflectance data. Surprisingly, the

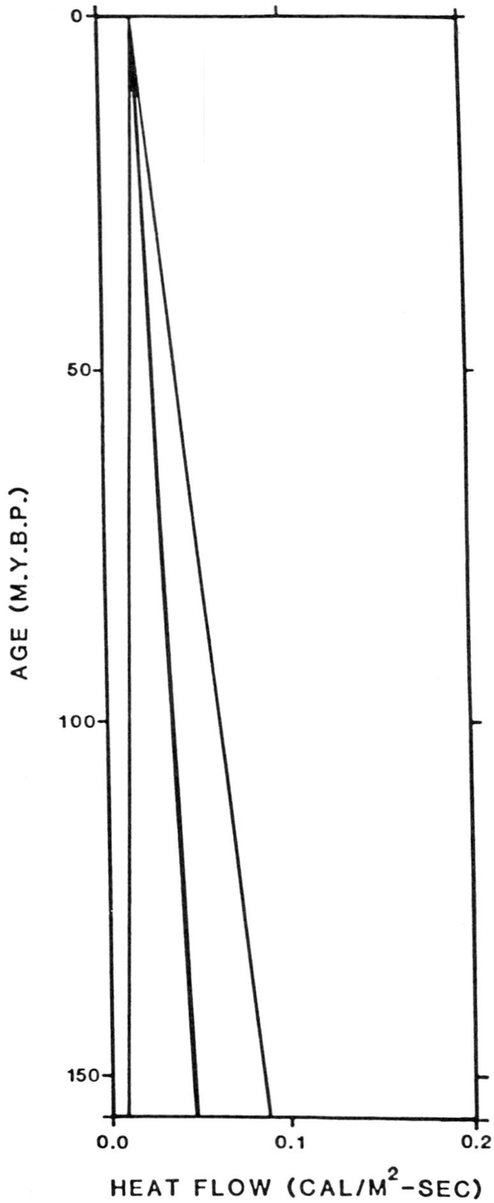

FIGURE 8.6 Linear paleoheat flow with time for well X determined from hopane isomer data using the inversion method given in the text. The ordinate is time (MYBP); the abscissa is heat flow (HFU). The central solid curve is the heat flow which produces the best fit (between the predicted isomer values) to the measured data, and the flanking solid curves are error assessment curves determined as described in the text.

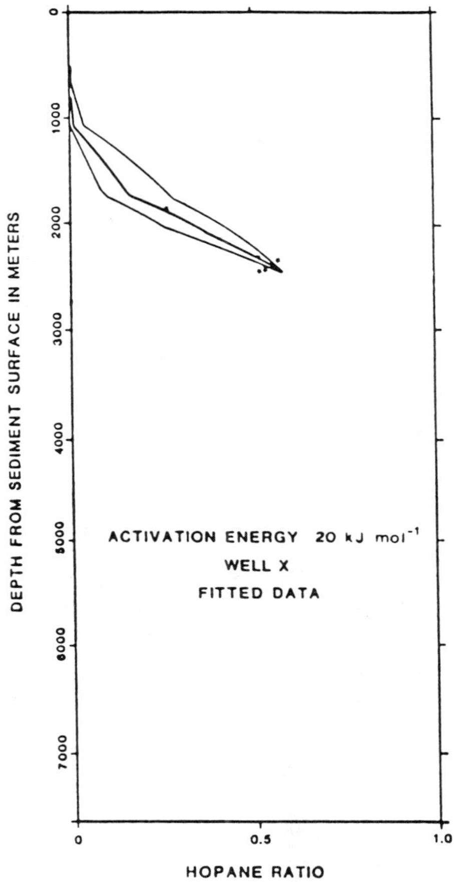

FIGURE 8.7 Hopane isomer data with depth for well X. The central solid curve is the heat flow which produces the best fit between the predicted isomer values and the measured data using the paleoheat flux inversion method given in the paper. The two enveloping solid curves are the error assessment curves.

paleoheat flux and its uncertainty, as shown in Fig. 8.9, appear to be well constrained. However, although a precise fit is obtained (as can also be seen in Fig. 8.10, showing the theoretical estimate of the sterane isomer, and its uncertainty, with depth superimposed on the sterane data), there are caveats and reservations associated with the sterane isomer determination of paleoheat flux. The point is that we really do not yet have sufficient data to determine the activation energy accurately, but even then the inversion scheme works well, emphasizing the importance of gathering more data.

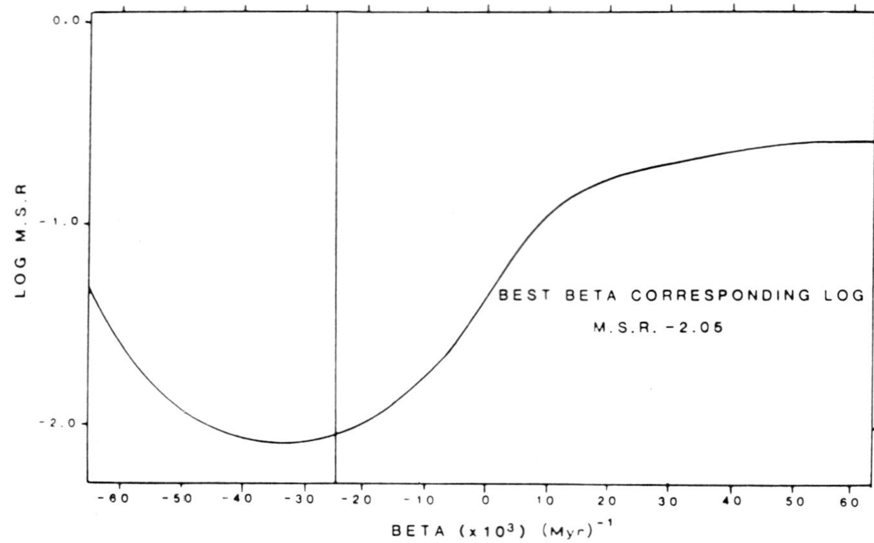

FIGURE 8.8 Plot of MSR (between predicted sterane isomer values and measured sterane isomer values) versus β for determining the best linear paleoheat flux using sterane isomer data from well X. The best fit occurs at the minimum MSR indicating the best β.

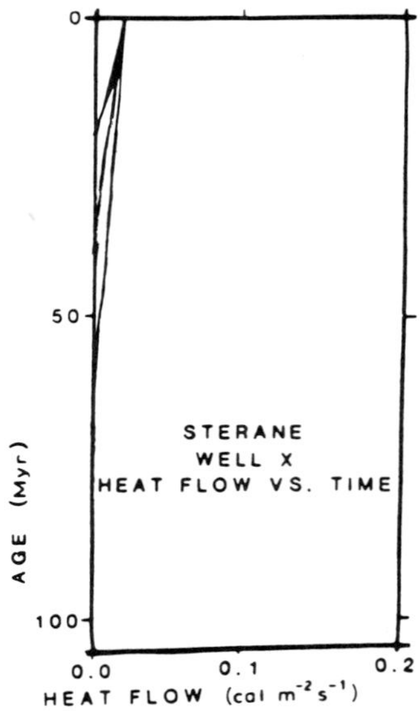

FIGURE 8.9 As for Fig. 8.6 but with sterane isomer data from well X used in place of the hopane isomer data.

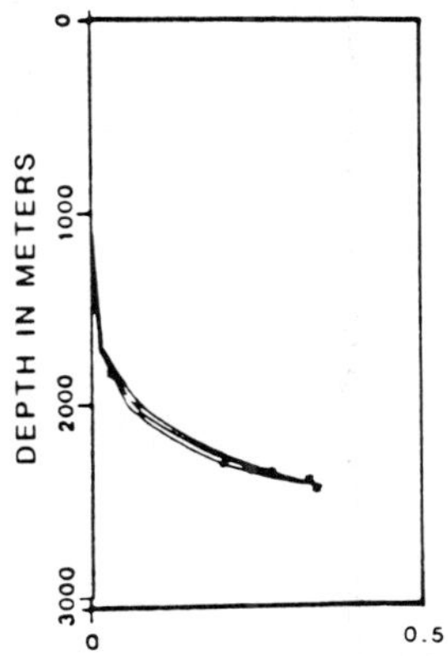

FIGURE 8.10 As for Fig. 8.7 but with sterane isomer data with depth from well X rather than hopane isomer data.

Running the inversion procedure for sterane isomers with an activation energy of 80 kJ/mol produced the determination of heat flow with time shown in Fig. 8.11. This determination, too, had a fairly good fit to the data (an MSR of −2.05). We also note that it, too, provides a fairly precise determination of paleoheat flow with time and fits the sparse sterane data fairly well (Fig. 8.12). However, there is a considerable degree of *inconsistency* between the two paleoheat flux determinations with the sterane isomer data. Basically the system is conveying the message that *if* we could pin down the activation energy to a precise value, the sterane isomer method would provide an accurate paleoheat flux with time.

Again, the paleoheat flux with time is predicted to be higher in the past—as also given by both the vitrinite reflectance and hopane inversions. But again we should view this fact as merely providing qualitative corroborative support for the more precisely determined paleoheat flux variations from the vitrinite reflectance data.

We record the β value suggested by the sterane isomer inversion using an activation energy of 70 kJ/mol. We have

$$\beta_s = 0.017(1 \pm 0.1)\ (\text{Myr})^{-1}$$

If we accept all three of the β values, β_{VR}, β_H and β_s at face value and naively average them, we would predict an average β, $<\beta>$, of $<\beta> = 7 \times (1 \pm 0.6) \pm 10^{-3}\ (\text{Myr})^{-1}$ where the standard error is based on the individual values of β_{VR}, β_H and β_s. If we include the error in

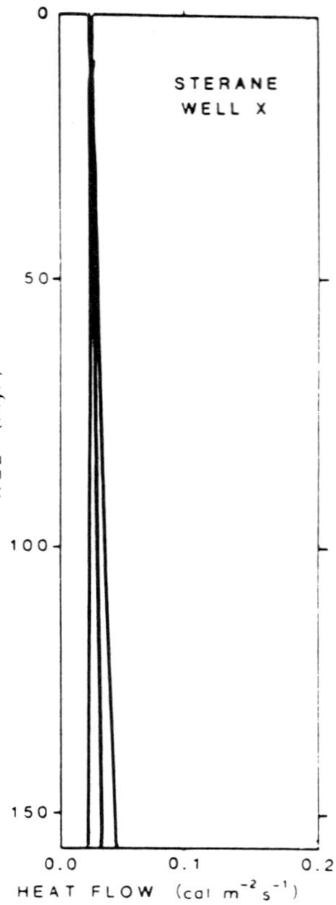

FIGURE 8.11 As for Fig. 8.9 but with the activation energy of steranes decreased to 20 kJ/mol.

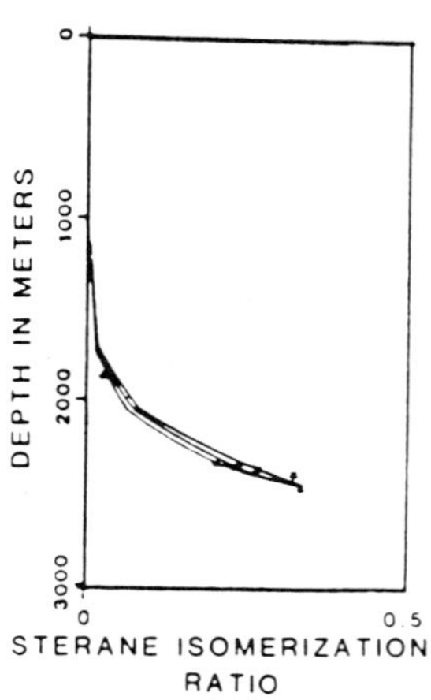

FIGURE 8.12 As for Fig. 8.10 but with the activation energy of steranes decreased to 20 kJ/mol.

determination of each of the three individual values of β, then we replace ± 0.6 as an error assignment by ± 1.1.

4. Confirming Evidence

The consistency we have obtained in paleoheat flux by inverting three very different thermal indicators can be used to argue strongly that the relationships proposed between heat flow, vitrinite reflectance, sterane isomerization, and hopane isomerization are valid, because if we were to be making a significant error in the underlying theoretical framework of any of the indicators, we would not expect consistency of the paleoheat flux results obtained from the three indicators—contrary to what has been obtained. So each indicator acts as corroborative support for the correctness of the theory underlying the remaining two indicators.

In addition, by exhaustive trial-and-error methods, we were able to obtain approximate values for the chemical activation energies of the hopane and sterane isomerization reactions based on well X.

One other well, well Y, in the same block as well X, also had an adequate enough distribution of hopane data with depth for an inversion to be performed and paleoheat flux variations with time and the best activation energy to be determined. The results are shown in Fig. 8.13, following the procedure laid down in the previous section. From Fig. 8.13 we see that an activation energy for hopane isomerization of $E = 60 \pm 40$ kJ/mol is indicated. This is the same as the value obtained independently from well X, indicating that this method of extracting chemical parameters (which should be independent of global, or at the least basinal, behavior) is an appropriate procedure. Unfortunately, well Y did not have any vitrinite reflectance data, nor did it have a sufficient number of sterane isomerization datum points that we could perform inversions. Accordingly, we cannot confirm the paleoheat flux variations obtained from the hopane isomerization for this well.

Well Z, a few blocks away, had about five sterane isomerization values. While hardly sufficient to perform an inversion, nevertheless we ran through the formal machinations in an attempt to see if an estimate could be obtained of the sterane isomerization energy, using the minimum MSR contour method of the previous section. The results are shown in Fig. 8.14. We see from the figure that, despite the paucity of the data in well Z, a rough approximation to the sterane isomerization activation energy is $E = 65 \pm 35$ kJ/mol, which is not discordant with the value of $E = 70 \pm 40$ kJ/mol obtained from well X.

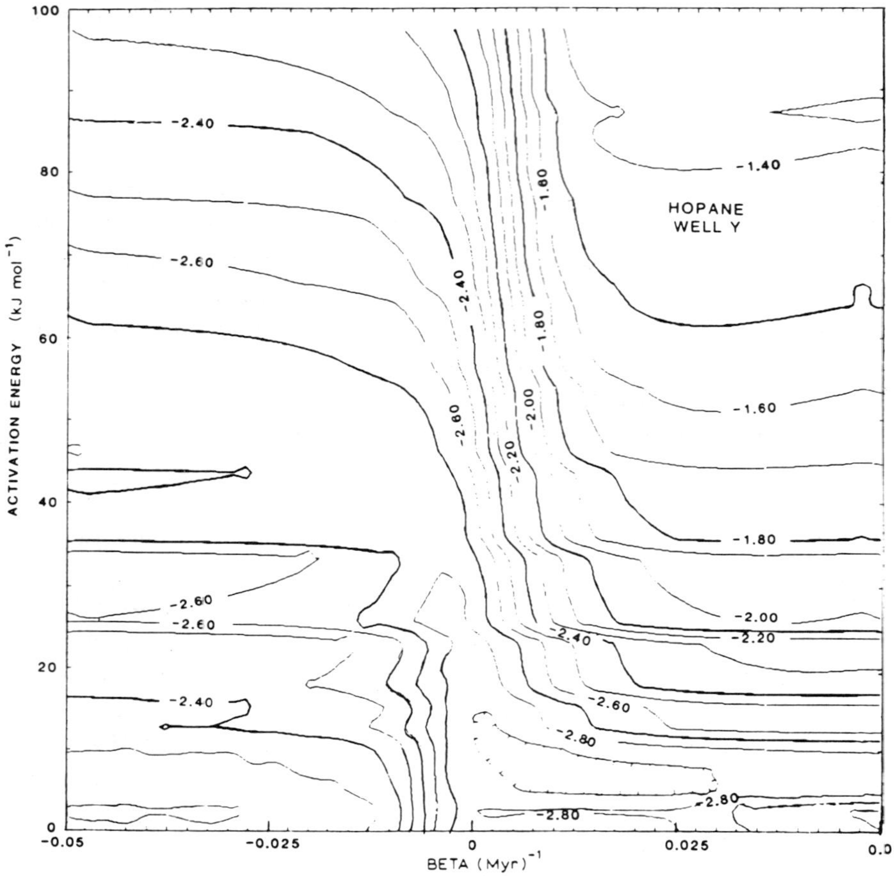

FIGURE 8.13 Contours of constant MSR between predicted hopane isomer values and measured hopane isomer values for variations in activation energy (E), β for well Y.

II. Paleoheat Flux and Stratigraphic Age

In this part of the chapter we show how thermal indicator data are used in an inverse mode to determine ages of stratigraphic horizons. Results from "blind" tests on wells with horizon ages ranging from Ordovician through Carboniferous and Jurassic to Miocene indicate that thermal indicator inversions are capable of resolving such ages to within about 10% uncertainty. Confirmation of the accuracy of the inversion procedure with one thermal indicator (vitrinite reflectance) was achieved by using another independent thermal indicator (sterane isomerization) in the same

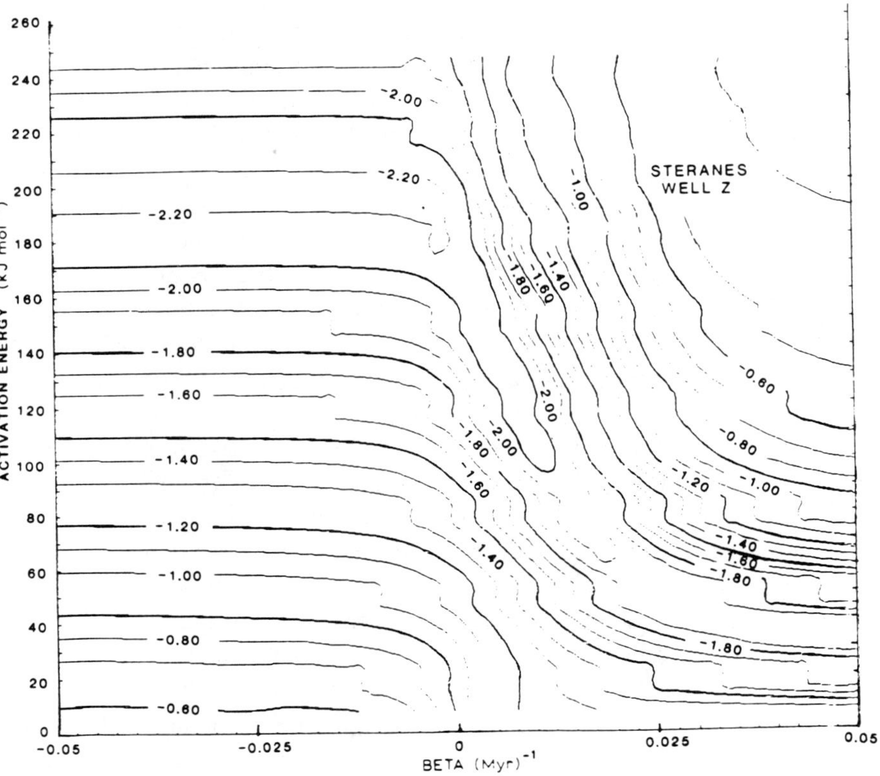

FIGURE 8.14 Contours of constant MSR between predicted sterane isomer values and measured sterane isomer values for variation in activation energy (*E*), β for well Z.

well, and the self-consistent activation energy for sterane isomerization was determined to be 30 ± 15 kJ/mol. In addition, (1) it was possible to determine simultaneously the age of a stratigraphic horizon, the thickness of eroded sediments at an unconformity, and the variation of paleoheat flux with time by thermal indicator inversion in a single well, and (2) two neighboring wells, less than 10 km apart, provided essentially identical ages for the same formation when tested using the inversion procedure.

The ability to determine stratigraphic horizon ages from inversion of thermal indicator data implies that we can determine sedimentation rates, and so basinal evolution, to a degree of resolution not previously obtainable from assumed interpolation methods applied to determine the age of formations between a limited set of stratigraphic horizons of known ages.

As we remarked earlier, when assessing an area for hydrocarbon potential, the petroleum geologist needs to determine quantitatively the timing and extent of hydrocarbon generation, as well as to provide a quantitative assessment of the migrational history of oil and gas. To achieve these objectives, the burial history of formations in the area is required. Before source rock maturity can be evaluated, the ages of formations, the amount of erosion of unconformities, and the timing of erosion need to be determined. Paleontologic studies are considered essential to determine the age of formations, length of time of unconformities, and paleobathymetry. Unfortunately, the paleontological data are not always available and, even when available, are often not accurate enough to use in the quantitative manner required by burial history calculations. Here a method is presented which uses the biostratigraphic marker information for most of the ages of formations and then uses thermal indicator data to recover the remaining formation ages.

When inverting thermal indicator data the canonical assumptions are that (1) the burial history of the sediments is precisely known; (2) the dominant source of heat to the sediments was the basement, and (3) the dominant mode of heat transport was thermal conduction.

Here we provide quantitative formulas which detail the influence on the thermal history of sediments when assumption (1) is not met. These formulas are then used to determine the temporal variation of paleoheat flux together with the age of a stratigraphic horizon. First we illustrate how the various factors violating assumptions (1), (2), and/or (3) influence the thermal history of sedimentary material.

Temperature information at depth z at time t is calculated from

$$T = T_s + \int_0^z \frac{Q(t)\,dz}{K(t, z)} \tag{8.1}$$

where T is temperature, T_s is the temperature at the sediment surface, $Q(t)$ is the heat flux at time t, and $K(t, z)$ is the thermal conductivity at time t and depth z. K is a function of lithology, depth, and time and is calculated from the relationship between porosity and conductivity discussed in Sclater and Christie (1980). The integral runs from $z = 0$ (corresponding to the surface) to the formation's depth z at time t.

Two major effects are apparent. First, any changes in the burial path of the sediments from some preassigned evolution means that, even if paleoheat flux was constant at its present-day value, there would be a direct change in the temperature history of the sediments resulting from the altered burial path. Thermal indicators would be directly influenced by this altered temperature history. Second, a change in temperature can also be caused by igneous intrusives (Dow, 1977) (i.e., a certain portion of the sedimentary sequence would have an additional heat source for a discrete period of time); by overthrusting (an increase in burial depth caused by the overthrust overburden as well as compaction of the underlying layers; in addition hotter layers are moved over cooler layers, and frictional heating may also play a role); by variable overpressuring with time [since formational thermal conductivity K depends on porosity, with lower K for higher porosity, and since porosity is high in overpressured formations, formations overpressured for any period of time would have a higher temperature gradient than normally pressured formations at the same depth (Fertl, 1976)]; by proximity of the sediments to dynamically evolving salt ridges, diapirs, and "tongues" (Vizgirda *et al.*, 1985); and so forth. Thus, although the burial history may be precisely known, both the thermal conductivity and the sedimentary temperatures are influenced directly by the foregoing geological events.

A. Results Based on Vitrinite Reflectance Measurements

An age must be ascribed to each formation in order to use isostatically compensated burial history models. However, for some formations we do not know the ages precisely or even at all. In those cases it has been a common practice to assume a constant depositional rate between two formations of known age and to interpolate linearly between them to assess the age of an intervening formation.

The cumulative record of thermal history carried by a thermal indicator (such as vitrinite reflectance or sterane/hopane isomer ratios) provides precisely the information needed to determine the age of a horizon. A thermal indicator, enmeshed in sedimentary layers deposited later than the formation of unknown age, has no means of recording the thermal events previously seen by that formation but does record the later thermal history. However, a thermal indicator occurring in the layer of unknown age, and in layers deposited earlier, records both the thermal history before that age and also thereafter to the present day. This is because the combination of thermal indicator theory together with burial history backstripping (Guidish *et al.*, 1985) allows a determination of paleoheat flux with time.

The degree of accuracy is provided by

$$\mathrm{MSR} = \log\left\{\left[\sum_{i=1}^{n}(P_i - M_i)^2\right]/(n-2)\right\} \tag{8.2}$$

where P_i is the predicted value and M_i the measured value. Here MSR is a measure of the degree of mismatch between the theoretically predicted thermal indicator values with depth and the observed values and is the logarithm (base 10) of the mean square residual mismatch. Using MSR, one can quantify the degree of fit for one predicted set of values based on a particular heat flux versus the observed values against another predicted set based on a different heat flux. As the rate of change of heat flux with time is varied, the predicted thermal indicator values change, so the MSR varies, and a minimum in MSR indicates the best fit (Fig. 8.15). By such a procedure, β (which characterizes the rate of change of heat flux) is tightly constrained and the fit of theoretical to observed vitrinite reflectance with depth for the "best" β value is determined.

Determining an age for a horizon adds another dimension to the problem. For an assumed age of a stratigraphic horizon we can find the best β and record the MSR. With either too young or too old an assumed age, the degree of mismatch between the predicted and observed vitrinite values with depth will increase. There is, then, a best age for the horizon,

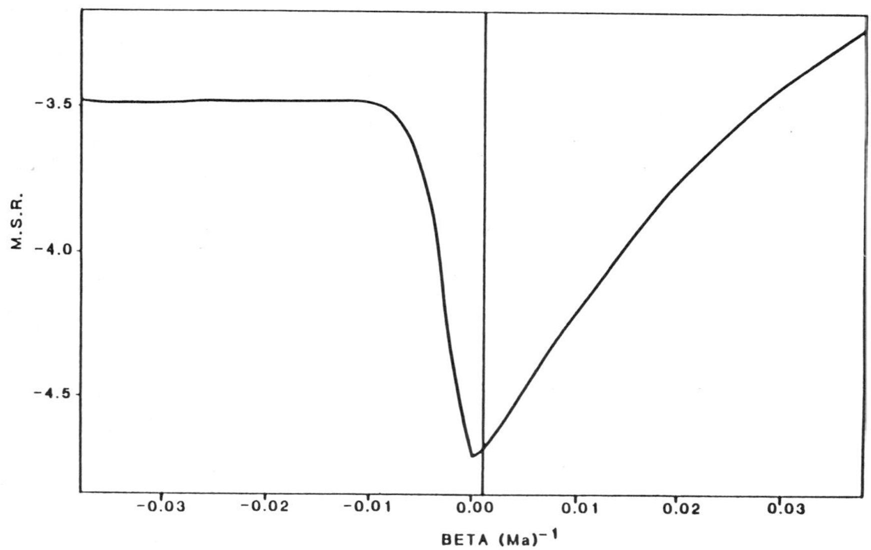

FIGURE 8.15 Plot of the logarithm of the mean square residual mismatch between theory and observations as a function of the linear heat flow coefficient β for vitrinite reflectance data from the well Yalcumma 1, treating the Triassic–Jurassic unconformity as a depositional hiatus. Note the increase in agreement of fit by more than an order of magnitude for β values in the range 0 to 2×10^{-3} (Ma)$^{-1}$.

which equates with both the smallest MSR and the best β value. The question we investigate here is the degree to which the inversion of thermal indicator data determines the age of a stratigraphic horizon, and with what degree of resolution (determined by the *relative* improvement in MSR). In this preliminary investigation we examine the impact of using "blind" tests on wells in which all the stratigraphic horizons have known ages, and then we assume that one of the dated horizons has an unknown age. We ask how well we can recover the correct answer. First, we carry through the inversion analysis using vitrinite reflectance measurements as the sole thermal indicator. We then ask how well the determined ages are substantiated by using another independent thermal indicator in the inverse mode. Satisfactory agreement provides strong corroborative support for the theoretical framework for each such thermal indicator.

The essence of inversion of any thermal indicator is as follows:

1. Approximate the physical and chemical mechanisms of the reaction being modeled.
2. Normalize out as many chemical and physical constants as possible, leaving as small a number as possible of "free" parameters.
3. Use the data to solve for these remaining free parameters such that a minimum is obtained for the difference between predicted values and observed data.
4. Determine the range of the free parameters such that the parameters give consistency with the observed data for each well, and determine further the acceptable range from many wells where the free parameters related to chemical and physical constants overlap.

Using this scheme, semiempirical relationships of thermal indicators to cumulative measures of temperature and time along burial paths are tested against measurements. We are, of course, limited by the accuracy and quality of data collected. Any disagreement between chemical and physical constants determined through inversion and laboratory experiments could be accounted for by the shift of time and temperature scales and the multimillion scale of geologic processes. In addition, we must remember that the conditions of the laboratory environment are by no means those of the subsurface, so again we have reason to expect a discordance between laboratory-determined values of "constants" and the values best suited to the geological record (Snowdon, 1979). With this in mind, the accuracy of the inversion scheme is continuously being tested and updated over the full spectrum of geological time with many wells and, by requiring the empirical relationship of the chemical reaction framework and its "constants" be consistent with downhole data, is independent of that time scale. We used several wells from different basins, of different geological ages and tectonic regimes, to show that this scheme is successful independent of the basin type.

1. An Ordovician Section

For well Yalcumma 1 from the Cooper Basin (Australia) the results of varying the age at total depth (TD) show that in a certain range of ages a consistent paleoheat flow coefficient was determined and that the MSR improved significantly. The Cooper Basin is a complex Permian–Triassic northeast-trending intracratonic basin which is overlain by the Eromanga (Jurassic–Cretaceous) Basin. Earlier Paleozoic basins and Carboniferous intrusives underlie the Cooper/Eromanga basins (Battersby, 1976).

Figure 8.16a shows an improvement of MSR (by 0.2) in a constrained age ($\approx$400–500 Ma) and β (7.5 $\pm$ 7.5) $\times$ 10^{-4} $(\mathrm{Ma})^{-1}$ range. Therefore the age at TD would be 450 $\pm$ 50 Ma from inversion of vitrinite data, consistent with Kantsler *et al.*, (1983), since TD is on the pre-Permian side of the unconformity, which is reported as lower Carboniferous to lower Proterozoic (350–700 Ma) (Battersby, 1976). A β of 7.5 $\times$ 10^{-4} $(\mathrm{Ma})^{-1}$ represents a slightly higher heat flux in the past than at the present, approximately 40% greater 500 Ma ago than today. Figure 8.16b, c, and d show the observed vitrinite reflectance data (asterisks) with depth and the solid lines show the predictions based on the thermal inversion procedure for assumed ages of 300 Ma (Fig. 8.16b), 450 Ma (Fig. 8.16c), and 600 Ma (Fig. 8.16d). Inspecting Fig. 8.16a–d further, one can see the underlying logic. If the age is too young (Fig. 8.16b) the heat flow coefficient β has to be increased to compensate for the shorter period of time in which the sediments can mature. As seen in Fig. 8.16a, the MSR is lowest with a β which would have the heat flow from times greater 300 Ma than it is today and does have the predicted reflectance values overmature relative to the measured values in the 2400–2800-m range and undermature relative to the last two measured points. With too old an age (Fig. 8.16d), the heat flow has to be reduced in the past (Fig. 8.16a) in order to have the best MSR which has the heat flow at 600 Ma less than the present-day value. Yet with more time and less heat the predicted values are still undermature relative to the deepest observed data. But β can be neither increased nor decreased without limit while maintaining a minimal MSR, because overlying, younger formations of known ages must also mature in a manner consistent with their observed thermal indicator behavior with depth. This fact constrains the range of β. So, with combinations of age and β, the optimal fit of predicted and observed vitrinite reflectance data is obtained (Fig. 8.16a and 8.16c).

2. A Carboniferous Section

To increase the precision of an age determined for a formation, two wells in close proximity to each other were used. The close proximity of these wells justifies the assumption that their burial and thermal histories were nearly the same.

Figure 8.17 illustrates age determination for the Merrimelia Formation for wells 3 and 8 in the Toolachee field of the Cooper Basin, Australia, within 10 km of each other. Both wells' burial histories and vitrinite reflectance data produced a consistent thermal history. To the same extent that the β's overlapped, the determined ages were similar. Indeed, for the age determinations we obtain well 3, age = 325 ± 20 Ma, and well 8, age = 325 ± 20 Ma. Paten (1969) and Williams and Wild (1985) have both assigned a palynological age of late Carboniferous–early Permian to the Merrimelia Formation.

3. A Cenozoic Section

We chose a Californian basin to test, and an interesting facet of the inversion algorithm using an isostatic burial history model was exemplified. In basins which are fairly young (less than about 20 million years old), where sedimentation rates are extremely high (of the order of 1000 ft per million years), the effective paleoheat flow determined from vitrinite reflectance was zero until the last several million years (Fig. 8.18). From a geological and/or physical point of view, this zero heat flow is absurd. If the heat flux is indeed purely from the basement through conduction, the correct interpretation of the model prediction of the equivalent heat flow seen by the sediments is as follows. The equilibration time for conductive heat flow through sediments about 5000 m thick is on the order of 10 million years (Carslaw and Jaeger, 1947). From inversion of the vitrinite reflectance data, the indication is that the sediments matured only in the recent past and that they indeed recorded zero heat flow prior to the last few million years. Thus the temperature of the sediments remained between the surface temperature and their equilibrium temperature from the time of deposition until close to the present day, with their temperatures closer to the surface temperature at deposition and ap-

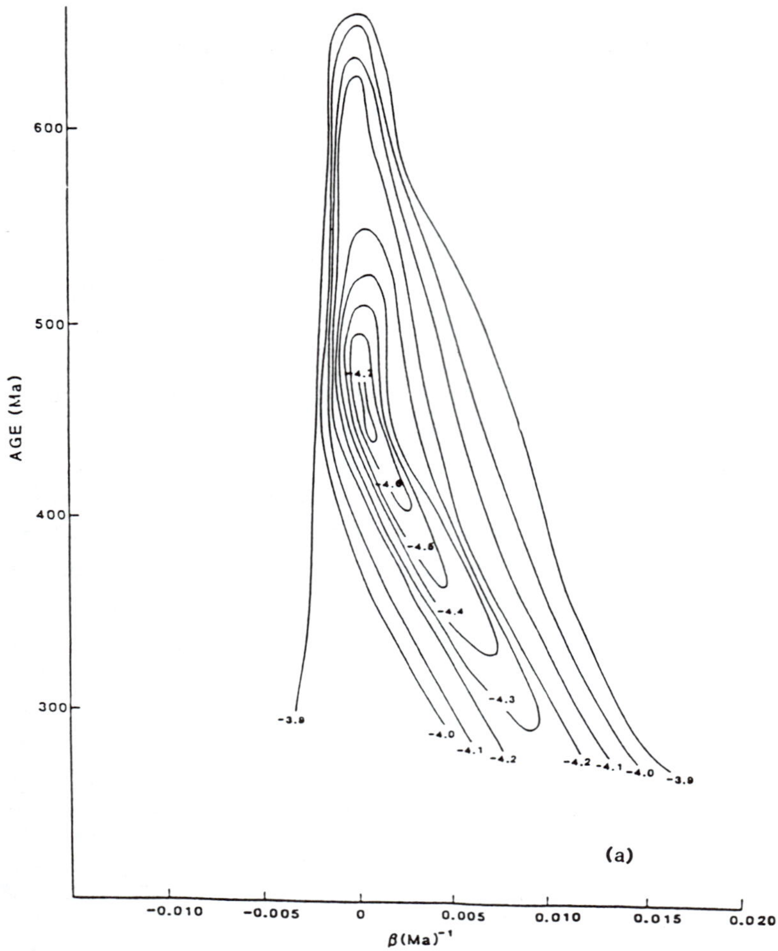
AGE (Ma)
600
500
400
300
-3.9
-4.0
-4.1
-4.2
-4.3
-4.4
-4.2
-4.1
-4.0
-3.9
(a)
-0.010
-0.005
0
0.005
0.010
0.015
0.020
β(Ma)⁻¹

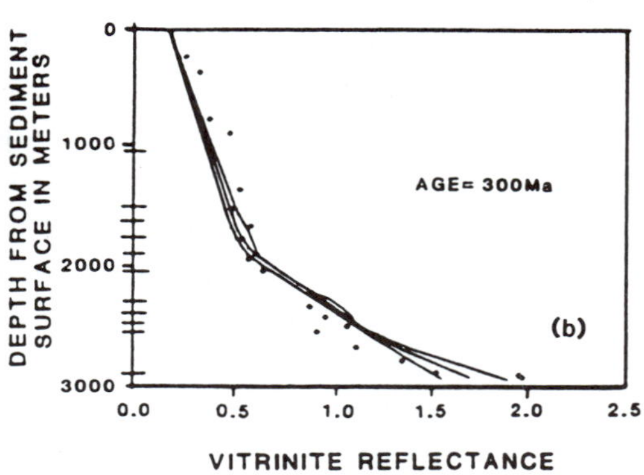
DEPTH FROM SEDIMENT SURFACE IN METERS
0
1000
2000
3000
AGE= 300Ma
(b)
0.0
0.5
1.0
1.5
2.0
2.5
VITRINITE REFLECTANCE

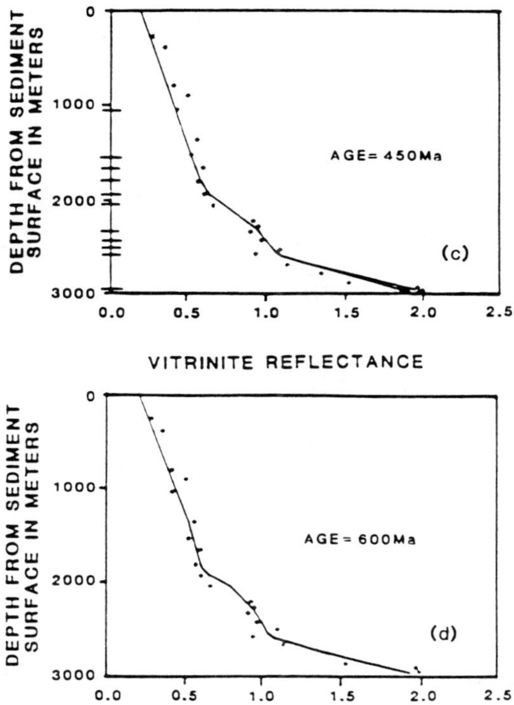

FIGURE 8.16 (a) Contours of constant MSR in an age–β space for vitrinite reflectance data from the well Yalcumma 1. Note that the best fit contours tightly cluster around the point age 450 Ma, $\beta = 10^{-4}$ (Ma)$^{-1}$. (b) Plot of observed and predicted vitrinite reflectance values with depth at the present day for an assumed TD formation age of 300 Ma. The three curves represent the best obtainable β value determined from the minimum MSR and also β values obtained from MSR values 10% larger than the minimum. (c) As for (b) but for an assumed formation age of 450 Ma. Note the significant improvement in fit to the data. (d) As for (c) but for an assumed age of 600 Ma. Note the worsening of fit to the data compared with that in (c).

proaching their equilibrium temperature as the system equilibrates. We were able to use this fact to constrain better the critical temperature T_c, depending on surface temperature input (Fig. 8.19).

The critical temperature T_c is that temperature at which vitrinite starts to mature. Since $T_c = 295 \pm 10$ K had been acceptable for the many previously tested wells, this extreme case was examined further so that the sensitivity of T_c could perhaps be resolved further. (This continuing inspection of sensitivity of the parameters is imperative to the inversion

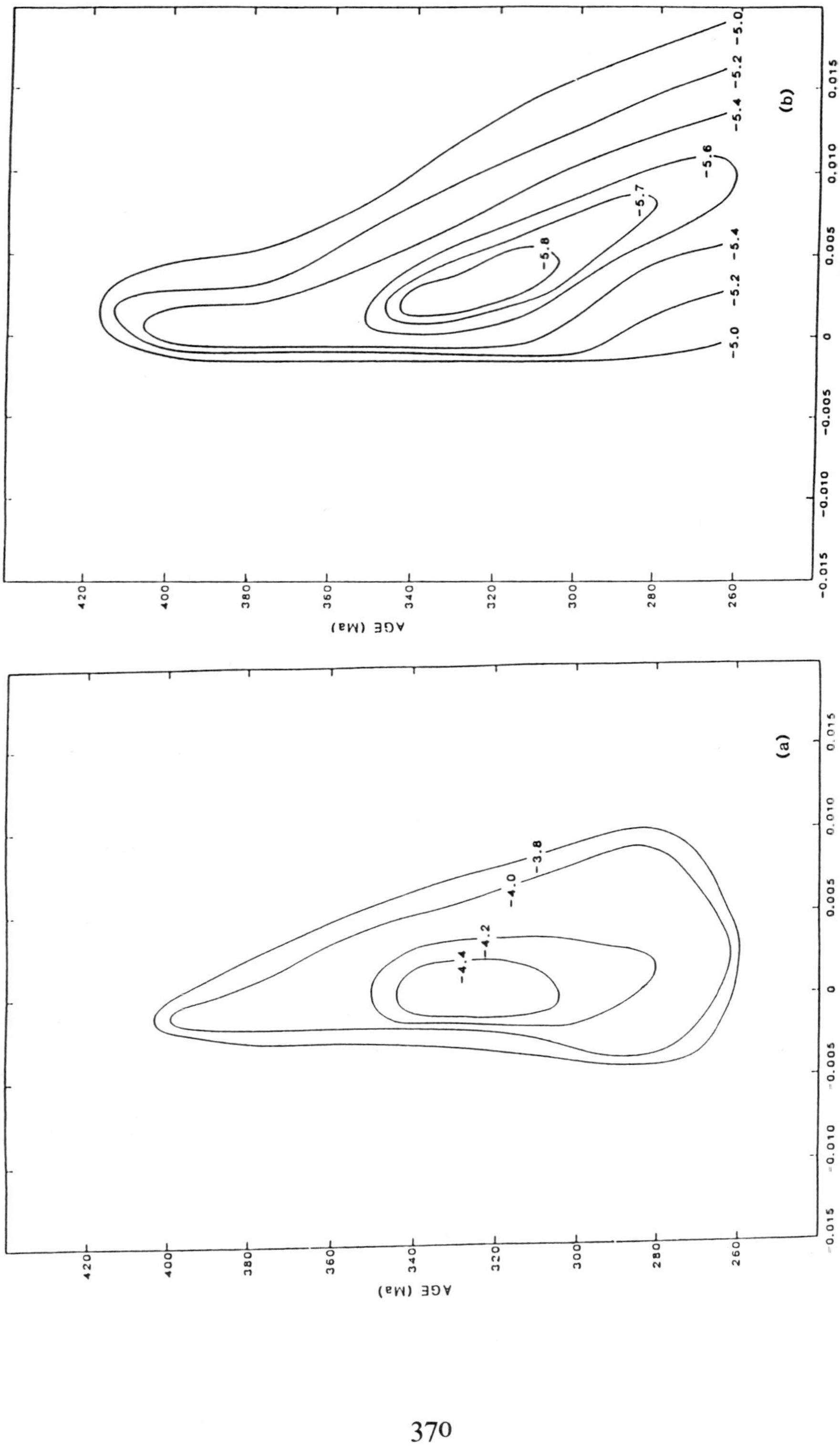
(a)
(b)
AGE (Ma)
β (Ma)⁻¹
420
400
380
360
340
320
300
280
260
-0.015
-0.010
-0.005
0
0.005
0.010
0.015
-3.8
-4.0
-4.2
-4.4
-5.0
-5.2
-5.4
-5.6
-5.7
-5.8

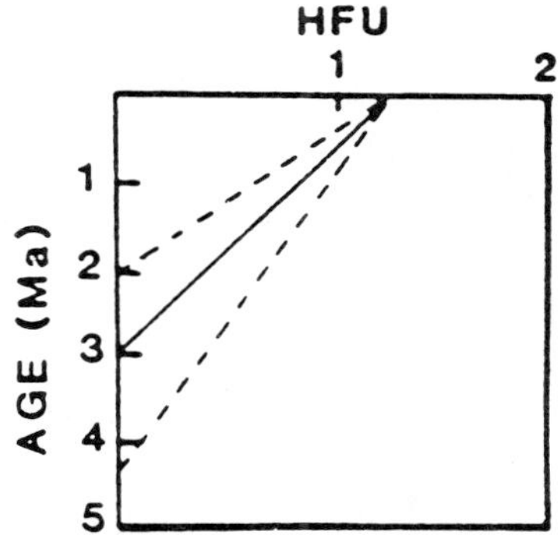

FIGURE 8.18 Plot of the nominal best basement heat flux (HFU) versus age for a California well. Note that the heat flux reaches zero between 2 and 4.5 Ma ago—an absurd result. The proper interpretation of this "observed" heat flux variation is provided in the text.

scheme in order to check for consistency of system parameters nominally independent of basin type and geologic process). For a young basin with a high sedimentation rate, a fixed surface temperature of 60°F (289 K), and low vitrinite reflectance values (indicative of some California basins), a poor correlation is found between predicted and measured reflectance values at low critical temperatures (≤287 K). However, when the critical temperature is in the range 290–305 K a significant improvement in MSR results (Fig. 8.19). To illustrate the interrelation of critical temperature T_c and surface temperature T_s in a different mode, we obtained curves of constant MSR in a T_c–T_s space (Fig. 8.20). As can been seen from Fig. 8.20, we trade determination of T_c for resolution of T_s. Therefore neither T_s nor T_c can be constrained beyond the limits already established.

However, it does not follow that the surface temperature has been constant over the lifetime of the basin as assumed. A more rigorous way to examine the interrelation of surface temperature variations with time

---

FIGURE 8.17 (a) Same as for Fig. 8.16a but for vitrinite reflectance data from well 3 in the Toolachee field of the Cooper Basin, Australia. Note that the best-fit contours tightly constrain the formation age to be 325 ± 20 Ma. (b) Same as for Fig. 8.16a but for vitrinite reflectance data from well 8 in the Toolachee field of the Cooper Basin. Note that the best-fit contours tightly constrain the formation age to be 325 ± 20 Ma. The formation is the same as that whose age was determined from well 3 data.

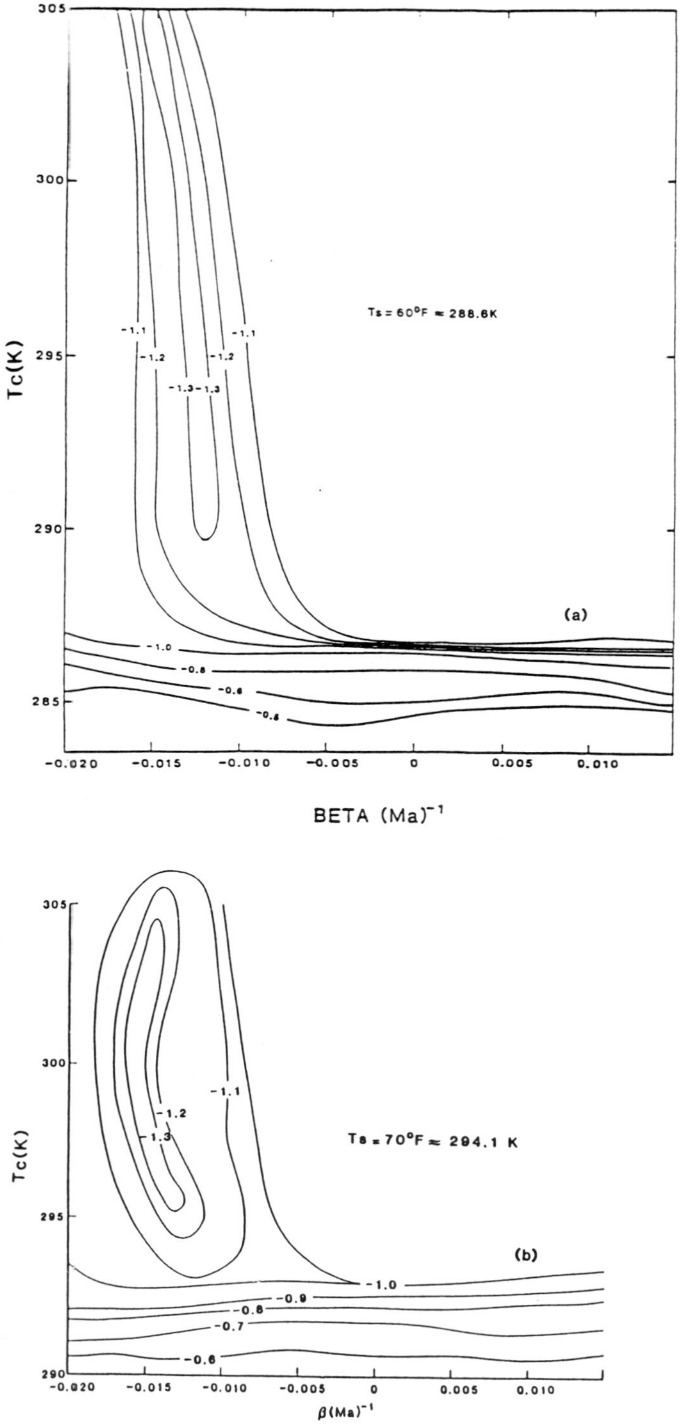

Tc(K)
BETA (Ma)⁻¹
Ts = 60°F ≈ 288.6K
(a)
Tc(K)
β(Ma)⁻¹
Ts = 70°F ≈ 294.1 K
(b)

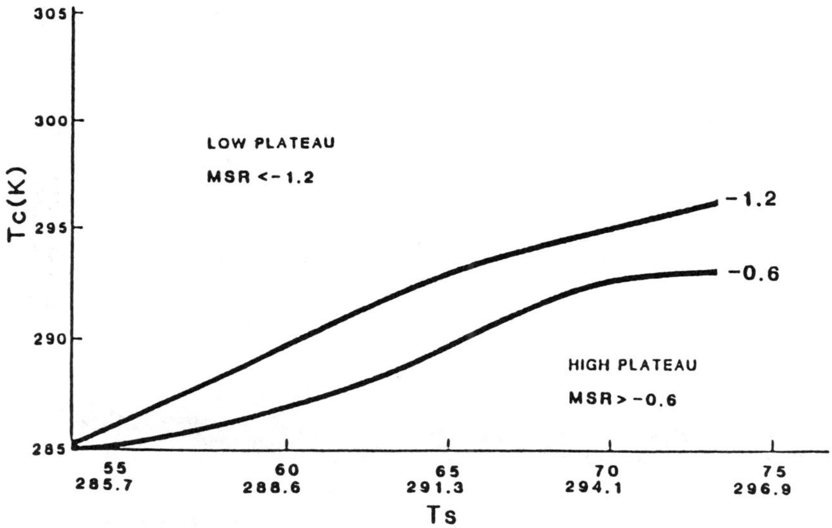

FIGURE 8.20 Contours of constant MSR for the California well in a T_c–T_s space with T_c in degrees kelvin and T_s in degrees Fahrenheit. Note that above the MSR = −1.2 curve the contours of MSR flatten out into a very broad valley so that resolution on T_c is being traded for resolution on T_s.

and the critical temperature is to allow T_s to vary from its present-day value. Thus with $T_s(t) = T_{s0}(1 + \gamma t)$, where T_{s0} is the present-day surface temperature and γ is the linear paleo-surface temperature coefficient, we can obtain plots of constant MSR in a T_c–γ space as shown in Fig. 8.21. Note that there is a very sharp boundary contour of MSR cleanly separating, by about an order of magnitude, poor fits (below $T_c \approx 287$ K) from very good fits (above $T_c = 287$ K). Unfortunately, the broad range of possible allowed γ values means that both the quality and quantity of the vitrinite data are not adequate to resolve the rate of change of surface temperature with time. A considerable body of very high-quality data would be needed because the requirement that the temperature be above

FIGURE 8.19 (a) Contours of constant MSR for the California well in a critical temperature T_c versus linear heat flux coefficient β space, for a fixed surface temperature T_s of 60°F (288.6 K). Note that a very poor fit obtains if $T_c < 285$ K, while a very good fit is found for $305 \text{ K} \geq T_c \geq 290$ K. (b) As for (a) but with the surface temperature T_s raised to 70°F (294 K). Note the very poor fit for $T_c \leq 293$ K with a very good fit for $305 \text{ K} \leq T_c \leq 295$ K.

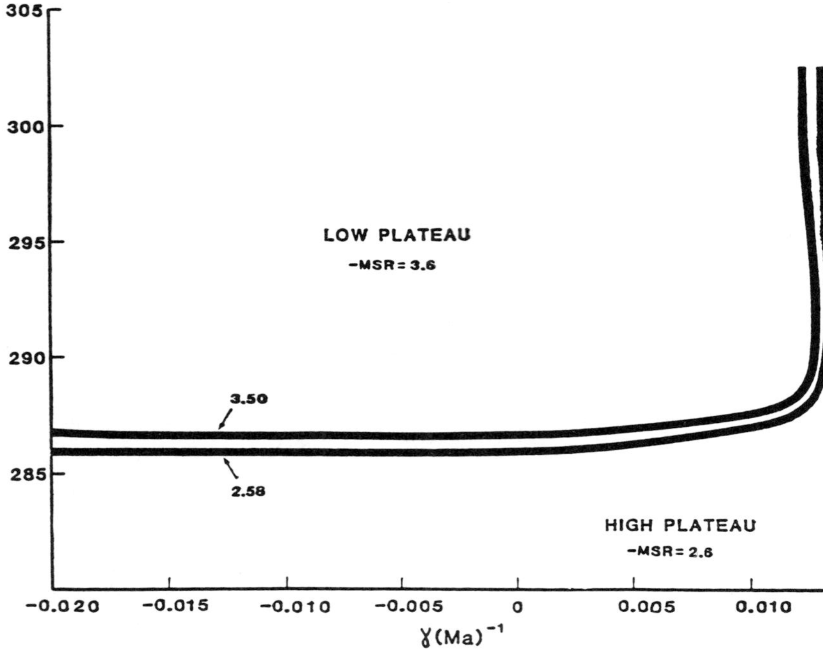

FIGURE 8.21 Contours of constant MSR for the California well in a T_c–γ space with T_c in degrees kelvin. Note that the contrast contour between an order of magnitude improvement is very steep so that above the contour line any range of γ is acceptable.

T_c in order to initiate thermal maturation means that the equality $T_c \approx T_s$ functions as an on–off switch: on for $T_s > T_c$, off for $T_s < T_c$, as shown by the contour nearly parallel to the γ axis separating an order-of-magnitude improvement in fit.

Effectively, then, we can hold the surface temperature at its present-day value of 60°F (289 K) or any other acceptable T_s because of the relationship of T_c and T_s (Fig. 8.20). The effective heat flux seen by the sediments will indeed head toward zero a very short time into the past. As a result, the sensitivity of the system to the unknown age, e.g., of the formation at TD, is no longer contained in an age–β space for contours of constant MSR.

Figure 8.22 shows the resulting contours of a different Cenozoic well, which has a nonzero effective heat flux extending over the total basinal history covered by the thermal indicators. Note the large spread in acceptable β values relative to any other well of older age and lower sedimentation rate, but note also from Fig. 8.22 that the thermal indicator

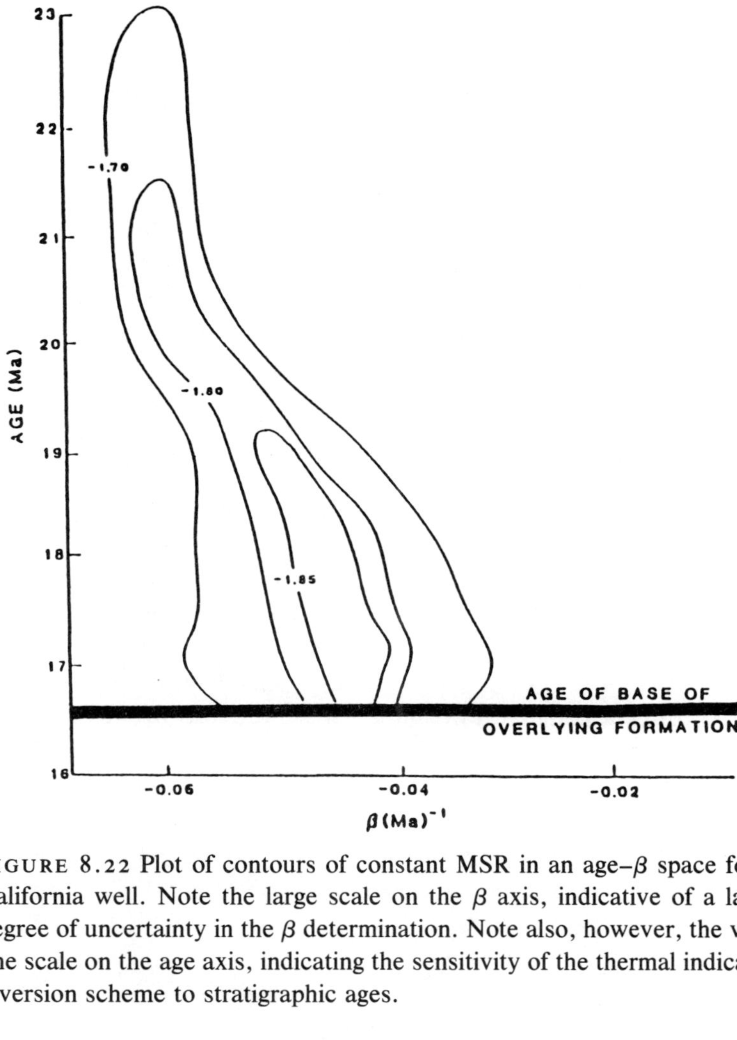

FIGURE 8.22 Plot of contours of constant MSR in an age–β space for a California well. Note the large scale on the β axis, indicative of a large degree of uncertainty in the β determination. Note also, however, the very fine scale on the age axis, indicating the sensitivity of the thermal indicator inversion scheme to stratigraphic ages.

inversion has indeed preserved the sensitivity in age determination, yielding an age of about 17.8 ± 1.3 Ma. The biostratigraphic age for this formation was assessed at 19 Ma. The MSR contours do not go below 16.5 Ma because the age of the base of the formation above this unit was nominally given as 16.4 Ma.

Now a finite-thickness formation does not have to be continuously deposited but can have different sedimentation rates during its history and, to complicate matters even further, unconformities can occur (either by erosion or depositional hiatus). Since changes in sedimentation rate can have a substantial influence on the thermal history, it is necessary to determine the burial history parameters as precisely as possible. Thus we assign an age to a given depth position in a formation (whose ages at the top and bottom of the formation are known from paleontological evidence). For vitrinite reflectance values measured throughout the well and, in particular, at the assigned depth in the formation, we proceed through the inversion scheme to determine the best age for the given depth. The rate of sedimentation and its variation across the formation can then be determined. It is important to realize that high-quality thermal indicator data are needed and that the burial history of the rest of the well must be known for this procedure to provide good resolution.

4. Unconformity Determination

Armagnac (1985) and Armagnac *et al.* (1988a) have shown that the thickness of an erosional unconformity can be determined using inversion of a thermal indicator through a burial history program. But when we wish to determine *both* the thickness of an erosional unconformity and the age of a given stratigraphic horizon, the problem arises of trying to determine simultaneously, by inversion, two parameters which affect the burial history paths of the sediments differently. To illustrate this, Fig. 8.23 shows contours of constant MSR in the space of β and h, the thickness of a surface unconformity, when a fixed age is assigned to TD. Figure 8.16a shows contours of constant MSR in the space of β and age of TD when the erosional thickness h is held fixed at the value h = 1000 m. From Figs. 8.23 and 8.16a we see that the paleoheat flux coefficient β is less sensitive to variations in the thickness of the unconformity than to variation in the age of the TD horizon, but the relative difference of MSR (improvement in degree of mismatch) does not appear to favor strongly either burial history parameter. Deposition and erosion of different amounts of sediments for an unconformity, along with the time at which deposition ends and erosion begins, influence all the underlying layers (which will be buried deeper during the deposition phase and then uplifted during the erosion phase, leading to further compaction due to the additional overburden; the thermal conductivity will increase, causing a shift in the geothermal gradient). Sensitivity of the thermal indicator data from well Yalcumma 1 to the size of the unconformity is disappointing. Acceptable answers from Fig. 8.23 are 1300 ± 1000 m, so we retain nearly 100% uncertainty in the determination of the size of the erosional event. However, the thermal data do bracket the unconformity size to the scale of a kilometer or two. Presumably with higher-quality thermal indicator

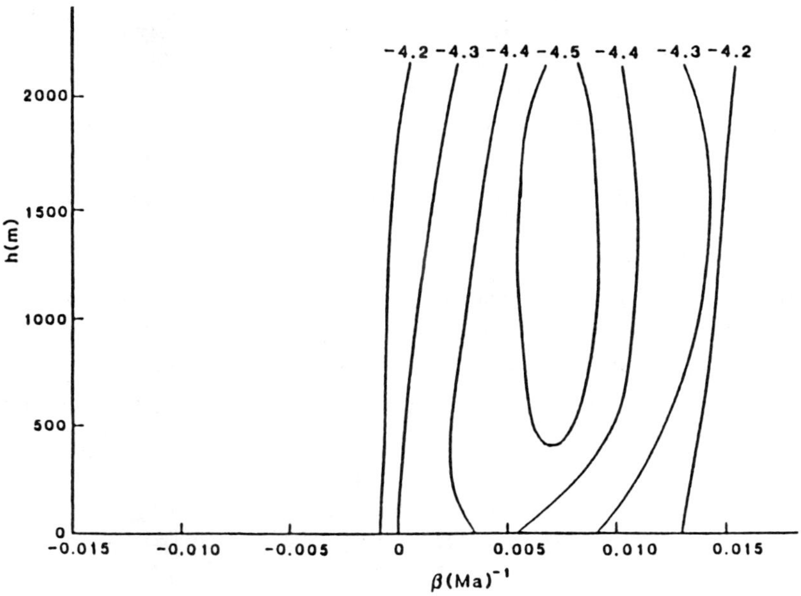

FIGURE 8.23 Contours of constant MSR for Yalcumma I in an unconformity thickness h and β space. Note that the unconformity thickness is not as well constrained as the β value, illustrating the less sensitive nature of the vitrinite reflectance to erosional unconformities than to heat flux changes.

data (or thermal indicator data of different sensitivity from a different maturation chemistry), this uncertainty on the unconformity size could be reduced.

B. Results Based on Biomarker Measurements

It might be argued that we have been singularly fortunate in our choice of wells from which we determined the ages of selected stratigraphic horizons by inversion of thermal indicator data. Concern can also be expressed that the theoretical formulation of vitrinite reflectance might be less than complete and so might be providing the known ages of the formations in an automatic mode. We think such concerns are illfounded, for we have been able to perform the type of analysis recorded here on a wide variety of wells from many basinal settings in a blind-test mode of operation with remarkably consistent results for all wells tested to date.

Nevertheless, there is always the slight, nagging, unresolved worry that we could be in error. In efforts to alleviate such worries and concerns we first note two points:

1. The general logic underlying the age determination of selected stratigraphic horizons by inversion of thermal indicators is independent of the type of thermal indicator (although an *accurate* determination of an age does, of course, depend on the sensitivity of a given thermal indicator, as well as on having available a significant body of high-quality data for the thermal indicator, roughly uniformly spread down the well).

2. Although the specific theory underlying the evolutionary behavior of a given thermal indicator (such as vitrinite reflectance) may indeed be less than complete, a large number of thermal indicators of different types have been investigated over the years; for several of them kinetic theories are available detailing evolution of the thermal indicators with time and temperature. In particular, we consider here sterane isomerization, which is known (MacKenzie and McKenzie, 1983) to be a useful thermal indicator. It has been shown (Armagnac *et al.*, 1988b) that variations of paleoheat flux determined from inversion of vitrinite reflectance data, and from hopane and sterane isomerization data in the same well, are consistent with one another, which strongly indicates the basic validity of the theoretical framework for each of the thermal indicators separately. Despite the sparseness of the sterane isomerization data in those wells, it was possible to show (Armagnac *et al.*, 1988b) that a rough estimate of the activation energy for the sterane isomerization reaction was in the approximate range 50 ± 40 kJ/mol, at the 1σ level, based on varying the activation energy to minimize the MSR.

In well P, of Cretaceous age, we had available not only vitrinite reflectance data with depth but also sterane isomerization information with depth. Therefore we again followed through the thermal inversion procedure for the sterane measurements to see how well we could determine the age of the TD formation. In addition, we required consistency with all the other information available. The following observations are pertinent:

1. Using vitrinite reflectance data (with $T_c = 295 \pm 10$ K and $T_D = 200$ K), we have been able to determine paleoheat flow change with time for over 300 wells, with consistent heat flux variations for neighboring wells and regional consistency across the basin.

2. The relationships of sterane isomerization ratios changing with

time and temperature have also been used to determine paleoheat flux variations with time, but with fewer than 20 wells. The apparent consistent activation energy from all the wells is 50 ± 40 kJ/mol.

3. Laboratory experiments and geophysical models (Abbott *et al.*, 1985; MacKenzie and McKenzie, 1983) have derived frequency factors, activation energies, and γ, the ratio of forward and backward rate constants. We use $\gamma = 1.174$, given by MacKenzie and McKenzie (1983). On the other hand, we normalize out the frequency factor partly because we can do so, thereby limiting the number of constants to be determined, and partly because Abbott *et al.* (1985) point out that the frequency factors measured in the laboratory are unlikely to be applicable to sediments. Also, Snowdon (1979) has shown the pitfalls of extrapolating laboratory-derived kinetic parameters to thermal alteration of sedimentary organic matter on the geological time scale. As shown in Fig. 8.24, the contours of constant MSR in a β–age plot demand an age of 130 ± 15 Ma for the formation—in accord with the value 137 ± 15 Ma obtained independently from vitrinite reflectance data (Fig. 8.25). Once again, the age of the base of the overlying formation was set as the lower boundary and searches were made only down to 115 Ma (age of the base). The biostratigraphic age provided was stated to be 125 Ma but without any assigned uncertainty.

In addition, if we demand that the rate of change of heat flux coefficient β be within the same range of uncertainty for both the vitrinite reflectance data and the sterane isomerization data, the domain shown by the dashed lines in Figs. 8.24 and 8.25 represents the common overlap domain of acceptable β values. If this constraint is used, we can further narrow the age of the TD formation to 135 ± 15 Ma—a value consistent with both the vitrinite reflectance data and the sterane data. With β and the age so determined, we can reexamine the resolution of the activation energy for the sterane isomerization.

In Fig. 8.26 we show contours of constant MSR in a β–activation energy space. Constraining β to lie between the dashed lines in order to obtain consistency between the vitrinite and sterane indicators, we see that we limit the activation energy to lie in the range 30 ± 15 kJ mol—a value lower than, but consistent with, the preliminary estimate 50 ± 40 kJ/mol previously obtained from other wells with sparse sterane data.

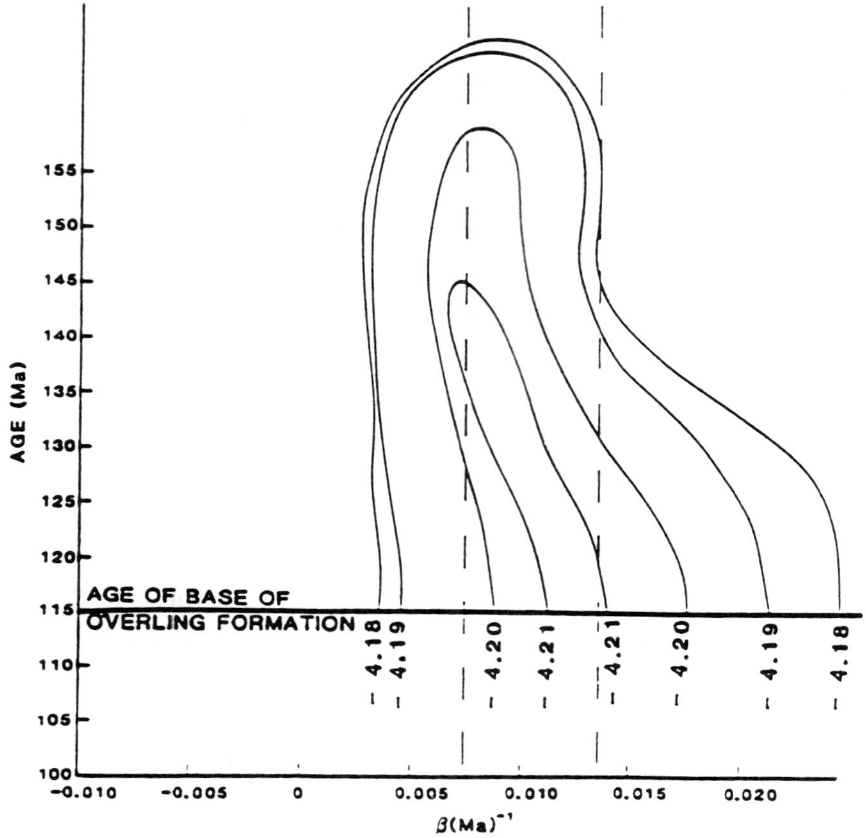

FIGURE 8.24 Contours of constant MSR for sterane isomer data from well P in an age–β space. The sterane data indicate an age for the formation of 135 ± 20 Ma. The vertical dashed lines show the domain of β values consistent with inversions of both sterane isomer data and vitrinite reflectance data in well P.

What is surprising is that the activation energy for the sterane isomerization reaction of 30 ± 15 kJ/mol, consistent with the downhole data, is considerably lower than the activation energy of 91 kJ/mol quoted by MacKenzie and McKenzie (1983). In attempting to obtain some understanding of this discrepancy, we have examined the work of MacKenzie and McKenzie (1983) (hereinafter MM83) in considerable detail—in particular with respect to their method of using sedimentary information to obtain an estimate of the activation energy.

Their basic approach is to plot the logarithm (base e) of the rate

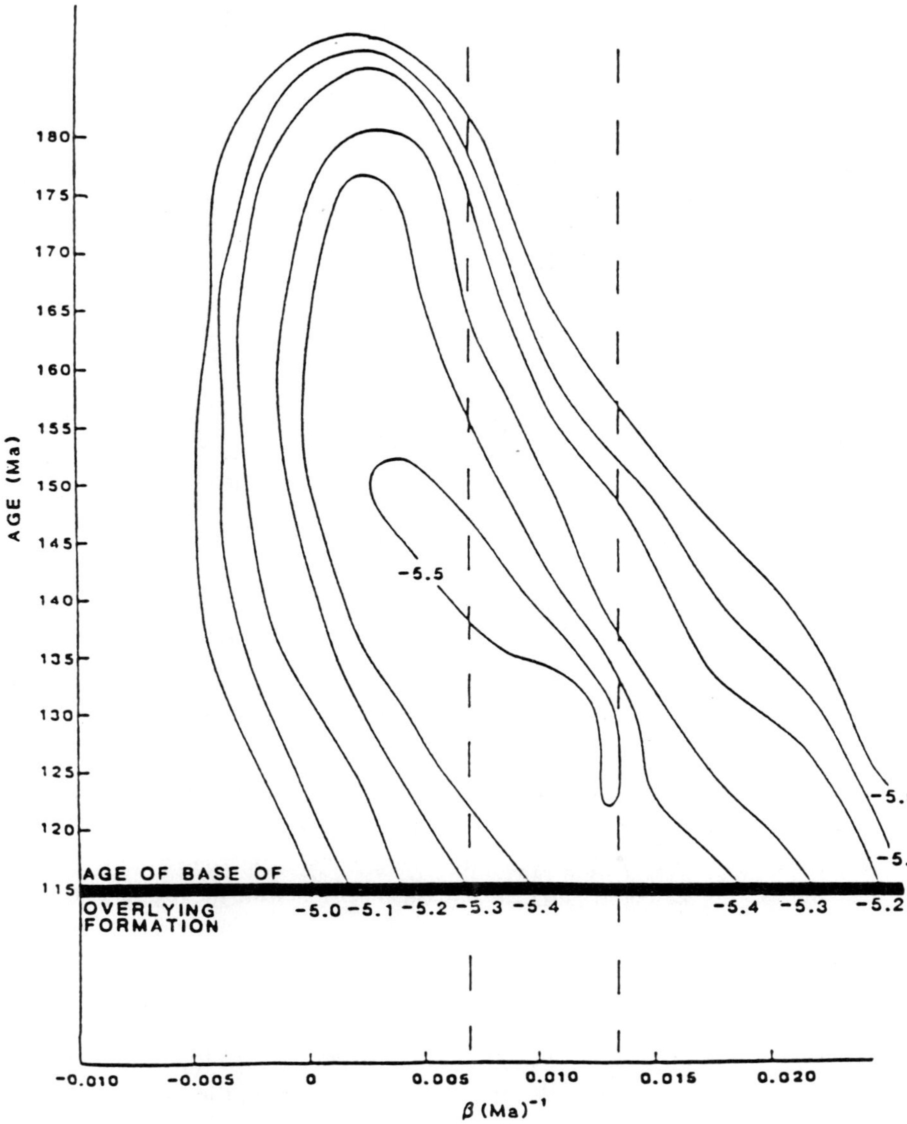

FIGURE 8.25 Contours of constant MSR for vitrinite reflectance data from well P in an age–β space. The vitrinite reflectance data indicate an age for the formation of 137 ± 15 Ma. The vertical dashed lines show the domain of β values consistent with inversion of both sterane isomer data and vitrinite reflectance data from well P.

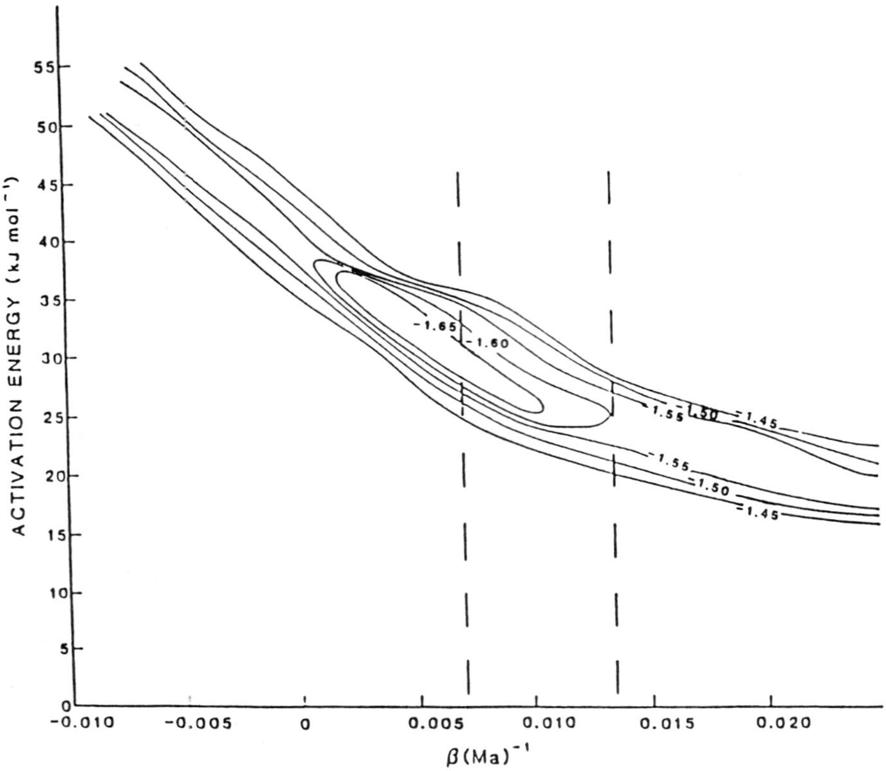

FIGURE 8.26 Plot of contours of constant MSR for sterane isomer data from well P in an activation energy–β space. Note that the "closure" contours of best fit indicate an activation energy for sterane isomerization of around 30 ± 15 kJ mol^{-1}. The vertical dashed lines indicate the regime of β values consistent with the inversion of both sterane isomer data and vitrinite reflectance data from well P.

constant k versus the inverse of the temperature T; the slope yields the activation energy divided by the gas constant. To obtain the rate constant k at each depth z_i, MM83 obtain the measured ratio [20S]/([20S] + [20R]) and equate this measurement to $\gamma(1 + \gamma)^{-1}[1 - \exp(-(1 + \gamma)k_it)]$. MM83 then insert the equilibration constant $\gamma = 1.174$, from which they determine the product k_it. They subsequently take the time t to be *fixed* at 40 Ma (40% of the time claimed since extension) for North Sea data and also take the temperature at each depth to be the present-day value. Then, by setting $k_i = A \exp(-E/RT_i)$, E is obtained as described. There are several difficulties with this procedure. First, it assumes that evolution of the sterane isomerization ratio started only some 40 Ma ago, and it does not allow for the continued evolution of the ratio since deposition of the steranes. Second, it makes the intrinsic assumption that the temperature with depth recorded at the present day was precisely the same in the past—no variation of basinal heat flux with time is permitted. Third, no allowance is made for burial history of the sediments, including changes in thermal conductivity caused by porosity variations as the sediments are buried deeper (even if the heat flux did not change with time, such effects would also cause a change in formation temperature with burial depth). Results for the Pannonian Basin, which MM83 use to bolster their North Sea results, suffer from the same assumptions.

By contrast, the results reported here permit isomerization from the time of deposition, incorporate compaction of the sediments during burial, and, most important, permit the heat flux to change with time. When these effects are allowed for, a search is undertaken for an activation energy most consistent with the downhole data, yielding the 30 ± 15 kJ/mol uncovered and consistent with the coarser value 50 ± 40 kJ/mol based on much sparser isomer data from other wells. In hindsight, given the crude assumptions in MM83, we are surprised that they obtain a value for the activation energy only two to three times larger.

C. Summary

Unknown ages for stratigraphic horizons ranging from near Cambrian through Carboniferous and Jurassic to Cenozoic can be determined with about 10% accuracy by inverting thermal indicator data at the present time. The general method of determining ages is independent of the thermal indicator used, as demonstrated by the results obtained in well P using either vitrinite reflectance data or sterane isomerization data. The fact that independent thermal indicators in the same and neighboring wells provide basically identical results implies that the theoretical framework underlying each such indicator is essentially correct, and the blind tests on sedimentary layers of known age confirm this beyond any further doubt.

When different thermal indicators in a single well are used in conjunction (as opposed to independently), they constrain very tightly the geological age and rate of change of heat flux in that well—much more so than either indicator will do on its own. In addition, the constrained values are then sufficiently precise that it is possible to reiterate the procedure to hone determination of the self-consistent chemical activation energy.

The rapid, and high, deposition rates in a Cenozoic well (of high present-day surface temperature) permit us to determine more accurately the critical temperature for vitrinite reflectance. Unfortunately, since resolution in critical temperature is traded for resolution in surface temperature, we would require a significant improvement in both the quantity and quality of thermal indicator data in order to make a determination of the evolution of the surface temperature with time.

While it is possible to determine two (sometimes three) of the parameters influencing burial histories from inversion of thermal indicator data (as shown by the example of the thickness of an erosional unconformity determination simultaneously with the age of a sedimentary layer and the paleoheat flow variation), to go beyond this point, to a determination of several parameters simultaneously, rapidly becomes unwieldly using a trial-and-error method.

To improve significantly on our current understanding of the degrees of uniqueness, resolution, and precision of stratigraphic age determination by thermal indicator inversion, a large number of wells with well-distributed, high-quality thermal indicator data of many types is needed so that the inversion scheme can be tuned to achieve the highest level of consistency with all data.

III. Paleoheat Flux and Unconformity

The stratigrapher tracking the history of basins is constantly confronted with the problem of determining the amount of sediment deposition and erosion at unconformities. Similarly, in some sedimentary basins, the geologist is asked to determine the timing of listric faulting, the emplacement of igneous intrusions and salt tongues, and the onset of overpressuring and overthrusting. Here we focus on unconformities.

The classical approach to determining when erosional events were initiated and estimating how much sediment was eroded is to extract this information from the regional stratigraphy. Some of these techniques are outlined by van Hinte (1978), who describes the use of burial history or geohistory algorithms. These techniques include establishment of the age of the youngest rocks beneath the unconformity at specific locations. The stratigrapher then locates the oldest rocks resting on the unconformity and notes their age. This approach does not handle the problem of how much erosion took place prior to the development of the present unconformity surface or when the erosion began. One possible solution is to assume that the rate of sedimentation below the unconformity continued into it and that the rate of deposition following the unconformity matched the rate of erosion (van Hinte, 1978; Guidish *et al.*, 1985). One can assume that erosion began at the time where the slopes of these lines intersect on a burial history diagram (Chapter 3). Another method is to (1) assume the erosion associated with the unconformity began at the most obvious major sea-level change on the coastal onlap chart of Vail *et al.* (1977), (2) extrapolate the rate of sediment accumulation prior to the unconformity to this time, and (3) erode to the age of the overlying sediment (Chapter 3). Undoubtedly, there are other ways of deriving the timing and erosion involved during an unconformity (Moshier and Waples, 1985), but all carry assumptions similar to those of the approaches listed above.

Once we have estimated the time of the onset of erosion, we need an independent way to establish how much sediment was eroded. Magara (1976b) used sonic logs to plot shale transit time with depth and estimated the amount of erosion at unconformities by extrapolating the rate of change in the transit time into the unconformity. The sediment removed from the unconformity is added back until the transit times on either side of the unconformity match. Dow (1977) outlined a method of estimating the amount of material removed at an unconformity by examining vitrinite maturation on a logarithmic scale. The method entails the development of a maturation profile, the slope of which is influenced by the geothermal gradient, the geologic age, and the sedimentation rate of the section under investigation. With the presence of a major unconformity, the slope of the maturation profile is offset and the thickness of the sediment removed during this event can be estimated by projecting the trend of the earlier vitrinite reflectance values into the slope of later values. We use a scheme similar to those of Magara (1976b) and Dow (1977) but, instead of invoking a discontinuity in a trend in transit time or vitrinite reflectance with depth, we tie thermal history to the vitrinite reflectance (Lerche *et al.*, 1984).

The method, like that of Dow (1977), uses the fact that if vitrinite reflectance is traced across an unconformity in a well or a measured section, the trend of the vitrinite reflectance data will show a jump at the unconformity. Thus, between unconformities there is a progressive upward decrease in vitrinite reflectance, reflecting a shorter temperature history. At the unconformity, this trend is interrupted and the vitrinite reflectance values above the unconformity are markedly less than those below the jump. The trend with depth of the younger vitrinite reflectances may or may not parallel the trend of the vitrinite reflectance values below the unconformity. We determine the heat flow during the unconformity period by using the vitrinite reflectance data above and below it. Then we vary the size of the unconformity, in terms of its timing and the amount of sediment removed, until the predicted vitrinite reflectance values follow the trend of the vitrinite reflectance measurements on either side of the unconformity. In other words, the cumulative thermal history is retained by the vitrinite reflectance throughout time, recording the thermal maturation of the sediments, including evidence of what happened during any unconformities. This information can be extracted to determine the total sediment thickness stripped off.

Vitrinite in sedimentary layers deposited later than the unconformity has no means of recording the presence of that unconformity. However, vitrinite occurring in sedimentary layers deposited earlier than the unconformity records not only the thermal history before the unconformity period but also that during and after. Usually one assumes a linear dependence for the paleoheat flux with time: $Q(t) = Q_0(1 + \beta t)$, where Q_0 is present-day heat flux, t is time, and β is the linear constant coefficient to be determined.

The total amount of material deposited and then eroded is unknown. However, for a given total amount of assumed erosion we find the best β and record the MSR. We then increase the amount of erosion in stepped increments and at each step we record the best β and the MSR. With too little or too much erosion the degree of mismatch between predicted and observed vitrinite reflectances with depth will be large. With too little erosion the discontinuity in the reflectance profiles at the unconformity surface will not be large enough to match the observed jump, while with too much erosion the predicted jump will be too large relative to observations. There is, then, a best amount of erosional removal during the unconformity, which equates with both the smallest MSR and the best β value.

One caveat needs to be mentioned. When the unconformity occurs near the present-day sedimentary surface, as occurs in several wells in the NPRA, we do not have vitrinite reflectance on the upper side of the unconformity to guide us. In this case we take the vitrinite reflectance to have a specific value at the depositional surface. An investigation of nearly 300 wells has narrowly constrained this reflectance value to be $R_0 = 0.2 \pm 0.05$. When this depositional value is so constrained, we can go through the previous sequence of operations with wells with present-day surface unconformities until the best thickness h and β value are determined. The next sections describe the results obtained with this procedure for a number of wells in the NPRA and for one well from Morocco.

A. Mid-Tertiary Unconformity in Alaskan Wells from the National Petroleum Reserve, Alaska

1. Regional Geology—NPRA

Using the scheme described above, we estimated the thicknesses of unconformities for a number of wells in the NPRA to see if we could improve our understanding of the regional geology. The wells used are in the southern, eastern, and northern NPRA and include Awuna-1, Inigok-1, and Ikpikpuk-1. In the northern area along the Barrow Arch, a mainly Cretaceous section is penetrated, but at Inigok-1 to the south, sediments of Mississippian age have also been drilled. The wells cut across several unconformities, including a major one in the Tertiary which removes sediments from the late Tertiary to late Cretaceous. As mentioned earlier, the classical way to determine the size of this unconformity is to examine the area's regional structure and stratigraphy. The NPRA has been studied by many geologists, using seismic cross sections, outcrop, and well information to interpret the structural and stratigraphic history (e.g., Grantz *et al.*, 1979; Churkin *et al.*, 1980; Mull, 1980; Churkin *et al.*, 1980; Molenaar, 1981b; Bushnell, 1981), and this provides an excellent data base.

The NPRA can be divided into four major east–west-trending structural provinces which generally correspond to the major physiographic terrains (Fig. 8.27). The structural complexity of the region decreases from the south in the Brooks Range to the Arctic coastal plain in the north (Fig. 8.28).

Uplift of the Brooks Range began in latest Jurassic to earliest Cretaceous in response to (1) the extension and rifting to the north which formed the Canada Basin and the Barrow Arch and (2) the compression and overthrusting to the south caused by the collision of a southern microplate of which the Brooks Range is a part. These events led to the termination of deposition from the northerly continental provenance which had existed from Mississippian to earliest Cretaceous time and switched the source terrain to the southerly Brooks Range. A renewed uplift of the Brooks Range in the middle to late Cretaceous increased sedimentation and subsidence in the NPRA and led to the progradation of sediments over the Arctic platform.

The province of the Brooks Range consists of a complex of folded and

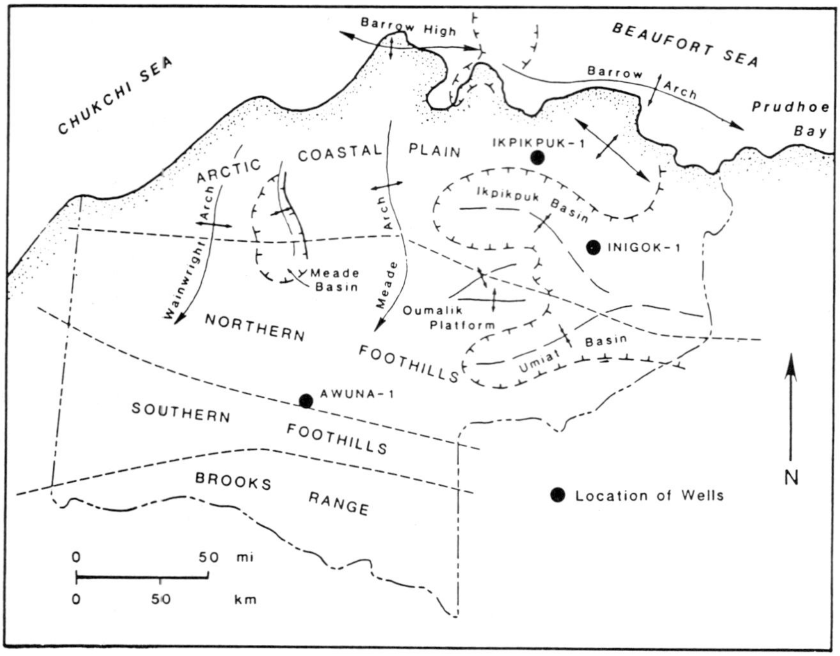

FIGURE 8.27 Location of Awuna 1, Inogok 1, and Ikpikpuk 1 in the National Petroleum Reserve of Alaska (NPRA) and of the major physiographic terrains described in the text. (Modified from Kirschner *et al.*, 1983.)

imbricated thrust sheets of mainly Ellesmerian (early Mississippian–early Cretaceous) rocks, which may have undergone 100 to 500 km of crustal shortening (Bird, 1981). The easterly trending Southern Foothills province is slightly less deformed than the Brooks Range but has undergone a greater degree of folding and faulting than the Northern Foothills province. Structures in the Northern Foothills include a series of thrust fault detachment zones with narrow asymmetric anticlines (Kirschner *et al.*, 1983). The Arctic coastal province, to the north of the foothills, is a relatively stable region covered with coastal to shallow marine sediments.

Important subsurface features are seen on the seismic sections. Beneath the foothills and coastal plain, a series of horsts and grabens disturb the basement physiography and form the Ikpikpuk, Umiat, and Meade basins. Along the northeastern coast of the NPRA, a major subsurface basement ridge, the Barrow Arch, trends and plunges to the southeast. Most of the wells drilled in the NPRA have been located along this arch, although no significant accumulations of hydrocarbons have

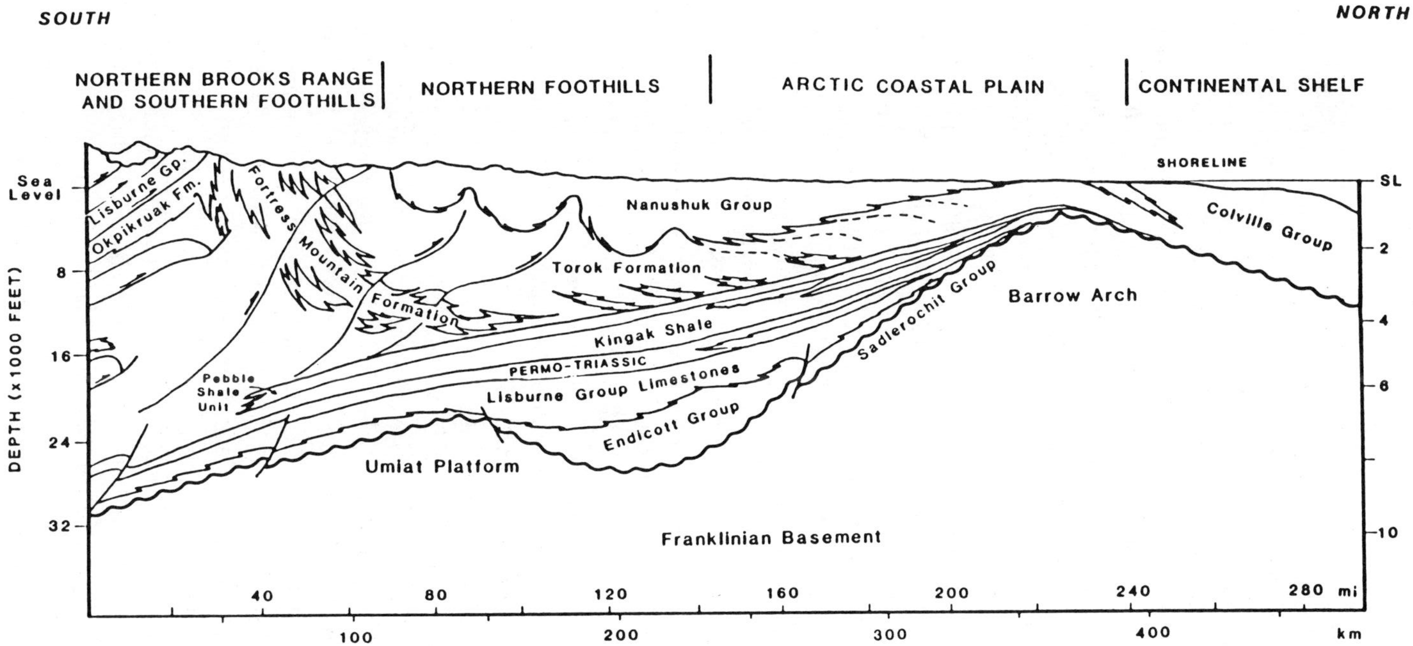

FIGURE 8.28 Schematic north–south cross section through the NPRA showing general stratigraphy and structure (modified from Bird, 1981). Note the decreasing complexity northward.

been discovered to date. Another important subsurface feature, located beneath the foothills and coastal plain, is the Colville Geosyncline, an asymmetric foredeep that formed just to the north and in front of the Brooks Range area. Sediments in this trough are thickest to the south near the Brooks Range and thin toward the Barrow Arch.

The stratigraphy of the NPRA can be divided into three major sequences: the Franklinian, the Ellesmerian, and the Brookian (Fig. 8.29). In the vicinity of the wells, the Franklinian is thought to consist of metamorphosed rocks of Devonian age and older and is considered economic basement in the NPRA. These folded and faulted Franklinian rocks have been encountered in most wells on the Barrow Arch but are buried under more than 20,000 ft of sediments in the Ikpikpuk, Meade, and Umiat basins (Kirschner *et al.*, 1983). A regional erosional unconformity separates the basement rocks from the overlying Ellesmerian sequence, composed of up to 20,000 ft of early Mississippian to early Cretaceous clastics and carbonates which probably formed part of an earlier extensional continental margin that now dips south. The clastics were derived from a source terrain to the north of this ancient margin and the present-day coastline (Kirschner *et al.*, 1983). Sediments are thinnest over the Barrow Arch and become finer toward the south, away from the continental source. The earliest Ellesmerian rocks are part of an early extensional nonmarine graben fill which was progressively overlain by slope to shallow shelf deposits of the old continental margin. Shales of the upper Ellesmerian sequence record the onset of the late Jurassic–early Cretaceous collision of the ancestral Brooks Range which overrode the old southward-dipping continental margin.

The Cretaceous to Tertiary Brookian sequence is formed by mainly prograding sediments which shed to the north and northeast from the ancestral Brooks Range to debouch into an asymmetric, northward-shallowing foredeep. Over much of the northern NPRA, the Brookian–Ellesmerian contact is sharp and uncomformable and is marked by the Pebble Shale Conglomerate, but in the deeper portions of the region to the south, the Pebble Shale undergoes a facies change and the sequences are thought to interfinger. The oldest Brookian unit, the Albian Fortress Mountain formation, is primarily a thick deepwater flysch deposit. The Fortress Mountain formation grades into shales of the lower Torok formation in the northern part of the southern foothills and the southern part of the northern foothills, and is overlain by the Torok foreset beds to the north and northeast (Molenaar *et al.*, 1981a). Foreset dip angles in the Torok increase in an easterly direction from less than 2°

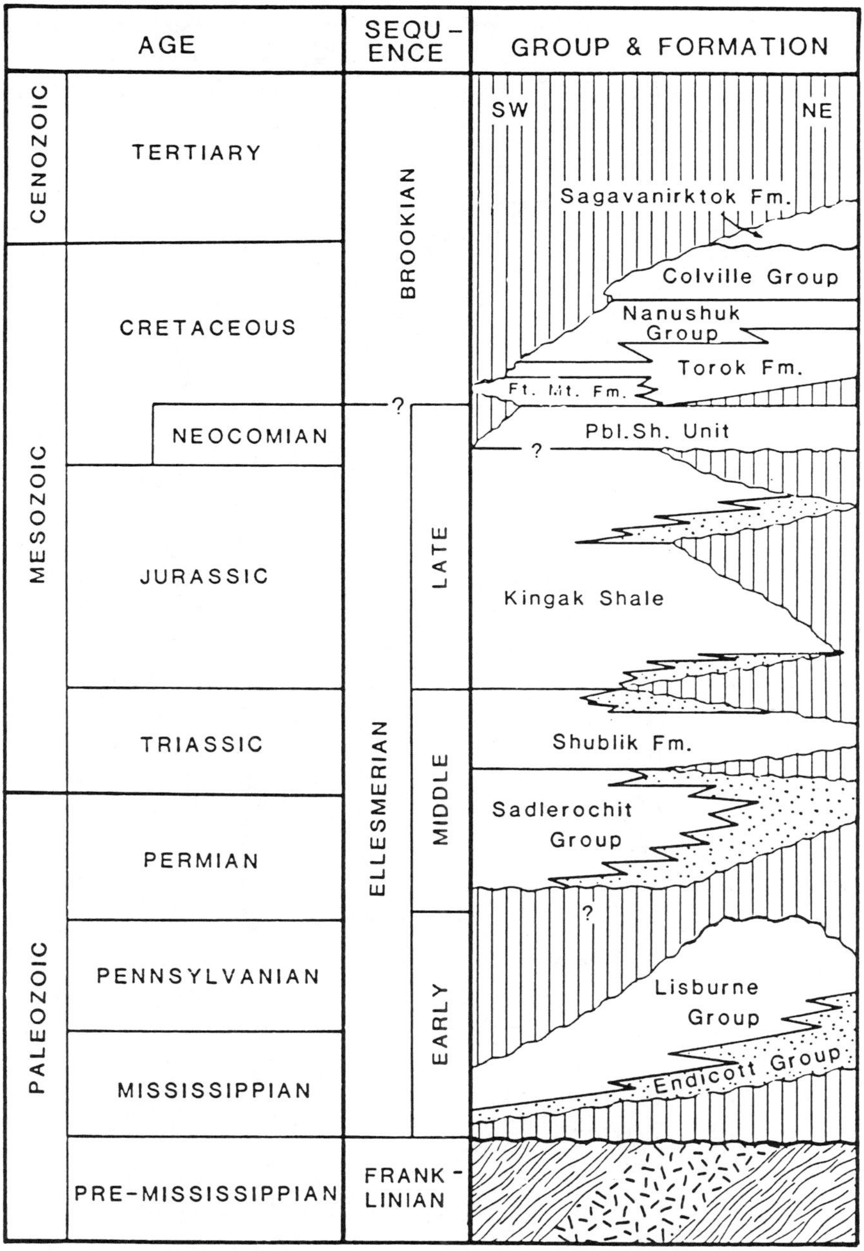

FIGURE 8.29 General stratigraphic terminology and relationships for the NPRA (after Kirschner *et al.*, 1983). Ft. Mt. Fm., Fortress Mountain Formation.

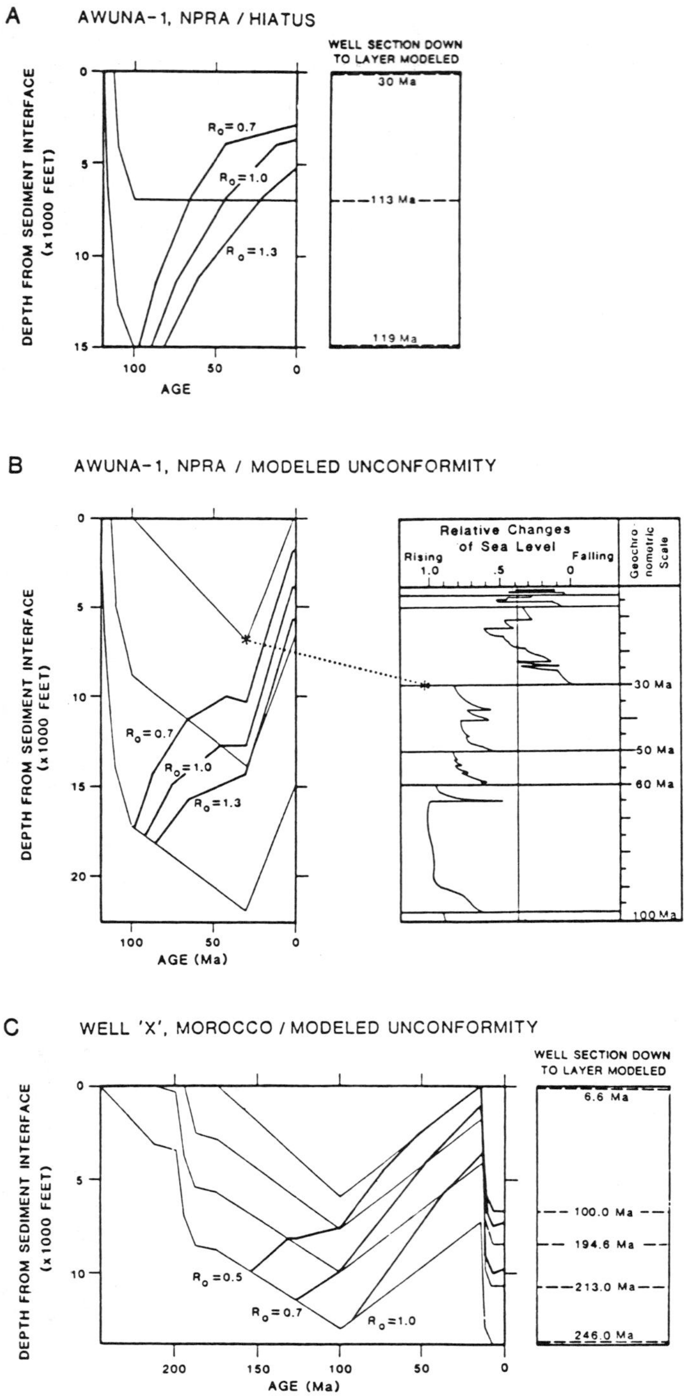
A
AWUNA-1, NPRA / HIATUS
WELL SECTION DOWN TO LAYER MODELED
30 Ma
113 Ma
119 Ma
DEPTH FROM SEDIMENT INTERFACE (x1000 FEET)
AGE
R₀=0.7
R₀=1.0
R₀=1.3
B
AWUNA-1, NPRA / MODELED UNCONFORMITY
Relative Changes of Sea Level
Rising
Falling
Geochronometric Scale
30 Ma
50 Ma
60 Ma
100 Ma
DEPTH FROM SEDIMENT INTERFACE (x1000 FEET)
AGE (Ma)
R₀=0.7
R₀=1.0
R₀=1.3
C
WELL 'X', MOROCCO / MODELED UNCONFORMITY
WELL SECTION DOWN TO LAYER MODELED
6.6 Ma
100.0 Ma
194.6 Ma
213.0 Ma
246.0 Ma
DEPTH FROM SEDIMENT INTERFACE (x1000 FEET)
AGE (Ma)
R₀=0.5
R₀=0.7
R₀=1.0

to approximately 6°, suggesting more rapid progradation as the Torok beds progressed eastward (Molenaar, 1981b).

Topset beds of the progradational sequence belong primarily to the Nanushuk Group of Albian to Cenomanian age, which outcrops in the northern foothills but is buried beneath much of the Arctic coastal plain. The Nanushuk rises stratigraphically to the east–northeast in the direction of the shelf progradation (Molenaar, 1981b). Two delta lobes have been recognized as being developed during Nanushuk time. The western lobe or Corwin delta was derived from a source to the southwest that presently may be in the Chukchi sea (Molenaar, 1981b), while the southerly lobe, the Umiat delta, was probably derived from rocks around the Delong Mountains, Tigara uplift, and Herald Arch (Mull, 1980). To the east and northeast of the NPRA, the prograding sequences of the Colville Group overlie the Nanushuk Group, but in the eastern NPRA shales of Turonian age overlie the formation (Molenaar, 1981a).

2. Unconformity Determination

A major unconformity erodes to Brookian rocks of middle Cretaceous age in the western NPRA and down to Tertiary-age rocks in the eastern NPRA. The beds beneath the unconformity have been dated from Eocene to Maastrichtian (Carter *et al.*, 1977; Wolfe, personal communication, 1979; both in Nelson, 1981), suggesting that a period of major erosion began after the Eocene.

The global coastal onlap curves of Vail *et al.* (1977) show a major drop in eustatic sea level approximately 30 million years ago, in the Oligocene (Fig. 8.30). This sea level change apparently halted deposition in the NPRA and coincided with the period of marked erosion. It is thought to have initiated erosion in the vicinity of all the wells we studied. Thus we use the time of this drop to begin a period of erosion in the computer runs we made of our burial history model for these wells. At the same time we vary the thicknesses of sediment deposited before the onset of erosion to

---

FIGURE 8.30 Combined geohistory plots for NPRA well Awuna 1 with the Tertiary unconformity shown (A) as a hiatus and (B) with the onset of erosion coinciding with the Oligocene sea level low from the Vail *et al.* (1977) coastal onlap chart. (C) Plot for Moroccan well "X" with the onset of erosion coinciding with the mid-Cenomanian sea level low set to 100 Ma; note that from Vail *et al.* (1977) it might be argued that the low is at 95 Ma.

test how much sediment could have been deposited in the late Cretaceous to early Tertiary. We plotted β, the fractional rate of change in paleoheat flux, versus unconformity thickness h (Fig. 8.31, 8.32, and 8.33). These plots were contoured for values of equal log mean square residual and show ranges of unconformity thickness centered around a minimum value of the log mean square residual for that best β as well as ranges in β for a given unconformity thickness. The "bull's-eye" area of the plot shows the best range of fit of unconformity thickness and of the linear variation in heat flux for that well.

B. Late Cretaceous–Early Tertiary Unconformity in Moroccan Well "X", Rharb Basin

1. Regional Geology

Using the methodology set out above for the NPRA, we estimated the size of the late Cretaceous–early Tertiary unconformity that can be seen

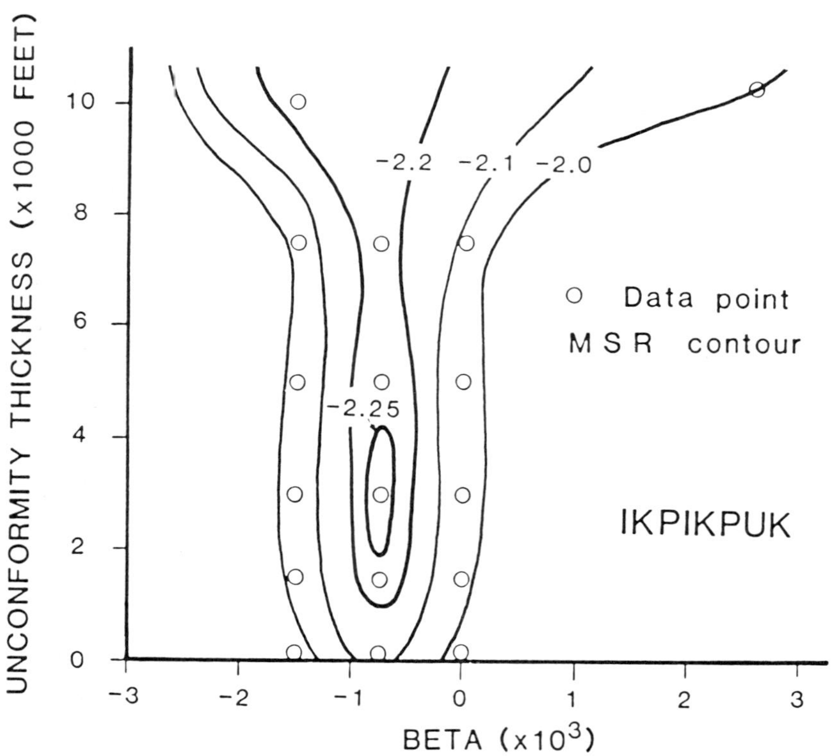

FIGURE 8.31 Plot of unconformity thickness versus best β for Ikpikpuk 1. The lowest MSR is at −2.25 which coincides with the Oligocene unconformity eroding 914 m (3000 ft) and a best β of −0.0007/Ma.

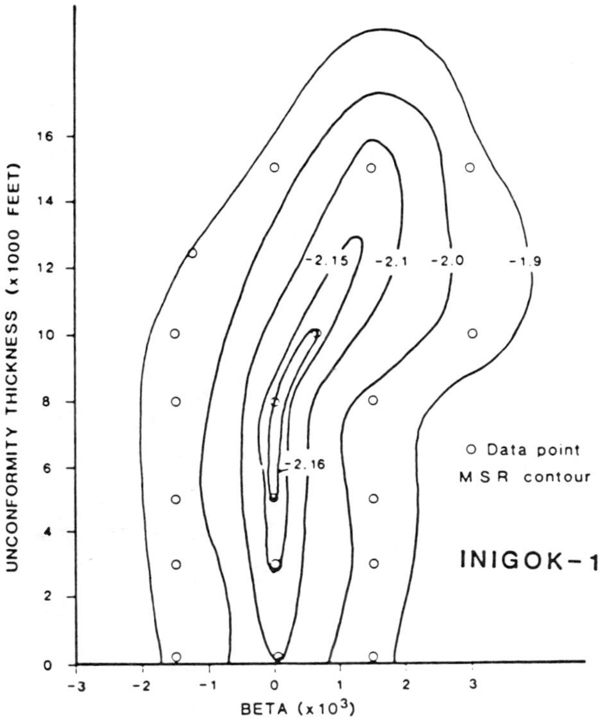

FIGURE 8.32 Plot of unconformity thickness versus best β for Inigok 1. The lowest MSR is at −2.165, which coincides with the Oligocene unconformity eroding 2440 m (8000 ft) and a best β of 0.

in well "X" from the Rharb Basin. This basin, as Bucci (1985) describes it, is located in Morocco just south of the Rif overthrust zone, at the north end of the Moroccan Mesata (Fig. 8.34). The Mesozoic and Tertiary rocks of the basin lie unconformably on a structurally deformed, slightly metamorphosed Hercynian basement with some intrusive granites (Fig. 8.35). The extension associated with the initiation of the Atlantic in the Triassic developed a series of normal faults which strike N 45°E in the Rharb Basin area, creating a series of horsts and grabens. The resulting basin is filled by a Mesozoic succession related to rifting and postrift sedimentation and is followed by a Tertiary section related to Alpine deformation and foredeep development.

The earliest rift fill is a presumed Triassic breccia composed of fragments of Paleozoic rock (too thin to mark on Fig. 8.35) which lies unconformably on the Hercynian basement. Overlying these breccias is a sequence of continental red beds containing anhydrite, gypsum, and salt

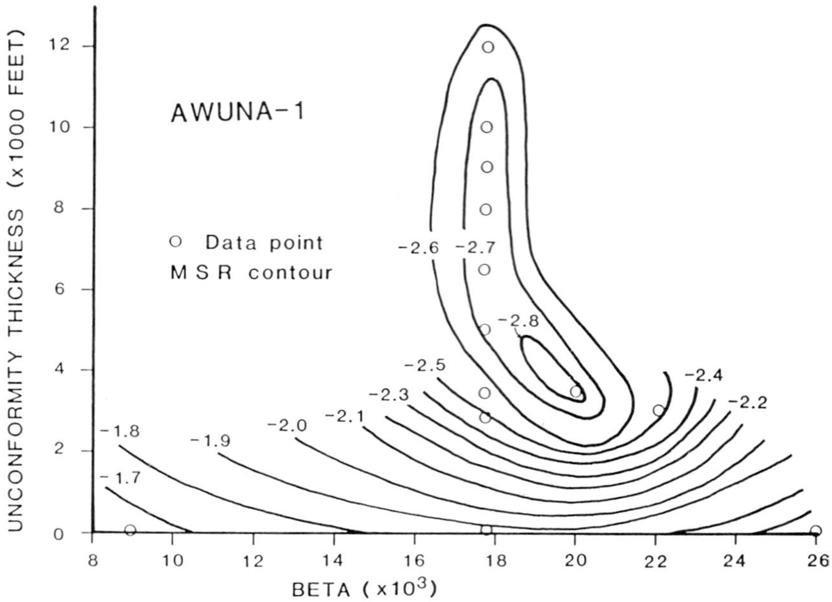

FIGURE 8.33 Unconformity thickness versus β for Awuna 1. The lowest MSR is at −2.6 which coincides with the Oligocene unconformity eroding 1070 m (3500 ft) and a best β of 0.02/Ma.

beds. Basalts, 100 m (330 ft) thick, are interbedded with the red beds. The Triassic red beds terminate with a thick succession of dolomites and anhydrites with minor shales. The postrift fill began with Liassic shallow-marine carbonates, followed by deep-water marls. The Jurassic deposition was terminated in the Dogger by arkosic sands that were probably derived from nearby horst blocks. No Upper Jurassic is present in the Rharb Basin, but thin onlapping Cenomanian and very thin Turonian shallow-water carbonates and marls are present on the western flank. A thick Cretaceous sequence may have been deposited in the basin itself, but, if so, it was eroded in the late Cretaceous to early Eocene (Arana and Vegas, 1974). Bucci (1985) discusses the possibility that the western Rharb Basin may have been a positive feature until the Tertiary, although no conclusive evidence for this is available. As we shall show, within the basin itself, the vitrinite reflectance data from well "X" in the Sidi Fili Field can be used to argue that an extra 1630 m (6000 ft) or more of Mesozoic sediment was present before erosion was initiated at a Cenomanian sea level low some 96–100 Ma ago (Vail *et al.*, 1977).

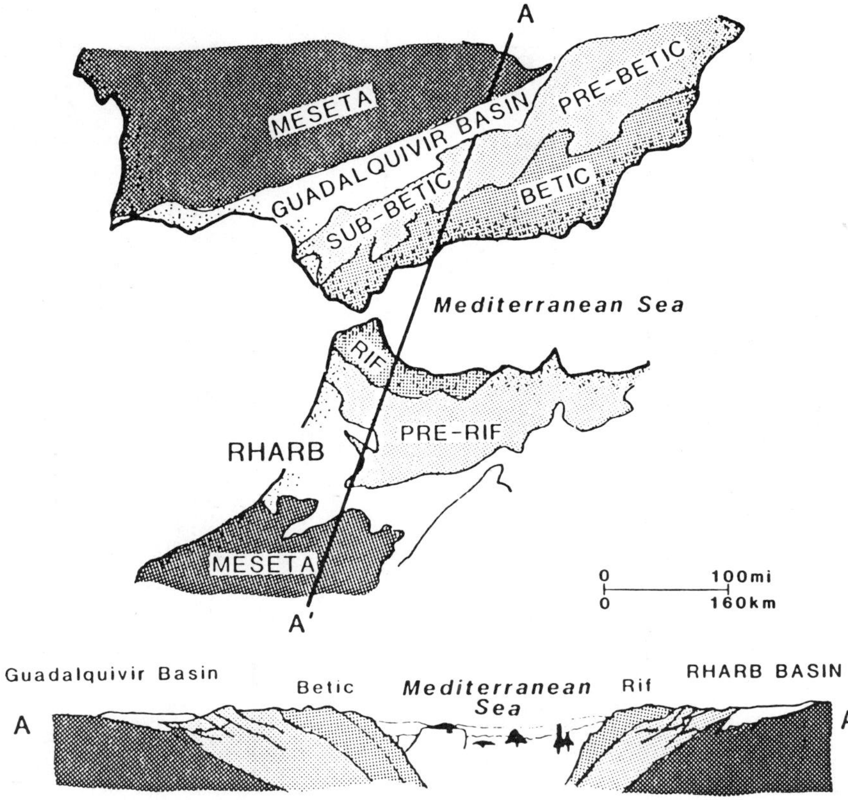

FIGURE 8.34 Location map for the Rharb Basin with a schematic crustal cross section across northern Morocco. (After Biju-Duval and Montadert, 1977).

In the early Miocene, overthrusting from the north in the Rif extended this nappe complex to the northern edge of the Rharb Basin. Deposition began with Aquitanian and Burdigalian sands, which were succeeded by shaley and marly sediments of Serravalian to Tortonian age. These deep-water sediments (Bucci, 1985) are capped by Pliocene shallow-water sands and marly sands.

Bucci (1985) reports an abrupt deflection in the basement and an increase in sediment thickness adjacent to the Rif nappe system, suggesting that basin subsidence in the Rharb Basin at the time of nappe implacement was at least partially related to crustal loading.

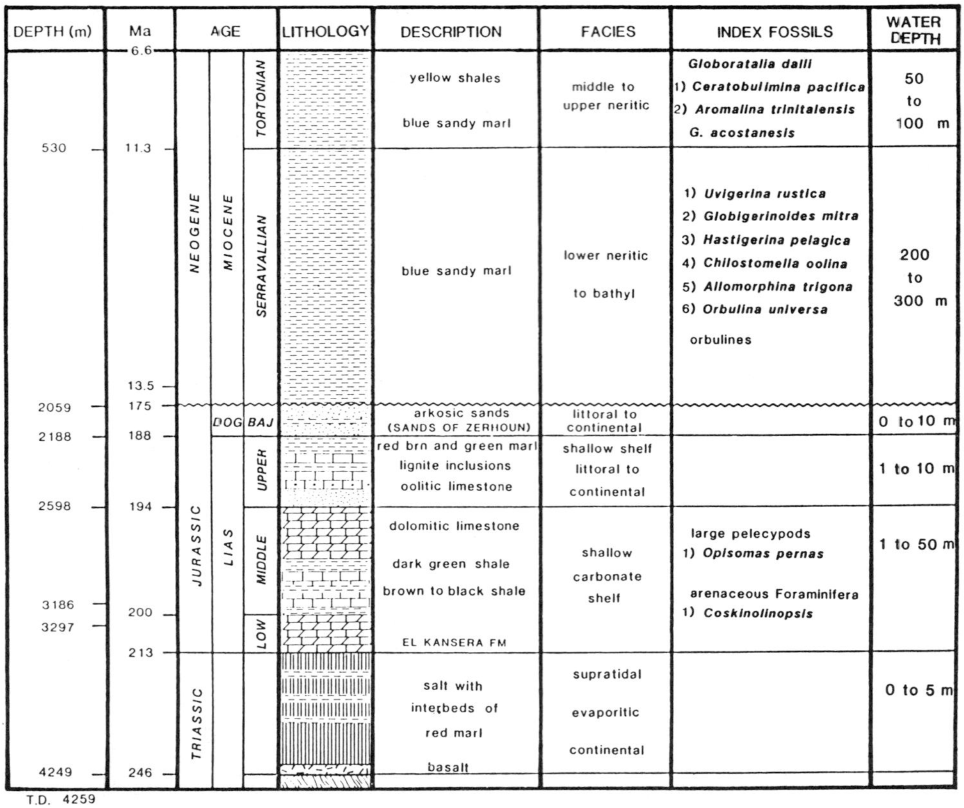

FIGURE 8.35 Stratigraphy of well "X" from the Sidi Fili Field of the Rharb Basin, Morocco. (After Bucci, 1985.)

2. Input Data for Well "X"

The input data used for modeling the unconformity were the same kinds as we used for the NPRA wells. The source of our information was Bucci (1985), who gathered elevation, Kelly–Bushing height, and temperatures from well-log headings. Biostratigraphy and lithologies came from final logs and well reports from the Bureau de Recherche et de Participations Minières, Rabat, Morocco. The vitrinite samples were assembled by James Bucci and their reflectances were measured by Iris Hubbard of the Earth Sciences Resources Institute at the University of South Carolina.

3. Runs and Results

On the basis of the work of Arana and Vegas (1974) and the coastal onlap charts of Vail *et al.* (1977), which show a Cenomanian sea level low, we

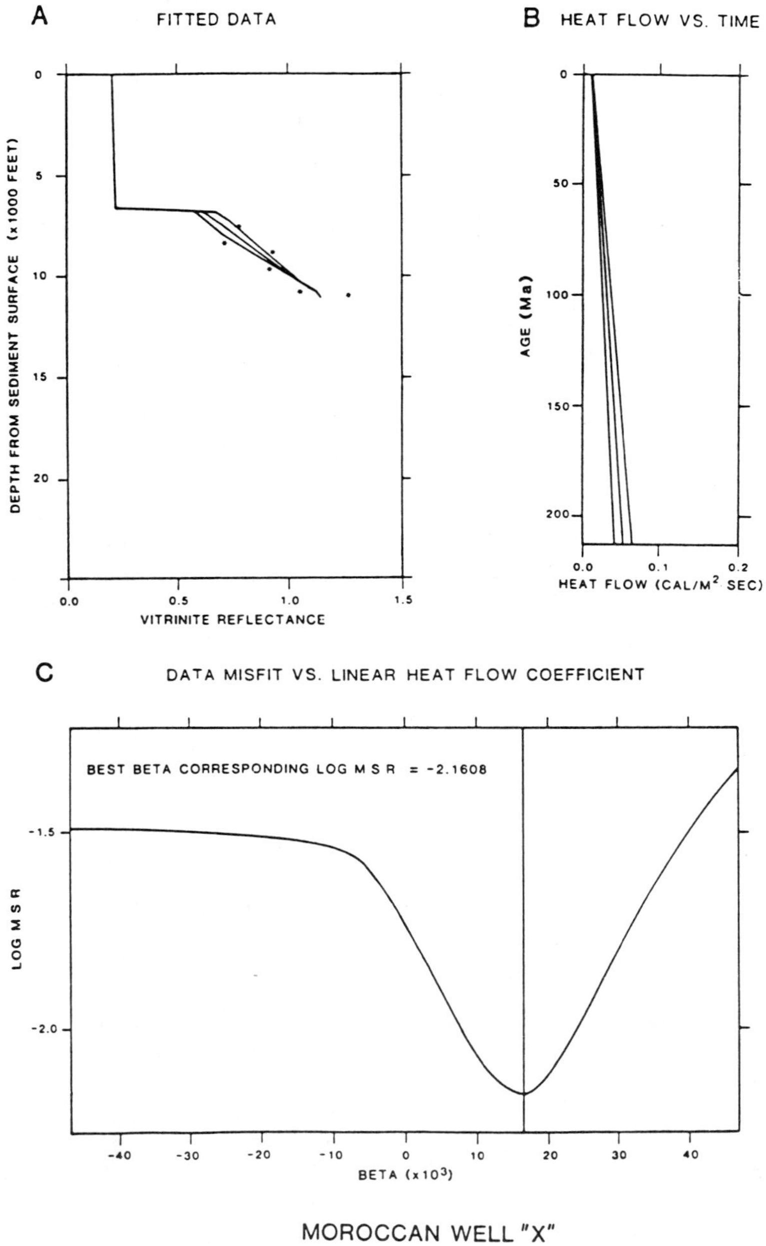

FIGURE 8.36 (A) Predicted and observed vitrinite reflectance as functions of depth for Moroccan well "X" when the unconformity is modeled with 1830 m (6000 ft) of deposition before the onset of erosion at 30 Ma. (B) Plot of best heat flow for best β for Moroccan well X with 1830 m (6000 ft) of deposition modeled for unconformity. (C) Plot of MSR against β for 1830 m (6000 ft) of deposition at 30 Ma unconformity for Moroccan well "X".

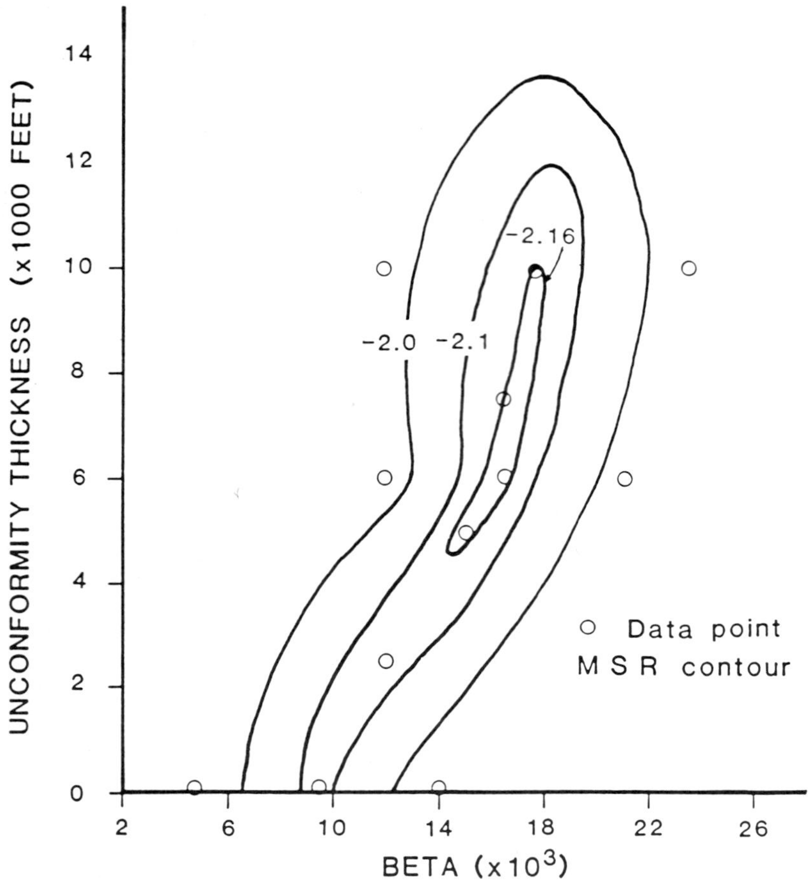

FIGURE 8.37 Unconformity thickness versus β for Moroccan well "X" from the Rharb Basin. The lowest MSR is at −2.16 which coincides with the Cenomanian unconformity eroding 1830 m (6000 ft) and a best β of +0.0164/Ma.

estimate that the unconformity which extends from 175 to 13.5 Ma in well "X" began at around 96 to 100 Ma. This sea level event coincides with an early Alpine orogenic event during which regional uplift is thought to have occurred (Arana and Vegas, 1974). Several examples with varying amounts of erosion beginning at 100 Ma were used. At the same time, the linear heat flux coefficient β was plotted as a function of the log MSR (Fig. 8.36). The best β values were then plotted as a function of 20 different thicknesses (h), and the log MSR values were contoured as they were for the Alaskan wells (Fig. 8.37). The minimum MSR contour occurs when approximately 1830 m (6000 ft) of section is eroded at the unconformity. The best range of h (smallest contour interval) occurs between 1470 and 2170 m (4800 and 7100 ft) with the β range of 0.0107 and 0.0078.

The implications of this study for the regional stratigraphy and for hydrocarbon generation modeling are important. In the case of the regional stratigraphy, this study suggests that the late Cretaceous–early Tertiary unconformity seen in the region was more than just a long-lived hiatus and that it actually removed some 1830 m (6000 ft) or so of Upper Cretaceous sediment. This result opposes the current stratigraphic model, which would have a long-lived hiatus instead. From the hydrocarbon generation viewpoint, the extra 1630 m (6000 ft) of rock means that potential source rocks from the Liassic could have entered the oil window earlier, remained there longer, and hence generated more oil.

9

Thermal Indicator Tomography

I. Theoretical Considerations

In earlier chapters we presented the general inversion schemes for a suite of different thermal indicators. The inversion procedures use present-day downhole measurements to determine paleoheat flux variations subject to several constraints:

1. The underlying thermal evolution process and temperature dependence controlling the maturation of each thermal indicator must be known.
2. The chemical and physical parameters associated with the evolution process must be provided. (To the extent possible, we normalized out of the inversion procedures as many as possible of such parameters so that paleoheat flux determinations were minimally beholden to the vagaries of uncertainties in the parameters.)
3. The burial history of each sedimentary layer was assumed to be known with precision in order that the inversion procedures could be used to extract the paleoheat flux variations with maximal sharpness.

With geological data, and their attendant "scruffiness," there is not too much improvement that can be made in the underlying thermal evolution processes. Such functional determinations are more properly determined by controlled laboratory experiments.

But the chemical and/or physical parameters of such processes determined in the laboratory, over human lifetimes and under pristine conditions, do not have to be the appropriate values for maturation of thermal indicators over geological time scales and in the uncontrolled environment of the evolving sedimentary basin. So it behooves us to regard such parameters as unknowns, also to be determined by the inversion procedures. In a similar vein, geological parameters related to unknown variations in burial history information should also be determinable from the inversion procedures.

Here we examine the consequences of relaxing, or abandoning, the second and third constraints listed above, and we ask whether the inversion procedures are capable, even in principle, of resolving parameters associated with paleoheat flux variations as well as chemical and physical parameters and geological parameters, at the same time on the same footing.

To maintain continuity, the notation follows as closely as possible that of Chapter 3.

A. Determination of Chemical and Physical Parameters

In each of the thermal indicators examined previously, two factors are apparent. First, we are able to normalize out of the problem a considerable degree of the dependence of the thermal indicators on scaling constants and to replace such a dependence by a dependence on data-related quantities. For instance, in the case of $^{39}Ar/^{40}Ar$ we removed the dependence on D_0/l^2, and similarly for each thermal indicator. Presumably, when we have determined the paleoheat flux variation with time which best provides a least-squares fit to the data, we should be able to reverse the process and determine that value of, say, D_0/l^2 which is most consistent with all of the data from all of the wells, since D_0/l^2 should at the least be a basinal, if not a global, constant. This determination will hinge not only on obtaining an accurate paleoheat flux but also on having an accurate burial history. A departure from isostatic evolution which allows for paleo-overpressuring changes both the burial history path with time and the temperature with time (since the thermal conductivity of an overpressured region is lower than it would be under isostatic burial). In turn, this alters the assessment of heat flux. In concert, these effects will limit the degree of accuracy with which we can determine the scaling constants for each thermal indicator.

Second, we note that it is not possible to normalize out of an indicator every physical or chemical parameter by replacement in terms of data-dependent quantities. For instance, in the case of vitrinite reflectance we are left with a doubling temperature T_D and a critical temperature T_c; in the case of $^{39}Ar/^{40}Ar$ with an activation energy E; in the cases of sterane and hopane isomerization and sterane aromatization with activation energies; in the case of apatite fission scar tracks with both an activation energy and an annealing rate constant; in the case of optical rotation with an activation temperature; and in the case of pollen translucency with a doubling temperature and a critical temperature.

One of the difficulties that arises with attempts to invert measurements of a thermal indicator to determine the paleoheat flux is that the respective time–temperature integrals depend on the physical and chemical parameters, so that these parameters must be known in order to perform the inversions. Two methods are available for determining the relevant physicochemical parameters. (1) We can provide laboratory measurements specifically tailored to extract the required parameters. But such methods suffer from two major disadvantages: (a) Laboratory measurements are geared to phenomena on a time scale of, at best, a human lifetime. The parameter values so determined are not necessarily representative of the values appropriate on the multimillion-year time scale of basin evolution. (b) The conditions of the laboratory experiments often do not represent the prevailing milieu of the subsurface in respect of temperature and pressure, salinity, mineral composition and relaxation, solubility in the subsurface brines, and so on—all of which can influence the relevant physical and chemical processes. Thus it is, perhaps, unwarranted to place a significant degree of confidence in such laboratory-determined parameter values. (2) We can use the downhole measurements of the thermal indicators themselves to determine the most appropriate values of the physicochemical parameters. The argument here is that inspection of the least-squares control function for each thermal indicator shows that the physicochemical parameters, as well as the paleoheat flux variations with time, enter in a highly nonlinear manner. Hence, for each well it is possible to do an exhaustive trial-and-error search to determine simultaneously both the heat flux variation with time and the physicochemical parameters which provide a global minimum for the mean square residual. By repeating the process with a particular thermal indicator for each well, a histogram of the best-fit parameters then can be determined. Since the physicochemical parameters are supposedly unrelated to geological evolution, they must take on the same constant value for each and every well. This method provides the parameter values most consistent with the downhole information for each thermal indicator.

Considerations of the above sort applied to vitrinite reflectance data from over 300 wells have narrowly constrained the critical temperature at $T_c = 295 \pm 20$ K and have provided the doubling temperature $T_D = 200 \pm 50$ K; similarly, the activation energy for sterane isomerization is found to be $E = 50 \pm 40$ kJ/mol, for hopane isomerization $E = 60 \pm 30$ kJ/mol, and so on.

In short, we can invert present-day downhole thermal indicator data to determine the paleoheat flux with time, and we can also capitalize on the nonlinear dependence of the relevant time–temperature integrals on the physicochemical parameters to determine these parameters as well.

B. Resolutions to Geological Problems Using Thermal Indicators

Apart from the intrinsic dependence of the various thermal indicators on paleoheat flux variations with time, it is also clear that the sundry time–temperature integrals depend implicitly on three other behaviors: (1) the burial path of each sedimentary unit with time, (2) the temporal variation of the thermal conductivity of the sedimentary layers along each and every burial path with time, and (3) the changes in temperature with time produced by sources other than the change of basement heat flux with time—such as igneous intrusions, salt domes, salt sheets and "tongue" emplacements, overthrusting, highly radioactive layers, and unconformities.

Quite often a degree of ambiguity is present in assessments of parameters associated with these various geological events. For example, in order to determine the amount of sediment eroded at an unconformity which occurred at some time in the past, we clearly need some measurement in a present-day geological setting which retains a quantitative memory of the event. Such a memory is carried in a cumulative manner by thermal indicators, as amply illustrated by Armagnac *et al.* (1988a) and discussed earlier.

Likewise with other geological events. Each event has associated with it a set of parameters (e.g., timing of overthrusting and amount of frictional heat generated; emplacement time of an igneous intrusion and its temperature at insertion). Each geological event can obviously provide an influence on the thermal maturity of sedimentary layers deposited before the event. Some events (e.g., the subsequent cooling of emplaced igneous intrusives) can also influence the thermal maturity of sedimentary layers after the event, while others (e.g., unconformities) do not influence the thermal maturity of postevent sedimentary units.

If we can provide models of the influence of the geological events on the temperature, thermal conductivity, and burial path of sedimentary units and if, further, the analytic expressions for the models contain unknown (*a priori*) geological parameters, then it should be possible to (1) incorporate such models into the relevant time–temperature integrals and (2) use the downhole observations of the thermal indicator in conjunction with the models to determine directly from the data the best parameter values which provide a minimum least-squares value. The disagreement between theory and observations is then minimal.

Such a scheme has already proved inordinately successful in determining amounts of erosion during unconformities (Bucci, 1985; Armagnac *et al.*, 1988a) and produces a regional well-defined trend to the amount of the unconformity consistent with the geological setting and inferred tectonic and sedimentologic history. In addition, a similar application to determine the unknown ages of stratigraphic horizons has met with considerable success in basins ranging in age from Cambrian through Cretaceous to Miocene, using multiple thermal indicators (Pantano and Lerche, 1988). In the latter case, the degree of veracity and accuracy has been directly tested by "blind" trials against sedimentary units of known biostratigraphic age. The high correlation obtained provides firm corroborative support for the generic logic of such manipulations, as well as confirming the essential soundness of the underpinning physics and chemistry of particular thermal indicators. The point here is that no such correlation would be expected in units ranging from 15 to 500 Myr old if some fundamental (but unknown) flaw were present in the underlying basic tenets of the theoretical development for each such thermal indicator.

In this section of the chapter we provide the basic theoretical development for a suite of geological problems so that it is made clear how they are incorporated within the inverse methods used to unravel the evolution of thermal indicators with time from present-day measurements. Section II will be concerned with some applications and case histories.

We exclude the cases of determination of the parameters for a *single* unconformity and of the unknown age of a *single* stratigraphic horizon, for these cases have already been dealt with extensively in earlier chapters and elsewhere (Bucci, 1985; Armagnac *et al.*, 1988a; Pantano and Lerche, 1988). We defer to later a discussion of the problems of determining the geological parameters for multiple unconformities and of determining the unknown ages of several stratigraphic layers.

1. The Influence of a Highly Radioactive Layer on Thermal Maturation

Most sedimentary layers are mildly radioactive, but the amount of heat liberated by radioactive decay is normally minor compared with that crossing the basement. Once in a while, however, a layer is produced, such as the Kimmeridge shale in the North Sea, which has a significant heat-generating capability compared with that crossing the basement.

Since the radioactive decay is mainly from the U–Th–K series, with half-life of order 10^{10} yr, and since the sedimentary basins are typically less than 10^9 yr old, the heat production from such layers can be considered to be constant after deposition.

Let such a layer be deposited between the time intervals t_2 and t_1 ($t_2 > t_1$) so that at a later time t the layer is sandwiched between the depths $z_1 < z < z_2$ and is of thickness $L(t) = z_2(t) - z_1(t)$. Since the burial history is known we can take the depths $z_1(t)$ and $z_2(t)$ to be known functions of time. Let the basement heat flux be $Q_B(t)$ and let the radioactive layer have a total heat production capacity of Q_R, with Q_R a constant.

Prior to deposition of the radioactive layer the temperature of a sedimentary layer is given by

$$T = T_s + Q_B(t) \int_0^z dz'/K(z') \qquad (9.1)$$

After deposition of the layer we have the sedimentary temperature of a layer at depth z at time t given by (Carslaw and Jaeger, 1947)

$$T = T_s + [Q_R + Q_B(t)] \int_0^z \frac{dz'}{K(z')} \quad \text{in } 0 \leqslant z \leqslant z_1(t) \qquad (9.2a)$$

by

$$T = T_s + Q_R[z_2(t) - z_1(t)] \int_{z_1(t)}^z \frac{dz'}{K(z')}[z_1(t) + z_2(t) - 2z'] + [Q_R + Q_B(t)] \int_0^z \frac{dz'}{K(z')} \quad \text{in } z_1(t) \leqslant z \leqslant z_2(t) \qquad (9.2b)$$

and by

$$T = T_s + [Q_B(t) - Q_R] \int_0^z \frac{dz'}{K(z')} + Q_R \int_{z_1(t)}^{z_2(t)} \frac{dz'}{K(z')}\{[z_1(t) + z_2(t) - 2z'][z_2(t) - z_1(t)]^{-1} - 2\} \text{ in } z \geqslant z_2(t) \qquad (9.2c)$$

where T_s is the temperature at the sediment–water interface. This formulation for the temperature can then be inserted into the appropriate time–temperature integrals, which, in turn, can be evaluated along the burial paths of the sedimentary units and a comparison made with the observed variations of the thermal indicator with depth at the present day. The major unknowns in equations (9.1) and (9.2) are $Q_B(t)$ and Q_R. Clearly, these unknown values can be determined, or at least bracketed, by running through the least-squares methods given earlier until, by exhaustive trial-and-error searches, the values of Q_R and $Q_B(t)$ are obtained which minimize the least-squares mismatch between theory and observations. A more powerful method of incorporating the determination of Q_R and $Q_B(t)$ within a tomographic framework is presented later.

2. *A Method for Approximating Paleo-overpressuring Using Thermal Indicators*

Within the framework of isostatic, equilibrium burial history there is no formal possibility of including the development with time of overpressure within a self-consistent framework. Therefore one of two choices exist. Either (1) we develop a fluid flow/compaction modeling capability and use it to replace isostatic burial history codes, as done in Chapter 6, or (2) we attempt to incorporate overpressuring within the isostatic burial history framework in the least *ad hoc* way possible. The advantage of the first formulation is obvious; the advantage of the second (if it can be quantified numerically in a ruggedly stable manner) is that it may enable us to obtain a relatively cheap and simple way of estimating paleo-overpressuring and paleofluid flow in a large number of geological settings without having to use more complex analyses of time-dependent fluid flow and compaction.

We start with the observation that the total overburden W in a sedimentary column must be supported by matrix frame pressure P_f plus water pressure P_w with

$$W = g \int_z^{z_s} [\rho_s(1 - \phi) + \rho_w\phi]\, dz$$

$$= g\rho_w(z_s - z) + g \int_z^{z_s} (\rho_s - \rho_w)(1 - \phi)\, dz \tag{9.3}$$

so that

$$P_w = \rho_w g(z_s - z) + \Delta P_w \tag{9.4}$$

with

$$P_f + \Delta P_w = g \int_z^{z_s} (\rho_s - \rho_w)(1 - \phi)\, dz \tag{9.5}$$

where ΔP_w is the excess fluid pressure (i.e., above hydrostatic), ρ_s the grain density, ρ_w water density, ϕ the porosity, z_s the sediment surface measured from the basement ($z = 0$), and g the acceleration due to gravity.

Typical frame pressure laws for shales (at least in the Gulf Coast region) give

$$P_f = P_*(\phi_*/\phi)^n \tag{9.6}$$

where P_*, ϕ_*, and n are constants, with $n \approx 4$.

Under *equilibrium* conditions for which the excess fluid pressure ΔP_w is zero, let the normal porosity be $\phi_n(z)$ (for which an exponential model with depth is often chosen). Then a finite fluid pressure ΔP_w will cause an increase in the porosity.

With $\phi = \phi_n + \Delta\phi$ and $|\Delta\phi| << \phi_n$ we have

$$P_f(\phi_n) = g \int_z^{z_s} (\rho_s - \rho_w)(1 - \phi_n)\, dz \tag{9.7}$$

For a given overburden weight we then have

$$W(1 - n\,\Delta\phi/\phi_n) = -\Delta P_w + W \tag{9.8}$$

so that

$$\Delta P_w/nW \cong \Delta\phi/\phi_n \tag{9.9}$$

Typically, significant overpressure is a fraction (0.1–0.3) of total lithological overburden, that is, $\Delta P_w \approx 0.1\text{–}0.3W$.

Hence with $n \approx 4$ we have

$$\Delta\phi/\phi_n \approx 0.03\text{–}0.08 \tag{9.10}$$

so that the range

$$0.2 \geqslant \Delta\phi/\phi_n \gtrsim 0 \tag{9.11}$$

That is, $\Delta\phi = \alpha\phi_n$ with $0.2 > \alpha \geqslant 0$, generously encompasses the porosity range. Now under isostatic equilibrium the balance of basement heat flux Q and temperature gradient is struck by

$$K(\phi)\, dT/dz = Q(t) \tag{9.12}$$

For normal shales we write (Lewis and Rose, 1970)

$$K_n = K_{sh}(K_w/K_{sh})^{\phi}$$

so that at 100% porosity ($\phi = 1$), $K = K_w$, the thermal conductivity of water, while for $\phi = 0$, $K = K_{sh}$, the thermal conductivity of fully compacted shale. In an overpressured region we have

$$K_{op} = K_n \exp[\Delta\phi \ln(K_w/K_{sh})] \cong K_n[1 - \Delta\phi \ln(K_{sh}/K_w)] \quad (9.13)$$

where $K_n = K_{sh}(K_w/K_{sh})^{\phi_n}$, so that the thermal conductivity *decreases* relative to normal values. Hence

$$(dT/dz) = (dT/dz)_n[1 + \Delta\phi \ln(K_{sh}/K_w)] \quad (9.14)$$

so that the temperature gradient *increases* in an overpressured region relative to that in a normally pressured section. The temperature in an overpressured region at time T at depth z is then given by

$$T = T_s + Q(t) \int_0^z \frac{dz'}{K_n(z')} [1 + \alpha\phi_n(z') \ln(K_{sh}/K_w)] \quad (9.15)$$

where $\alpha = 0$ if the layer is not in the overpressured domain.

The problem is to find the best α ($0.2 \geqslant \alpha \geqslant 0$) and the timing and depth range of overpressure.

To illustrate the general philosophy, suppose for the moment that below a line $z = wt + z_0$ a significant overpressure exists and above the line we have a normally pressured section.

We then have to determine z_0, w, and α. For a well which is overpressured at the present day ($t = 0$), z_0 is the start of overpressuring with depth and so is known; hence α and w are to be found. For a fully paleo-overpressured situation we have to determine z_0 as well from the present-day downhole thermal indicator measurements.

If we have determined z_0, w, and α the paleo-overpressure is then given by

$$\Delta P_w = n\alpha g \int_z^{z_s} (\rho_s - \rho_w)(1 - \phi_n)\, dz' \quad \text{in } z > wt + z_0 \quad (9.16)$$

In general, we see from equation (9.15) that the theoretical prediction for the thermal indicator is a function of z_0, w, α, and the basement heat flux variation with time, $Q(t)$. Again, we have the problem of fitting theoretical predictions to observations of the thermal indicator with depth in order to determine the best parameter values.

More generally, if we relax the constraint of a straight line with time, we can argue that paleo-overpressure occurs below a curve

$$z_1(t) = \left[w_1 t + \sum_{k=1}^{m} a(k) \sin(\pi k t / t_m) \right]^2 \tag{9.17}$$

but above a curve

$$z_2(t) = z_1(t) + \left[w_2 t + \sum_{j=1}^{m} b(j) \sin(\pi j t / t_m) \right]^2 \tag{9.18}$$

(so that $z_2 \geqslant z_1$ everywhere and $z_1 \geqslant 0$ everywhere), where t_m is the age of the deepest formation for which we have thermal indicator information, w_1 and w_2 allow for a bulk trend of the curves with time, and the sets of coefficients $\{a(k)\}$, $\{b(j)\}$ and w_1, w_2 are to be determined by matching theoretical predictions to observations in a least-squares manner, as is the parameter α. Thus we see that as well as the paleoheat flux variation with time, we also have a suite of geologically related parameters to determine from observations with depth of a thermal indicator at the present day. Clearly, with more than two or three parameters to determine it becomes difficult, if not impossible, to do an exhaustive trial-and-error search to determine all of the parameters. Some more organized and systematic procedure of parameter determination is called for, as provided later in the chapter.

3. *Determination of the Timing and Insertion Temperature of Igneous Intrusions Using Thermal Indicators*

The thermal model here is appropriate for rapid emplacement of an igneous intrusion regarded as a slab emplaced parallel to bedding. Emplacement causes significant and abrupt changes in the thermal regime because the intrusive is much warmer than the sediments (initial intrusive temperatures of order 600–1000°C are typical). As time progresses the sediments warm and the intrusive cools.

Let the emplacement of an intrusive slab of thickness L and temperature T_I occur at time t_I at a sediment depth $z_I(0)$. The equation describing the later evolution with time of the temperature regime is (Brewer, 1981; Carslaw and Jaeger, 1947)

$$\begin{aligned} T(z, t) = {} & T_s + Q(t) \int_0^{z(t)} \frac{dz'}{K(z')} + [4\pi k(z)(t - t_I)]^{-1} \\ & \times \int_{z_I(t) - L}^{z_I(t)} \left[T_I - T_s - Q(t) \int_0^{y} \frac{dz'}{K(z')} \right] (\exp\{-(z - y)^2 [4k(z)(t - t_I)]^{-1}\} \\ & - \exp\{-(z + y)^2 [4k(z)(t - t_I)]^{-1}\}) \end{aligned} \tag{9.19}$$

where $z_I(t)$ marks the evolution of the igneous layer for $t > t_I$, z is the depth of a chosen sedimentary layer at time t, and k is the thermal diffusivity, which is related to thermal conductivity K, density ρ, and specific heat C_v through $k = K/\rho C_v$.

The igneous temperature influence has two unknown parameters: the initial temperature T_I and the insertion time t_I.

The influence of igneous intrusives is clearly seen in vitrinite reflectance measurements with depth (Dow, 1977; Hunt, 1979; Reeckman and Mebberson, 1984). We again note that the theoretical time–temperature integrals for a thermal indicator (e.g., vitrinite reflectance) depend on the insertion parameters for the igneous body. It follows that the best values of T_I and insertion time t_I as well as the paleoheat flux variations with time can be determined by minimizing the degree of mismatch of theoretical predictions of the thermal indicator with depth against observations.

4. Determination of Time of Emplacement and Frictional Heating of Overthrust Models

Furlong and Edman (1984) have argued that the overthrusting of sedimentary layers is very rapid on a geologic time scale so that the overthrust slab can be considered as being emplaced on top of the prevailing sedimentary layers, with the overthrust slab's temperature profile with depth maintained at the time of emplacement. Thereafter the excess heat in the overthrust slab is drained off by diffusion into the underlying sediments. The evolution of any thermal indicator is then changed both in the overthrust slab and in the underlying sediments by (1) the change in basement heat flux with time and (2) the "excess" heat provided by the overthrust slab. In addition, the thermal indicator in the slab already had achieved some degree of maturity before being overthrust. To handle these problems quantitatively we proceed as follows. At time $t = t_i$ emplace an overthrust slab of thickness L. Let the temperature profile in the slab be $T_i(z, t_i)$ at the instant of emplacement. Then the later development of temperature both in the slab and in the underlying sedimentary layers due solely to the entrapped heat is (Carslaw and Jaeger, 1947)

$$T_+(z, t) = T_s + Q(t) \int_0^z \frac{dz'}{K(z')}$$

$$+ [2\pi(t - t_i)^{1/2}]^{-1} \int_0^\infty dz' k(z')^{-1/2} (\exp\{-(z - z')^2[4(t - t)k(z')]^{-1}\}$$

$$- \exp\{-(z + z')^2[4(t - t_i)k(z')^{-1}\})$$

$$\times \left[T_i(z', t_i) - T_s - Q(t) \int_0^{z'} \frac{dz''}{k(z'')} \right] \quad \text{in } t > t_i \tag{9.20}$$

and

$$T(z, t) = T_s + Q(t) \int_0^z \frac{dz'}{K(z')} \quad \text{in } t < t_i \tag{9.21}$$

For times prior to $t = t_i$, and bearing in mind that we treat with single-well information so that we are restricted to a one-dimensional model, we treat the slab as having undergone a different thermal history than the overlain sedimentary material so that the temperature distribution at any depth z in the slab at emplacement is given by

$$T(z, t_i) = T_s + Q(t_i) \int_0^z \frac{dz'}{K(z')} \quad \text{in } t < t_i,\ L \geqslant z \geqslant 0 \tag{9.22}$$

The maturity of any thermal indicator is then described by two parts: that prior to emplacement of the slab and that postemplacement. For sediments under the slab ($z > L$) we have a continued evolution of the indicator with the temperature given by equations (9.20) and (9.21).

For the same thermal indicator in the slab we have to allow for the maturity that the slab was subjected to prior to emplacement. Thus as far as the slab is concerned, in a one-dimensional burial history framework it is as though the slab were deposited in a very short time on top of the sediments with the values of the thermal indicator "frozen" at their preemplacement stage.

Perhaps the most pragmatic way to treat this problem numerically is to consider the emplacement taking place over a short time interval Δt in $t \geqslant t_i$ at a constant rate $w \equiv L/\Delta t$. Then the depositional value of the thermal indicator, say $\mathrm{TI}_{\mathrm{dep}}$, in the slab is given by $\mathrm{TI}_{\mathrm{dep}} = \mathrm{TI}[L - w(t - t_i)]$ in $t_i \leqslant t \leqslant t_i + \Delta t$ and $0 \leqslant z \leqslant L$. Thus we replace the (usually constant) depositional value of the thermal indicator by a time-evolving value determined from the prior thermal history. Thereafter, thermal evolution proceeds under the influence of the evolving heat flux of the basement and the diffusive heat input from the overthrust. This process can be repeated for each overthrust slab, with its own thickness and its own deposition time. The upshot is clear. The final present-day values of the measured thermal indicator with depth are nonlinearly related to the emplacement time t_i of the overthrust slab, as well as to the evolution of basement heat flux with time; or, in the case of more than one overthrust slab, to the time of emplacement of each slab. If, in addition, the overthrusting generates an amount of frictional heating in the amount $q(t)$ throughout a thickness range ΔL centered on $z = L$ in the domain between the slab's lower surface and the underlying sedimentary units, we can model this frictional heating as in the manner of a radioactive layer, treated previously, taking over those formulas in their entirety with replacement of the constant heat flux Q_R by the frictional heat $q(t)$, and with the radioactive layer thickness being replaced by ΔL. With a model dependence for the frictional heating of the general form

$$q(t) = q_0 \exp[-(t - t_i)/\tau] \tag{9.23}$$

in $t > t_i$, with the two new free parameters q_0 and τ, again it is possible to invert the observed thermal indicator data with depth to determine these new parameters in addition to those already associated with emplacement of the overthrust.

5. *Determination of the Time of Salt Tongue Emplacement*

Within the framework of a one-dimensional burial history the emplacement of a salt tongue or sheet [of high thermal conductivity relative to the surrounding sediments by a factor of 3 to 5 (O'Brien and Lerche, 1984; Lerche and O'Brien, 1987)] appears as an increasing thickness of material deposited not at the sediment surface, but internally. The temperature in the salt sheet changes from that in the sediment due to the higher thermal conductivity, so that at any subsurface depth z at time t, for $t > t_i$, where t_i is the insertion time of the salt tongue, we have the temperature given by

$$T_1(z, t) = T_s + Q(t) \int_0^z \frac{dz'}{K(z')} \quad \text{in } 0 \leqslant z < L_s(t) \tag{9.24a}$$

$$T_2(z, t) = T_1(L, t) + \frac{Q(t)}{K_{\text{salt}}}[z(t) - L_s(t)] \quad \text{in } H(t) > z > L_s(t) \tag{9.24b}$$

$$T_3(z, t) = T_2(H, t) + Q(t) \int_H^z \frac{dz'}{K(z')} \quad \text{in } z > H(t) \tag{9.24c}$$

Here $L_s(t)$ $(t > t_i)$ measures the position of the top of the salt layer from the sediment–water interface, $H(t)$ measures the position of the bottom of the salt and includes both thickening of the salt and the effect of the increased overburden as time progresses.

Once more, then, we see that the sedimentary temperature in the post-salt emplacement stage depends not only on the evolving basement heat flux with time but also on the emplacement time of the salt layer. Clearly, the thermal maturation indicators are sensitive to some extent to this effect. Once again we are faced with the task of determining the emplacement time from present-day observations with depth of the thermal indicators.

6. Listric Fault Models

Listric sliding and slumping behaves in exactly the same way as an overthrust slab as far as the one-dimensional burial history is concerned. Basically, partially matured sediments are overlaid on sedimentary strata in a very short geological time period. The mechanics of handling the thermal maturity pre- and postfaulting then follows precisely the same pathway as already spelled out for the overthrust. We have, once more, the task of determining the emplacement times of the material from present-day measurements of the thermal indicator.

C. Tomographic Inverse Methods of Solution to Basin Analysis Problems Using Thermal Indicators

In previous sections of this chapter we have seen three basic types of parameters entering the methods used for thermal indicators:

1. Parameters such as activation energies, critical temperatures and rate constants directly associated with the underlying physics and/or chemistry peculiar to each thermal indicator.
2. Parameters such as paleoheat flux variations with time or radioactive heating with time directly associated with the changing thermal maturity of the sedimentary layers.
3. Parameters such as erosional unconformities, paleo-overpressuring, etc., which directly change the burial path of the sedimentary layers with time, and so the temperature, or which directly change the thermal conductivity and so the temperature even if the paleoheat flux does not vary with time. They therefore have a role to play in changing the time–temperature integrals and so the thermal maturity of individual thermal indicators with depth at the present day.

Out of this array of parameters it is relatively easy to determine one or two such parameters in a given geological scenario by trial-and-error searches. But if more than a few parameters are involved (such as would happen with multiple unconformities where the erosional thicknesses and/or onset time of erosion *and* paleoheat flux variations with time are to be determined), exhaustive trial-and-error searches are not only exhausting but also extremely difficult to implement.

The difficulties are clear: How do we determine simultaneously a large number (say four or more to fix ideas) of a mixture of chemical, thermal, and geological parameters by inversion of downhole thermal indicator data? And how do we know when we have reached resolution capability of the data so that it is pointless to search for further parameters?

Fortunately, precisely this type of parameter determination problem has arisen in a variety of other scientific disciplines over the years and a substantial body of information has been culled on appropriate methods of analysis to tackle such problems. In general, the most powerful and accurate techniques are those based on some form of tomographic analysis. Here we provide the relevant form of tomography appropriate to the inverse problems of thermal indicators.

Thermal indicator tomography is a nonlinear inverse problem. The nonlinearity arises because the thermal maturity along the paths taken by the sediments and the path themselves are affected by the parameters one is trying to determine (e.g., paleoheat flux, insertion times, unconformity erosion thicknesses). One generally has enough gross information about the paleostructure of a region to approximate roughly the thermal indicators—to this extent thermal indicator tomography is "almost linear." As usual in nonlinear problems, we linearize around an initial guess for the parameters, solve the linear inversion for an improved set of parameters, and iterate. Experience from many disciplines suggests that if the (linear) inversion has been properly regularized (stabilized), and if the intermediate solutions are constrained to be reasonably smooth, this procedure converges satisfactorily in a few iterations (typically 5 to 10—although extreme cases are known when many tens of nonlinear iterations are necessary).

The essence of thermal indicator tomography, then, is to (1) construct a model which approximates past thermal behavior the subsurface, (2) compute theoretical thermal indicators for each sedimentary layer by forward tracing in time through the model, (3) measure the actual thermal indicators from the appropriate depths, and (4) modify the thermal model through which the theoretical thermal indicators traveled in time to minimize the difference between measured and computed maturations. The model is modified by inverting a linearized version of the least-squares equations. Finally, the entire process is iterated by tracing through the nonlinearly modified model and proceeding as before.

Conceptually, this method is closely related to the iterative inversion method used by Hawley *et al.* (1981) for simultaneously determining hypocenters of earthquakes and three-dimensional velocity variations. Their method is a generalization of the one-step general inverse method proposed by Aki and Lee (1976). In the form outlined here, the technical development is closely allied to that of travel-time seismic reflection tomography (Cutler *et al.*, 1986).

The model is specified in a one dimension in time and one dimension (in depth) burial history of the earth. This "universe" is divided into boxes by a rectangular grid. Each box is labeled by the rectangular coordinates of its center (t, z). The heat flux field is parametrized in each box by a single value of rate of change of heat flux at the center of the box. The depth of the ρth layer, $z_\rho(t)$, is parametrized at the edges of the boxes.

Certain points about the model definition are worth emphasizing:

1. Layers can cut through the interior of boxes.
2. Rate of change of heat flux Q can vary both vertically with depth and temporally in an arbitrary (but reasonably smooth) manner. The reason for including a *spatially* varying paleoheat flux is to allow for internal sediment heating, for example, increased radioactivity in particular layers.
3. Layers are not necessarily associated with any changes in the heat flux field.

As we have seen in the previous sections, it is also possible to define other parameters which describe the thermal response of the earth, such as unconformities or igneous intrusives, overthrust slabs, or salt tongue intrusions. All that is necessary is to find the relationship between such parameters and the measured thermal indicator at depth. (In many cases, as we have shown, these relationships are trivial.) In the next section we will show how forward tracing (in time) of thermal indicators connects the parameters to measured thermal indicators. For convenience, we denote the parameters describing the model by a vector p. Given a model described by parameters p, forward modeling will predict a thermal maturity with depth at the present day. For N depths at which the thermal indicator is measured denote the predicted thermal indicator for a model by a vector $\tau_{\mathrm{m}}(p)$ of length N and the actual observed thermal indicators by a vector τ.

This notation simplifies the equations but obscures the fact that parameters have different dimensions and typical values—for example, the different magnitudes and units for heat flux and eroded unconformity thickness. We rescale all parameters according to the typical effect of that *type* of parameter on the thermal indicator. The rescaling is done using the elements of an A matrix, which will play a central role in the later development of the linearized and nonlinear tomographic inversion procedure. For the moment we write the formal rescaling relations

$$P_j \rightarrow \left(\sum_l \sum_k {}' A_{lk}^2 \right)^{1/2} P_j \tag{9.25a}$$

$$A_{ij} \rightarrow \left(\sum_l \sum_k {}' A_{lk}^2 \right)^{-1/2} A_{ij} \tag{9.25b}$$

where the primed sums are performed only over indices denoting parameters of the same type as j; for example, the primed sum done over heat flux parameters is used to provide a rescaling factor which multiplies only heat flux parameters. Similarly, the primed sum over eroded unconformity thickness derivatives rescales thickness parameters, and so forth. This rescaling will be suppressed in all subsequent formulas.

1. Linearized Inversion

We solve the nonlinear tomography problem by linearizing around an initial guess, solving the linear inverse problem for an improved set of parameters, and iterating (Menke, 1984; Cutler *et al.*, 1986). If the parameters describing the model are denoted as a vector p and the N expected thermal indicators resulting from tracing through the model by a vector $\tau_m(p)$, then the change in thermal indicators, relative to the observed values τ, $\Delta\tau = \tau - \tau_m(p)$, produced by a change in the parameters, Δp, may be approximated linearly by

$$\Delta\tau = A\ \Delta p \tag{9.26}$$

where A is an $N \times M$ matrix. Note that M and N, in a practical problem, may be of order 10 to 100.

The matrix A is the *linearized forward* modeling matrix. The components of the A matrix are derived by taking the appropriate functional derivatives.

The linearized inverse problem may be solved by various well-known techniques. Two solutions to the inverse problem are (Cutler *et al.*, 1986)

$$\Delta p = (A^T A + \alpha I)^{-1}\ A^T \Delta\tau \tag{9.27}$$

which is the overdetermined formulation, and

$$\Delta p = A^T (A A^T + \alpha I)^{-1}\ \Delta\tau \tag{9.28}$$

which is the underdetermined formulation. These two forms are equivalent if the factor α is zero, although the matrices to be inverted are of different size ($M \times M$ and $N \times N$, respectively). The first form is overdetermined because its square matrix $A^{\mathrm{T}}A$ has the smaller dimension if the number of datum points N is greater than the number of parameters M. Not only is the matrix smaller, but also it is less singular (in the sense that the larger matrix AA^{T} has at least $N - M$ zero eigenvalues). For these reasons, the first formulation is usually preferable if the number of datum points exceeds the numbers of parameters. (Similarly, the underdetermined matrix AA^{T} is smaller and less singular if there are more parameters than datum points.)

We choose our box size in time–depth coordinates so that there are considerably more datum points than parameters; accordingly, one might expect the overdetermined formula to be applicable. In fact, following Cutler *et al.* (1986), we use the underdetermined equation because it can be recast in a form which is efficient to calculate. The damping constant α, necessary to make the matrices invertible, is the usual route to regularizing this sort of inverse problem (Menke, 1984). A small value for α, usually 5% or so of a "typical" (r.m.s.) diagonal element of $A^{\mathrm{T}}A$, is used. To obtain an acceptable solution to the inverse problem, the box size is chosen so that there are typically three to five times as many datum points as parameters.

With Menke (1984) and Cutler *et al.* (1986) we note that to solve the linearized inverse problem we invert

$$\Delta p = A^{\mathrm{T}}(AA^{\mathrm{T}} + \alpha I)^{-1}\,\Delta\tau \qquad (9.29)$$

or, defining the vector X (of length N), by the pair of equations

$$\Delta p = A^{\mathrm{T}}X \qquad (9.30a)$$

$$X = (AA^{\mathrm{T}} + \alpha I)^{-1}\,\Delta\tau \qquad (9.30b)$$

we can then use the reliably convergent Gauss–Seidel iteration to solve for X from

$$(AA^{\mathrm{T}} + \alpha I)X = \Delta\tau \qquad (9.31)$$

Since the matrix A is relatively sparse, write the iteration formula for X in terms of A,

$$X_\kappa(\text{new}) = \frac{\Delta\tau_\kappa - \sum_{j=1}^{M_\rho} A_{\kappa j} \sum_{i=1}^{\kappa-1} A_{ij}X_i(\text{new}) + \sum_{i=\kappa+1}^{N_\iota} A_{ij}X_i(\text{old})]}{\left[\alpha + \sum_{j=1}^{M_0} A_{\kappa j}^2\right]} \tag{9.32}$$

where X_κ (old) is immediately updated to X_κ (new) as each component is calculated. The quantity in square brackets in the numerator of (9.32) is almost equal to Δp_j:

$$\Delta p_j = \sum_{i=1}^{N} A_{ij}X_i \tag{9.33}$$

except that some of the components of X are "old" and some are "new." The efficiency and speed of computation are now improved considerably by invocation of a very clever "trick" developed extensively in Cutler *et al.* (1986). Define a new quantity Δp(new) by $\Delta p(\text{new}) = A^{\mathrm{T}}X$, where each component of X takes on its most current value, that is,

$$\Delta p_j(\text{new}) = \sum_{i=1}^{\kappa-1} A_{ij}X_i(\text{new}) + \sum_{i=\kappa}^{N_\iota} A_{ij}X_i(\text{old}) \tag{9.34}$$

then

$$X_\rho(\text{new}) = \frac{\Delta\tau_\rho - \sum_{j=1}^{M} A_{\kappa j}\Delta p_j(\text{new}) + X_\kappa(\text{old}) \sum_{j=1}^{M} A_{\kappa j}^2}{\alpha + \sum_{j=1}^{M} A_{\kappa j}^2} \tag{9.35}$$

Essentially one of the sums is hidden in Δp(new). But, as noted by Cutler *et al.* (1986), each time a new X_κ is calculated, one can subtract the contribution of the *single* old component and add in the new one,

$$\Delta p_j(\text{new}) = \Delta p_j(\text{old}) + A_{\kappa j}[X_\rho(\text{new}) - X_\rho(\text{old})] \tag{9.36}$$

This procedure eliminates the second sum entirely!

Although convergence of the Gauss–Seidel algorithm can be slow, the algorithm above is so efficient that the linearized inversion step tomography uses considerably less computing resources than the burial path tracing required to calculate the vector $\tau_{\mathrm{m}}(p)$ prior to each linearized inversion.

2. Nonlinear Iterations

The linear inversion is a very clear procedure. But the nonlinear portion of the tomographic inversion scheme is less clear in general in all scientific disciplines—basically because an understanding of nonlinear equations is not as well developed as for linear equations. To date, our numerical experiences on synthetic and real data problems has shown that

1. The nonlinear inversion generally converges rather rapidly in 5–10 iterations.
2. The model arrived at after the nonlinear iterations is a significantly better fit to the data than the results of the first linearized inversions.
3. Pragmatically, when the "best" model parameters are determined they are then (a) reversed in sign, (b) halved, (c) doubled, (d) halved and reversed in sign, and (e) doubled and reversed in sign (except for intrinsically positive quantities, which are made to keep the same sign). The tomographic procedure then runs these new five "initial" guesses in both the linear and then the nonlinear modes to determine the "absolute" best-fit set of parameters. While even more time-consuming, this procedure ensures that the tomographic inversion is much more likely to find the global minimum than become stuck in a local minimum. In most cases steps (a)–(e) are unnecessary in hindsight, but in about 10% of cases investigated to date this more time-consuming procedure has proved to be very much worthwhile in improving the model parameters.

3. Comments on Thermal Indicator Tomography

We have applied thermal indicator tomography in case studies where temporal changes in heat flux were expected or known to exist. In each of these cases the object was to develop a tomographic model for one of the following goals:

1. Image the heat flux anomaly in time to verify or suggest a possible interpretation consistent with known geological events, such as a dolerite intrusion or a rift phase of a basin.
2. Resolve eroded unconformity thickness ambiguities for key geologic stages and interpret the resulting burial history. Conventional methods are not capable of correcting for these variations in detail.
3. Use a tomographic model to improve the processing of conventional thermal indicator data. Because temporal heat flux variations are determined, the thermal maturity field determined by tomography is a better estimate of where and when hydrocarbons

mature in the source rock. Tomography can also be used as an iterative modeling technique in complex geologic areas to test alternative interpretation hypotheses.

In summary, thermal indicator tomography has proved to be a useful technique where temporal anomalies cause burial path distortions and thermal variations thereby influencing maturity estimates.

Geologic parameters (unconformity thicknesses, emplacement time of overthrusts, insertion temperature and time of igneous intrusives) can be determined in an inverse framework from measurements of thermal indicators with depth at the present day. Clearly, the more thermal indicators we have which are capable of being inverted tomographically, the better we can narrow down the degree of uncertainty in the precise numerical values for the geologic parameters. In addition, we can simultaneously use thermal indicator tomography to determine physical and chemical parameters as well as, of course, paleoheat flux variations with time.

We have spelled out the theoretical development underpinning not only methods of determining paleoheat flux variations with time from inversion of present-day downhole thermal indicator data but also methods for assessing parameters of geological processes which affect the evolution of sedimentary basins in both a local and regional sense. The detailed technical development, when more than a handful of such parameters is to be determined, is best accomplished through the use of methods based on tomographic procedures.

The combined determination of quantitative paleoheat flux variations and associated structural and dynamical properties of petroliferous basins, together with quantitative measures of their uncertainties consistent with downhole data, offers the greatest probability of accurately assessing the petroleum generation and migration aspects of basins in respect of the timing of hydrocarbon production from kerogen and its relationship to the timing of the development of structural and stratigraphic traps which serve as oil reservoirs in a basin.

Underlying the development of the multiple thermal indicator methodology is also the idea that multiple thermal indicator information downhole acts as a very powerful constraint and provides checks and balances of one thermal indicator to another. The point here is that different thermal indicators spanning the same depth range in the same well must have been exposed to precisely the same burial path behavior and to the same heat flux variations with time. The inversion of each such indicator must then produce the same value for the rate of change of heat flux with time as well as for any parameters tied to geological events, such as the amount of erosion involved at an unconformity. By demanding such agreement, and since each thermal indicator has its own peculiar dependence on time and temperature, it is possible to consider the inverse mode of operation as one from which physical and chemical parameters (such as diffusion coefficients and activation energies) can be extracted in order to achieve consistency with the downhole data of the particular thermal indicator and also with the geological parameters determined by inversion from other downwhole indicators.

In the next part of this chapter we present several case histories illustrating and exemplifying the application of the theoretical development given here.

II. Applications and Case Histories

A. Inigok-1, Alaska

The well Inigok-1 is located in the Ikpikpuk basin of the National Petroleum Reserve of Alaska (NPRA). The well was drilled through a thick sequence of coastal to shallow marine sediments which have undergone several episodes of uplifting and erosion (Armagnac, 1985). Vitrinite was sampled from 400–19,900 ft and reflectance values ranged from approximately 0.42 to 4.80%. Three unconformities were dated in the well, with ages estimated as 1.8–69 Myr, 131–140 Myr, and 278–285 Myr, respectively, with unknown amounts of erosion. As a first-order estimate of the paleoheat flux and the surface unconformity, two-dimensional grid searching (in an unconformity amount, β, space with the other two unconformities modeled as hiatuses) yielded an erosion amount of 2900 m for the surface unconformity and a β of 0.6×10^{-3} $(\text{Myr})^{-1}$ (Figs. 9.1 and 9.2). The "best" χ^2 is 0.092. Inspection of the number of vitrinite data points and the systematic offset between the predictions and measurements in the shallow section (approximately between 1000 and 9000 ft), as shown in Fig. 9.2, led to the suspicion that the assumptions of (1) a linear variation of heat flux with time and/or (2) a fixed time of 30 MYBP to start the erosion were not adequate. Therefore, the next step taken was to include the age when the erosion began and for the heat flux the form

$$Q(t) = Q_0 \exp\left[\beta t + \sum_{k=1}^{3} a_k \sin(\pi k t / t_{\max})\right] \tag{9.37}$$

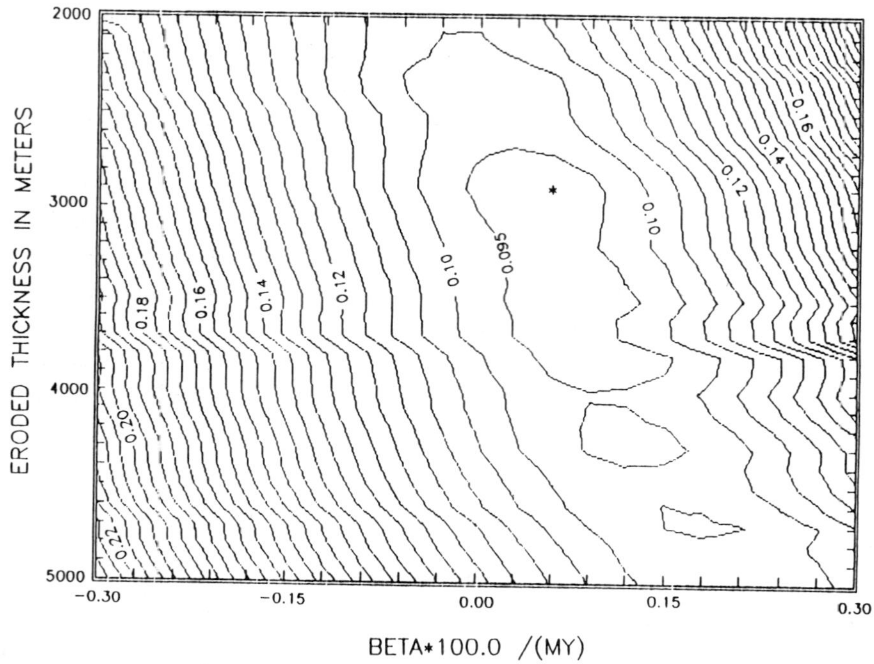

FIGURE 9.1 χ^2 contoured in a β–eroded thickness parameter space for Inigok-1, NPRA. Star denotes minimum of χ^2.

both to be varied within the thermal indicator tomography (TIT) framework. A χ^2 of 0.074 was obtained (Fig. 9.3), which is a 20% improvement over the best values from the linear search. The final heat flux variation derived (Fig. 9.4) suggests that the heat flow was higher during the late Cretaceous. This timing for the higher heat pulse corresponds to the period of rifting which separated the NPRA from its northern source and formed the Canada Basin (Grantz and May, 1983). The increase in deposition rate and basement depth seen on the geohistory plot (Fig. 9.5) from approximately 120 to 90 MYBP may be the result of both rifting and an increase of sedimentary influx from the Brooks Range. Continued subsidence during the early Tertiary may be caused by postrift cooling (Armagnac, 1985). The best parameters determined by TIT are listed in Table 9.1. An age of 30 MYBP for the onset of erosion was suggested by the tomography. This timing of uplift is younger than the youngest-dated sediments (37 MYBP) immediately below the same unconformity in a neighboring well, while the youngest rocks immediately below the unconformity in Inigok-1 are dated at about 69 Myr in age. When the

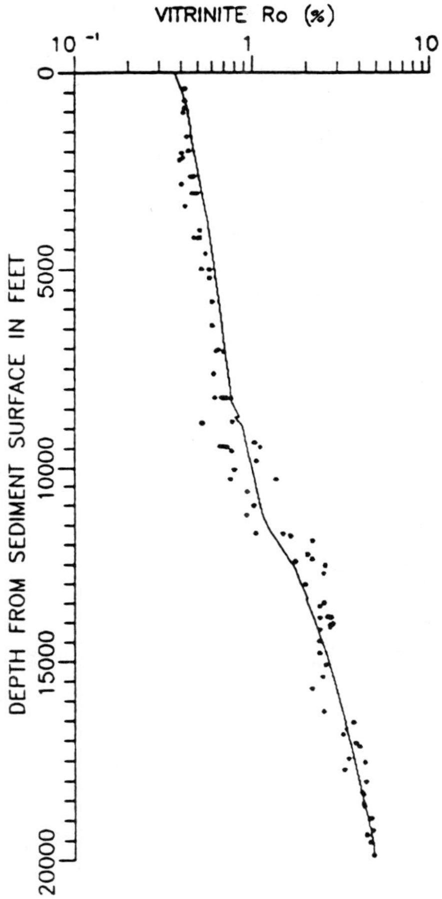

FIGURE 9.2 Reflectance versus depth, including predicted reflectance line, for Inigok-1, Alaska. Asterisks are measured vitrinite values; dark line is predicted vitrinite reflectance derived using $\beta = 0.6 \times 10^{-3}$ (Myr)$^{-1}$ and a surface erosion of 2900 meters derived from grid searching shown in Fig. 9.1.

Table 9.1 Tomography Results for Inigok-1, NPRA

χ^2	T_D (K)	β (Myr)$^{-1}$	a_1	a_2	a_3	Eroded thickness (m)	Age (Myr)
0.078	192	-0.84×10^{-3}	−0.015	0.36	−0.44	2500	24

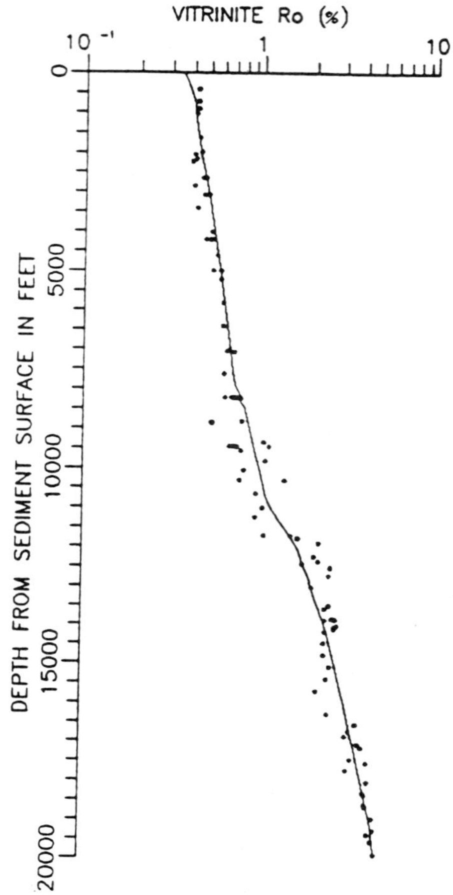

FIGURE 9.3 Same as Fig. 9.2, except predicted reflectance line is derived using the parameters listed in Table 9.1.

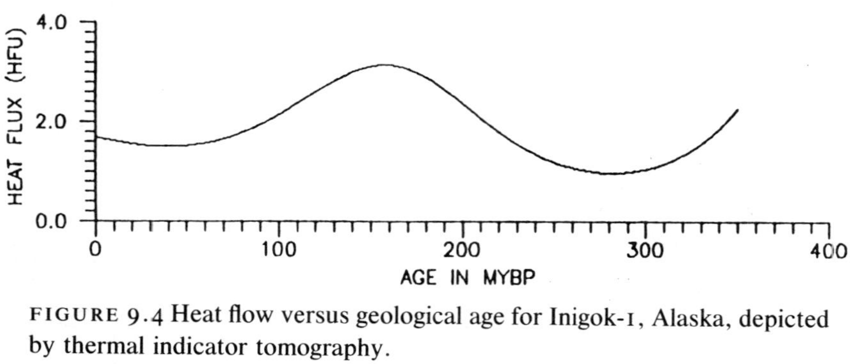

FIGURE 9.4 Heat flow versus geological age for Inigok-1, Alaska, depicted by thermal indicator tomography.

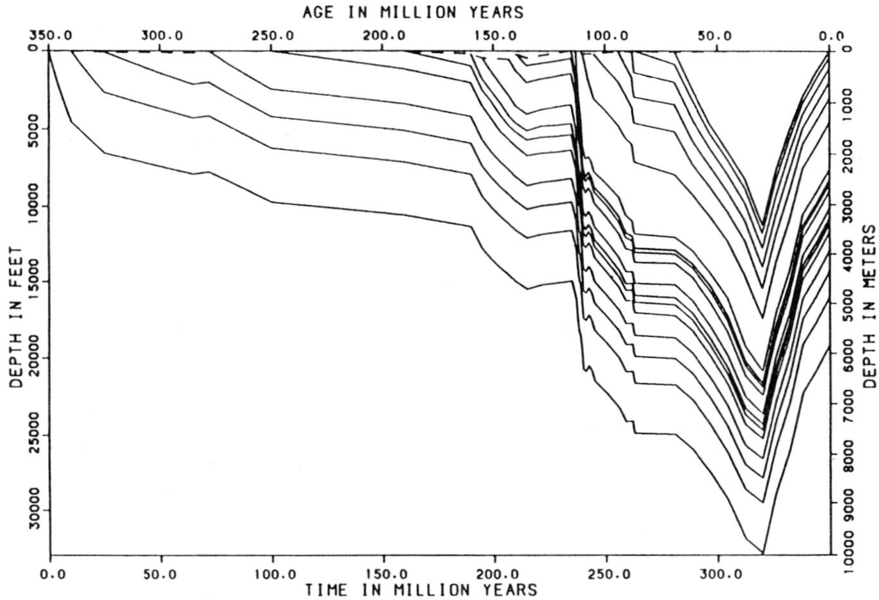

FIGURE 9.5 Geohistory plot for Inigok-1, Alaska. Erosion amount and timing are determined using vitrinite reflectance as a thermal indicator in tomography.

amounts of erosion for the two other possible unconformities at depths of 9040 feet and 13,890 ft were included as parameters to be determined, the tomography favored a hiatus for both unconformities. It can be seen from the data (Figs. 9.2 and 9.3) that the resolution limit has been reached for the determination of these unconformities. If there was any erosion, the "noise" in the vitrinite data is obscuring any determination of the amount of erosion.

B. NW Canning Basin, Australia

Three methods were investigated by Huntsberger and Lerche (1987) for determining paleoheat flux from fission scar tracks in detrital apatite collected from wells. For the distribution of track length method, at each depth z_i one measures the number of tracks $N(l_j, z_i)$ in $0 \leq l_j \leq L_{max}$, with l_j the track length (in micrometers) at the center of the histogram "bin" (for 1-μm bins we have $l_1 = 0.5$, $l_2 = 1.5$, . . .). What is usually plotted at each depth is the probability distribution (frequency of occurrence) of track lengths observed in each range of l:

$$P(l_j, z_i) = \frac{N(l_j, z_i)}{\sum_{j=1}^{J} N(l_j, z_i)} \tag{9.38}$$

where J is the total number of bins. The theoretical expression directly comparable to $P(l_j, z_i)$ is given by

$$p(l, t_i) = (\pi^{1/2} t\Delta)^{-1} \int_0^{t_i} dt' \exp\left[\int_{t'}^{t_i} \alpha(t'')\, dt''\right]$$

$$\times \exp\left\{-\Delta^{-2}\left[L_{av} - l \exp \int_{t'}^{t_i} \alpha(t'')\, dt''\right]^2\right\} \tag{9.39}$$

where $\alpha(t) = k_1 \exp[-T_c/T(t)]$, with k_1 the rate constant and T_c the critical temperature corresponding to an activation energy of $E = T_c R$ (R is the gas constant). For a detailed description of the formulation and constants the reader is referred to Huntsberger and Lerche (1987); see also Green and Duddy, 1988; Lerche, 1988c). The point here is that the predicted distribution $p(l, t_i)$ for an apatite sample deposited at time t_i and currently at depth z_i depends not only on the temperature along the burial path $T(t)$ but also on two "chemical" constants T_c and k_1. Huntsberger and Lerche (1987) demonstrated that the apatite fission track data could be inverted to determine the paleoheat flux variations. However, because of the limitations of the grid searching method then available, they were unable to improve very much on the values of k_1 and T_c suggested from laboratory experiments of Gleadow *et al.* (1983), as can be seen from the relatively poor fit to the data (see Huntsberger and Lerche, 1987, pp. 384–385).

Using the tomographic inversion method, an attempt was made to determine the heat flow history simultaneously[1] with k_1, T_c, and other variables such as the original track length distribution parameters Δ and L_{av}. The test was carried out on four wells from NW Canning Basin and four other wells in two other basins. In order to maximize the stress on tomography of this chapter, here we present only the result from Curringa-1 and Kambara-1, which were studied by Huntsberger and Lerche (1987).

Igneous intrusions dated as late Carboniferous to Jurassic, but thought

---

[1] For ease in numerical convenience we rewrite $k_1 \exp(-T_c/T)$ in the form $k_* \exp[T_c/(1/273 - 1/T)]$ with $k_* = k_1 \exp(-T_c/273)$. This procedure minimized overflow and allowed us to better determine T_c and k_*.

to be Permian in age, have been intersected by six wells in the northwestern Canning Basin. Both seismic and magnetic interpretations suggest that these intrusions are common over a large area. Regionally, the mafic intrusions are restricted to the northwest edge of the Canning Basin adjacent to the continental margin of northwest Australia. Igneous intrusions of a similar age are known from the northern Carnarvon Basin and these, together with the Canning Basin intrusions, are thought to be related to the initiation of tensional tectonics at the beginning of rifting (Reeckmann and Mebberson, 1984).

The final best values of the parameters determined using tomography are listed in Table 9.2. Despite the similarity in the burial histories of the two wells, the heat flow variations are somewhat different (Fig. 9.6). A heat pulse was needed between 250 and 350 MYBP for Kambara-1. This is coincident with the dating of the igneous intrusions. Although Kambara-1 is 3 km away from the nearest mapped igneous intrusion, the existence of a widespread heating event was suggested by Reeckmann and Mebberson (1984). For Curringa-1, which is 26 km from the nearest mapped intrusion (Gleadow and Duddy, 1984), this thermal event is not evident in the apatite tracks even with the use of thermal indicator tomography. Whether the thermal event was not felt by Curringa-1 or whether it is beyond the resolution of the technique is unknown. Results from two other wells in the basin which are penetrated by intrusive bodies have shown very pronounced effects of the heating event (He and Lerche, 1988).

Nevertheless, the combination of the parameters (Table 9.2) determined from thermal indicator tomography gave very good fits to the data (Figs. 9.7 and 9.8). The χ^2 values for Curringa-1 and Kambara-1 are 0.335×10^{-2} and 0.80×10^{-2}, which, when converted to the same scale, are improvements by about two to three orders of magnitude over the

Table 9.2 Tomography Results for Curringa-1 and Kambara-1, NW Canning Basin

χ^2	$T_c \times 10^{-3}$ (K)	$k_* \times 10^5$ (Myr)$^{-1}$	β (Myr)$^{-1}$	a_1	a_2	a_3
Curringa-1, 0.349×10^{-2}	7.3	0.14	0.66×10^{-3}	0.091	0.007	−0.020
Kambara-1, $0.7996 \times ^{-2}$	7.0	0.13	0.35×10^{-5}	0.69×10^{-2}	−0.20	0.59×10^{-1}

Note: $k_* = k_1 \exp(-T_c/273)$.

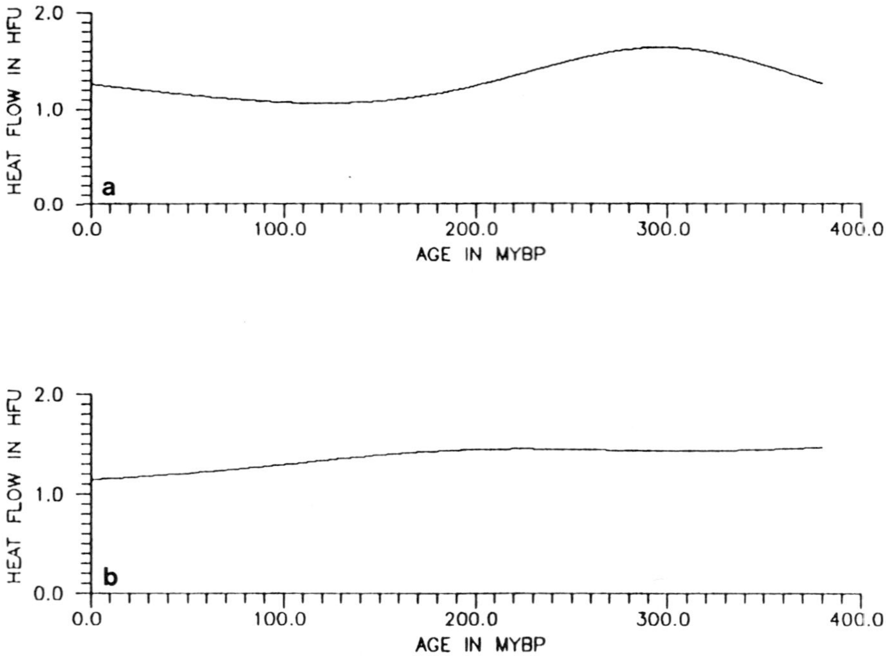

FIGURE 9.6 Heat flow versus geological age for (a) Curringa-1, and (b) Kambara-1, determined using length distribution of fission scar tracks as a thermal indicator in tomography.

results obtained by Huntsberger and Lerche (1987). However, much of the improvement is due largely to the better-defined chemical constants that describe the thermal behavior of fission tracks. As a result, the heat flow history determined using better-defined parameters is certainly more reliable, as described in the next section.

C. Resolution

It is obvious that the resolution limit from the data has not been reached when a systematic mismatch is observed. As discussed in the case of Inigok-1, varying only the unconformity size and the β value could not make up the mismatch. Some other parameters must have caused the offset (unconformity age and/or a heat pulse) and could be determined. It is also clear that the data do not provide enough information for a parameter to be examined when an expected change (such as a "kick" in the trend of vitrinite data caused by an unconformity) is not observed or is buried in the scatter of the data. But the question still exists: when the

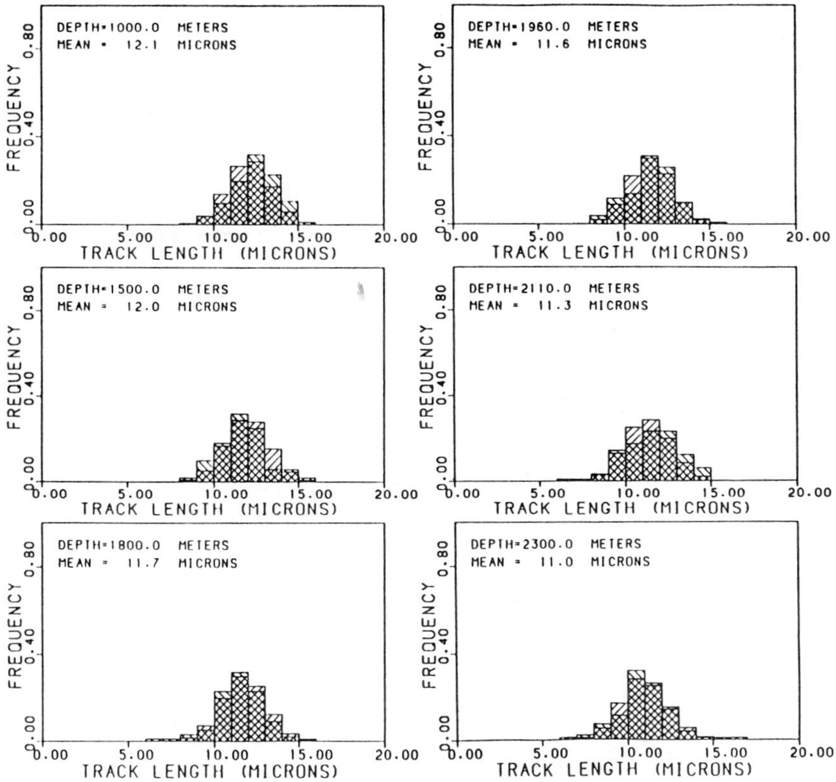

FIGURE 9.7 Theoretically predicted (hashed at 45°) and observed (hashed at −45°) track length distribution with depth at the present day for Curringa-1. The heat flux variation is shown in Fig. 9.6a.

"best" parameters are found through tomography and no further improvement is possible by adding more variables, what is the uncertainty associated with each parameter? To examine this problem, we vary one of the variables step by step while holding all others fixed at their optimal values to see how much the mismatch (χ^2) will increase when the single parameter deviates from its best value. The results of such an investigation are shown in Figs. 9.9 through 9.11 for Inigok-1. The χ^2 versus T_D curve gives a best T_D of 204 K, consistent with the findings of Lerche *et al.* (1984) on about 300 wells, which yielded a T_D of 200 ± 50 K. However, the T_D value is less (more) confined to the higher (lower) side of 200 K in this well. It appears that all the heat flow parameters are well confined (Fig. 9.10), except that a broad range of a_2 is observed where χ^2

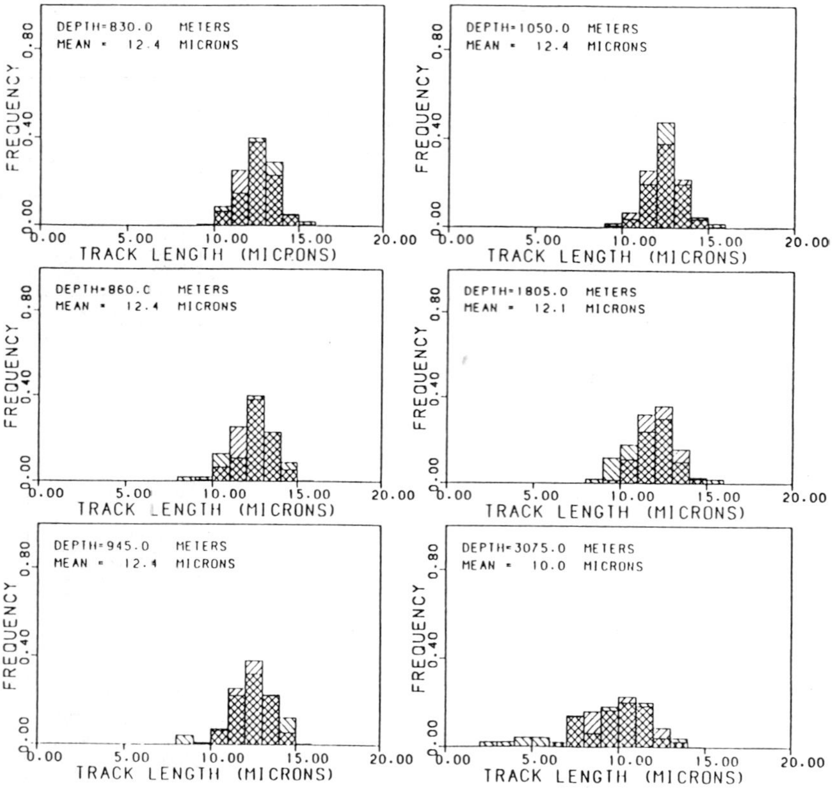

FIGURE 9.8 As for Fig. 9.7 but for track length distribution data for Kambara-1. The heat flux variation is shown in Fig. 9.6b.

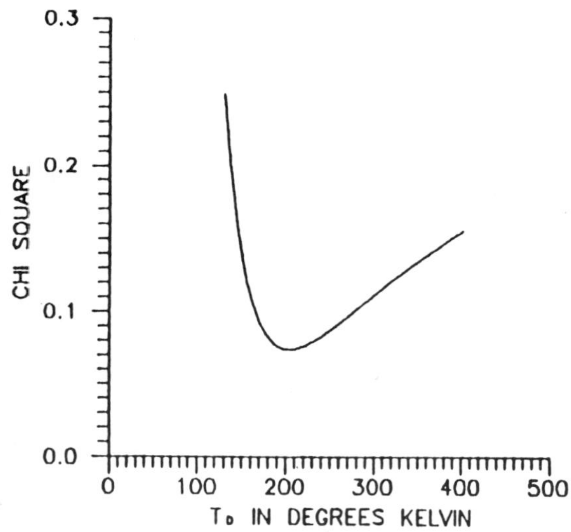

FIGURE 9.9 χ_2 versus T_D for Inigok-1, Alaska.

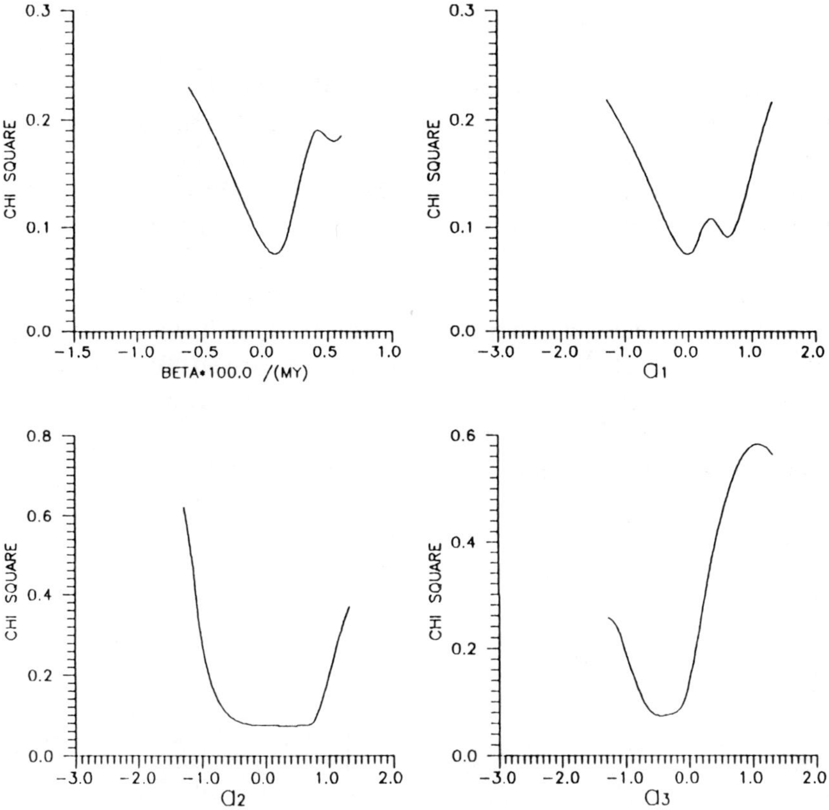

FIGURE 9.10 χ^2 versus heat flow parameters in equation (9.37) for Inigok-1, Alaska.

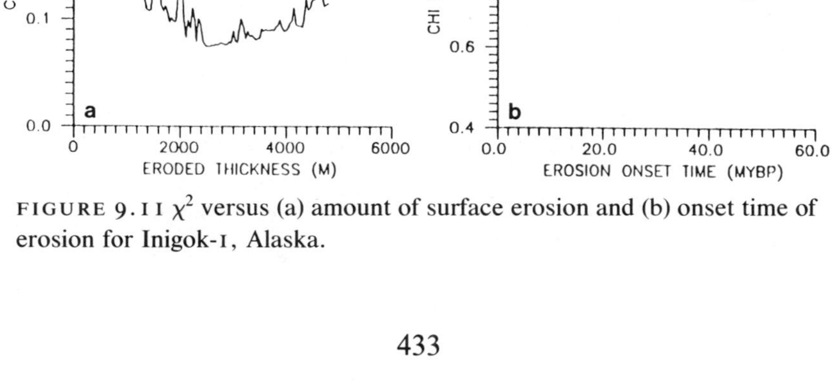

FIGURE 9.11 χ^2 versus (a) amount of surface erosion and (b) onset time of erosion for Inigok-1, Alaska.

has a "flat" valley. But an examination of the heat flow equation with different values of a_2 within the valley does not change the overall shape of the heat flow curve (Fig. 9.5). Therefore the resolution of the data is high enough to define the heat pulse. As for the burial history parameters, the curve of χ^2 versus unconformity size (Fig. 9.11a) indicates that a larger or smaller amount of erosion will not only increase χ^2 but also influence the stability of the fluid flow/compaction process. By contrast, when the time of onset of the erosion is varied around its best value (Fig. 9.11b), only an age of about 20–50 Myr could be assessed. Because an age of 30 Myr was chosen when applying the linear search run, the improvement of fit was due mostly to a more precise depiction of the paleoheat flux variation compared with a linearly varying pattern.

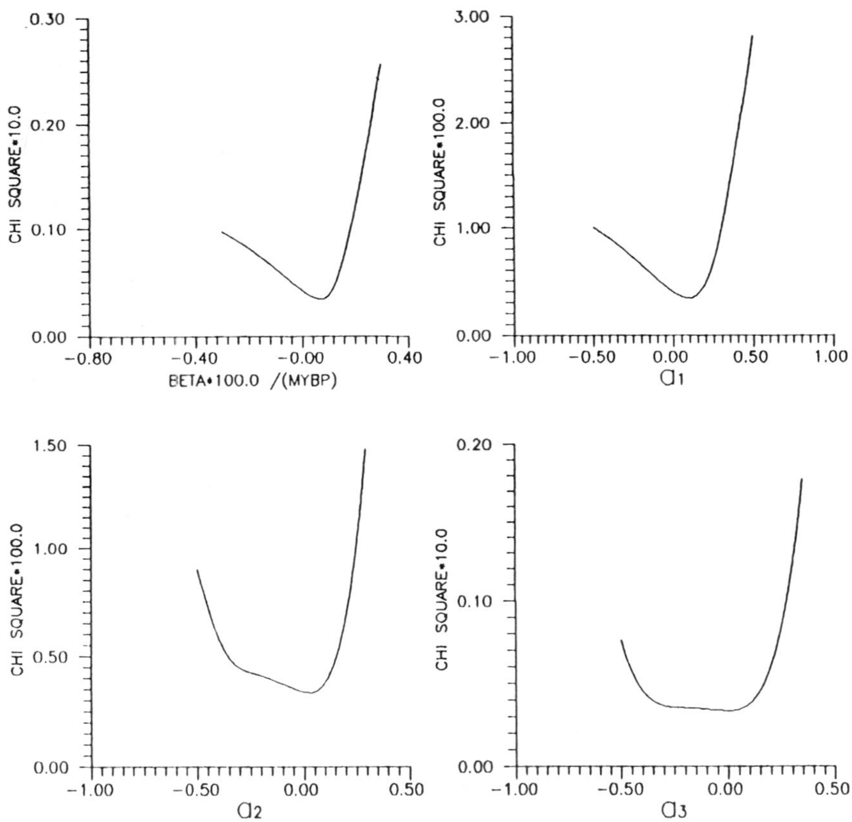

FIGURE 9.12 χ^2 versus heat flow parameters in equation (9.37) for Kambara-1, NW Canning Basin.

When this uncertainty analysis was applied to the apatite distribution method, the uncertainties in the heat flow parameters were even smaller (Figs. 9.12 and 9.13; note the finer scales of the abscissa as compared with those in Fig. 9.10) compared with those for Inigok-1 and almost every parameter has a well-developed minimum, suggesting a higher sensitivity of the apatite track distribution as a thermal indicator. Furthermore, the significant resolution in the chemical parameters k_1 and T_c (Figs. 9.14 and 9.15) and their consistency over eight wells in three basins (see also Table 9.3) imply a smaller uncertainty in the paleoheat flow and geological parameters determined by thermal indicator tomography. The strong suggestion is that the apatite track distribution is one of the best thermal indicators available at the present day. Fundamentally, this suggestion is made because the change of track length distributional shape with depth

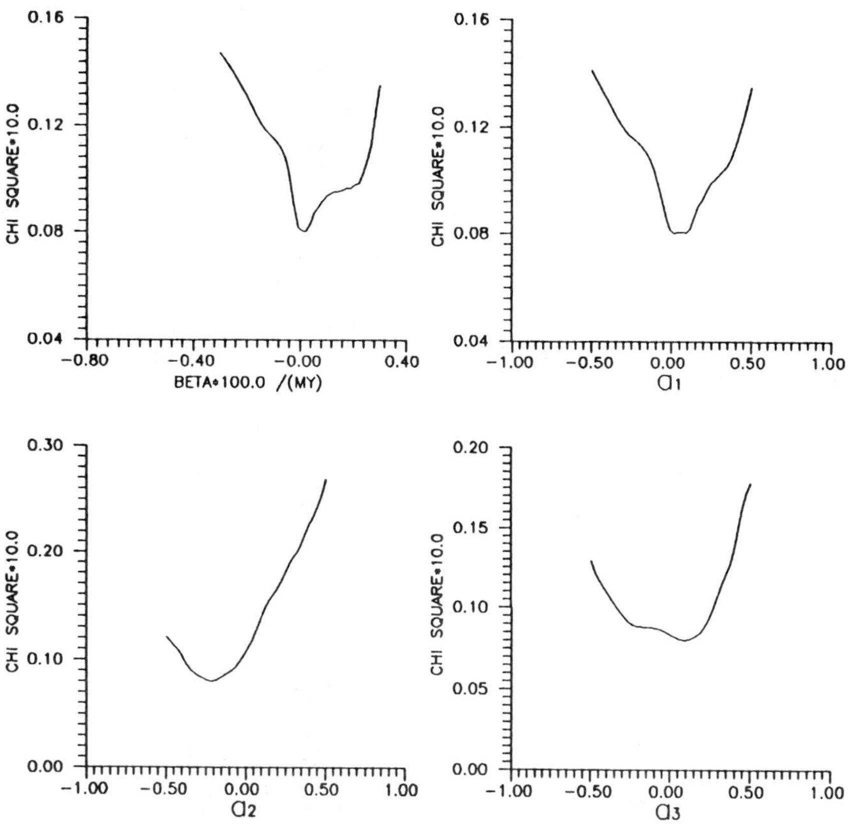

FIGURE 9.13 χ^2 versus heat flow parameters in equation (9.37) for Curringa-1, NW Canning Basin.

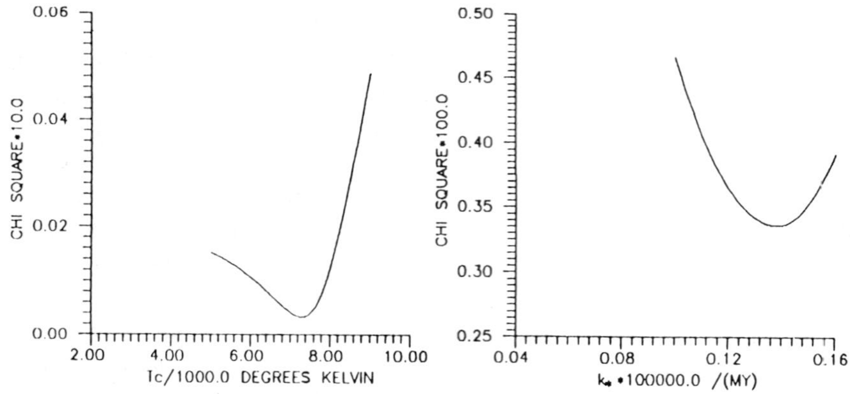

FIGURE 9.14 χ^2 versus T_c and k_* for Kambara-1, NW Canning Basin.

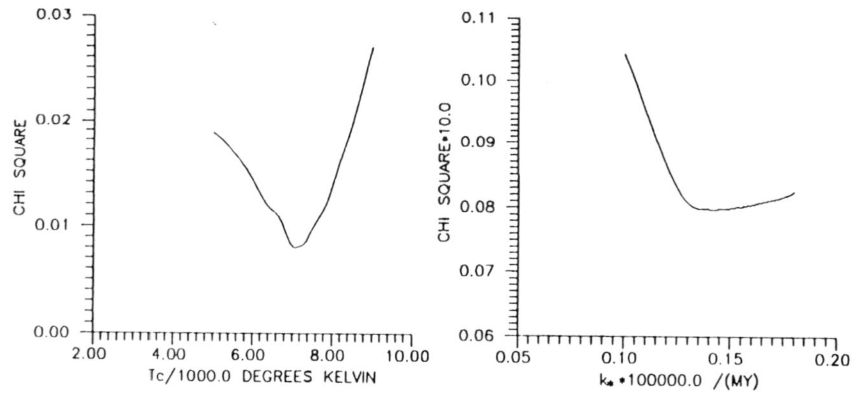

FIGURE 9.15 χ^2 versus T_c and k_* for Curringa-1, NW Canning Basin.

Table 9.3 Optimal k_* versus T_c for Different Wells

Well name	$k_* \times 10^5$ (Myr)$^{-1}$	$T_c \times 10^{-3}$ (K)
Curringa-1	0.136	7.3
Kambara-1	0.134	7.0
Fraser River-1	0.144	7.3
Perindi-1	0.145	7.5
x_1	0.141	7.1
x_2	0.137	6.7
x_3	0.143	7.4
x_4	0.144	7.4

Note: $k_* = k_1 \exp(-T_c/273)$.

contains a degree of sensitivity sufficient for us to determine multiple parameters, leading to a more accurate assessment of the geological and thermal history, and thus a more accurate prediction of maturation of organic matter when these parameters are used.

D. Summary

We have exhibited applications of a tomographic method which is capable of determining multiple geological parameters simultaneously, with the use of vitrinite reflectance and apatite fission track length distribution as thermal indicators. In the cases presented here, six or seven parameters were determined, including several chemical parameters which characterize the thermal and temporal behavior of the thermal indicators.

In the case of Inigok-1, the amount of erosion at the surface was bounded by about 2900 ± 1000 m, while the age when the erosion began is in the range 20–50 MYBP. However, the better fit to the data compared with earlier results can be attributed partly to the improved delineation of the heat pulse during the late Cretaceous.

The chemical parameters involved in fission track annealing were better defined with $T_c = (7.2 \pm 0.5) \times 10^3$ K and $10^5 k_* = 0.14 \pm 0.01$ $(\text{Myr})^{-1}$ from the data available. The heat flow variations with time determined for the two NW Canning Basin wells are consistent with the findings of earlier investigations concerning the igneous intrusions in the basin. However, it should be noted that the uncertainties associated with each parameter may actually be larger than determined by the simple sensitivity analysis given here.

While the results reported are encouraging, the sensitivity analysis needs further development and more resolution studies thermal indicator tomography as a sharp investigative tool. In addition, further applications and developments involving multiple indicators in the same well, "blind" tests, igneous intrusives, and overthrust timing needed to be investigated.

10

Morphologic Development of a Sedimentary Basin: Sea Level Variations and Clastic Sediment Loading Models

I. The Rise and Fall of Eustatic Sea Level

The evolution of a sedimentary basin's gross morphology depends on three principal factors: basement movement, sediment accumulation and compaction, and variations of eustatic sea level with time. Here we explore some of the various ways that have been proposed to determine eustatic sea level variations from present-day observations. We find that, without fail, each and every proposed scheme, including our own work, *assumes* some behavior for two of the three underlying processes: tectonic movement of the basement, sedimentary accumulation, and eustatic sea level. Each scheme then determines the third process relative to the assumed model behavior of the other two.

We quantify the problem of attempting to determine eustatic sea level by a detailed discussion of (1) the coastal onlap technique, (2) the stacked crustal subsidence technique (3) Pleistocene sea level movements based on reef terrace positions, and (4) the deep-ocean $\delta^{18}O$ isotope technique and (5) by showing mathematically that, at best, it is possible to obtain only the *sum* of tectonic basement subsidence and sea level variations from onlap methods and, at worst, not even that simple a combination, and (6) by building a numerical simulation of basin fill and using previously proposed techniques to "invert" the problem—with surprising results.

In short, all proposed techniques to date for estimating the rise and fall of eustatic sea level with time have enjoyed brief rises of popularity before their flaws were noticed sufficiently to cause their fall and wane.

It would seem that, like King Canute, we cannot yet obtain a control on the sea.

A. Importance of Eustatic Sea Level to Stratigraphy and Hydrocarbon Exploration

In this part of the chapter we discuss the significance of the use of eustatic sea level in unraveling basin stratigraphy and predicting hydrocarbon potential. We show how sediment fill of basins responds not only to eustacy but also to tectonic movement and to rates of sediment accumulation. We then review a variety of different systems for measuring the size of eustatic sea level excursions and discuss why none of these systems work, since they require the assumption of the size of other independent variables such as rates of subsidence and sediment accumulation. We conclude the section with a discussion of the mathematical relationships of the different variables needed and why no unique dimension can be determined for the size of eustatic events.

Eustatic sea level events are important to understanding basin history for a number of reasons. They represent unique singular events which, because they can be recognized in basins and worldwide, can be used by the geologist to date not only the sedimentary fill of specific basins but also the fill from basin to basin. They are also in part responsible for modifying depositional environments through time, and so changes in fauna (Hallam, 1984). Thus, while sea level changes do not always trigger evolution, they probably punctuate it, so that a smooth change in fauna through time is not always seen because the sedimentary record is interrupted by its response to eustacy, improving the role of fauna as good chronostratigraphic markers. Identification of eustatic events can then be used to separate them from *relative* sea level changes caused by other effects, such as local tectonic events and variations in the earth's rotational axis. Also, should the rate of change of sea level be more rapid than thermotectonic movements, the size of the resulting accommodation for sediment fill can be ascribed to that sea level change and can be used to predict the location, size, and geometry of source rock, reservoir, and seal. Finally, if the size of eustatic event were known, it would help us to separate and determine how much, if any, of the crust's flexural response is related to sea level change and how much is caused instead by thermotectonic subsidence.

Even if we cannot determine the exact dimensions of eustatic sea level excursions, sediment compaction, crustal flexural strength, and thermotectonic behavior, we can set limits on these variables and so construct an approximate simulation of the sedimentary fill of basins. Such simulations have been attempted on only a limited basis to date, but the results show that simulation is useful to exploration, not as a direct locator of prospects but as a stratigraphic tool which enhances our knowledge of the seismic stratigraphy of potential acreage. For instance, such a simulation can duplicate sediment geometry during periods of over-the-shelf-edge sedimentation; it can maximize the integration of geological, geophysical, and paleontological data; and it can reinforce interpretations made from internal seismic reflections. The data required can be derived from conventional sources such as biostratigraphic analysis, regional mapping, examination of core samples, well logs, and seismic crosssections. Such a simulation can be tied directly to oil and gas exploration and is a dynamic framework on which to superimpose depositional models to account for unconformities and lithological variation in vertical sequences and to highlight the far-reaching effects of worldwide sea level events.

B. Major Variables Responsible for Gross Sediment Geometry

The stratigraphy of the sediments that fill a basin is dependent on the source and depositional setting of the sediments, the role tectonism plays in controlling basin fill, and the influence of eustasy. It is the difficulty in separating these three influences that has plagued sedimentary stratigraphic interpretation and makes measuring the magnitude of eustatic sea level so difficult. Suess (1906), Stille (1924), Grabau (1936), Kuenen (1939), Umbgrove (1939), Bally (1981), Watts (1982), and Hallam (1984) are among many geologists who have confronted the problem and have catalogued how, over the last 80 years, different people have tried to solve it.

The most pervasive component of stratigraphic sections appears to be the "recurrent asymmetric sedimentary sequence." The fining-upward "red bed" cyclothems of Allen (1964), the carbonate tidal flat cycles of James (1979), the clastic tidal flat cycles of Klein (1970) the deltaic cycles Horne and Ferm (1976) and Walker and Harms (1971), and submarine sequences are perceived as examples of this behavior. The problem is how one interprets these cycles. Some, like Horne and Ferm (1976), have argued that the cyclicity is largely a response to the lateral migration of depositional settings; others, like Goodwin and Anderson (1985), favor eustacy; and still others, like Fischer (1964), favor a tectonic cause.

The problem of resolving these different controls on sedimentary geology can best be demonstrated by trying to establish how a basin margin developed. For example, in Fig. 10.1 we have taken a slightly modified cross-sectional diagram by King (1948) of the sediment geometry displayed at the margin the Delaware Basin by Permian Upper Guadalupian carbonate and clastic sediments in west Texas. We have cut up this

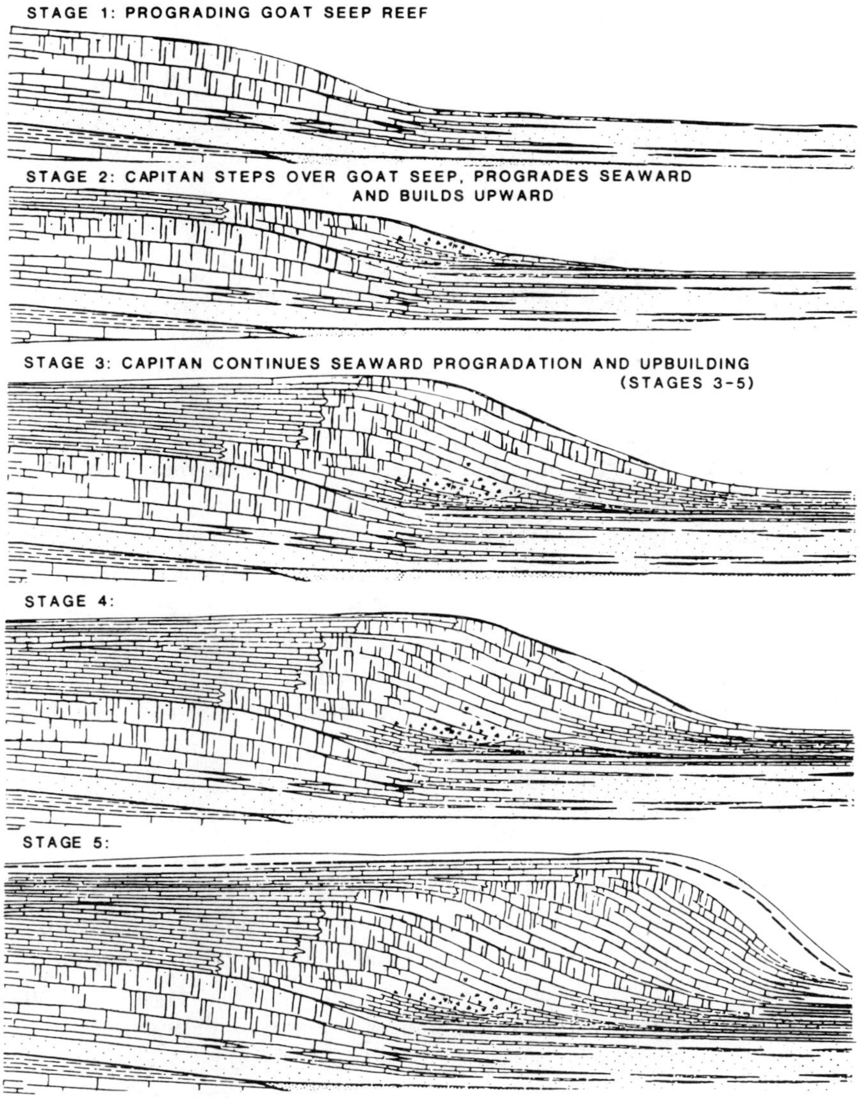

FIGURE 10.1 Sediment geometry displayed by the Permian Upper Guadalupian margin of the Delaware Basin in west Texas and its evolution through time.

cross section in an attempt to see how the sediment geometries evolved. At stage 1 the margin began as a gentle ramp on which a massive fine-grained carbonate bank edge progressively built up and moved seaward over itself, while simultaneously interfingering seaward with basin clastics and shelfward with shoaling-upward cycles of carbonates and clastics that are dominantly carbonate close to the margin but become more clastic landward. At stage 2 the margin suddenly steps basinward with backreef carbonates lying abruptly over what was once the basin edge, so that while the width of the belt in which the massive basin margin sediments were deposited narrowed, the margin prograded gently seaward. The backreef maintained its general position while the margin prograded over a wedge of basin clastics. The same steplike jump seaward can be seen in stages 3, 4, and 5.

The problem facing the stratigrapher is to explain the geometries and facies we see in this geological cartoon. Consider three different hypotheses that account for the observations.

1. Hypothesis 1

If we subscribe to the belief that depositional environments migrated along depositional strike, the cause of both the shoaling-upward cycles and the steplike progradation could be a response to the site of active carbonate sedimentation on the shelf laterally shifting its position, while away from localized sites of active carbonate sedimentation, clastics bypass into the basin is achieved through dense lagoonal brines carrying the clastics across the shelf into the basin. Subsequently, the carbonate factory spasmodically returns to its original position, producing shoaling-upward cycles and local progradation. This hypothesis may strike some as apt, but how do we prove it?

2. Hypothesis 2

Another possible hypothesis is that the tectonic subsidence which created the "accommodation" for the sediment of the shelf and immediate margin did not change at an even rate but instead accelerated and decelerated periodically. Slow acceleration of tectonic subsidence would initially have caused the shelf carbonates at stage 1 to maintain their general slow progradation, but a rapid slowing in subsidence would have resulted in a reduction of accommodation for sediment by stage 2. Rapid progradation would have occurred until another abrupt acceleration phase in subsidence slowed the progradation and so on. This also provides a hypothesis which might be advocated but, again, how do we prove it?

3. Hypothesis 3

A third hypothesis is that of a eustatic control. In this case, at stage 1 the observed progradation can be explained as taking place during a moderate eustatic sea level rise, while the adjacent basinal clastics were deposited during a rapid sea level fall which also provided the conditions for eolian transport of the sands over the shelf into the basin. The following rapid seaward stepping of the shelf carbonates of stage 2 could have been the result of a rapid sea level rise during which the coast retreated, carbonate sediments onlapped the shelf, and a condensed sequence formed offshore. This rapid event was followed by a very slow sea level rise which produced little accommodation, so that seaward stepping of the shelf resulted, and so on. Again, this provides an attractive hypothesis, but how do we prove it?

Supporting evidence can be found for each of the models to explain the geometries observed in the Delaware Basin margin, but no conclusive proof seems to be available for any one of them. The causes of the cycles we see in sediments in most cases remain unresolved; interpretation is very much a response to geologic fashion rather than to definitive data. It would seem that no matter how persuasive geologists are with their interpretation of eustacy, their hypotheses which claim to resolve the size of eustatic events cannot be proved. If stratigraphy is to become more interpretative and is to move further from the stage described so eloquently by Krynine's statement, "Stratigraphy can be defined as the complete triumph of terminology over facts and common sense" (circa, 1950; J. Ferm personal communication, 1985), at the least we need to be able to resolve the size of (1) sea level excursions, (2) tectonic movement, and (3) rates of sedimentary accumulation. Once we have determined the size of one of these variables, we stand a chance of establishing the size of them all. Then we could determine how a basin filled and what causes sedimentary cyclicity. From there we could predict more accurately the character of the basin fill away from described areas. An accurate eustatic sea level variation chart would be a boon to frontier basin exploration in the oil industry, where seismic cross sections and occasional stratigraphic wells and/or wildcats are the only data sources. Using eustatic sea level charts tied to wells, we could project eustatically related sedimentary sequences across a basin on seismic cross sections. We could then assess the potential of finding, say, sand-prone versus shale-prone versus carbonate-prone sections, the objective of so many hydrocarbon exploration companies.

C. Some Methods for Measuring the Magnitude of Eustatic Sea Level Excursions

Several methods for determining the size of eustatic sea level have been proposed by different authors. These include:

1. Sediment Aggradation and Onlap Curves from Seismic Data

Vail *et al.* (1977, 1984) and Hardenbol *et al.* (1981) emphasized the importance of recognizing eustatic events and developed one of the best-known systems for attempting to determine the size of these events. They provide two different methods: (1) examination of seismic cross sections and determination of the position of onlapping sedimentary wedges of marine, coastal, and alluvial sediments on the continental margin as a function of time, and (2) measurement of the size of perturbations on crustal subsidence curves produced from well data.

Vail *et al.* (1977) showed how to recognize the effects of eustatic sea level events by using seismic cross sections. Their technique for identifying seismic sequences assumes that continuous seismic reflectors on acoustic geophysical cross sections are close matches to bedding planes and unconformities. The unconformities that bound the sequences can be identified, as can be seen on Fig. 10.2, by recording where the seismic reflectors onlap and terminate either against the lower unconformity surface or against each other. The argument here is that the position of the onlapping seismic reflectors is controlled by the base level of the mean high-water mark. Thus a sediment encroachment chart can be drawn that shows how far the wedge of submarine, coastal, and alluvial sediment has onlapped the basin margin (Vail *et al.,* 1977). A sediment aggradation chart can also be constructed which shows the amount by which the vertical component of each onlapped seismic reflector has climbed or fallen (Fig. 10.2c) (Vail *et al.,* 1977). Vail *et al.* are careful to point out that in using this method, "The measurements of coastal aggradation are made as closely as possible to the underlying unconformity to minimize the effect of differential basinal subsidence." Vail *et al.* (1977) then correlate the regional cycles of relative changes of sea level at specific locations with those at other locations, constructing charts that record the occurrence of global sediment onlap cycles. Using the aggradational measurements from the seismic results, Vail *et al.* (1977) estimated the size of the relative sea level excursions. However, as Hardenbol *et al.* (1981) pointed out, "quantifying eustatic sea-level changes from measured changes in coastal onlap does not provide an accurate measure, because of variations

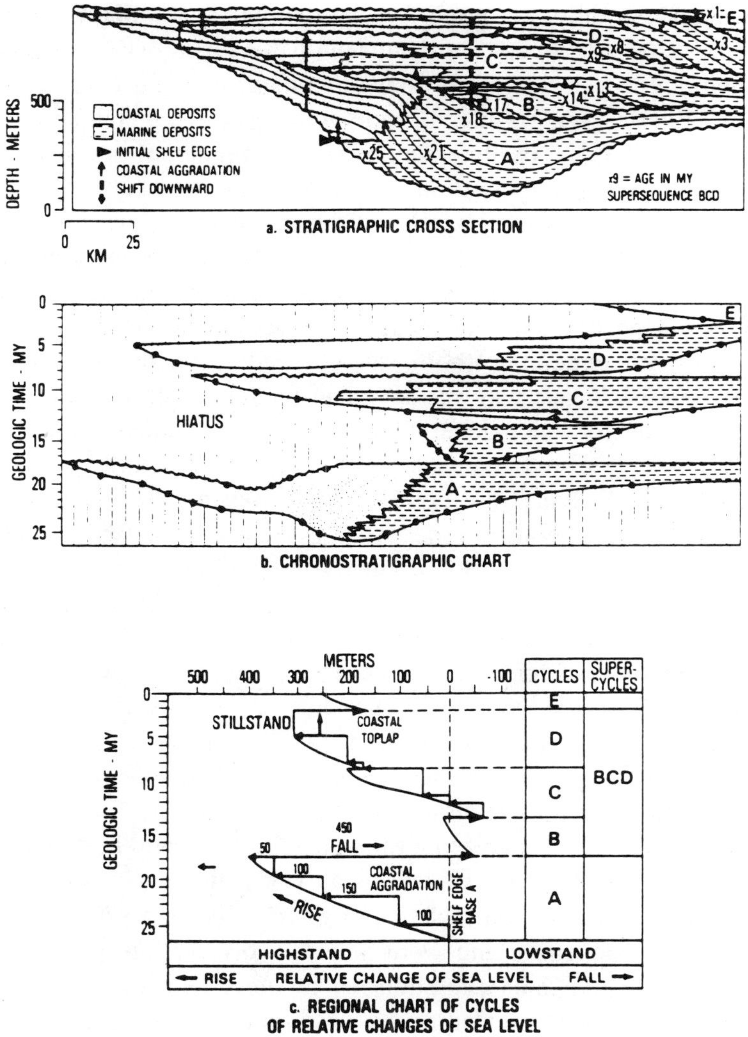

FIGURE 10.2 (a) Diagrammatic stratigraphic cross section across a hypothetical seismic cross section, showing distribution of sequence boundary types, downlap surfaces (condensed sections), and general facies of idealized sequences in depth. (b) Chronostratigraphic chart or Wheeler diagram identified in part a, showing nature of sequence boundaries, downlap surfaces and condensed sections, and facies. (c) Chart of cycles of relative changes of sea level constructed from part a. Note arrows of coastal aggradation transferred to part c. (After Vail *et al.*, 1984.)

in subsidence in different basins.'' As we show later in this chapter, not only do crustal subsidence and compaction vary from basin to basin but also their effects can exaggerate sediment onlap curves by as much as five or more times. Thus, while this method appears to identify eustatic events, it cannot be used to determine magnitude, or position, with respect to the sea level variations that bound the events.

Hallam (1984) questions whether seismic reflectors are close enough approximations to bedding plane surfaces and so to chronostratigraphic surfaces or isochrons that extraction of eustatic events can be undertaken. Although it is true that bedding planes and so seismic reflectors must transgress time, they both can be used to subdivide strata into relative time packages, with the sediment lying beneath a bedding plane or seismic reflector being generally older than the sediment above. This concept has been tested and used by hundreds of oil company geologists and geophysicists every day tying synthetic seismic traces constructed from sonic logs (Sengbush *et al.*, 1961; Gardner *et al.*, 1974) to seismic cross sections. They identify chronostratigraphic horizons on the sonic logs and tie these to reflectors on the seismic cross sections, tracing the stratigraphy from well to well (Vail *et al.*, 1977). Gairud *et al.* (1978) used such a system to determine the Tertiary stratigraphy across the Jan Mayen Ridge in the Norwegian–Greenland sea, using multichannel seismic and Deep Sea Drilling Project (DSDP) wells. In the same way, Sichlee and Fritsch (1983) used seismic lines and COST wells to begin unraveling the seismic stratigraphy of the Georges Bank Basin. Though they were unable to recognize all the unconformities and hiatuses on the seismic lines that they recognized in the wells, they were able to trace some. Von Rad and Exon (1983), using well data and seismic results, dated seismic sequences and were able to work out the paleoenvironmental and geodynamic evolution of offshore Northwest Australia. Similarly, Willumsen and Cote (1983), using many wells tied to regional seismic lines, worked out the Tertiary sedimentary history of the southern Beaufort Sea. Harding and Lowell (1983), Hamberg (1983), Curnelle and Marco (1983a,b), Phelps and Roripaugh (1983), Kirchner *et al.* (1983), Suzuki (1983), Rafavich *et al.* (1984), Dimian *et al.* (1983), Erxleben and Carnahan (1983), Thomson (1983), Stone (1983a,b), Pieri (1983), Bachmann and Koch (1983), Norton (1983), Fox (1983), Harding *et al.* (1983), Gamboa *et al.* (1985), Harding (*1985*), and Crain *et al.* (1985), also use the same method and tie wells to seismic reflectors to interpret the seismic stratigraphy and structure of subsurface. Van Hinte (1983) accepts, in general, that seismic reflectors can be equated to isochrons and goes even further to use well data to produce synthetic seismic sections by projecting biostratigraphically controlled markers or isochrons from well to well, as if they were seismic reflectors, as "isochrons" to model "pseudo wells" at common depth points (CDPs). The "dated" reflectors provide the pseudowell stratigraphy so that "seismic" geohistory analyses of basins can be carried out by backstripping the sedimentary fill through time.

We believe that Hallam's (1984) claim that evidence is unavailable to test the concepts of Vail *et al.* (1977) is not substantiated by the record, since seismic and well data sets are available from the Norwegian Oil Directorate for Norwegian waters, from the U.S. government for the National Petroleum Reserve of Alaska, and from the American Association of Petroleum Geologists (AAPG) for the East Coast of the United States (Buffler *et al.*, 1978) and the Gulf Coast (Watkins *et al.*, 1976) and are routinely handed out at AAPG seismic stratigraphy schools. By using these sets, most of the ideas suggested by Vail *et al.* (1977) can be examined. In fact, Thorne and Watts (1984) use just such data to show that some of the concepts in Vail *et al.* (1977) are not always valid.

2. *Differences between Crustal Subsidence and Thermotectonic Curves*

Hardenbol *et al.* (1981) and Vail and Todd (1981), from estimates of "changes in coastal onlap and from paleontologic studies," report that the magnitude of low-frequency eustatic events can be assessed by measuring the difference between crustal subsidence curves calculated from wells and the theoretical thermotectonic subsidence curves postulated for that location using the approach outlined in Royden *et al.* (1980).

Essentially these crustal subsidence curves are obtained for a well by (1) determining its burial path from the datum of sea level (Fig. 10.3) (the geohistory plot of van Hinte, 1978) and then (2) removing the effects of compaction and the effect of sediment and water weight on the crust in the manner of Watts and Steckler (1979).

To illustrate, let us consider a sedimentary section of height s and matrix grain density ρ_s, in which porosity depends on depth of burial z in a manner described by the function $\phi(z)$. The sedimentary section is overlain by a water column of height W_d (Fig. 10.4). In general, the quantities ρ_s and $\phi(z)$ are lithology dependent. We now ask what the depth of the basement, Y, below sea level would be if the sedimentary

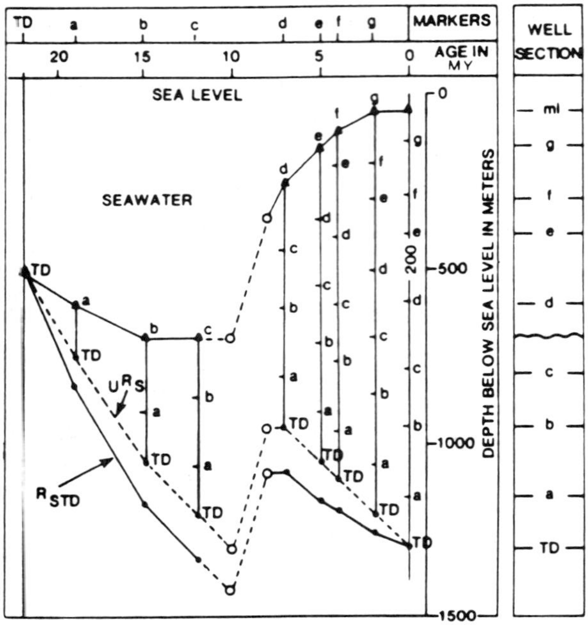

FIGURE 10.3 Geohistory diagram for hypothetical well (after van Hinte, 1978). Upper curve shows paleowater depth at this locality. The dashed lower curve $_uR_s$ is the uncorrected burial path of TD through time, measured cumulatively from the water–sediment interface; the solid lower curve R_{STD} is the corrected burial path of TD which incorporates the progressive compaction of the overlying sediment as TD is buried.

column were removed. By requiring balance of the weights of the overlying columns at the depth of compensation, Airy (1855) isostasy yields the following expression for Y:

$$Y = W_d + \int_0^s \frac{\rho_m - \rho_s}{\rho_m - \rho_w}\, dz + \int_0^s \frac{\rho_s - \rho_w}{\rho_m - \rho_w}\phi(z)\, dz \qquad (10.1)$$

where ρ_m represents the mantle density. We use this equation to estimate basement subsidence compensated for the influence of sedimentation but ignore the effect of sea level fluctuations on this subsidence. We also tacitly ignore nonisostatic components of response such as flexural plate effects.

In equation (10.1) the quantity s represents the total height of the sedimentary column at a given geologic time. However, information for the entire column is commonly not available and reconstructions are confined to the part of the column penetrated by the borehole. A

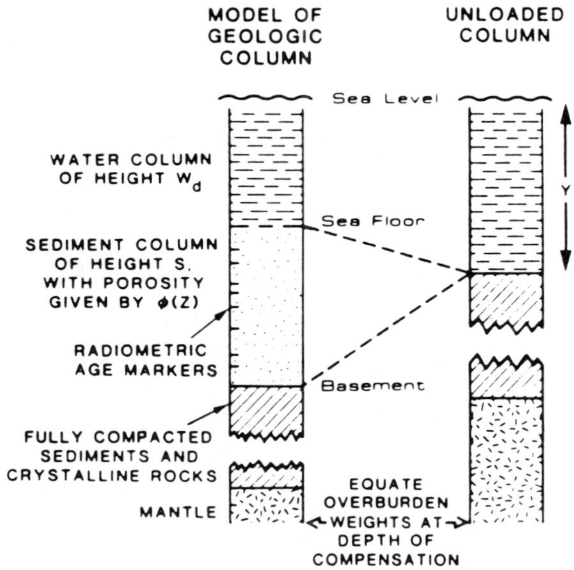

FIGURE 10.4 Watts and Steckler (1979) backstripping diagram.

consistent prediction of depth to basement may be achieved by extrapolating the porosity function ϕ below the total depth of the well to the depth at which the rocks are considered to be fully compacted. In our work, for instance, we used a model in which porosity decreases exponentially with depth, and we define "effective basement" as the sedimentary body that has a calculated porosity of 1% at the depth at which it is buried, when all sediments down to the total depth of the well have been backstripped. Thus, effective basement always has a porosity less than or equal to 1%.

Lack of information about sedimentary formations below this total depth results in an uncertainty in the decompaction of the sedimentary column during the backstripping calculation. This, in turn, yields an uncertainty in the estimate of basement subsidence, compensated for by sedimentation. Model calculations show that this effect typically changes basement subsidence integrated over the total depositional time, by about ± 70 m (± 230 ft), which is significantly less than typical basement subsidence after compensation for sedimentation but comparable to estimates for the rise and fall of eustatic sea level!

Results obtained from the burial history calculation are illustrated in Fig. 10.5, which shows the reconstructed depositional history and basement subsidence curve of a sample North Sea well. These basement subsidence curves are then to be matched to the model thermotectonic

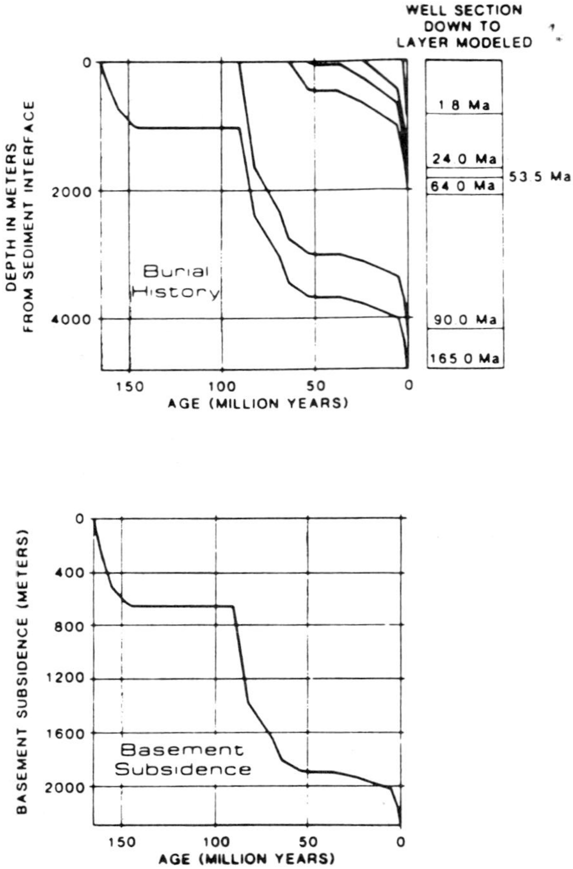

FIGURE 10.5 Burial history of North Sea well plus crustal subsidence.

curves and the difference is described as providing the size of the sea level excursion.

However the hypothetical thermotectonic curves differ from author to author. For instance, McKenzie (1978) models subsidence curves by assuming that when rifting occurs the continental lithosphere stretches rapidly horizontally and thins the lithosphere in a predictable way. In response, the hot asthenosphere wells up and then cools conductively and predictably, causing subsidence until thermal equilibrium is reached.

In McKenzie's model the lithosphere is assumed to be a single slab with a constant temperature at its base and the system is assumed to be in isostatic equilibrium at all times. The surface of the continent is taken to be at or below sea level, so that water occupies the entire volume created by subsidence. Contributions from radioactivity of crustal rocks are ignored and two-dimensional effects are neglected entirely. The contribution of sediment loading to overall subsidence is removed. The equations for this model and the physical parameters used are given in McKenzie (1978).

Hellinger and Sclater (1983) have a more complex model which uses many of the assumptions of McKenzie (1978) but involves modeling the stretching and thinning of two layers instead of one. As before, isostatic compensation is maintained by the upwelling of the hot asthenosphere, which then cools and subsides until thermal equilibrium is reached. The thermotectonic subsidence curve produced is different from that of McKenzie's (1978) model.

In contrast, the large-scale dike intrusion model of Royden *et al.* (1980) is constructed by assuming hot material from the asthenosphere intrudes into the continental lithosphere through a series of vertical dikes like those exposed in swarms in east Greenland. Here the density of the intrusions ranges from 5 to 10 per kilometer to more than 50 per kilometer. Even if dikes were present at a spacing of 5 km, the thermal effects of intrusion are assumed to have equilibrated horizontally within a million years, which allows Royden *et al.* (1980) to neglect the horizontal component of heat conduction and consider only a simple one-dimensional problem. They assume the intruding dikes have predictable thermal properties and the effects of intrusion can be modeled so that as the fraction of the lithosphere composed of dike material from the asthenosphere cools, the crust subsides.

Models have also been developed by Falvey and Mutter (1981), Nunn *et al.* (1984), and Beaumont (1984). Each model is different from the next and was developed to find ways around the weaknesses in other models. If we are to measure the difference between crustal subsidence derived from a well and thermotectonic subsidence, then because the subsidence curves predicted by McKenzie (1978), by Royden *et al.* (1980), by Hellinger and Sclater (1983), and Falvey and Mutter (1981) are so different, we are forced to choose one. Once we do, the difference between the crustal subsidence curve derived from a well and the arbitrarily chosen thermotectonic subsidence curves is given by a least-square-difference system and this in turn leads to an estimate of the size of the sea level excursion.

The results of such a calculation are given in Hardenbol *et al.* (1981), who choose the model of Royden *et al.* (1980) and apply it to an example from northwest Africa. They state that "the stratigraphic resolution of these changes rarely allows exact quantification of their magnitude, but a minimum rate of change of sea level can often be determined." Even though Hardenbol *et al.* (1981) restrict themselves to low-frequency events, the assumption of a smooth thermotectonic subsidence curve is at best an approximation. In fact, many of the high-frequency perturbations may be tectonic in origin. Also, the fact that the parameters in the Royden *et al.* (1980) subsidence curves are chosen to provide a good fit to the crustal backstripped curves does not mean that Royden *et al.* (1980) model is the most appropriate model.

The error limits in this system are not estimated by the various authors but are thought to be sizable. Such uncertainties are related not only to the constants used in the different models but also to the compaction history during subsidence above and below total (target) depth (TD) and to the flexural behavior of the crust.

Bally (1981), Watts (1982), and Thorne and Watts (1984) demonstrate how matters can be arranged so that the coastal onlap curves of Vail *et al.* (1977) are produced by eustacy or widespread regional tectonic movement or both. The existence of eustatic events seems noncontroversial. However, whether one eustatic event stood higher on the continent than another is complicated by the effects of tectonic subsidence (Bally, 1981; Watts, 1982; Thorne and Watts, 1984) and explains why Hallam (1984) differs with Vail and Todd (1981) as to the relative position of some sea level stands.

3. Stacked and Averaged Crustal Subsidence Curves

Guidish *et al.* (1984) stacked and then averaged crustal subsidence curves to determine the occurrence and magnitude of eustatic events. However, when their global average rate of basement subsidence is compared with Vail's eustatic curves, there is little similarity in detail between them. This suggests that high-frequency tectonic movements show up more distinctly in crustal subsidence curves than do eustatic events because the sizes of the sea level excursions are too small to register much of an effect on the crustal subsidence. Even if the crust did respond to the higher-frequency sea level events, we could at best recognize their occurrence but not measure their size. This is because the limits in this system are related (1) to assuming an equilibrium isostatic burial history, (2) to modeling isostatic and flexural response to sediment and water weight, (3) to assuming how sediments compact below TD, and (4) to the stacking "error" produced by the problem of averaging. Once again, we have a system that cannot determine the size of eustatic changes.

4. Determinations of the Size of Eustatic Sea Level Excursions Associated with Pleistocene Glacial Stages

Fillon (1984) provides a very complete review of continental and marine Pleistocene stratigraphy and of how the two can be correlated and tied to eustacy. As Fillon points out, Agassiz (1840) probably inspired geologists to begin looking at the Pleistocene in terms of glaciation. Since then, geologists have recognized a series of different major and minor glaciations through the Pleistocene and have been using three principal geological sources to unravel Pleistocene stratigraphy and, in particular, the response to glacioeustatic events. These are the terrestrial tills and interglacial sediments dated by ^{14}C and K/Ar, isotopic records from marine benthic and planktonic microfossils dated by paleomagnetic stratigraphy, and geomorphologic evidence of sea levels based on U/Th-dated corals and ^{14}C dated in place peats (Fillon and Williams, 1983).

Perturbations on plots of $^{18}O/^{16}O$ ratios derived from microfossils from the deep sea can be dated and have been related to glacial eustatic events (Broecker and Van Donk, 1970; Shackleton and Opdyke, 1973; and Fillon and Williams, 1983). Similarly, Mesollella *et al.* (1969) and Steinen *et al.* (1973) have shown that eustatic sea level changes have left their record in the coral reef terraces that enveloped Barbados during its tectonic uplift in the Pleistocene. These terraces were dated using the U/Th method. Evidence for the same eustatic events can be seen in New Guinea (Chappell, 1974) and Indonesia (Chappell and Veeh, 1978) where, as in Barbados, tectonic uplift has exposed a series of reef terraces that mark these sea level events. As with Barbados, these reefs were dated by using U/Th.

The causes of the Pleistocene glacial events are unknown, but the events can probably be ascribed to the fluctuations in solar radiation received at the northern and southern hemispheres through time, as controlled by the periodic change in the tilt of the earth's axis every 41,000 years, the precession of the equinoxes every 26,000 years, and the changes in eccentricity of the earth's orbit every 90,000 years (Milankovich, 1941). Broecker (1966) and Berger (1976) have modified the original relationships described by Milankovich (1941) to emphasize the effect of precession on solar radiation. Ruddiman and McIntyre (1981) and Fillon and Williams (1983) used these modifications to model ice sheet formation in the Arctic Ocean and over continental land masses and believe that they can account for the enrichment of the $\delta^{18}O$ record. However, it is the size of the eustatic sea level excursions that we are concerned with here, not the recognition of glacial eustatic events or their causes.

Moore (1982) published a late Pleistocene eustatic sea level curve, showing sea level change dimensions through time, based on the dating of coral reef terraces in Barbados (Mesollela *et al.*, 1969), the Far East (Chappell, 1974; and Chappell and Veeh, 1978), and several other locations. The size of the sea level excursions is dependent on assuming that one of the events, that at 125,000 BP, was 2 to 10 m above present sea level and that the rate of tectonic uplift was constant at this datum between 125,000 BP and today. Chappell (1974) and Chappell and Veeh (1978) assume constant rates of uplift of 0.5 to 3 m/1000 years. In Barbados, Mesolella *et al.* (1969) assumed a rate of uplift of 0.2 m/1000 years. Both groups were then able to separate out the changes in sea level by using an assumed tectonic uplift model. However, as Ward and Chappell (1975) pointed out, it is impossible to determine the size of the tectonic movement, and to do so we first need to know the size of the sea level excursions. Thus, while the dated coral terrace positions are signals produced by the eustatic events, we cannot determine the size of the sea level events because we have to assume a model of tectonic behavior.

We face problems if we try to determine the size of sea level excursions from $\delta^{18}O$ variations in deep-sea cores. Fillon and Williams (1983) show two curves for the size of sea excursion, one from dated coral reef terraces and the other from $\delta^{18}O$. The two curves have the same frequency of signal but the amplitude of the events is different. Broecker and Van Donk (1970) were among the first to argue that sea level position, ice volume, and ^{18}O change are related. Shackleton and Opdyke (1973) assumed that a change of 0.1 per mil in $\delta^{18}O$ was equivalent to a 10-m sea level change and produced a glacioeustatic sea level curve that matched that of Mesolella *et al.* (1969). Since then, Williams *et al.* (1981) and Fillon and Williams (1984) have examined this relationship more closely. The problem is to predict the volume of sea ice versus continental ice and to know how much ^{18}O is trapped in the different ice bodies. Fillon and Williams (1984) produce figures that match the original generalization of Shackleton and Opdyke (1973), but the results are slaved in order to match those derived from reef terrace dating with assumed tectonic uplift models. The dimensions of the variables involved are too unconstrained and can only be guessed at and certainly not measured (such as the rate of tectonic uplift over the last 125,000 years). Even if we knew the ice volume involved, we would still not be able to predict the isostatic response of the crust to the changes in weight of seawater since the crustal flexural properties change from one setting to the next and also, presumably, with time.

II. Clastic Sediment Loading Models

A. Marine Sediment Simulation

Hardenbol *et al.* (1981) have argued that the sedimentary aggradational curves (the coastal aggradational curves of Vail *et al.,* 1977) are not an accurate means of determining the size of a eustatic event because the measure of sedimentary aggradation is also influenced by compaction, the rate of crustal subsidence, and, importantly, the extent of alluvial plain onlap, and these latter variables may not be accurate. For instance, the compaction algorithms which are commonly used are based on the assumption that fluid escapes uniformly during burial and that the sedimentary section compaction is in equilibrium with this escape (Guidish *et al.,* 1985). Fluid escape is much more complex and the compaction is not being handled accurately by these algorithms. Also, using a thermotectonic model to predict a crustal subsidence curve, as a means of determining the size of the eustatic excursion, is dependent on the choice of the model used, as well as on parameters entering the model. These model choices are not grounded in empirical fact, no matter how well and cogently the model is argued and presented. The same problem with tectonic models occurs in using coral reef terraces to measure sea level changes. Similarly, attempts to measure high-frequency sea level changes from stacked subsidence curves do not work since the crust does not respond noticeably to such small sea level events. Lastly, although the oxygen isotope ratio in the sea is controlled by the ocean volume, we cannot determine the size of eustatic excursions because we cannot accurately model the crust's flexural response to changes in ocean volume except in direction and to a very rough order of magnitude. Hallam (1984) reviews other methods for determining the size of eustatic sea level events, including Pitman's (1978) ridge spreading, hypsometric curves of continental elevation (Bond, 1978), and his own (Hallam, 1977). However, all these methods, like those we have discussed, require making basic assumptions about the size of a number of variables that cannot be measured. Exciting and imaginative concepts arise, but so far without supportive, unequivocal evidence.

1. Simulation Model

Because of this intrinsic dissatisfaction with previously discussed systems, we developed a forward (in time) model to handle sea level, tectonics, and sedimentation, with the ultimate goal of *inverting* the forward model so that we could use it to see to what extent we could determine the sizes of the different variables needed to produce the observed sedimentary geometries within given basins.

Here we describe a means of simulating marine clastic deposition independent of the alluvial wedge. We first deal briefly with the sequential simulation of geologic processes such as deposition, compaction, and subsidence; the simulation of the individual processes; and the required inputs. The second part describes suggested sources of data for input to the simulation. The last part presents an evaluation of the significance of the simulation with respect to our current understanding of shelf sedimentation and what outputs of the simulation are available. An Appendix describes details of how we handle deposition, compaction, and subsidence.

We developed a simple system of modeling designed to generate a two-dimensional representation of the sediment geometry that can be observed on dip-trending seismic lines that cross buried shelf margins (Fig. 10.6). The geometries seen on seismic lines are the result of the interaction of changes in sea level, rates of sedimentation, and rates and directions of tectonic movement. The simulation varies sedimentation, subsidence, and position of eustatic sea level change and plots the resulting geometries. If a favorable match between the simulation plot and the seismic cross section could be seen, it was our hope that some specific statements could be made concerning sediment accumulation rates, subsidence rates, and sea level changes. These data, in turn, could then be applied to the risk evaluation of hydrocarbon exploration plays in terms of burial history, stages of diagenesis, timing of generation and migration of hydrocarbons, definition of trends, and so forth. The applications to hydrocarbon play evaluation could thus be significant.

The simulation is based on the concept that important gross features of the depositional geometry are controlled by macroprocesses such as eustatic sea level variation, crustal subsidence, compaction, and sediment deposition, precisely the processes which are needed to interpret the stratigraphy of seismic sections. No attempt is made to simulate the physical processes involved. Rather, we attempt to reproduce the sedimentary geometries which are the results of such processes, averaged over long periods of time.

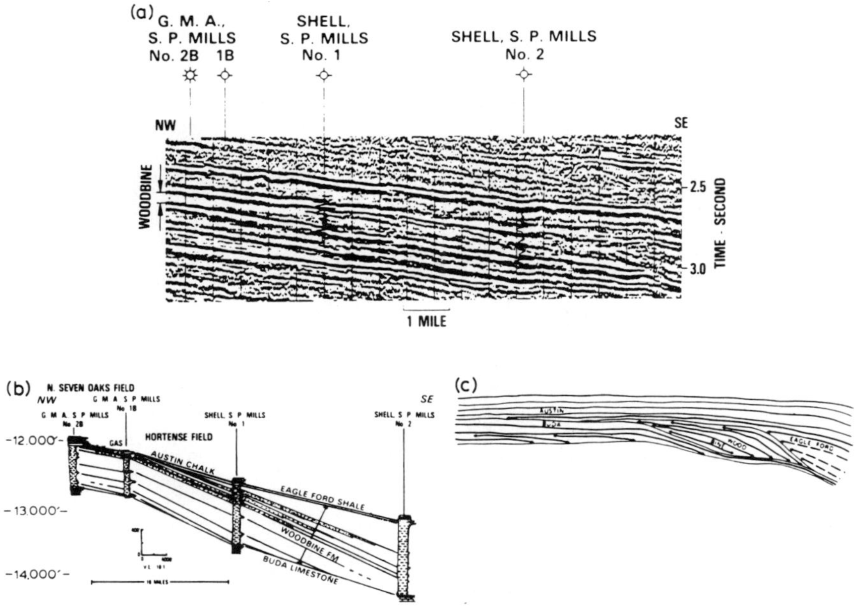

FIGURE 10.6 (a) Seismic cross section through the Woodbine delta, Polk County, Texas. (b) Electric log cross section showing distribution and geometry of Woodbine delta sandstone beds. (c) Diagrammatic interpretation of (a) and (b). (After Vail *et al.*, 1977.)

We simulate the deposition of packages of sediment over fixed time intervals, which we define as sediment triangles of specified length and thickness (see Fig. 10.8). The distribution of these packages for each time interval is controlled by

1. The configuration of the original depositional surface (Fig. 10.7A).
2. The position of the sea relative to that surface (Fig. 10.7D).
3. The quantity of sediment (shale and/or sand) deposited during that time interval (Fig. 10.7B).
4. The subsidence behavior of the depositional surface (including hinged tectonic or thermal subsidence, isostatic loading, and sediment compaction in response to dewatering) (Fig. 10.7C).

The simulation includes functions that allow the sediment to build to sea level, after which erosion and bypass take place.

The simulation inputs, such as eustatic sea level variation with time,

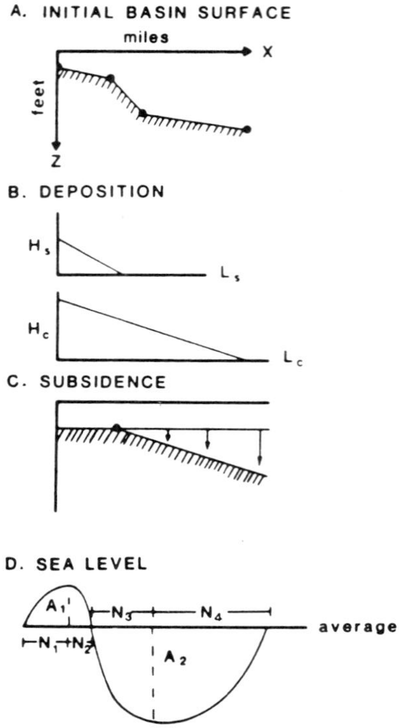

FIGURE 10.7 Definition of the size of the input parameters used in the graphic simulation.

are described in the section that follows. For a given time each input is described by an array corresponding to the vertical position of the surface between evenly spaced horizontal points, giving all surfaces a staircase-like appearance (Fig. 10.8). Figure 10.9 schematically illustrates the depositional geometry after five time steps. The basic sequence of processes in each time step is as follows.

1. Locate the point of greatest onlap (POGO), x', that is, the intersection of sea level and the highest sediment surface (Figs. 10.8 and 10.9).

2. Erode between x'', the point of landwardmost previous sediment deposition, and x' (Figs. 10.7 and 10.8).

3. Deposit sediments, including those eroded in step 2 and those to be deposited for that time step.

4. Compact all sediment layers according to the weight of overlying rock.

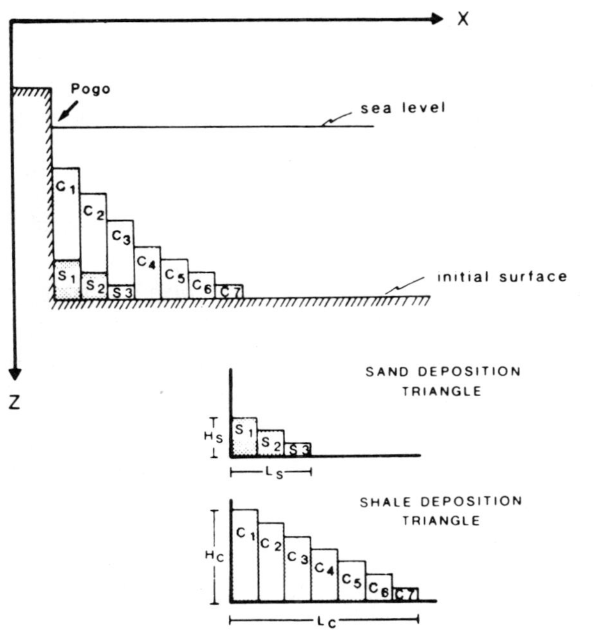

FIGURE 10.8 Depositional surface defined by an array of evenly spaced horizontal points, giving the surface a staircaselike appearance.

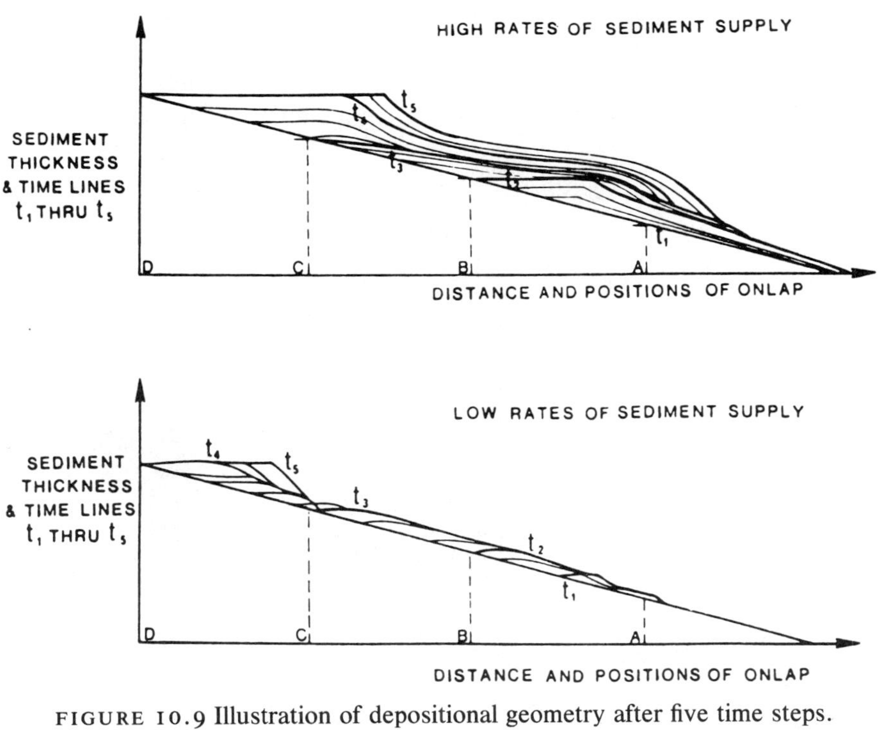

FIGURE 10.9 Illustration of depositional geometry after five time steps.

5. Subside all surfaces in response to fixed tectonic dependences and the isostatic accommodation due to sediment loading and change in water depth.

This sequence of erosion, deposition, compaction, and subsidence is carried out for each time step.

Deposition involves the creation of new surfaces at the same height as, or above, the highest previous surface. Both erosion and compaction involve reduction in the thickness of layers. Tectonic subsidence causes all surfaces to drop vertically by the same amount at a given horizontal position.

Detailed descriptions of erosion, deposition, compaction, and subsidence as well as the simulation inputs and outputs can be found in the Appendix. Figure 10.10 provides a simplified flowchart of the simulation.

2. Results of the Simulation

The purpose was to simulate marine clastic deposition alone, ignoring the alluvial wedge landward, and so reproduce the gross sedimentary geometries of the marine sediments showing the location of sand-prone and shale-prone portions of the section (Fig. 10.11). The area between the dark lines represents the sediment deposited during a designated time interval. The proportion of shale and sand within the time step is displayed in the following way. The area immediately below the dark lines

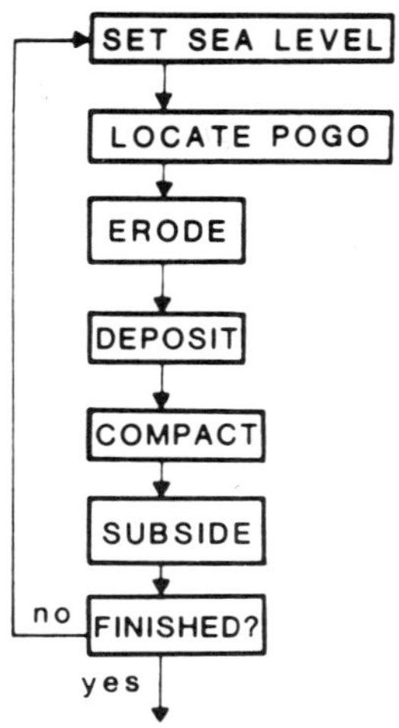

FIGURE 10.10 Simplified flowchart of simulation.

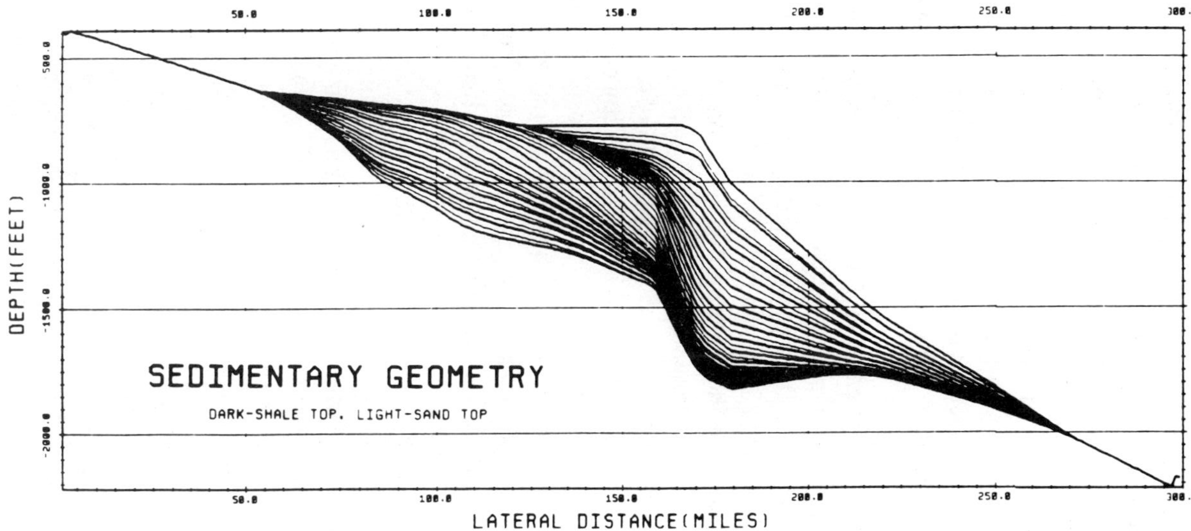

FIGURE 10.11 Cross section of sedimentary geometry produced by simulation of eustasy, compaction, tectonic movement, isostasy, and sediment fill, assuming no alluvial wedge landward and sediment fill to sea level.

represents the shale portion of the sediment, and the area below the light lines represents the sand portion. Individual sand and shale bodies, such as point bars, filled channels, or deltas, are not specifically identified, although areas where sand or shale bodies are likely to occur are shown. The primary objective of this output is to aid in the interpretation of seismic data by simulating the general, rather than specific, depositional history of the rock sequence. In particular, the simulation can be utilized to help identify probable zones of good source rocks and seals (shale-prone areas) and the general region in which potential stratigraphic traps (pinch-outs of sand-prone areas within and/or beneath shale-prone areas) may occur.

Figure 10.12 shows the progressive evolution of the shelf and basin margin geometry. At stage 1 the progressive onlap of the shelf in response to a sea level rise is seen; in stages 2 and 3 we see top-lapping progradation in which sea level fall is not rapid enough to drop the coastal sedimentary wedge below the shelf margin. In stage 4 we see the effect of the sea level rise and the onlapping sedimentary wedges. One of the interesting responses we see in the simulation is the effect of the weight of the sediment on the crust. In the first stage the crust is bowed down close to the shore, while at the basin margin the crust has a positive area which is ridged up through a lack of subsidence. In the third and fourth stages, the crust is bowed down seaward of the positive ridge. The resulting "basement" geometry is a product of sediment weight alone, assuming a perfect elastic Airy response of the crust.

In Fig. 10.13 we plot the eustatic curve (the lower curve) plus a modified Vail aggradational curve (without the alluvial plain, which is not included in this model) constructed after the deposition of the whole sedimentary package using precisely the same criteria as outlined in Vail *et al.* (1977). The result is at least a fourfold exaggeration, as well as a distortion, of the sea level excursion. The magnification and distortion would be even larger if we added the alluvial plain sediments to the model. This magnification effect explains why the sea level excursions are so large and yet the crustal response seen by Guidish *et al.* (1984) is so small. It appears that small sea level excursions have a marked effect on the sedimentary response and that compaction and tectonic movement exaggerate the size of the excursion if one tries to measure them on seismic cross sections. The last diagram shows the position of POGO, the shelf break, and the farthest basinward position of the sediment wedge (Fig. 10.14). The sinusoidal shape of these curves and of the coastal onlap curve is due to the fact that we are not modeling the alluvial plain

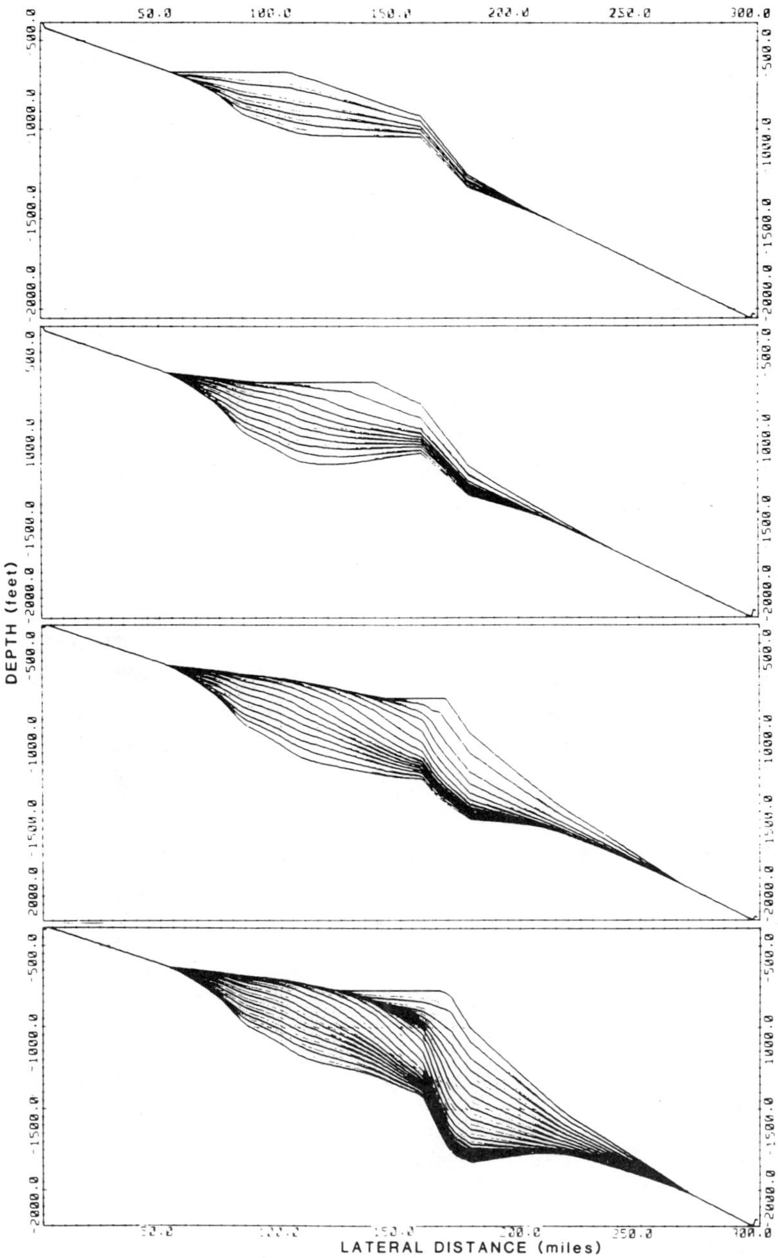

FIGURE 10.12 Five steps involved in sediment fill to produce the geometry of Fig. 10.11.

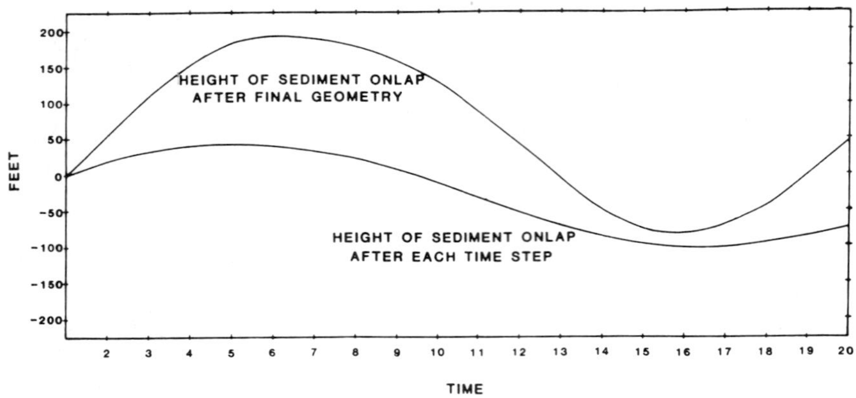

FIGURE 10.13 Vail onlap curve versus eustatic sea level variation.

sediments, which, as Hardenbol *et al.* (1981) point out, are responsible for the sawtooth appearance of their "coastal" or sediment onlap curves.

We still have a considerable amount of work to do before we can assess quantitatively the possibilities of using sea level variation in risk evaluation of plays (e.g., a major element yet to be added to the simulation reported here is the alluvial wedge that onlaps landward of the coastline; see Fig. 10.2). Nevertheless, such simulations give a general sense of the response to sedimentation, sea level, and tectonism in the offshore marine and coastal sediments. Systematic testing of any simulation's reliability and its effectiveness as an aid to interpretation seems not to have been undertaken. The principles for choosing test criteria should involve

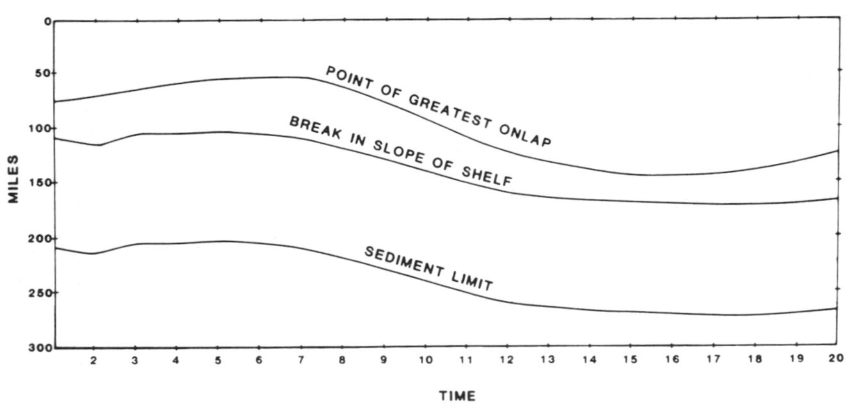

FIGURE 10.14 Position of POGO, shelf break, and basinward position of sediment.

well-studied areas which provide good geologic control and which can be used to make reasonable estimates of the program inputs. Such areas should lack geologic phenomena not included in the programs, such as faulting, carbonate buildups, and salt movement. Until such tests are done, the applications to less well known areas cannot be made with confidence.

B. Mathematical Structure Relating Sea Level Variations to Basin Fill under Isostatic Burial History

Having developed the simple modeling system just described, which showed how the sediment fill of a basin margin responds to sea level variation, tectonic movement, and rate of sediment input, we are still no nearer to being sure that we are providing the correct size for the variables we input. As a result, we tried to see if there was a mathematically unique answer. The results are educational and make transparent the assumptions involved, but once again we found we could not determine the size of the variables.

The purpose of this portion of the chapter is to spell out the detailed mathematics of how laterally varying onlap sequences are related to sediment fill, sea level rise and fall, tectonic movement, and water level variations in a basin.

We start at some instant of time t with a basin filled as shown (Fig. 10.15). In the time interval t to $t + dt$, add the solid grammage

$$dm_s = \rho_L(x)[1 - \phi_L(x)]\, dz_{sed} \qquad (10.2)$$

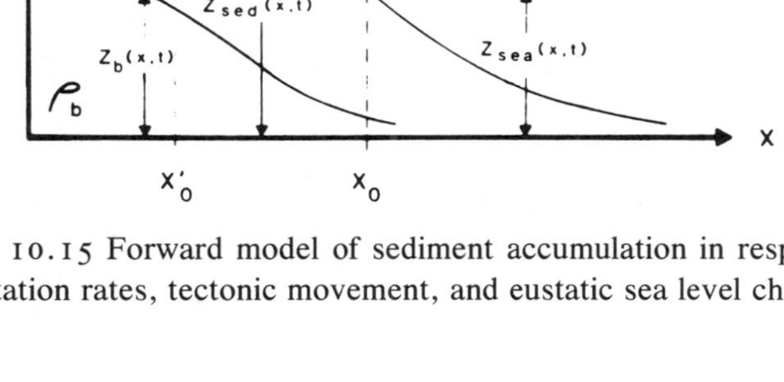

FIGURE 10.15 Forward model of sediment accumulation in response to sedimentation rates, tectonic movement, and eustatic sea level changes.

the water grammage

$$dm_w = \rho_w \phi_L(x)\, dz_{sed} \tag{10.3}$$

and the sea level grammage

$$dm_{sea} = \rho_w\, dz_{sea} \tag{10.4}$$

with $dz_{sed} = 0$ to the left of x_0. Here $\rho_L(x)$ is the lithology-dependent grain density, $\phi_L(x)$ the surface porosity, dz_{sed} the increase in thickness of sediment at position x, dz_{sea} the increase in water depth (sea level), which is independent of x (for the moment) to the right of the onlap point x_0 and shoals to the left as sketched.

Since we assume that all our calculations are isostatically compensated in order to establish the levels of sea level, sediment thickness, etc. at the next time stage of deposition, we break the calculation into three steps:

1. The compaction of the sediments underlying the added material, holding the basement fixed in position.
2. The vertical compensation movement of the basin.
3. The motion of the point of onlap as a consequence of 1 and 2.

Then proceed to the differential limit as $dt \to 0$, appropriate for continuous deposition.

1. Compaction

As a consequence of increased solid grammage deposited, the underlying sediments are compacted such that an element of sediment of lithology l which was at coordinates (z, x) at time t, moves to coordinates $(z + \Delta z, x)$ at time $t + dt$ and an element of lithology l at $(z + H, x)$ moves to $(z + H - \Delta H, x)$ at $t + dt$.

Skeletal matter being conserved, we have

$$\int_z^{z+H} [1 - \phi_l(z', x)]\, dz' = \int_{z-\Delta z}^{z-\Delta z+H-\Delta H} [1 - \phi_l(z', x)]\, dz' \tag{10.5}$$

with $\Delta z = 0$ for an element sited on $z = z_b$ at time t.

For a given law of porosity with depth (we use an exponential variation of porosity with depth throughout, but the calculations can be redone for any specified porosity-with-depth variation)[1] we write

$$\phi_l(z, x) = \phi_{lx}(x,t) \exp[-(z_{\text{sed}} - z)/a_l] \tag{10.6}$$

so that

$$dH_l = dz\, \phi_{l\text{b}} \exp[-(z_\text{b} - z)/a_l][\exp(H_l/a_l) - 1] \times\{ 1 - \phi_{l\text{b}} \exp[-(z_\text{b} - z - H_l)/a_l]\}^{-1} \tag{10.7}$$

where

$$\phi_{l\text{b}} = \phi_{l\text{s}} \exp[-(z_\text{s} - z_\text{b})/a_l] \tag{10.8}$$

The base of the Lth layer is at $z_L = z_{L-1} + H_{L-1}$ at time t and at $z_L - \Delta z_L = z_{L-1} + dH_{L-1}$ at time $t + dt$.

The total *increase* in sediment thickness at $t + dt$ at position x is

$$dh = dz_{\text{sed}} - \sum_{l=1}^{N} dH_l \tag{10.9}$$

where N is the number of sedimentary layers deposited prior to time t. The height of the sediments above the basement at z_b at time $t + dt$ is

$$h = dh + \sum_{l=1}^{N} H_l \tag{10.10}$$

The grammage of water lost from the Lth layer in time t to $t + dt$ is

$$\Delta m_{\text{w}l} = \rho_\text{w} dH_l \tag{10.11}$$

so that the total lost grammage of water is

$$\Delta m_\text{w} = \rho_\text{w} \sum_{l=1}^{N} dH_l \tag{10.12}$$

The net balance of pore water in t to $t + dt$ is then

$$\delta m_\text{w} = \rho_\text{w} \phi_\text{L}(x) dz_{\text{sed}} - \Delta m_\text{w} \tag{10.13}$$

[1] Note that this assumption retains a *fixed* porosity law with depth for all time; that is, fluid flow and overpressuring are assumed to be ignorable—which assumptions are known to be untrue in general.

In an infinite ocean this pore water change provides no shift in sea level; it corresponds to a shift in weight on the underlying mantle. The increase in mass in t to $t + dt$ at x acting on the mantle is

$$dw = dm_s + \delta m_w + \rho_w \, dz_{sea} \tag{10.14}$$

while the sediment column height above basement increases by dh.

2. Basement Motion

As a consequence of the increase in weight, the isostatic basement position $z_b(x, t)$ will move. At x let the basement move in time t to $t + dt$ from z_b to $z_b - dz_b$. Then at time $t + dt$ the column sedimentary height is $h + dh$ above basement, while the water height above sediment is

$$dz_{sea} + dz_b = dl \tag{10.15}$$

Thus, measured from the basinal coordinates sketched in Fig. 10.15, we have PE_b (= potential energy of basement material) given by

$$PE_b = g\rho_m \int_0^{z_b - dz_b} z' \, dz' \tag{10.16}$$

$$PE_{sed} = g \int_{z_b - dz_b}^{z_b - dz_b + h + dh} z'\{\rho_l(z')[1 - \phi_L(z')] + \rho_w \phi_L(z')\} \, dz' \tag{10.17}$$

$$PE_{water} = \rho_w g \int_{z_b - dz_b + h + dh}^{z_b - dz_b + h + dh + dl} z' dz' \tag{10.18}$$

The total potential energy PE is then

$$PE = PE_b + PE_{sed} + PE_{water} \tag{10.19}$$

Isostatic equilibrium is reached when the change of PE with respect to dz_b is minimum, that is, when

$$\partial PE / \partial (dz_b) = 0 \tag{10.20}$$

This occurs at $t + dt$ when

$$\begin{aligned} &\rho_b(z_b - dz_b) + \{\rho_L(x, t + dt)[1 - \phi_L(x, t + dt)] + \rho_w \phi_L(x, t + dt)\} \\ &\quad \times (z_b - dz_b + h + dh) \\ &\quad - \{\rho_L(z = z_b - dz_b, x)[1 - \phi_L(z_b - dz_b, x)] + \rho_w \phi_L(z_b - dz_b, x)\} \\ &\quad \times (z_b - dz_b) + \rho_w \, dl = 0 \end{aligned} \tag{10.21}$$

Now prior to increasing the load, at time t we have

$$\rho_b z_b + \{\rho_L(x, t)\,[1 - \phi_L(x, t)] + \rho_w \phi_L(x, t)\}(z_b + h) - \{\rho_L(z = z_b, x)[1 - \phi_L(x, t)] + \rho_w \phi_L(z_b, x)\}z_b = 0 \quad (10.22)$$

Since the basement porosity changes as a result of added load but the basement lithology does not, we have

$$\rho_L(z = z_b, x) = \rho_L(z = z_b - dz_b, x) \quad (10.23)$$

Then by subtraction we obtain

$$-\rho_b\, dz_b + \rho_w(dz_b + dz_{sea}) + dh\, \rho_{sed}(x, t + dt) - dz_b \rho_{sed}(x, t + dt) + [\rho_{sed}(x, t + dt) - \rho_{sed}(x, t)](z_b + h) + dz_b\, \rho_{sed}(z = z_b, x) - [\rho_L(z = z_b, x) - \rho_w]\, \frac{\partial \phi_L(z_b, x)}{\partial z_b}\, dz_b = 0 \quad (10.24)$$

that is,

$$dz_b \left\{ \rho_b - \rho_w + \rho_{sed}(x, t + dt) - \rho_L(z = z_b, x, t) + z_b[\rho_L(z = z_b, x) - \rho_w]\, \frac{\partial \phi_L(z_b, x)}{\partial z_b} \right\} = \rho_w dz_{sea} + dh\, \rho_{sed}(x, t + dt) + (z_b + h)[\rho_{sed}(x, t + dt) - \rho_{sed}(x, t)] \quad (10.25)$$

Hence we determine the basement motion dz_b in the time interval to $t + dt$ due to sediment and water load. If, in addition, there are tectonic forces at play causing a downshift in basement position dz_{tect}, then the total basement response in t to $t + dt$ is

$$dz_{total} = dz_{tect} + dz_b \quad (10.26a)$$

with

$$z_b(t) = \int_0^t dz_{total} \quad (10.26b)$$

so that

$$z_b(t + dt) = z_b(t) - dz_{total} \quad (10.27)$$

3. Sediment Thickness and Onlap

At the end of time interval t to $t + dt$ the basement at x is now at position $z_b(t + dt)$, the sedimentary column is of height $h' = h + dh$, and the water depth above the sediment is increased by $dz_{sea} + dz_{total}$.

At time t the height of sea level above the basement (at z_b) is $H_{sea}(\text{old}) = z_{sea} - z_b$. At time $t + dt$ the height of sea level above the new basement (at $z_b - dz_{total}$) is

$$
\begin{aligned}
H_{sea}(t + dt) &= z_{sea} + dz_{sea} - (z_b - dz_{total}) \\
&= H_{sea}(t) + dz_{sea} + dz_{total}
\end{aligned}
\tag{10.28}
$$

The height of sea level above the new sediment column (of thickness $h + dh$) is

$$H(t + dt) = H_{sea} + dz_{sea} + dz_{total} - (h + dh) \tag{10.29}$$

The base of a sedimentary layer, layer L, at z, x at time t is now at $z' = z - dz - z_b(t + dt)$ above the new basement position. At time t the point of greatest onlap, x_p, was where $H_{sea}(t) = H(t)$; that is, $z_{sea} - z_b = H(t)$.

At time $t + dt$ the point of greatest onlap has moved to $x_p + dx_p$, where the sea level height above basement equals the sediment column height above basement, such that

$$H(t + dt, x_p + dx_p) = 0 \tag{10.30}$$

Now

$$H(t + dt, x) = H_{sea}(t, x) + dz_{sea}(t, x) + dz_{total}(t, x) - h(t, x) - dh \tag{10.31}$$

and

$$h(t, x_p + dx_p) = h(t, x_p) + dx_p \frac{\partial h(t, x_p)}{\partial x_p} \tag{10.32}$$

while

$$H_{sea}(t, x_p + dx_p) = 0 \tag{10.33}$$

so that the shift in point of greatest onlap in t to $t + dt$ is

$$dx_p = \left[\frac{dh - dz_{\text{sea}} - dz_{\text{total}}}{\partial h(t, x_p)/\partial x_p} \right]_{x=x_p+dx_p} \tag{10.34}$$

Note that motion of the onlap point is determined not only by sediment increase, sea level shift, and basinal drop under sediment plus tectonics but also by the *slope* of the compacted sedimentary layer surface at the onlap point.

The whole process is very nonlinear. To determine sea level change from present-day measurements of sedimentary strata and their onlap points by inverting this forward modeling procedure would seem difficult.

Even if we could invert, we have two major problems:

1. All we can deduce is *the sum of sea level and basin motion.*
2. We do not know where the points of *greatest* onlap are in an onlap sequence without the complete basin history, and that requires (a) *complete* knowledge of sedimentary fill laterally in a basin together with the basement response and (b) knowledge of how overpressure with time influences the sequence and onlap effects, since basins are not isostatically controlled for all time.

In short, sedimentary geometries which we can see on seismic cross sections and in outcrop are the products of variations in rates of sediment accumulation, eustatic sea level change, and tectonic movement. To date, the innovative concepts of Vail and his co-workers, the forward modeling presented here, ties to stacked crustal subsidence (Guidish *et al.*, 1984), and systems for measuring Pleistocene sea level movement using either reef terrace positions or deeper-ocean oxygen isotope ratios have not provided a unique and unequivocal key to determining the relative size of these processes. The major conclusion from all of this work and from the formal mathematical development is that at the very least two of the three processes (sediment accumulation, eustatic sea level, tectonic subsidence) must be specified in order to determine the third. Presumably, this lack of knowledge underlies the disparite nature of the many equally good hypotheses often invoked to account for observed sedimentary geometries. The information just is not present to decide which hypothesis represents truth.

Appendix

A. Erosion

Erosion is simulated by stripping off rock to a constant depth between the present POGO (x') and the landwardmost of previous POGOS (x''). The erosion depth is determined by dividing the maximum area of sediment which can be eroded per time step, by the distance $x' - x''$.

Major constraints are placed on erosion. Only previously deposited sediments above sea level are available for erosion; no erosion into the initial surface can occur because the simulation input does not define sediments below the initial surface.

The layers of sediment available for erosion, both sand (coarse) and shale (fine), are increasingly compacted with depth. Following the erosion of these variously compacted rocks to a depth D, these eroded sands and shales are uncompacted to their original porosity, increasing the thickness of any removed sediment $d_2 = d_1(1 - \phi_1)/(1 - \phi_2)$, where d_1 is the compacted thickness and ϕ_1 and ϕ_2 are the compacted and original porosities (the original fractional porosities of sands and shales at deposition are 0.4 and 0.75 in our model).

The total eroded quantities of uncompacted sands and shales are then added to the sand and shale already available for deposition for that particular time step.

B. Deposition

Following erosion, coarse (sand) and fine (shale) sediments are deposited seaward of POGO (x') subject to two principal constraints: the sediment functions specify the amount of each sediment to be added to the top of the last surface as a function of distance from x', and the sand is deposited first, followed by the shale. The resultant cross section (Fig. 10.11) of the sand–shale distribution represents the relative proportions of each, with distance, without providing any insight into their specific geometries; that is, the probability of sand bodies or percentage of sand at a given location can be inferred, but the size and location of individual sand bodies cannot.

The thickness of deposited sediments, both sand and shale, is assumed to decrease seaward of POGO for that time step. For simplicity, the sand–shale deposition functions are triangles defined by height H and length L, both parameters specified by the user (see Fig. 10.8).

Prior to deposition, the quantities of uncompacted sand and shale accumulated by erosion are added to the input deposition triangles by increasing only the height of the respective triangles but not the length.

The rationale is that the distance of sediment deposition from POGO depends primarily on properties such as particle size and currents rather than volume of sediment.

The sediment deposition simulated consists of columns of blocks of varying thickness to which additional sand and shale blocks may be added at each time step. The sediment deposition functions are like a set of blocks of linearly decreasing height. Deposition begins by stacking first the largest sand block and then the largest shale block onto the column corresponding to POGO. Shifting seaward one column, the next largest sand and shale blocks are added, and so on until the blocks run out (Fig. 10.8).

This simple picture is modified by two constraints. First, deposited sediment is stable over long periods where the slope of the depositional surface is below the critical angle. Above the critical angle, even sediments that are temporarily stable are assumed to be removed by sporadic turbidites. Using an input critical angle for stable deposition, the vertical drop between adjacent columns determines whether deposition is allowed to occur on a given column. If the vertical drop is more than that allowed by the critical angle, the simulation skips that column without depositing and goes on to test the next column.

Second, sediment cannot be deposited above sea level. If the sand and shale blocks to be piled onto the column actually reach above sea level, the thickness of both blocks is reduced so that they just come to sea level. The relative sizes of the two blocks are reduced in proportion to the sizes by which they exceeded sea level at that position. The sand and shale left over are added to the sediment triangles by increasing the height of the remaining triangles and not their length; that is, the number of remaining blocks is unchanged by the increase in sediment.

Finally, the simulation tests to see if deposition has been completed, that is, there are no more blocks. When deposition is completed, the program moves on to compaction.

C. Compaction

As with the decompaction of sediments performed in the erosion segment, compaction requires knowledge of the thickness of layers (the sediment blocks) and their present porosities, i.e., their present state of compaction. For each block, the weight of overlying blocks determines the porosity. We have modified expressions relating porosity and depth for sands and shales into expressions for porosity and the overlying weight.

Uncompacted porosity for shale is taken as 75% while the porosity at 10,000 ft is about 20%. For sandstones, the void ratio (ratio of pore space to rock space) is inversely proportional to depth. Uncompacted porosity for sands is 40%, while the porosity at 10,000 ft is about 27%. With depth *d,* given in feet, porosity for shale is defined here as

$$\phi = 0.75 - 0.06 \ln d$$

and for sands

$$\phi = 0.27a/(a + d)$$

where $a = 2.1 \times 10^4$ ft.

In these empirical expressions, d represents thickness of overlying sediment. The above expressions were converted to porosity versus weight. When the actual overburden weight is less than the computed value, nothing is done. The latter can occur when erosion takes place and reduces the depth of the sediment. When sediment is eroded, the remaining sediments do not decompact. Thus, when weight is again added, compaction does not occur until the newly incremented weight exceeds that previously eroded.

D. Subsidence

Three independent types of subsidence are included: (1) a global subsidence on the scale of the model, taken to be a constant independent of horizontal location; (2) a hinged subsidence increasing linearly seaward of an input hinge point (Fig. 10.9); and (3) isostatic accommodation to sediment loading and change in water depth. The first two are independent of sedimentation. The three components are summed at each horizontal position and the column or block shifted down accordingly.

The isostatic response to sediment loading is found from an equation stating that the weight of water, sediment, and mantle to some fixed depth is conserved;

$$\rho_w \Delta_w + \rho_s \Delta_s - \rho_m \Delta_m = 0$$

where ρ_w, ρ_s, and ρ_m are the densities of water, sediment, and mantle and Δ_w, Δ_s, and Δ_m are the changes in water, sediment, and mantle thickness. The isostatic response is

$$-\Delta_m = \rho_w \Delta_w / \rho_m + \rho_s \Delta_s / \rho_m$$

The densities are taken to be $\rho_w = 1$ g cm^{-3}, $\rho_s \cong 2.7$ g cm^{-3}, and $\rho_m = 3.4$ g cm^{-3}, and Δ_s is the newly added sediment.

E. Summary of Simulation Inputs

The following are descriptions of the simulation inputs. As mentioned above, the derivation of some of these inputs is from readily available geologic information. To run the simulation, the following are required.

1. Initial surface, defined by a set of points with linear interpolation between them; vertical units are feet; horizontal units are miles. In the example we show in the figures, the number of vertices cannot exceed 60.
2. Horizontal range in miles, with start and end values used in output plots.
3. Range of time in millions of year, with start and end values used in output usually determined from paleodata.
4. Number of time steps.
5. Stable deposition angle, assumed to be 2° unless changed.
6. Maximum area eroded per time step, assumed to be 0 unless changed.
7. Length and height of sand and shale deposition.
8. Global subsidence per time step, which is kept 0 unless changed.
9. Hinge point and slope of tectonic or thermal subsidence seaward from hinge point.
10. Density of matrix rock and shale at deposition, assumed to be 2.7 g cm^{-3}.
11. Sea level variation, defined by average, initial phase, and amplitude per half-cycle and time steps per quarter-cycle.
12. Time steps after which geometry plots are desired.

F. Output Plots

1. Geometry of deposited surfaces at the end of each user-chosen time step (Fig. 10.11). The initial surface is plotted as a thick line. The subsequent alternations of sand and shale surfaces are shown in alternating thin and thick lines. The horizontal scale is in miles and the vertical scale in feet.

2. Vail *et al.* (1977) aggradation versus time (Fig. 10.13). Vail aggradation, although carrying at least a fourfold exaggeration and some distortion, is a means of estimating sea level variation from the rock geometry. Two curves are plotted; the thicker curve represents the true aggradation history, while the thinner curve is derived from the final geometry, simulating the estimate of sea level variation, by Vail's method. The technique associates the vertical change in successive POGOs with variations in sea level, although this neglects the different subsidence histories of the various POGOs with time. This neglect affects even the true aggradation curve, since the previous POGO has always subsided between time steps, and therefore the difference in the two POGOs includes subsidence of the first POGO with the change in sea level.

3. Horizontal location of POGO, break in slope, and last point of deposition versus time (Fig. 10.14). The break in slope is defined as the last point at which sediment filled to sea level.

G. Suggested Sources of Data Input for Simulation

This section describes the sources of data which can be input to the simulation, and Fig. 10.12 shows an example of this for clastic deposition and the resulting gross sedimentary geometries (Fig. 10.11).

1. Number of Time Steps

The simulation described runs for an arbitrary maximum of 30 equal time steps. The duration of each step is determined by the length of the simulation (as defined by the start and end times) divided by the number of time steps. To simulate accurately the sediment geometry, which results in part from sea level variations, the time sampling must be no less than the time between maxima and minima of the sea level oscillations with, preferably, several samples between peaks and troughs. Thus the length of a time step must be less than the minimum time between sea level extremes and the number of time steps is just the total time desired (say 25 Myr, giving 25 time steps).

The length of time taken for the sea level cycles may be determined from paleontological dating. The various charts available include those for the Jurassic and Cretaceous by van Hinte (1976a and b), those for the Tertiary by Vail and Hardenbol (1979), and a time scale for the Phanerozoic and Archaen, which provide the relationship between geological ages and a chronometric time scale. An individual sea level cycle may last anywhere from at least 2 million to 12 million years, so the length of time steps will vary according to the length of these sea level cycles.

In our simulation, plots can be made after any time step to display sedimentary geometry (Fig. 10.12). After the final time step, any of the plots shown in Fig. 10.11, 10.12, 10.13, and 10.14 can be displayed including amplitude of eustatic sea level variation versus time. Two shoreline positions of the latter chart are found differently. One is determined before compaction and is the true shoreline position, and the other is determined after compaction and is the position identified by Vail *et al.* (1977). Figure 10.14 shows the horizontal location of the point of greatest onlap, break of slope of shelf, and farthest extent of sediment deposition.

2. Number of Horizontal Points

The simulation model described has a finite length with a maximum of 300 equally spaced horizontal locations. The distance between points, the horizontal resolution, is determined by the width of shelf and shelf-derived sediment in the basin and matches the length of the model, as defined by the start and end locations, divided by the number of horizontal positions desired.

3. Height and Length of Sediment Triangles

The quantities of sand and shale deposited in a particular time interval and their vertical and lateral distribution can be determined in the following way.

1. Use at least two wells along the line of the simulation, in conjunction with a seismic cross section, to determine the amount of sediment in terms of the cross-sectional area of the sediment body (Fig. 10.6).

2. Calculate the percentage of sand and the percentage of shale within this sediment volume. Using these percentages and the total volume of the sediment, one can determine the volume and/or area of sand (A_s) deposited and the volume and/or area of shale (A_c) deposited per time step. For instance the volumes of sand and shale are divided by the number of time steps. The sand and shale should be uncompacted after this last calculation. The calculation is as follows:

$$d_1 = d_2(1 - \phi_2)/(1 - \phi_1)$$

where d_1 is the uncompacted thickness, d_2 the compacted thickness, ϕ_2 the compacted porosity, and ϕ_1 the uncompacted porosity.

3. Derive the length of the triangle representing the volume of sand and shale deposited within the time interval of the simulation. The shale triangle length (L_c) is a fraction of the length of the total sediment body being simulated. The exact fraction depends on the configuration of the depositional surface and the amplitude of sea level variations and is determined by iterating from an initial estimate. An initial estimate of the shale length might be half the length of the entire sediment body with height $h_s = 2A_c/L_c$. The sand triangle length (L_s) might be less than half the shale triangle, with height $h_s = 2A_s/L_s$.

4. Determine the subsidence rate and the location of the hinge point about which thermal subsidence takes place. The hinge point is a position at which no thermal subsidence takes place. Seaward of the hinge point, subsidence increases linearly. This subsidence is assumed to be due to thermal cooling that causes the crust to contract and so subside. The position of the hinge point is determined from a seismic section at the location where sediments cease to thin rapidly landward. The subsidence rate, however, is determined from an analysis of the geohistory curves (Fig. 10.5).

5. Choose the bypass angle. The angle beyond which sediment is not deposited was fixed at 2°, but this may be changed to suit the operator. We chose 2° because modern continental slopes seldom exceed this.

6. Account for erosion. At sea level lows the shelf is exposed above sea level. The maximum area or volume of sediment eroded for each time interval while sea level is low and below the shelf is set at zero unless changed by the operator. Evidence from modern coastal plains suggests that unless the coastal plain is uplifted and acquires angles in excess of 5°, erosion to all intents and purposes tends to be nonexistent, and deposition may even occur.

4. Geometry of the Initial Surface

This is identified by using vertical and horizontal coordinates, which are determined from seismic data by making a time-to-depth conversion or from a combination of seismic and well data. Clearly, the present configuration will not serve except as a first guess. We suggest iterative trial and error.

5. Eustatic Sea Level

It was our hope that eustatic sea level variations could be derived from the integration of data from many wells from all over the world (Guidish *et al.* 1984). However, as discussed, we are unable to separate eustatic sea level changes from sediment accumulation effects and/or tectonic effects. Initially, the curves of Vail *et al.* (1977) and Vail and Hardenbol (1979) or Pitman (1978) or, alternatively, curves based on the operator's initial geologic interpretation of seismic data can be used to estimate sea level changes. A different sea level curve can then be found by varying the initial guess (along with the other program inputs) in the simulation geometry. The derivation of sea level from burial history by the approach outlined by Vail *et al.* (1977) is discussed in Guidish *et al.* (1985).

11

Sensitivity Analysis

GEOLOGIC PROCESSES related to petroleum generation, migration, and accumulation are very complicated and none of the models can simulate these processes exactly. Sensitivity analysis of a model is a comprehensive examination of how the results of the model are influenced by the change in the assumptions and parameters of the model and by errors in the data. From sensitivity analysis, we can determine what error range in the input data permits accurate results to be obtained and also how strongly the model depends on the assumptions, and we can thereby make the model a more effective tool in basin analysis.

The purposes of this chapter are (1) to examine how sensitively models depend on variations in the input data, the equation parameters, and the assumptions of the model; (2) to estimate the probability of the results produced corresponding to those variations; and (3) to provide constraint conditions for variations in the input data, the equation parameters, and the assumptions of a model.

In order to provide a comprehensive investigation of behavior patterns, throughout this chapter we restrict our discussion to the one-dimensional fluid flow/compaction model, although analyses similar to those presented here can, of course, be performed for any model.

I. Précis of Integrated Model

The fluid flow model consists of three parts: a geohistory model, a thermal history model, and a hydrocarbon generation model. Since the model simulations are one-dimensional, the input data for the model are those commonly used: geological and geochemical data from a single well, which makes the simulations useful both in frontier basins, where only a few wells are available, and in well-developed basins.

I. Précis of Integrated Model

A. Geohistory Model

Through a simulation of the fluid movement in sediments caused by the compaction of the sediments, the geohistory model reconstructs the burial history, basement subsidence, vertical fluid flow, and changes of porosity, permeability, pressure, and fluid flow rate with both time and depth. Also, the evolution of cementation, dissolution, and fracturing caused by abnormally high pore pressure is simulated in terms of the change in formation permeability. The input data required to run the geohistory model are the depth and age of each formation base and the lithology and paleowater depth of each formation.

B. Thermal History Model

Based on the burial history created in the geohistory model, the thermal history model reconstructs the thermal history by (1) comparing predicted thermal indicator values (such as vitrinite reflectance) with measurements down a borehole and (2) adjusting the paleoheat flux to minimize discordances. The outputs are a heat flow change with time, a temperature change with time and depth, and vitrinite reflectance and time–temperature integral (TTI) changes with time and depth. The input data required to run the thermal history model are the temperature at the sediment surface, bottom-hole temperature (or thermal gradient/present-day heat flow), and some thermal indicator measurements with depth.

C. Hydrocarbon Generation Model

The hydrocarbon generation model is based on the kinetics of kerogen degradation and the general scheme of evolution sketched in earlier chapters. Two mathematical models are used here to simulate hydrocarbon generation: Tissot's model (Tissot and Welte, 1978), which simulates the formation of oil from kerogen in six parallel reactions (first stage) and the formation of gas from oil in a single reaction (second stage), and a modified model (Cao *et al.*, 1986) (see Chapter 5), which adds the major gaseous products from both kerogen degradation and oil cracking. Based on the burial history and temperature history produced in the model, the generation model gives the absolute amount of hydrocarbons generated (per gram of kerogen) with time and depth. The input data required to run the generation model are the contents of different kerogen types of each formation.

At each instant of time, spatial contouring of results from a number of wells in a study area permits an investigation of temporal and spatial regional studies such as basin subsidence, paleostructure evolution, heat flow evolution, source rock maturity, and hydrocarbon generation, migration, and accumulation (Cao *et al.*, 1986).

Results that need to be examined in the sensitivity analysis include

- formation thickness
- formation porosity
- formation permeability
- formation pressure
- heat flow with time
- temperature with time and depth
- R_0 and TTI with time and depth
- "oil window" in terms of time and depth
- amount of hydrocarbons generated

Three groups of variables must be examined: input data, equation parameters, and intrinsic assumptions.

1. Input Data

Very commonly there are errors in the measurements of the geological and geochemical data. The variables which need to be tested in the input data are

- depth and age of each formation base
- lithology and paleowater depth of each formation
- unconformity time and eroded thickness
- temperature at sediment surface
- bottom-hole temperature
- heat flow at present day
- thermal gradient
- vitrinite reflectance
- kerogen type and content of each formation

2. Equation Parameters

Most of the parameters and constants in the equations used in the models are based on empirical data; for example, depositional porosity of shale, 0.62, is used in the model. The parameters which need to be tested are

- depositional porosity, permeability, and frame pressure
- viscosity of the fluid in the sediments

- parameter A in the void ratio–frame pressure function
- parameter B in the void ratio–permeability function
- critical temperature and doubling temperature in the R_0 equation
- thermal conductivity of each lithology
- activation energies and frequency factors in the generation model

3. Assumptions

The following assumptions have been used in the model and will be tested here.

1. Change the power law functions of void ratio–frame pressure and void ratio–permeability to exponential functions.
2. Change constant sediment surface temperature with time to a variable sediment surface temperature.
3. Change linear heat flow function to a nonlinear heat flow function.

Not tested in this chapter are *intrinsic* assumptions of the model, such as the one-dimensional nature versus a two-dimensional problem, the replacement of complete isostatic movement of basement motion by a partial flexural compensation, and so forth. Examination of these assumptions would make for a very long text indeed, and on this ground they are not included here, although we recognize the pressing need for such a development in order to ascertain the conditions under which a one-dimensional model is adequate.

According to Yukler and Kokesh (1984), the models estimating hydrocarbon resources can generally be grouped into three categories: physical models, symbolic models, and mathematical models. Physical models are used to study one variable at a time, such as faulting under different stress distributions or subsidence with varying heat flow.

Symbolic models are used for graphical presentation of the interrelationships among the components of a system, as in maps, cross sections, and block diagrams.

Mathematical models are used to simulate complex processes with one or more variables and thus they are essential in the assessment of hydrocarbon resources. Mathematical models are applied either as statistical models or as deterministic models to reconstruct and predict geological processes, assuming that these processes are deterministic. The statistical models are mainly used in estimation of hydrocarbon resources because they cannot directly analyze the dynamic processes of hydrocarbon generation, migration, and accumulation. A Monte Carlo simulation is usually employed in the statistical models to construct various probability curves (Lumley, 1970). The most commonly used assessment methods are geological analogy (Weeks, 1952; Conybeare, 1965; Bally, 1975; Pitcher, 1976; Warren, 1979), delphi (Miller *et al.*, 1975), areal and volumetric yields (Stoian, 1965; Walstrom *et al.*, 1967; Smith, 1968; Jones, 1975; Newendorp, 1975; Roadifor, 1975), geochemical yields (Conybeare, 1965; Halbouty *et al.*, 1970; McDowell, 1975; Tissot and Welte, 1978), and field distribution (Atwater, 1956; Roy *et al.*, 1975).

II. Simulation Models

The simulation of sedimentary basin development was established in the 1960s (Chorafas, 1965; Griffiths, 1967). The initial models were basically process–response models (Krumbein and Grayhill, 1965) and they were followed by simple static and deterministic models which evolved into more complex static models and then into dynamic models (Harbaugh and Bonham-Carter, 1970).

In the past 10 years, more comprehensive models have been developed to simulate the dynamic processes related to hydrocarbon generation, migration, and accumulation (Yukler *et al.*, 1978; Sclater and Christie, 1980; Nakayama and Van Siclen, 1981; Welte and Yukler, 1981; Ungerer *et al.*, 1984; Guidish *et al.*, 1985; Cao *et al.*, 1986; Nakayama, 1986). The simulation can be divided into four general parts: (1) reconstruction of the geohistory, which includes basement subsidence, sediment deposition, and changes of porosity, permeability, fluid pressure, and fluid flow with time and depth; (2) reconstruction of the thermal history, which includes heat flux evolution with time, temperature (thermal gradient) changes with time and depth, and thermal maturation history in terms of thermal indicator evolution with time and depth; (3) reconstruction of the hydrocarbon generation history, thereby modeling the change of the amount of hydrocarbons generated with time and depth and determining the time and depth of peak hydrocarbon generation; and (4) reconstruction of hydrocarbon migration and accumulation history, including the amount of hydrocarbons migrated, the time and depth of peak hydrocarbon migration, the direction of hydrocarbon migration and relation among the accumulations (traps), the migration, and the "kitchen" where the hydrocarbons matured.

Because of the differences in level of understanding of geological, geochemical, and geophysical processes and in the mathematical methods used to describe these processes, different models have different features in reconstructing the geohistory, the thermal history, the hydrocarbon generation history, and the hydrocarbon migration and accumulation history. For example, for the geohistory reconstruction there are available (1) backstripping isostatic models, which remove the layer of sediment younger than the age considered and permit the sediments below to decompact by moving them vertically up using an exponential relationship between porosity and depth (Sclater and Christie, 1980; Ungerer *et al.*, 1984; Guidish *et al.*, 1985); and (2) "dynamical fluid flow" models, which simulate the dynamic interaction of deposition, compaction, and fluid migration (involving the changes of sediment thickness, porosity, permeability, pressures, and fluid flow rate with time and depth) by solving a time-dependent partial differential equation that describes the fluid flow in sediments (Welte and Yukler, 1981; Ungerer *et al.*, 1984; Cao *et al.*, 1986; Nakayama, 1986). The dynamical fluid flow models can be grouped further into one-, two-, and three-dimensional models which allow for fluid flow during sediment compaction in one, two, and three dimensions.

In choosing a mathematical model, the following aspects can be considered important:

1. The "reasonableness" of the model: Is the model quantifiable in terms of the principles on which it is based? Can one set up the geological, geophysical, and geochemical factors considered in the model and find a satisfactory method to use in the solution technique?

2. The practicality of the model: Is the model of practical use considering the input data, computer time and storage, and so forth?

3. The accuracy of the model: How well do the model results compare with the observed geological, geophysical, and geochemical data?

No model can simulate exactly the geological, geophysical, and geochemical processes related to hydrocarbon generation, migration, and accumulation. A simulation model usually has three pitfalls: (1) assuming that the input data are correct without any error or with negligible error, (2) assuming that the equation parameters in the model are independent of the input data, and (3) assuming that simplifications made in the model do not affect the accuracy of the simulation. Corresponding to these pitfalls, the users of simulation models should ask the following questions: (1) How do the errors in the input data affect the modeling results? For example, if the age of the bottom layer is uncertain (no age data available) and a guessed age has to be given, say, 100 MYBP with ±10% error, it is important to know how this error affects the maturity determination of the bottom sedimentary layers in order to give a reasonable interpretation of the modeling results. (2) How do equation parameters in the models affect the model results? For example, the sediment surface porosity (ϕ_*) for shale in the porosity law with depth (or frame pressure) ranges from 40 to 80% (Magara, 1978a) and it is necessary to know how the change in the sediment surface porosity affects the compaction of the shale in terms of porosity, permeability, pressure, and the like. (3) If one of the assumptions in the models is relaxed, how are the modeling results affected? For example, most models assume a constant sediment surface temperature with time (Welte and Yukler, 1981; Ungerer *et al.*, 1984; Cao *et al.*, 1986; Nakayama, 1986), but we know that the sediment surface temperature changes with time. Therefore it is necessary to know how a change in sediment surface temperature affects the maturity calculation in the models.

III. Methodology of Sensitivity Analysis

Sensitivity analysis is a study of the sensitivity of a system's response to various disturbances in the system. These disturbances may have widely different characters. They may be small or large, transient or permanent; they may be related to initial conditions, to coefficients and parameters, and so on. (Yukler, 1979). In this study, the following disturbances to the one-dimensional model are considered: (1) errors in the input data, such as the depth, age, and lithology of individual formations; (2) disturbances in the equation parameters, such as the surface porosity and the permeability; and (3) disturbances in the assumptions, such as constant sediment surface temperature with time.

There are usually two ways to handle parametric sensitivity: perturbation methods and direct methods. The perturbation methods are traditional methods which derive the system's sensitivity to a single parameter by making small increments in the parameter and determining the resulting change in output. The disadvantage of perturbation methods is the excessive amount of computer time involved. Because of this disadvantage, direct methods have been developed and widely used by many people (Tomovic, 1963; Anderson, 1965; Yukler, 1979; Fiacco,

1984). The direct methods first derive a sensitivity equation by taking the partial derivative of the equation used in a system and then solve the sensitivity equation by computer.

For this study, a perturbation-type method is used, for the following reasons:

1. The one-dimensional model is a complex system which involves more than one partial differential equation. In some cases, it is impossible to devise a closed-form sensitivity equation by the direct method.
2. Sensitivity testing of the assumptions made in the model is not the simple problem of parametric sensitivity; rather it is a complicated problem. A perturbation-type method has more flexibility to handle this problem.

A. Procedure of Sensitivity Analysis

Generally, we deal with two types of variables in sensitivity analysis: input variables and output variables. The input variables (independent variables) are the input parameters in the model which the sensitivity of the model is to test. These variables are independent of the model and are input data.

The output variables (dependent variables) are the output results of the model. These variables are model dependent and are the responses of the model's sensitivity to the input variables—the perturbation of input parameters.

The general procedure of sensitivity analysis of the one-dimensional model is to (1) determine the relationships between the input variables and the output variables, (2) estimate the probability relationships between the input variables and the output variables, and (3) define restrictive conditions for the input variables.

B. Determination of the Relationship between Output Variables and Input Variables

Let x be the input variable and y the output variable. For n given values of x, $x_1, x_2, \ldots, x_n$, we have n values of y, $y_1, y_2, \ldots, y_n$ by running the model n times. By plotting the y variable against the x variable, numerical and functional relationships between x and y can be determined by least-squares techniques if we assume that (1) there is a relationship between x and y, (2) the influence of the other input variables of the model on y can be neglected, or (3) all the other input variables have at least approximately constant values during the calculation of the model (Brownlee, 1960; Shchigolov, 1965). Assumptions 2 and 3 are satisfied in this study because we vary only one input variable and keep all other input variables constant in the sensitivity analysis. Assumption 1 is checked by graphic distribution of the points in an x–y plot and by the correlation coefficient r between x and y variables from

$$r^2 = \frac{\sum_{i=1}^{n}(x_i - \bar{x})^2(y_i - \bar{y}]^2}{\sum_{i=1}^{n}(x_i - \bar{x})^2\sum_{i=1}^{n}(y_i - \bar{y})^2} \tag{11.1}$$

where r is the linear correlation coefficient, $\bar{x}$ the mean of the x_i's, and $\bar{y}$ the mean of the y_i's. The value of r lies between -1 and 1, inclusive. A value of r near zero indicates that the variables x and y are uncorrelated (Press *et al.*, 1986).

A fitting procedure, called FIT, was developed in the study to establish the functional relationship between x and y (see Fig. 11.1). FIT consists of three parts:

1. Determination of Parameters

The parameters in the function of fitting n points can be estimated by chi-square minimization

$$\chi^2 \equiv \sum_{i=1}^{n}[y_i - y(x_i, a_1 \ldots a_m)]^2 \tag{11.2}$$

where n is the number of points to be fitted and m is the number of the parameters in the fitting function $y(x_i; a_l \ldots a_m)$. Seven fitting functions are used in FIT. They are

$$y = a_1 \exp(-a_2 x)$$
$$y = a_1 \exp(a_2 \ln x)$$
$$y = a_1 + a_2 x$$
$$y = a_1 + a_2 x + a_3 x^2$$
$$y = a_1 + a_2 x + a_3 x^2 + a_4 x^3$$
$$y = a_1 + a_2 x + a_3 x^2 + a_4 x^3 + a_5 x^4$$
$$y = a_1 + a_2 x + a_3 x^2 + a_4 x^3 + a_5 x^4 + a_6 x^5$$

The best-fitting function is determined by choosing the smallest mean-square-root error of the seven functions. A singular-value decomposition method is used in solving equation (11.2).

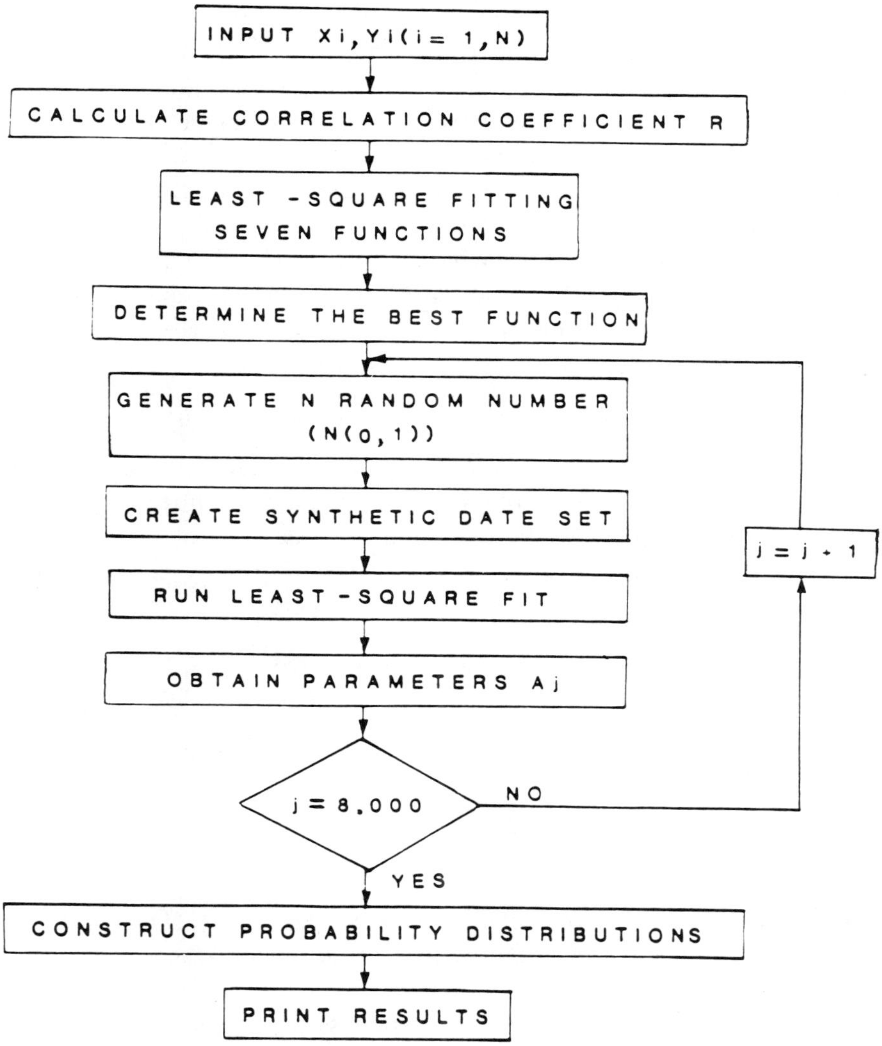

FIGURE II.1 Flow diagram of FIT program.

2. Error Estimates on the Parameters

Error estimates on the parameters are used to determine the confidence in fitted parameters, that is, how much trust there is in a parameter determined by any least-squares fit. For a set of M estimated parameters A, there is some underlying true set of parameters A_{true}. Since there are errors in the measurements (x_i, y_i; $i = 1,n$) there would be a slightly different set of fitted parameters, A, from a least-squares fit. Knowing the probability distribution of $A(i)$ would allow us to determine the quantitative uncertainties in the estimated parameters. Press *et al.* (1986) give an algorithm for simulating the probability distributions $A(i)$, which can be summarized as follows. First, simulate N sets of synthetic data by Monte Carlo realization; second, apply a chi-square fitting procedure to these N sets of synthetic data by using the best-fitting function, thus obtaining the N sets of parameters $A(i)$ $(i = 1, \ldots, N)$; finally, construct the probability distributions for each parameter a_i $(i = 1, m)$ from these N sets of parameters. In this study 8000 sets of synthetic data are created and 8000 sets of parameters are used to construct the probability distribution for each parameter.

3. Statistical Measure of Goodness of Fit

In equation (11.2), the measurement errors are assumed to be normally distributed. Therefore the probability distribution for different values of χ^2 is the chi-square distribution for $n - M$ degrees of freedom. The probability of a particular value χ^2 being exceeded by chance can be calculated from

$$Q(\chi^2 \mid \nu) = Q(\nu/2, \chi^2/2) \tag{11.3}$$

where ν is the number of degrees of freedom and Q is the incomplete gamma function defined by

$$Q(a, x) \equiv \frac{\Gamma(a, x)}{\Gamma(a)} = \int_x^\infty e^{-t} t^{a-1}\, dt / \Gamma(a) \tag{11.4}$$

with the limits $Q(a, 0) = 1$, $Q(a, \infty) = 0$, and $a > 0$. The computed probability Q gives a quantitative measure of the goodness of fit of the fitting procedure.

If Q is larger than, say, 0.1, the goodness of fit is acceptable; if it is larger than, say, 0.01, the fit may be acceptable if the errors are nonnormal or have been moderately underestimated. If Q is less than 0.001, the fitting procedure can rightly be called into question (Press *et al.*, 1986).

C. Estimate of the Probability Relationship between Input and Output Variables

When the functional relationship between input variable x and output variable y is established, it is not difficult to determine the probability relationship between x and y.

Suppose we already have the functional relationship between x and y from the least-squares technique, $y = f(x)$, and we are given

$$P\{a \leqslant x < b\} = \int_a^b B(c)\, dc \qquad (11.5)$$

where $B(c)$ is the probability density function of the input variable x.

We wish to determine

$$P\{d \leqslant y < e\} = \int_d^e B'(c)\, dc \qquad (11.6)$$

where $B'(c)$ is the probability density function of the output variable y.

If $y = f(x)$ is monotone, we can write

$$a \leqslant x < b <==> f(a) \leqslant y < f(b) \qquad (11.7)$$

and immediately

$$\int_{f(a)}^{f(b)} B'(c)\, dc = \int_a^b B(c)\, dc \qquad (11.8)$$

By a simple change of variable, $c = f(c')$, we can write

$$\int_a^b B(c)\, dc = \int_a^b B'(f(c')) \frac{df(c')}{dc'}\, dc' \qquad (11.9)$$

which gives

$$B(c)\, dc = B'(f(c))\, df(c) \qquad (11.10)$$

If the relationship is not single-valued in both directions, the statement (11.7) must be examined; for example, suppose that $y = x^2$. Then

$$\begin{aligned} a \leqslant x < b &<==> a^2 \leqslant y < b^2 \\ -b < x \leqslant -a &<==> a^2 \leqslant y < b^2 \end{aligned} \qquad (11.11)$$

and we have

$$\int_{-b}^{-b} B(c)\, dc + \int_a^b B(c)\, dc = \int_{a^2}^{b^2} B'(c')\, dc \qquad (11.12)$$

which, in a manner similar to (11.8), gives

$$B(c) = B'(f(c)) \mid df(c)/dc \mid \tag{11.13}$$

(Brownlee, 1960; Lumley, 1970).

In this study, we assume that the error in the input variable x is normally distributed with the probability density function

$$B(x) = (2\pi\sigma^2)^{-1/2} \exp[-(x - \xi)^2/2\sigma^2)] \tag{11.14}$$

where ξ and σ are the mean and standard deviation of x. If we know the mean ξ and the standard deviation σ of the input variable x in equation (11.14), the probability of the output variable y can be calculated through equations (11.6)–(11.13).

D. Restrictive Conditions for Input Variables

A restrictive condition for an input variable is defined here as an allowable error range in the input variable corresponding to a given allowable error range in a particular output variable. For example, the allowable error in the depth of the bottom layer (input variable) may be ± 100 ft, and we call the error of ± 100 feet the restrictive condition for the depth of the layer—the input variable.

There are two reasons why it is impossible to set a general and absolute restrictive condition for an input variable:

1. The quantitative relationship between an input variable and an output variable depends on the model and the data used in the sensitivity analysis, and for different models and different data the restrictive condition for the input variable may change.
2. A restrictive condition for an input variable corresponding to an output variable is based on the assumption that all other input variables are kept constant or have negligible effect on the output variable. In practice, we know that this assumption does not always hold.

Therefore the restrictive conditions for the input variables determined in this study are relative, limited to the one-dimensional model and the particular data used in the study.

Figure 11.2 illustrates the method used in this study to set restrictive conditions for an input variable x corresponding to an output variable y. For a given allowable error in the output variable y, the allowable error in the input variable x can be determined by the functional relationship between x and y, $y = f(x)$.

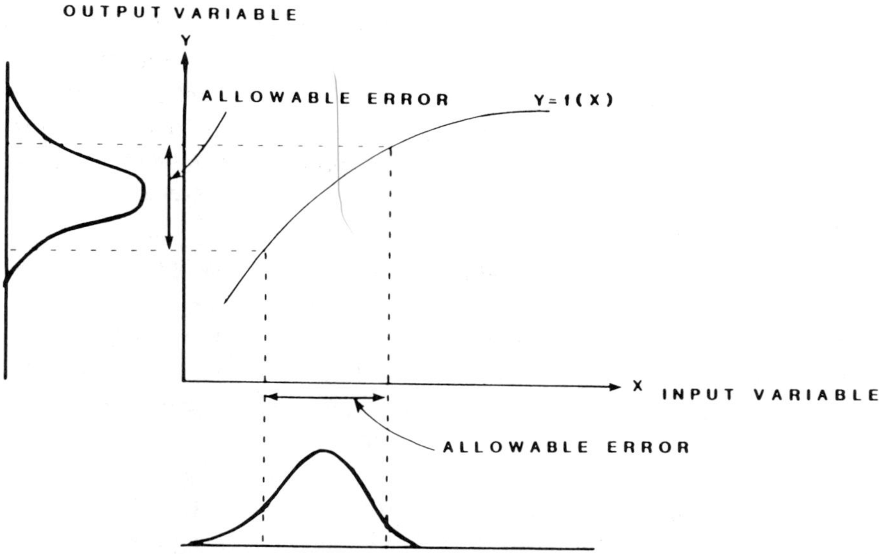

FIGURE 11.2 Determination of restrictive conditions for input variable.

IV. Applications and Case Histories

A. Data Sets of the Study

Three wells from different basins are used in this study: well NWTEST301 in the Norwegian sector, North Sea; well COST-1 in Bering Sea, Alaska; and well Inigok-1 in the National Petroleum Reserve of Alaska (NPRA).

In view of space limitations, we cannot do an exhaustive sensitivity analysis here of each of the variables listed above. Such an investigation would constitute a major thesis (Cao, 1987). Instead, we shall give selective illustrations representative of different behaviors and responses.

B. Sensitivity to Input Data

In earlier chapters we have seen analyses illustrating the sensitivity of determination to unconformity amounts and stratigraphic ages. Here we illustrate response behaviors resulting from variability in formation depth, present-day heat flux, geothermal gradient, and kerogen type.

1. Formation Depth

The data used to test the sensitivity to the depth of the input layers is from well NWTEST301, North Sea. We first examine the depth of the second layer from the bottom and then the depth of the first layer.

Table 11.1 gives the sensitivity results for the depth of the second layer with BUGG OFF.[1] The depth of the second layer is the input variable, which is changed from 5065 m to 4916 m. The output variables are the total depth; the formation porosities, permeabilities, and pressures of the first and second layers; the present-day heat flow; the β value for the heat flow function; and MSR, the logarithm of the mean-square-root error of the vitrinite reflectance fit. From Table 11.1 we see that only the formation pressure of the second layer and the MSR are sensitive to the change in depth of the second layer. The sensitivity of the formation pressure to the depth is expected because we know that the formation pressure increases with increasing depth. The sensitivity of the MSR to the depth suggests that the MSR could be improved by adjusting the depth of an individual formation or that a relatively accurate depth of the formation could be determined by choosing the best MSR value.

Table 11.2 gives the sensitivity results for the depth of the second layer with BUGG ON. Compared with Table 11.1, the model with BUGG ON gives a better fit in terms of the total thickness (the input total thickness is 5055 m), and the total thickness is more sensitive to the change in the depth of the second layer than in the case of BUGG OFF. Because of the sensitivity of the total thickness to the depth of the second layer, the present-day heat flow and β values are affected slightly. Table 11.2 tells us that the model is more sensitive to the depth of the second layer with BUGG ON.

Table 11.3 gives the sensitivity results for the total depth (depth of the bottom layer) with BUGG ON. The total depth (the input variable) is changed from 5287 to 5079 m. From Table 11.3, we see that all the output variables are sensitive to the total depth except the formation permeability and porosity. The best-fit functional relationship between the total depth (x) and the formation pressure (y) of the first layer is $y = 12x^{0.49}$ (with a correlation coefficient r of 0.68), where y is the formation pressure in atmospheres and x is the total depth in meters. The best-fit functional relationship between the total depth (x) and the present-day heat flow (y) is $y = 24.5x^{-0.35}$ (with a correlation coefficient r of -0.83), where y is the present-day heat flow in HFU and x is the total depth in meters. The

[1] BUGG is a subroutine in the 1-D model, whose function is to adjust automatically equation parameters A or B (see Chapter 6) to obtain the best thickness and porosity fit. BUGG ON runs the BUGG subroutine. BUGG OFF does not run the BUGG subroutine.

Table 11.1 Sensitivity of Burial History, Q_0, β, and MSR to Depth of Second Layer (BUGG OFF) (NWTEST301)

Depth of layer 2 (m)	Total depth (m)	Average values of the bottom layer: Permeability (md)	Porosity	Pressure (atm)	Layer 2 depth (m)	Average values of the second layer: Permeability (md)	Porosity	Pressure (atm)	Heat flow Q_0 (HFU)	β	MSR
5065.0	4219	5.90E − 05	0.067	575.8	4244	6.20E − 05	0.069	562.4	1.530	−0.0029	−1.9785
5040.5	4219	5.90E − 05	0.068	573.9	4205	6.30E − 05	0.069	560.5	1.530	−0.0029	−1.9812
5015.7	4219	5.90E − 05	0.068	573.9	4205	6.30E − 05	0.069	560.5	1.530	−0.0029	−1.9812
4990.8	4219	6.00E − 05	0.068	572.0	4165	6.30E − 05	0.069	558.6	1.530	−0.0029	−1.9714
4978.0	4219	6.00E − 05	0.068	572.0	4165	6.30E − 05	0.069	558.6	1.530	−0.0029	−1.9733
4971.0	4219	6.00E − 05	0.068	570.1	4125	6.40E − 05	0.069	556.7	1.530	−0.0029	−1.9741
4966.0	4219	6.00E − 05	0.068	570.1	4125	6.40E − 05	0.069	556.7	1.530	−0.0029	−1.9745
4961.0	4219	6.00E − 05	0.068	570.1	4125	6.40E − 05	0.069	556.7	1.530	−0.0029	−1.9749
4953.8	4219	6.00E − 05	0.068	570.1	4125	6.40E − 05	0.069	556.7	1.530	−0.0029	−1.9754
4941.2	4219	6.00E − 05	0.068	570.1	4125	6.40E − 05	0.069	556.7	1.530	−0.0029	−1.9764
4916.3	4219	6.10E − 05	0.068	568.2	4085	6.40E − 05	0.069	554.9	1.530	−0.0029	−1.9763

Table 11.2 Sensitivity of Burial History, Q_0, β, and MSR to Depth of Second Layer (BUGG ON) (NWTEST301)

Depth of layer 2 (m)	Total depth (m)	Average values of the bottom layer			Layer 2 depth (m)	Average values of the second layer			Heat flow Q_0 (HFU)	β	MSR
		Permeability (md)	Porosity	Pressure (atm)		Permeability (md)	Porosity	Pressure (atm)			
5065.0	5045	5.60E − 05	0.066	746.7	5007	5.96E − 05	0.068	730.5	1.263	0.0000	−1.9396
5040.5	5066	5.60E − 05	0.066	745.5	4979	6.01E − 05	0.068	726.9	1.262	−0.0014	−1.9712
5015.7	5061	5.60E − 05	0.066	748.4	5023	5.90E − 05	0.068	732.1	1.261	−0.0010	−1.9960
4990.8	5051	5.70E − 05	0.067	748.9	4966	6.10E − 05	0.068	732.6	1.261	−0.0010	−1.9598
4978.0	5070	5.60E − 05	0.066	748.2	4984	6.00E − 05	0.068	731.9	1.261	−0.0010	−1.9913
4971.0	5049	5.70E − 05	0.067	747.6	4964	6.10E − 05	0.068	731.4	1.262	0.0000	−1.9691
4966.0	5067	5.60E − 05	0.067	750.2	4981	6.00E − 05	0.068	733.9	1.261	−0.0010	−1.9777
4961.0	5026	5.70E − 05	0.067	738.7	4894	6.10E − 05	0.068	722.5	1.268	−0.0014	−1.9670
4953.8	5057	5.70E − 05	0.067	747.3	4924	6.10E − 05	0.068	730.9	1.260	−0.0010	−1.9793
4941.2	5057	5.70E − 05	0.067	747.5	4924	6.10E − 05	0.068	731.1	1.260	−0.0010	−1.9808
4916.3	5041	5.70E − 05	0.067	742.5	4908	6.10E − 05	0.068	726.2	1.264	−0.0014	−1.9868

Table 11.3 Sensitivity of Burial History, Q_0, β, and MSR to Total Depth (NWTEST301)

Input total depth (m)	Total depth (m)	Average values of the bottom layer: Permeability (md)	Porosity	Pressure (atm)	Layer 2 depth (m)	Average values of the second layer: Permeability (md)	Porosity	Pressure (atm)	Heat flow Q_0 (HFU)	β	MSR
5286.7	5149	5.70E − 05	0.067	760.8	4963	6.20E − 05	0.068	739.0	1.258	−0.0010	−1.9927
5260.8	5137	5.70E − 05	0.067	756.1	4952	6.20E − 05	0.068	734.4	1.258	−0.0005	−1.9824
5234.8	5098	5.70E − 05	0.067	751.9	4962	6.10E − 05	0.068	732.9	1.262	−0.0005	−1.9621
5208.9	5087	5.70E − 05	0.067	749.1	4952	6.00E − 05	0.068	732.5	1.266	−0.0014	−1.9596
5193.4	5041	5.70E − 05	0.067	740.1	4908	6.10E − 05	0.068	723.7	1.264	−0.0010	−1.9785
5183.0	5067	5.70E − 05	0.067	750.2	4981	6.00E − 05	0.068	733.9	1.261	−0.0010	−1.9770
5172.6	5011	5.70E − 05	0.067	738.8	4927	6.10E − 05	0.068	722.6	1.272	−0.0014	−1.9895
5157.1	5061	5.70E − 05	0.067	751.1	4975	6.00E − 05	0.068	737.2	1.260	−0.0010	−1.9592
5131.2	5004	5.70E − 05	0.067	744.5	4968	6.00E − 05	0.068	730.8	1.273	0.0010	−1.9615
5105.3	5004	5.70E − 05	0.067	744.1	4968	6.00E − 05	0.068	730.4	1.273	−0.0005	−1.9852
5079.3	4996	5.70E − 05	0.067	745.5	5007	5.90E − 05	0.068	734.2	1.275	0.0034	−1.9638

probability density function of the present-day heat flow has the form

$$p(y) = (2\pi\sigma^2)^{-1/2} \exp[-(g(y) - x^2/2\sigma^2)]\, dg(y)/dy$$
$$\equiv (2\pi\sigma^2)^{-1/2} \exp[-((y/a_1))^{1/a_2} - \bar{x})^2/2\sigma^2](a_1 a_2)^{-1}(y/a_1)^{-1+1/a_2} \qquad (11.15)$$

where $\bar{x} = 5183.0$, $\sigma = 109.3$, $a_1 = 24.5$, and $a_2 = -0.35$. No strong relationships exist between the β value and the total depth and between the MSR value and the total depth because the β value and MSR value are influenced more by the present-day heat flow than by the total depth.

To examine the effect of varying the total depth on the hydrocarbon maturation and generation, runs were done with a *fixed* heat flow function with time chosen in the form

$$Q(t) = 1.26(1.0 - 0.001t) \qquad (11.16)$$

where $Q(t)$ is in heat flow units and t is in million years before present. (This one-parameter heat flow is maximally consistent with the vitrinite reflectance inversion.)

Table 11.4 gives the sensitivity results, in which the output variables are the maximum R_0 value,[2] maximum TTI value, maximum temperature, time and depth at which TTI = 15, time and depth at which $R_0 = 0.6$, kerogen transformation ratio (KTR), and total amount of hydrocarbons generated for the bottom layer.

All the output variables except the depth at TTI = 15 increase with an increase in the total depth. Table 11.4 also gives the relative errors in the changes of the output variables corresponding to a relative error of 2% in the input variable, indicating that the maximum TTI is most sensitive to the total depth, with a relative fractional error of 19%!

The maximum R_0 is less sensitive to the total depth, with a relative fractional error of 7%. The generation variables, kerogen transformation ratio, and total amount of hydrocarbons (HC) have a relative fractional error of 2.5% for the bottom layer and of 0.5% for the second layer, suggesting that the change in the total depth does not significantly influence the generation potential of the second layer. The depth to TTI = 15 has a variable range of 3634–3682 m (48-m difference), corresponding to a variable range of 5079–5287 m (208-m difference) in the total depth.

[2] In this chapter, the maximum R_0, maximum TTI, and maximum temperature mean the values of R_0, TTI, and temperature of the bottom layer at present.

Table 11.4 Sensitivity of Maturation and Hydrocarbon Generation to Total Depth (NWTEST301)

Total depth (m)	Maximum			TTI = 15		R_0 = 0.6		Layer 1		Layer 2	
	R_0	TTI	Temp. (°C)	Time (MYBP)	Depth (m)	Time (MYBP)	Depth (m)	KTR	THC (mg)	KTR	THC (mg)
5286.7	2.093	1480	161.9	39.5	3634	119.3	1303	0.81	245.8	0.75	236.0
5260.8	2.087	1306	161.5	39.9	3643	119.0	1284	0.80	245.8	0.75	235.8
5234.8	1.947	1191	159.7	39.2	3646	117.5	1278	0.79	245.9	0.75	235.8
5208.9	1.939	1117	158.7	38.8	3651	117.4	1250	0.78	245.9	0.75	235.5
5193.4	1.928	1066	157.9	38.1	3648	117.3	1228	0.78	245.9	0.75	235.5
5183.0	1.928	1101	158.8	38.1	3665	117.2	1203	0.78	245.9	0.75	236.8
5172.6	1.916	1025	157.0	37.5	3648	117.1	1205	0.77	245.9	0.75	235.0
5157.1	1.918	1071	158.6	37.3	3628	117.1	1198	0.78	246.0	0.75	236.5
5131.2	1.899	1018	156.9	36.4	3682	116.9	1157	0.77	246.0	0.75	236.1
5105.3	1.833	1017	156.9	35.7	3674	114.1	1143	0.77	246.0	0.75	236.6
5079.3	1.829	979	156.6	36.0	3681	113.9	1133	0.77	245.8	0.75	237.1

From the above discussion, we have the following summary of the sensitivity of the model to the formation depth:

1. The BUGG subroutine is necessary to obtain a better fit in thickness and porosity.
2. The model has no significant sensitivity to the depth of the second layer in terms of the total depth, formation pressure, and present-day heat flow.
3. The model is sensitive to the total depth. The most sensitive output variables with the total depth are the formation pressure of the bottom layer, the maximum R_0, and the maximum TTI.

2. Heat Flow at Present Day

Heat flow with time is a major factor controlling the temperature determination and so the maturity estimates. A present-heat flow value (Q_0) is constant in the heat flow function of either linear form

$$Q(t) = Q_0(1 + \beta t) \tag{11.17}$$

or exponential form

$$Q(t) = Q_0 e^{\beta t} \tag{11.18}$$

The present-day heat flow Q_0 can be determined from

$$Q_0 = \frac{\text{BHT} - T_s}{\int_0^{z_i} dz/K(z)} \tag{11.19}$$

where BHT is the bottom-hole temperature at depth z_i, T_s is the sediment surface temperature, and $K(z)$ is the thermal conductivity of the sediments at depth z at the present day. Also the Q_0 value can be supplied by users as an input parameter.

Table 11.5 gives the sensitivity of the maturation and generation calculations of the model to variations in the present-day heat flow, using the data of well NWTEST301. In Table 11.5 the present heat flow (the input variable) varies from 0.6 to 2.6 HFU, assuming a constant heat flow through time ($\beta = 0.0$). The output variables are the maximum R_0, maximum TTI, maximum temperature, ages and depth at TTI = 15 and $R_0 = 0.6$, kerogen transformation ratio, and total amount of HC for the bottom layer. Figures 11.3 to 11.6 show plots of these output variables

Table 11.5 Sensitivity of Maturation and Generation to Present-Day Heat Flow (NWTEST301)[a]

Present-day heat flow (HFU)	Maximum			TTI = 15		$R_0 = 0.6$			
	R_0	TTI	Temp. (°C)	Time (MYBP)	Depth (m)	Time (MYBP)	Depth (m)	KTR	THC (mg)
0.6	1.305	6	75.5	—	—	96.6	1294	0.06	0.02
0.8	1.534	25	100.7	13.0	5119	108.2	1243	0.23	8.6
1.0	1.797	147	125.9	27.0	4351	117.2	1203	0.64	199.2
1.2	2.032	784	151.1	37.9	3672	120.7	1187	0.75	237.5
1.4	2.439	4216	176.3	47.6	3256	135.5	1137	0.91	286.5
1.6	2.831	25436	201.4	55.3	2925	145.2	901	0.91	287.8
1.8	3.200	1.4E06	226.4	63.3	2594	147.0	856	0.92	289.5
2.0	3.622	8.2E06	251.8	70.6	2289	148.7	814	0.99	313.0
2.2	4.108	4.8E07	277.0	77.0	2018	150.8	762	1.00	315.0
2.4	4.667	2.7E08	302.2	86.4	1624	155.6	646	1.00	315.0
2.6	5.314	1.0E09	327.3	96.6	1294	160.2	535	1.00	315.0

[a] $\beta = 0.0$ and $\alpha = 0.0045$.

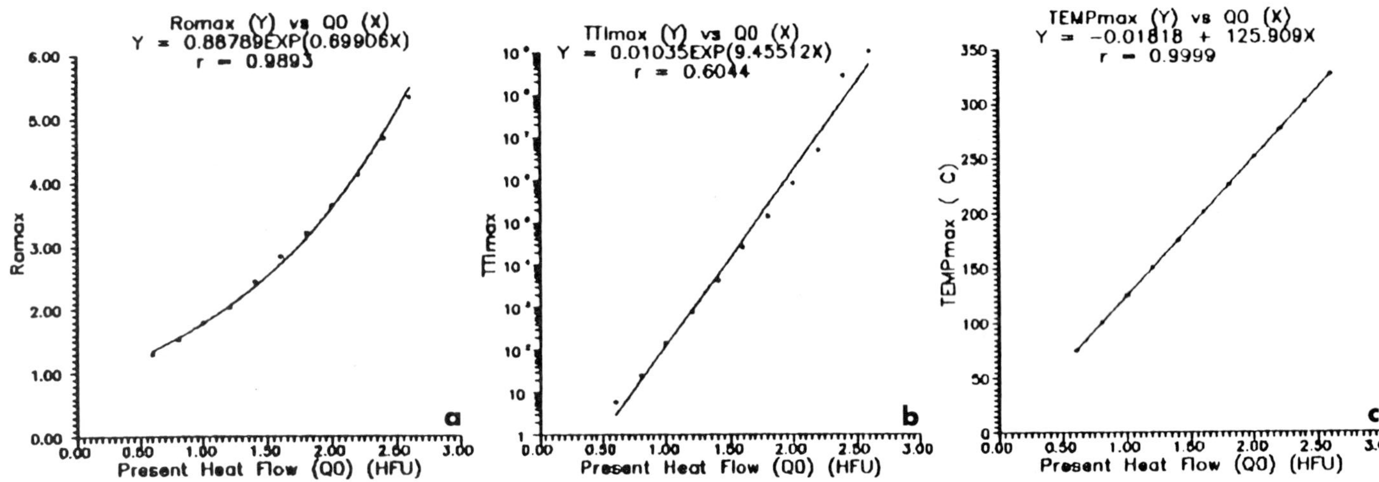

FIGURE 11.3 Plots of maximum R_0, TTI, and temperature versus present-day heat flow.

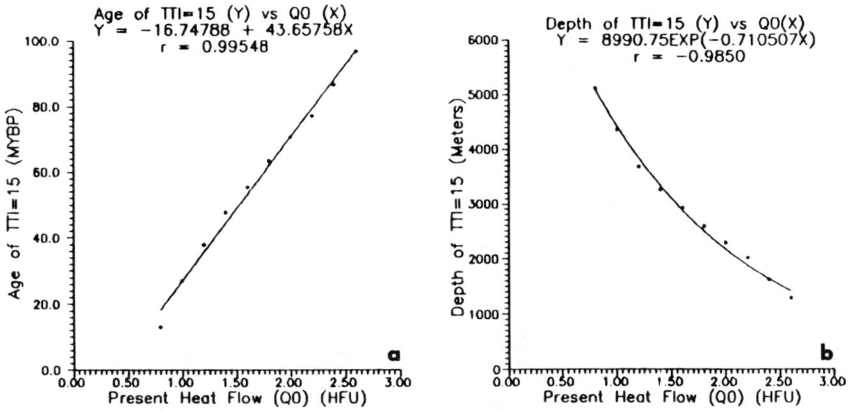

FIGURE 11.4 Plots of age of TTI = 15 and depth of TTI = 15 versus present-day heat flow.

versus the present heat flow, illustrating that the heat flow is a major factor in the maturation and generation calculations. All these output variables have a strong relation to the present-day heat flow, indicated by a linear correlation coefficient higher than 0.9, except for the maximum TTI, the KTR, and the total amount of HC. From Fig. 11.3b we see that the maximum TTI has a strong exponential relation to the present heat flow, even though the linear correlation coefficient is as low as 0.6. The reason for the relatively low correlation coefficients for KTR and the total amount of HC is that after the present heat flow exceeds 2.2 HFU, the KTR and total amount of HC reach their limiting values (1.0 for the KTR and 315 mg for the total amount of HC). From Fig. 11.6a and b we see that

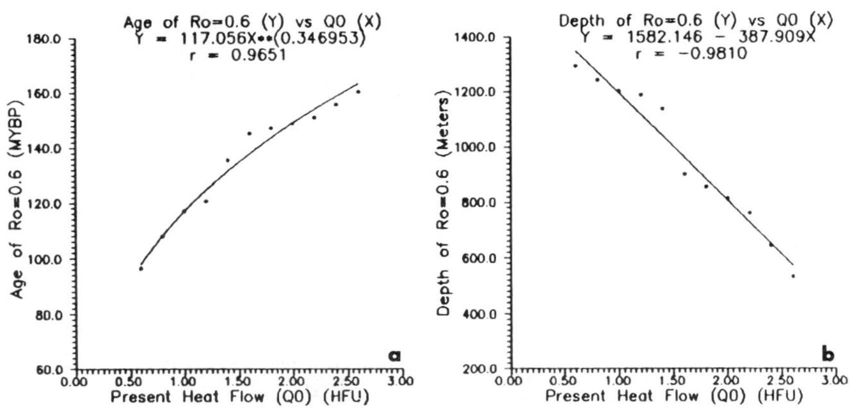

FIGURE 11.5 Plots of age of R_0 = 0.6 and depth of TTI = 15 versus present-day heat flow.

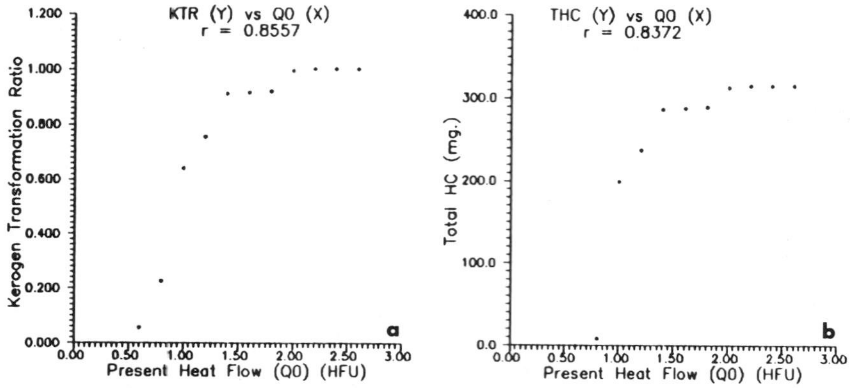

FIGURE 11.6 KTR and total hydrocarbon (THC) versus present-day heat flow.

the KTR and the total amount of HC increase quickly to their limits with the present heat flow changing from 0.6 to 2.0 HFU and then stay constant. Higher heat flow gives a higher thermal gradient, which make sediments mature at earlier times of deposition, as seen by the times at which TTI = 15 and R_0 = 0.6 (Figs. 11.4 and 11.5), and at shallow depths with small values in the depths at which TTI = 15 and R_0 = 0.6 (Figs. 11.4b and 11.5b).

3. Thermal Gradient

The thermal gradient (TG) is used as an alternative to the present heat flow. Equation (11.19) can be rewritten as

$$Q_0 = K \cdot \mathrm{TG} \tag{11.20}$$

with

$$\mathrm{TG} = (\mathrm{BHT} - T_s)/z_i \tag{11.21}$$

where K is the *average* thermal conductivity of the sediments from the surface to the BHT depth z_i.

It is easy to see that the sensitivity coefficient of the present heat flow to the thermal gradient is just the average thermal conductivity K. For given input data K is a constant. Since K is a positive number, the present-day heat flow will increase linearly with an increase in the thermal gradient at a rate equal to the average thermal conductivity K, as shown in the following table with K = 0.7073:

Input (TG, °F/100 ft)	Output (Q_0, HFU)
0.5	0.354
1.0	0.707
1.5	1.061
2.0	1.415
2.5	1.768
3.0	2.122
3.5	2.476
4.0	2.829
4.5	3.183

4. Kerogen Type and Content

There are basically three factors controlling the hydrocarbon generation of a given source rock. They are the burial history, the thermal history of the source rock, and the organic matter composition of the source rock. The organic matter composition of a source rock can be described by the kerogen type and content of the rock. Here we examine the sensitivity of the model to the kerogen type and content for a given burial history and thermal history, using data from well COST-1, Bering Sea, Alaska. Sensitivity tests on the kerogen type and content are encompassed by 13 different cases. Table 11.6 gives the sensitivity results.

From Tissot and Welte (1978), the genetic potential of kerogen, X_0, is 0.895 for type 1, 0.695 for type 2, and 0.313 for type 3. The kerogen transformation ratio is defined as the ratio of the kerogen already transformed to the genetic potential:

$$\mathrm{KTR} = (X_0 - \Sigma X_i)/X_0 \qquad (11.22)$$

KTR is zero at shallow depths and progressively increases to unity, which is reached when all labile organic material has been expelled, leaving a carbonaceous residue (Tissot and Welte, 1978).

Comparing cases 1, 2, and 3 in Table 11.6, we see that for the same burial history and thermal history, different kerogen types give different evolutions of kerogen degradation and hydrocarbon generation. Kerogen type 3 gives the highest value (0.575) and kerogen type 1 gives the lowest value (0.177) of the kerogen transformation ratio. Kerogen type 2 produces the most hydrocarbons (320.3 mg) in terms of the total amount of hydrocarbons generated. Kerogen types 2 and 3 have already passed the "oil window," and the produced oil has been cracked to gas. Kerogen

Table 11.6 Sensitivity Analysis for Kerogen Type and Content (Well COST-1)

	Kerogen content			Bottom layer			
Case	Type 1	Type 2	Type 3	KTR	Oil (mg)	Gas (mg)	THC (mg)
1	1.0	0.0	0.0	0.177	157.7	0.3	158.0
2	0.0	1.0	0.0	0.461	2.8	317.5	320.3
3	0.0	0.0	1.0	0.575	3.3	178.0	181.3
4	0.0	0.8	0.2	0.472	2.9	289.6	292.5
5	0.0	0.6	0.4	0.487	3.0	261.7	264.7
6	0.0	0.5	0.5	0.497	3.0	247.7	250.7
7	0.0	0.4	0.6	0.507	3.1	233.8	236.9
8	0.0	0.2	0.8	0.535	3.2	205.9	209.1
9	0.8	0.0	0.2	0.209	126.8	35.8	162.6
10	0.6	0.0	0.4	0.252	95.9	71.4	167.3
11	0.5	0.0	0.5	0.280	80.5	89.1	169.6
12	0.4	0.0	0.6	0.314	65.0	106.9	171.9
13	0.2	0.0	0.8	0.410	34.1	142.4	176.5
14	0.8	0.2	0.0	0.223	126.7	63.7	190.4
15	0.6	0.4	0.0	0.273	95.7	127.2	222.9
16	0.5	0.5	0.0	0.301	80.2	158.9	239.1
17	0.4	0.6	0.0	0.329	64.7	190.6	255.3
18	0.2	0.8	0.0	0.329	33.8	254.0	287.8

type 1 is still in the oil window and produces only a little gas from cracking oil.

Therefore, we can conclude that (1) kerogen type 1 takes more energy to break down to oil than kerogen types 2 and 3; (2) kerogen types 2 and 3 enter the oil window earlier than kerogen type 1; and (3) for a given thermal evolution history, different input data in kerogen type and content will give different kerogen degradation and generation histories in terms of the kerogen transformation ratio, the amount of oil and gas, and the total amount of hydrocarbons generated.

Cases 4 through 8 show the difference in hydrocarbon generation for the bottom layer with different contents of kerogen type 2 and type 3, indicating that with an increase in the content of kerogen type 3, the kerogen transformation ratio increases and the total amount of hydrocarbons decreases because of the decrease in the content of kerogen type 2 with higher genetic potential.

Cases 9 through 13 show the difference in hydrocarbon generation for the bottom layer with different contents of kerogen types 1 and 3, indicating that with an increase in the content of kerogen type 3, the kerogen transformation ratio increases, the amount of oil decreases, and the amount of gas increases with nearly a constant total amount of hydrocarbons.

Cases 14 through 18 show the difference in hydrocarbon generation for the bottom layer with different contents of kerogen type 1 and type 2, indicating that with an increase in the content of kerogen type 2, the kerogen transformation ratio increases, the amount of oil decreases, the amount of gas increases, and the total amount of hydrocarbons increases.

C. Sensitivity to Equation Parameters

1. Sediment Surface Porosity

The sediment surface porosity (ϕ_*) is an input parameter of the model. It is used (1) in the porosity–depth equation describing the present-day porosity change with depth

$$\phi(z) = \phi_* e^{-cz} \tag{11.23}$$

where $\phi(z)$ is the porosity at depth z and c is a lithology-dependent constant, and (2) to calculate the void ratio at the sediment surface

$$e_* = \phi_*/(1 - \phi_*) \tag{11.24}$$

where e_* is the sediment surface void ratio used as a constant in the constitutive equations connecting void ratio–frame pressure and permeability–void ratio.

Table 11.7 shows the sensitivity of the burial history to the sediment surface porosity of shale using data from COST-1 Bering Sea, Alaska, with BUGG OFF. The sediment surface porosity changes from 40 to 80%. Figures 11.7 and 11.8 show sensitivity plots for the output variables (total depth, formation permeability, and porosity and pressure of the bottom layer) versus the input variable (ϕ_*). From Table 11.7 and Fig. 11.7 and 11.8 we see that the total depth, formation permeability, and porosity and pressure of the bottom layer are all sensitive to the sediment surface porosity, especially when ϕ_* is larger than about 65%.

Figures 11.9, 11.10, and 11.11 show the profiles of the porosity, permeability, and formation pressure versus depth at present, respectively, for the well COST-1, with the sediment surface porosity varying from 50 to 80%. These three profiles suggest that (1) the change in the

Table 11.7 Sensitivity to Surface Porosity for Shale (COST-1)

Surface porosity (%)	Total depth (m)	Average values for the bottom layer: Permeability (md)	Porosity	Pressure (atm)
40.0	3318.8	0.139E − 03	0.050	372.2
45.0	3319.3	0.142E − 03	0.061	376.7
50.0	3317.4	0.146E − 03	0.074	383.5
55.0	3332.9	0.152E − 03	0.090	395.0
60.0	3343.0	0.161E − 03	0.110	412.3
65.0	3364.5	0.178E − 03	0.136	440.6
70.0	3403.8	0.209E − 03	0.172	485.3
75.0	3474.8	0.275E − 03	0.224	547.9
80.0	3601.5	0.415E − 03	0.302	620.6

sediment surface porosity of shale would have little effect on the porosity, permeability, and formation pressure of other lithologies, and (2) the change in the permeability of shale is less than the changes in the porosity and pressure of shale, and the permeability is less sensitive to the sediment surface porosity than are the porosity and pressure.

2. Sediment Surface Permeability

The sediment surface permeability (k_*) is used in the permeability–void ratio equation

$$k/k_* = (e/e_*)^B \qquad (11.25)$$

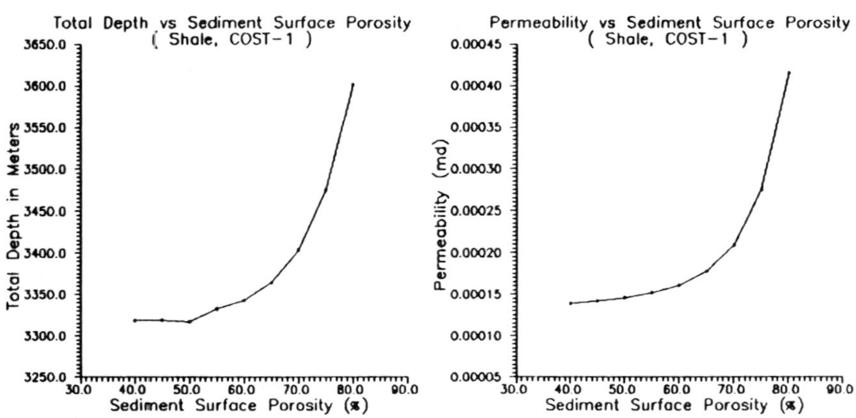

FIGURE 11.7 Plots of total depth and formation permeability versus ϕ_*.

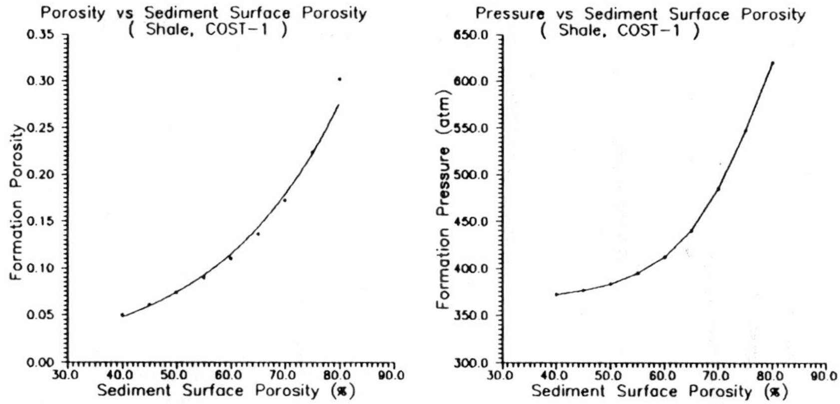

FIGURE 11.8 Plots of formation porosity and pressure versus ϕ_*.

where B is a constant. From this quasi-empirical equation of state we can calculate the permeability (k) for a given void ratio (e).

Table 11.8 shows the sensitivity of the total depth, formation permeability, porosity, and pressure of the bottom layer to the sediment surface permeability of shale for the data from well COST-1 Bering Sea, Alaska, with BUGG OFF. The sediment surface permeability changes from 0.001 to 100.0 md and, accordingly, the total depth changes from 4116 to

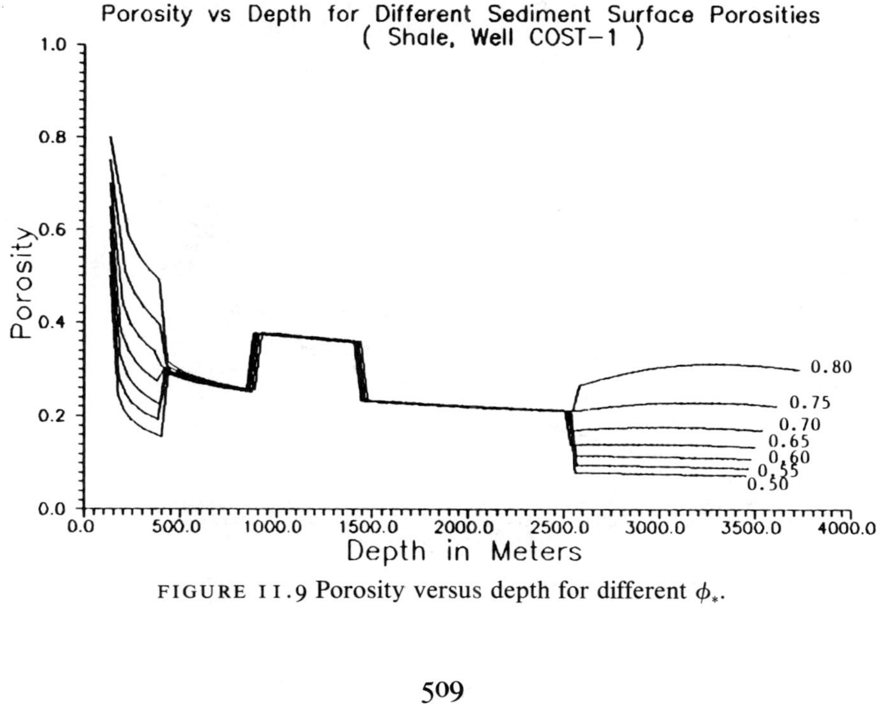

FIGURE 11.9 Porosity versus depth for different ϕ_*.

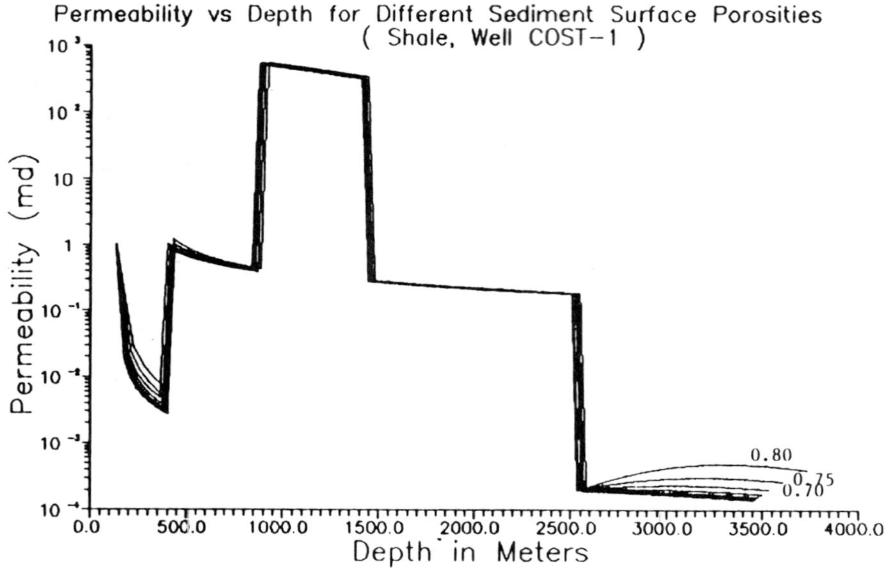

FIGURE 11.10 Permeability versus depth for different ϕ_*.

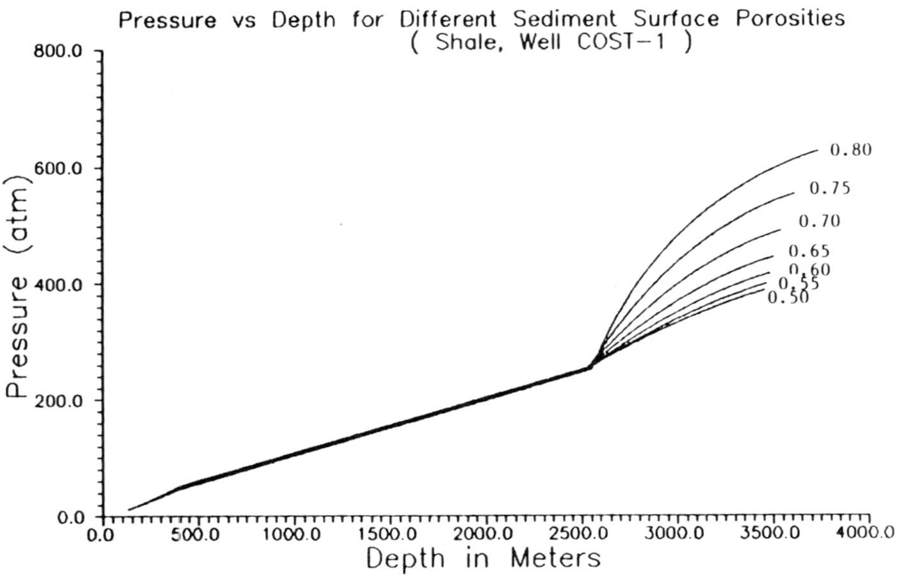

FIGURE 11.11 Formation pressure versus depth for different ϕ_*.

Table 11.8 Sensitivity to Sediment Surface Permeability of Shale (COST-1)

Surface permeability (md)	Total depth (m)	Average values of the bottom layer		
		Permeability (md)	Porosity	Pressure (atm)
0.001	4166.3	0.984E − 04	0.468	807.9
0.01	3825.5	0.680E − 04	0.306	773.5
0.1	3533.3	0.117E − 03	0.212	707.3
1.0	3364.5	0.178E − 03	0.136	440.6
10.0	3338.7	0.131E − 02	0.126	335.6
100.0	3335.4	0.128E − 01	0.125	326.6

3335 m, the deepest-layer permeability from 9.83×10^{-4} to 1.28×10^{-2} md, the porosity from 0.47 to 0.13, and the pressure from 808 to 327 atm. Figures 11.2 and 11.13 show the variation of total depth, formation permeability, porosity, and pressure of the bottom layer versus the sediment surface permeability of shale.

A higher sediment surface permeability gives a higher permeability (Fig. 11.12b). With increasing sediment surface permeability, the fluid in the shales moves out easily because of higher permeability. The shales are then more quickly compacted, which results in decreases of the total depth, the formation porosity, and the pressure of the bottom layer (Fig. 11.12a and 11.13a and b).

Figures 11.14, 11.15, and 11.16 show profiles of the porosity, perme-

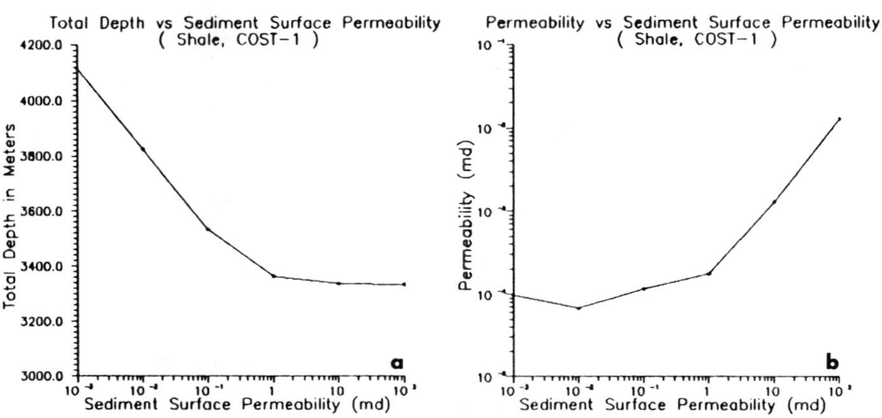

FIGURE 11.12 Plots of total depth and formation permeability versus k_*.

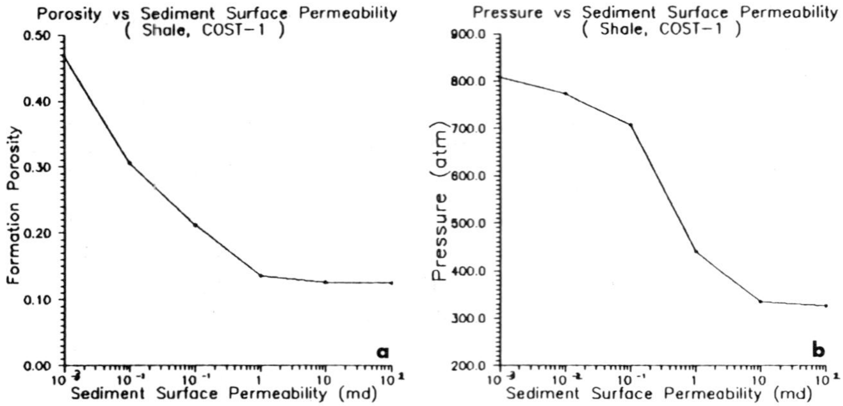

FIGURE 11.13 Plots of formation porosity and pressure versus k_*.

ability, and formation pressure versus depth for different values of the sediment surface permeability of shale. From these profiles, we see that (1) when the sediment surface permeability is larger than 1.0 md, the shales at the bottom of the sedimentary sequence are compacted like sand with no overpressure (Fig. 11.16), and a linear decrease in porosity with depth (Fig. 11.15) results; and (2) when the sediment surface permeability

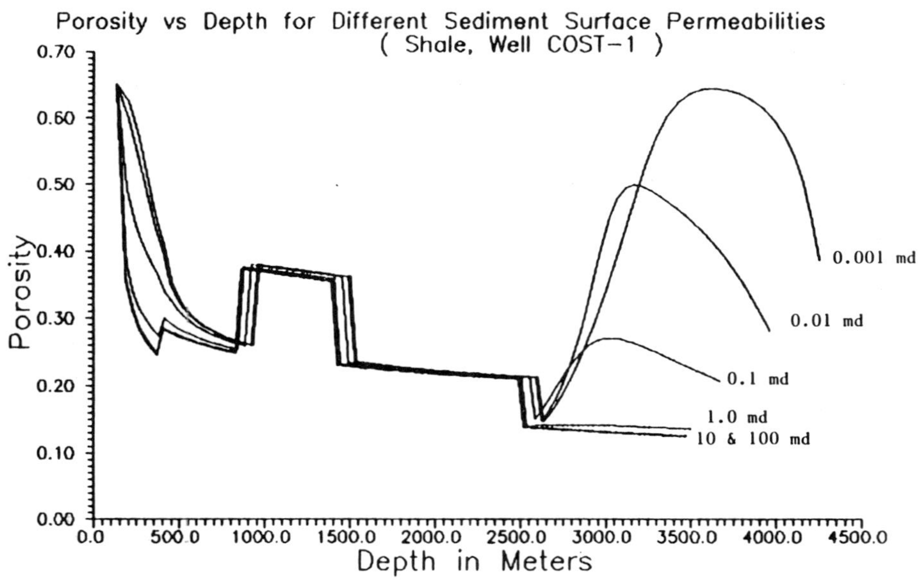

FIGURE 11.14 Porosity versus depth for different k_*.

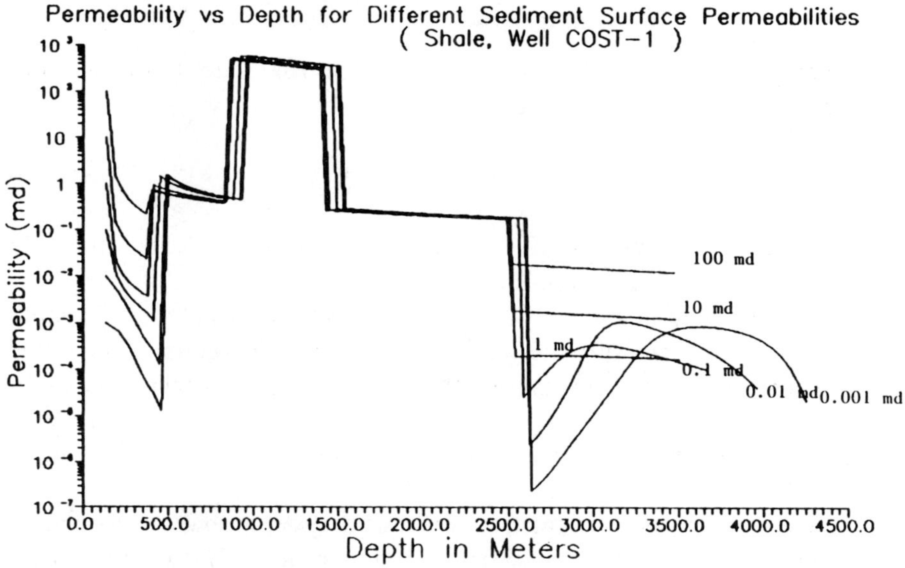

FIGURE 11.15 Permeability versus depth for different k_*.

is less than 0.1 md, the shales are abnormally noncompacted with unrealistically high values of porosity. Therefore, a choice of sediment surface permeability of shale between 0.1 and 1.0 md provides a reasonable compaction behavior of shales, in accord with observed values.

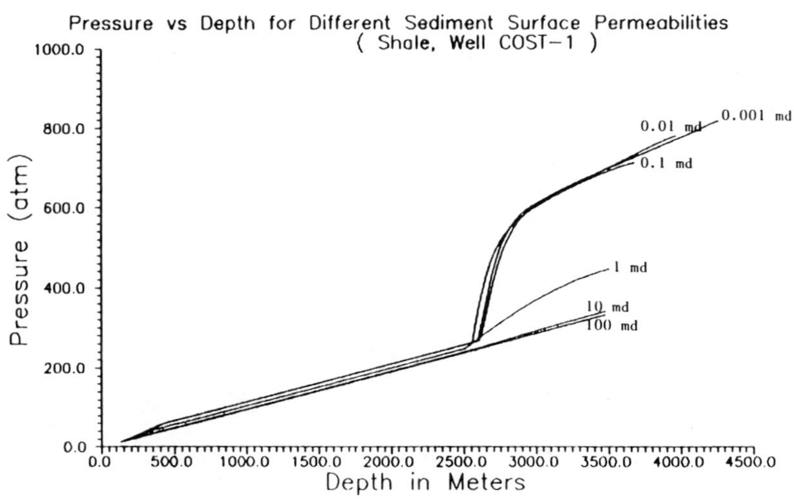

FIGURE 11.16 Formation pressure versus depth for different k_*.

3. Scaling Value of Frame Pressure

The frame pressure constant (P_{f*}) is used in the void ratio–frame pressure equation

$$(e/e_*) = (P_f/P_{f*})^{-1/A} \tag{11.26}$$

where A is a constant. From equation (11.26) we can calculate the void ratio for a given frame pressure.

A larger value of P_{f*} gives a higher frame pressure and a correspondingly lower fluid pressure (because the total pressure is equal to the frame pressure plus the fluid pressure). Therefore, a higher porosity is associated with a large value of P_{f*} because P_f is larger [equation (11.26)], and therefore a higher permeability [equation (11.25)] also results from a large P_{f*}.

Table 11.9 gives the sensitivity of the total depth and the formation permeability, porosity, and pressure of the bottom layer to the sediment surface frame pressure of shale for the data of Well COST-1 with BUGG OFF. The sediment surface frame pressure changes from 6.9 to 6.9×10^8 dyn. Figures 11.17 and 11.18 show plots of the above four output variables versus the input variables, demonstrating that with an increase in P_{f*} the total depth and the formation porosity and permeability of the bottom layer increase. However, the formation pressure of the bottom layer decreases first, reaches a lowest value of 364 atm at $P_{f*} = 6.9 \times 10^5$

Table 11.9 Sensitivity to Sediment Surface Frame Pressure of Shale (COST-1)

Surface frame pressure (dyn)	Total depth (m)	Average values of the bottom layer: Permeability (md)	Porosity	Pressure (atm)
6.89476E00	3244.1	0.287E − 05	0.045	724.9
6.89476E01	3272.2	0.586E − 05	0.056	720.8
6.89476E02	3296.3	0.176E − 04	0.075	703.3
6.89476E03	3318.2	0.665E − 04	0.106	656.0
6.89476E04	3365.1	0.179E − 03	0.136	440.9
6.89476E05	3526.1	0.145E − 02	0.223	364.1
6.89476E06	3841.5	0.146E − 01	0.357	377.1
6.89476E07	4286.5	0.135E00	0.512	414.9
6.89476E08	4610.2	0.726E00	0.629	602.3

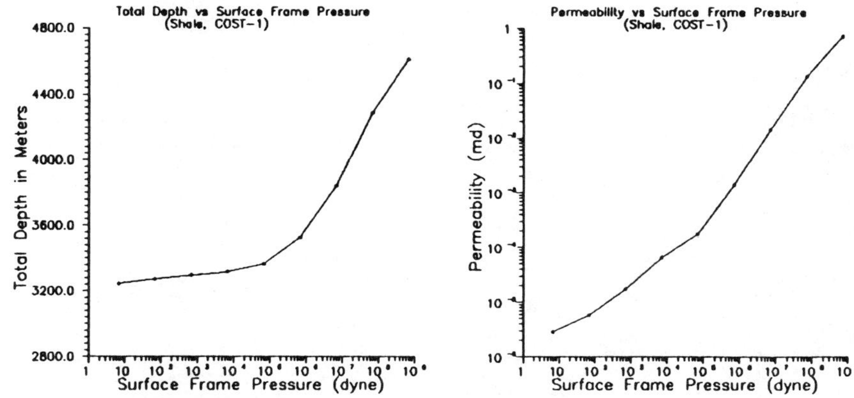

FIGURE 11.17 Plots of total depth and formation permeability versus P_{f*}.

dyn, and then increases with increasing P_{f*} (Fig. 11.18b). This turning point of pressure at $P_{f*} = 6.9 \times 10^5$ dyn may be an upper limit of P_{f*} for shale for the data from well COST-1, because when P_{f*} is larger the model gives an unrealistic relationship between the frame pressure and the formation pressure (fluid pressure). Table 11.9, Figs. 11.19, 11.20, and 11.21 also show that the total depth and the formation porosity and pressure are not so sensitive to P_{f*} when P_{f*} is less than 10^3 dyn, which may be a lower limit of P_{f*} shale.

4. *Parameter* A *in the Void Ratio–Frame Pressure Equation*

The parameter A is a constant in equation (11.26) describing the relationship between void ratio and frame pressure. For a given lithology, A

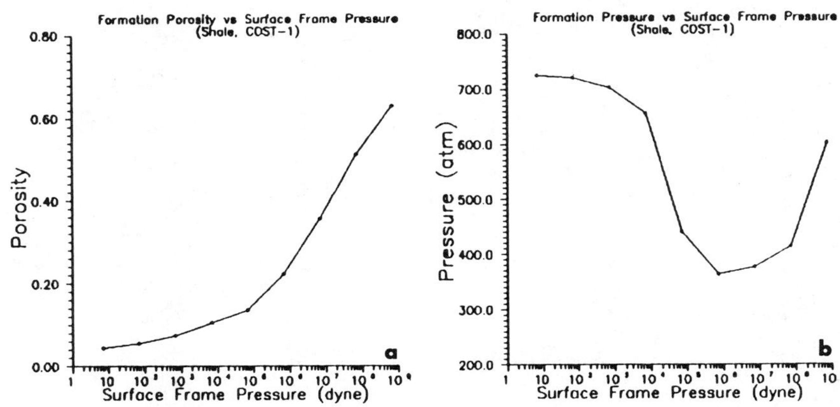

FIGURE 11.18 Plots of formation porosity and pressure versus P_{f*}.

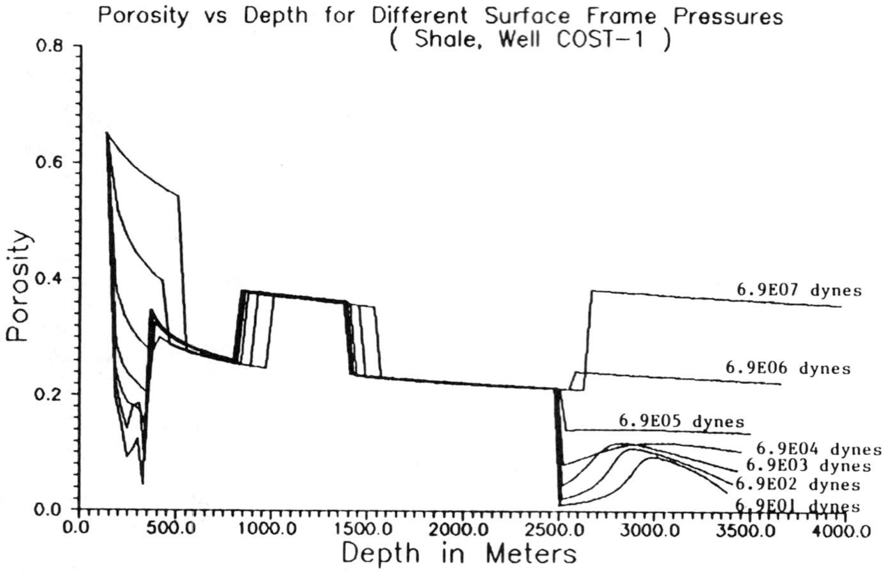

FIGURE 11.19 Porosity versus depth for different P_{f*}.

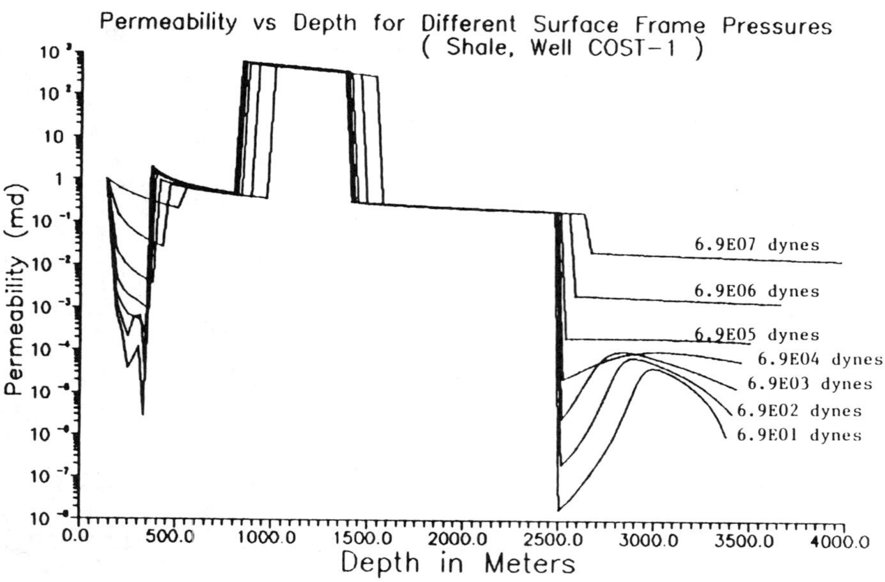

FIGURE 11.20 Permeability versus depth for different P_{f*}.

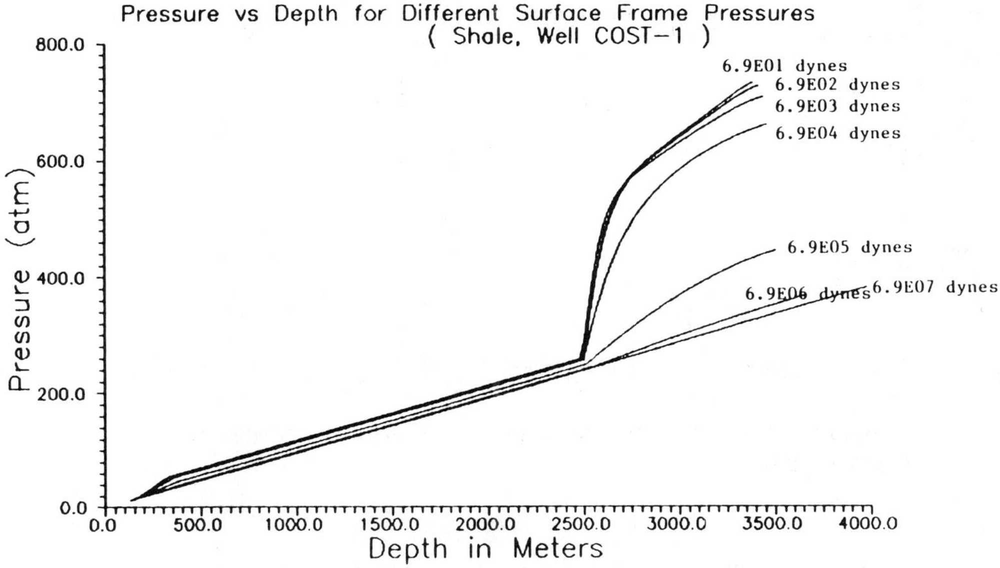

FIGURE 11.21 Formation pressure versus depth for different P_{f*}.

should be different with different burial histories; that is, *A* reflects the nature of the deposition and compaction. Usually *A* ranges from 1.0 to 6.0, with an average around of 3.0 (Lerche and Glezen, 1984).

Table 11.10 gives the sensitivity of the total depth and the formation

Table 11.10 Sensitivity to Parameter *A* for Shale (COST-1)

Parameter *A*	Total depth (m)	Average values of the bottom layer: Permeability (md)	Porosity	Pressure (atm)
1.0	3232.1	0.214E − 05	0.022	722.3
1.5	3247.8	0.764E − 05	0.035	704.3
2.0	3259.6	0.321E − 04	0.056	662.9
2.5	3273.2	0.753E − 04	0.073	546.0
3.0	3308.1	0.195E − 04	0.097	386.6
3.5	3360.5	0.606E − 03	0.136	354.0
4.0	3418.1	0.149E − 02	0.175	346.8
4.5	3477.2	0.305E − 02	0.212	347.1
5.0	3535.5	0.543E − 02	0.246	350.0
5.5	3591.9	0.871E − 02	0.276	353.7
6.0	3645.9	0.129E − 01	0.303	357.6

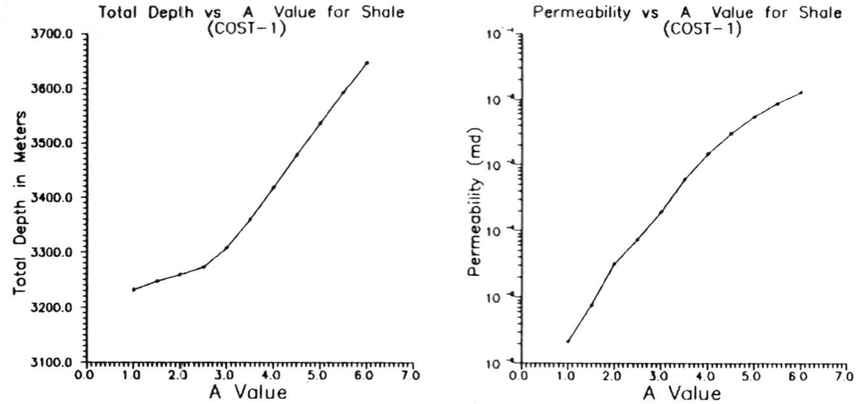

FIGURE 11.22 Plots of total depth and formation permeability versus parameter *A*.

permeability, porosity, and pressure of the bottom layer to the parameter *A* for shale, using the data of well COST-1 with BUGG OFF. The parameter *A* changes from 1.0 to 6.0 in steps of 0.5. Figures 11.22 and 11.23 show plots of the total depth, formation permeability, porosity, and pressure versus *A* and indicate that the total depth, formation permeability, and porosity increase with increasing *A* value. The increasing trends can be explained from equations (11.25) and (11.26) as follows: when *A* increases, the void ratio will increase for a given frame pressure (P_f) from

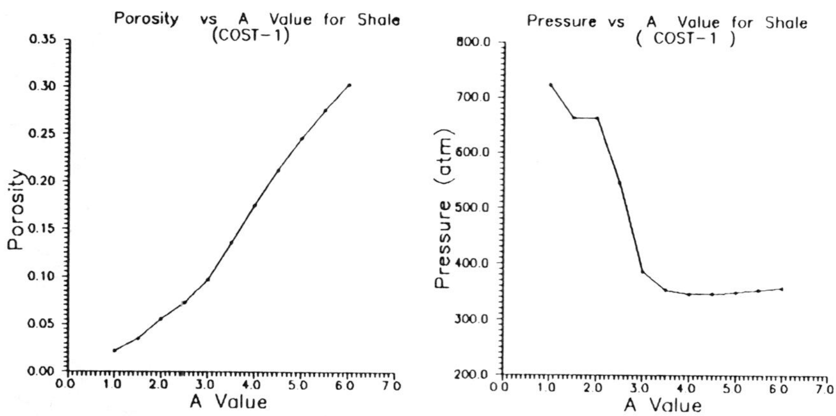

FIGURE 11.23 Plots of formation porosity and pressure versus parameter *A*.

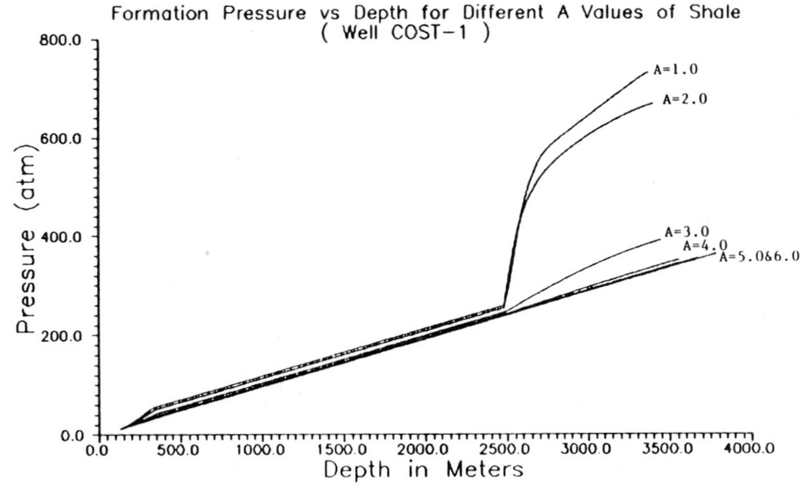

FIGURE 11.24 Formation pressure versus depth for different values of parameter *A*.

equation (11.26), which results in an increase in the permeability [equation (11.25)] and an increase in the total thickness.

Figure 11.23 shows that the formation pressure of the bottom layer decreases with an increase of *A* because the increase in the formation

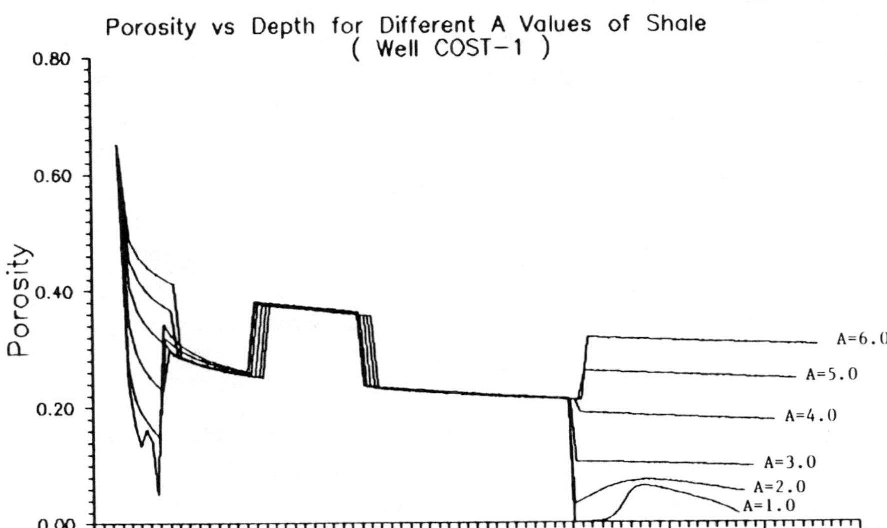

FIGURE 11.25 Porosity versus depth for different values of parameter *A*.

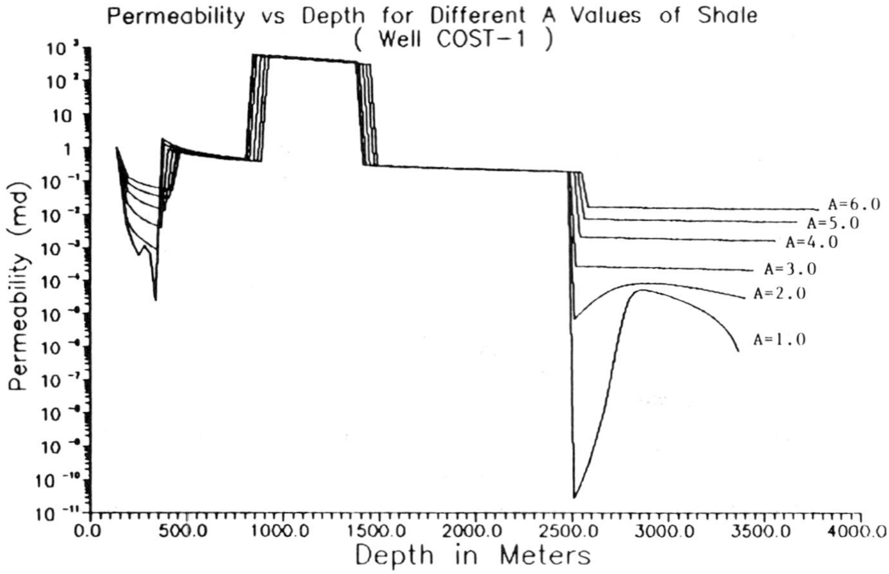

FIGURE 11.26 Permeability versus depth for different values of parameter *A*.

permeability makes the bottom layer more compacted. However, when the *A* value increases to around 4.0, the formation pressure stops decreasing and keeps a nearly constant pressure equal to hydrostatic pressure, which suggests that for this sedimentary sequence the parameter *A* of shale has its upper limit at around 4.0, beyond which the compaction of shales behaves like that of sand. This feature can be seen clearly from Fig. 11.24, showing that the pressures for $A = 4.0$, 5.0, and 6.0 are hydrostatic in the whole sequence. Figures 11.24, 11.25, and 11.26 suggest that there would be a lower limit of *A* of about 2.0, because when *A* is less than 2.0, the porosity and permeability of the bottom layer are too low compared with the observed values. Therefore the effective range of the parameter *A* of shale for well COST-1 is from 2.0 to 4.0.

5. *Parameter* B *in the Permeability–Void Ratio Equation*

Parameter *B* is a constant describing the relationship between permeability and void ratio. Like parameter *A*, *B* varies with different burial histories for a given lithology. In equation (11.25), we see that the permeability decreases with an increase in the *B* value for a given void ratio because the ratio e/e_* is always less than or equal to unity. When the permeability decreases, it is less easy for the fluid to escape from the

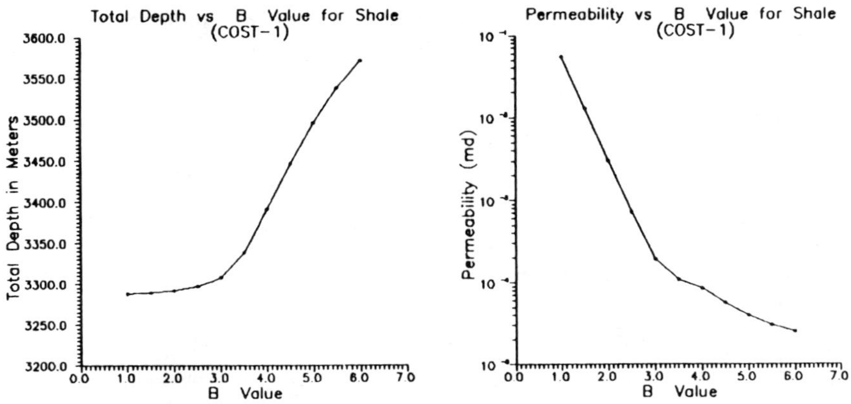

FIGURE 11.27 Plots of total depth and formation permeability versus parameter *B*.

sediments and formation pressure buildup occurs, which makes the sediments undercompacted with higher formation porosity.

Figures 11.27 and 11.28 show plots of the total depth, formation permeability, porosity, and pressure versus parameter *B* for data from well COST-1 with BUGG OFF. Parameter *B* was varied from 1.0 to 6.0 in steps of 0.5. With an increase in the *B* value, the total depth, formation porosity, and pressure increase while the formation permeability decreases. From the plots we see that the total depth, formation porosity, and pressure do not change too much when the *B* value is less than 3.0,

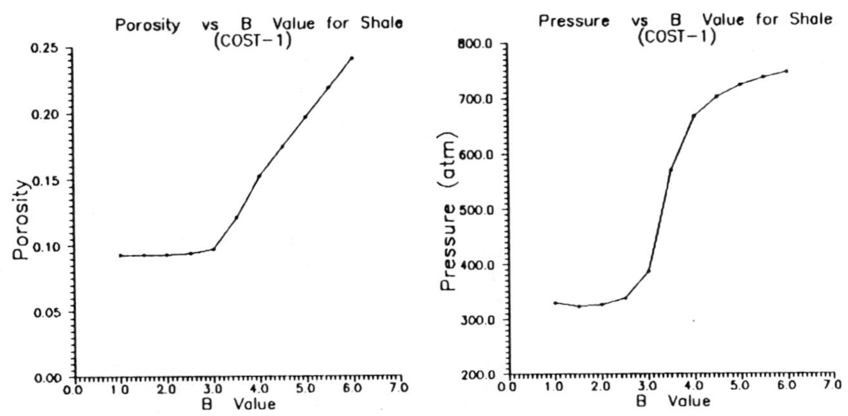

FIGURE 11.28 Plots of formation porosity and pressure versus parameter *B*.

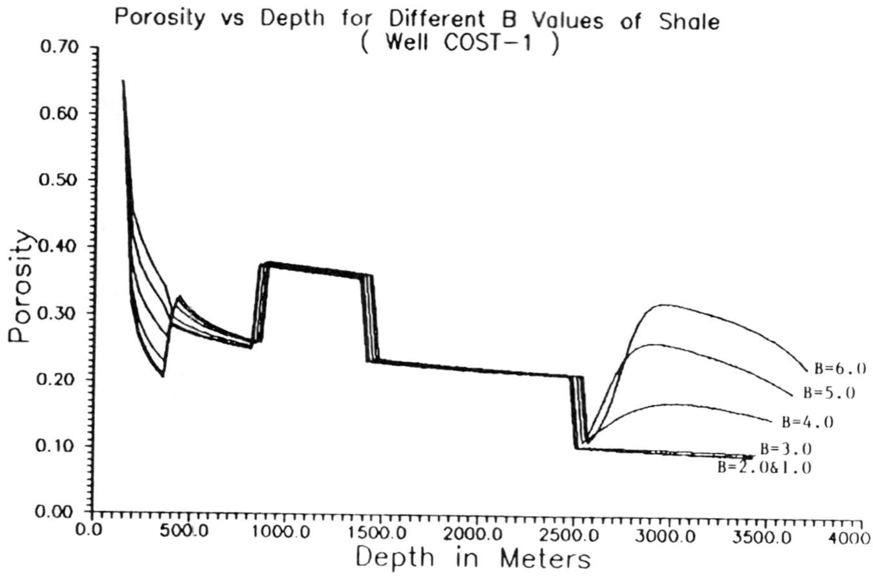

FIGURE 11.29 Porosity versus depth for different values of parameter *B*.

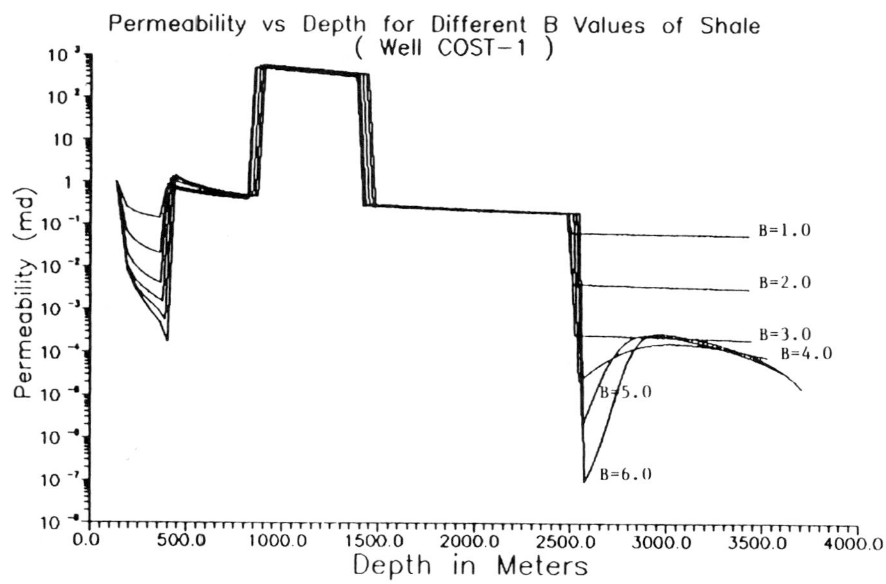

FIGURE 11.30 Permeability versus depth for different values of parameter *B*.

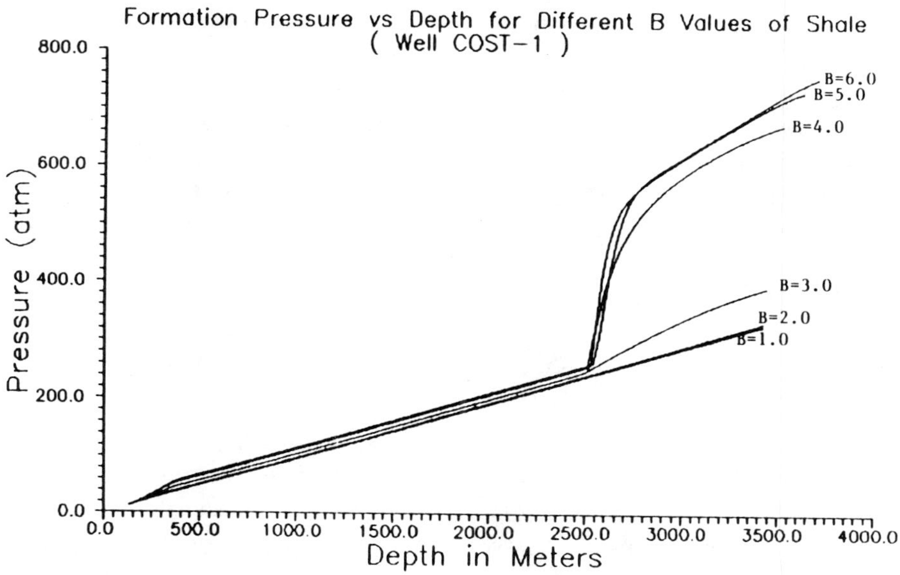

FIGURE 11.31 Formation pressure versus depth for different values of parameter *B*.

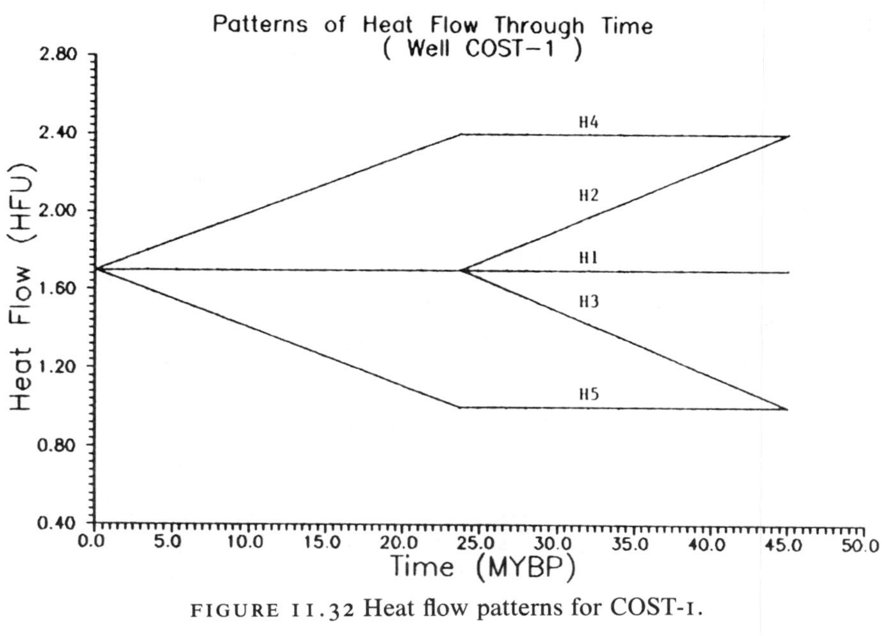

FIGURE 11.32 Heat flow patterns for COST-1.

but they are sensitive when the *B* value is larger than 3.0, which suggests that the *B* value of 3.0 may be the lower limit of shale for the COST-1 well.

Figures 11.29 to 11.34 give profiles of the porosity, permeability, and formation pressure versus depth for different values of parameter *B*, showing that when *B* is lower than 3.0, the bottom shales are in a hydrostatic condition; that is, no overpressure is developed. The pressure profile (Fig. 11.31) also indicates that when *B* takes the value of 6.0, the formation pressure is close to the overburden weight—the upper limit of the formation pressure—which suggest that a *B* value of 6.0 may be the upper limit of shale for the COST-1 well.

Parameters *A* and *B* are important in the simulation of sediment compaction from models because they define the relationship between void ratio and frame pressure and between permeability and void ratio. By choosing the best *A* and *B* values, models can give more accurate results when compared with the observed data in terms of thickness, porosity, permeability, and pressure. Hence a more appropriate burial history can be reconstructed, which is the critical factor in the determinations of thermal history and hydrocarbon generation and migration history.

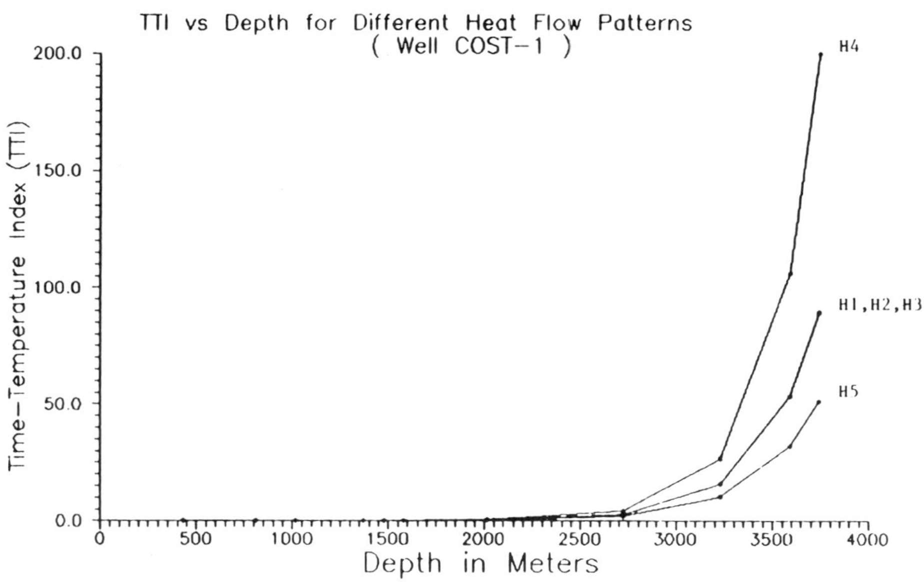

FIGURE 11.33 TTI versus depth for different heat flow patterns for COST-1.

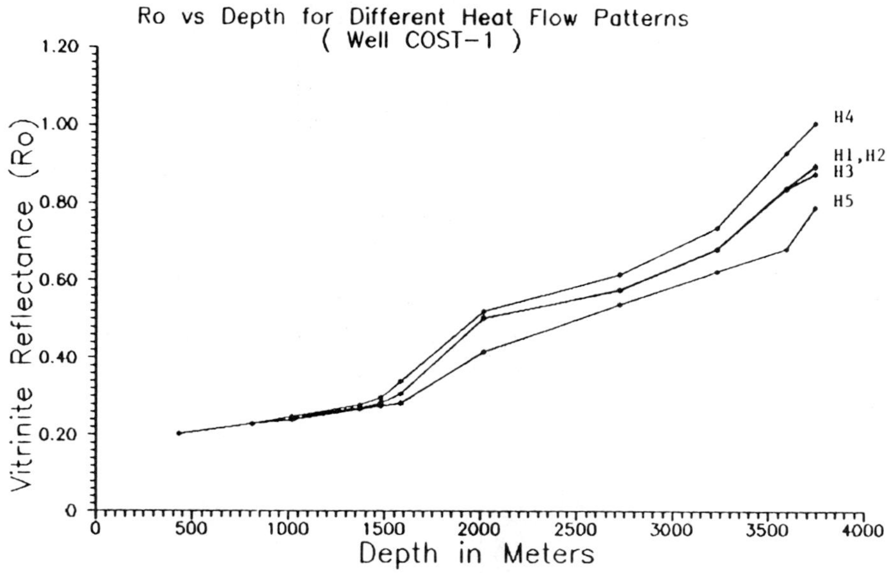

FIGURE 11.34 R_0 versus depth for different heat flow patterns for COST-1.

D. Sensitivity to Assumptions

1. Constant Sediment Surface Temperature

In the temperature calculation we assume that the sediment surface temperature (T_s) is constant through geological time to obtain the temperature at time t_i and depth z_i for a given heat flow $Q(t_i)$. Since the sediment surface temperature may vary with time, a sensitivity analysis is necessary.

Three types of sediment surface temperature variations are designed to run the sensitivity tests: (1) constant sediment temperature with time, (2) a linear decrease of sediment temperature with time, and (3) a linear increase of sediment surface temperature with time.

For well COST-1 with a present heat flow of 1.7 HFU, a β value of -0.017, an α value[3] of 0.01, and $T_c = 295$ K, the sediment surface temperature was chosen to (1) decrease from 41.0°F at present to 32.0°F at 45.0 million years ago, and (2) increase from 41.0°F at present to 80.0°F at 45 million years ago. Table 11.11 gives the sensitivity of the maturation

---

[3] α is a constant in the calculation of vitrinite reflectance based on the inversion technique (see Chapter 4).

Table 11.11 Sensitivity to Sediment Surface Temperature through Time (COST-1)

Surface temperature	Maximum		$R_0 = 0.6$		Bottom layer, TTI = 15		KTR	Oil (m)	Gas (mg)	THC (mg)
	TTI	R_0	Age (MYBP)	Depth (m)	Age (MYBP)	Depth (m)				
Decrease	51.7	0.695	13.3	2662.7	2.9	1437.5	0.498	0.68	288.7	289.4
Constant	52.9	0.708	13.6	1656.7	3.1	3421.4	0.499	0.67	289.1	289.9
Increase	58.9	1.063	24.5	2050.1	4.0	3280.2	0.502	0.63	290.8	291.4

and generation calculations to the sediment surface temperature. From Table 11.11 it can be seen that type 1 (constant) and type 2 (decrease) are nearly identical, and only a type 3 (increase) temperature variation has an effect on the R_0 and TTI calculations. The generation variables basically keep the same values for all three types of surface temperature variations. The insensitivity of the TTI and generation variables to the surface temperature variations can be explained as follows: these three types of surface temperature variations influence only the temperature variation with depth in the past (around 25 MYBP), and during that time the bottom layer was not deep enough to enter the maturation temperature range (above 90°C, Waples, 1980) for hydrocarbon generation. Therefore the temperature variations have little effect on the TTI and generation calculations. The R_0 calculation is slightly more sensitive to the sediment surface temperature because the effective maturation for R_0 starts from the critical temperature T_c in the time–temperature integral. For COST-1, the critical temperature is 22°C, much lower than the effective temperature for the TTI and generation calculations.

Therefore the sediment surface temperature variation through time slightly affects the maturation and generation of source rocks. It maybe necessary to consider this sediment surface temperature effect in basin analysis models.

2. Linear and Exponential Heat Flow with Time

Heat flow variation with time is very important in the reconstruction of temperature history, which, in turn, is the key factor controlling the maturation and hydrocarbon generation of source rocks. In the simplest models, heat flow changes are described by a one-parameter variation, which we take to be either linear or exponential with time. This section examines the sensitivity of the model to a variable heat flow with time (nonlinear and/or nonexponential). Five heat flow patterns were tested: pattern 1 (H1), heat flow constant with time; pattern 2 (H2,) heat flow constant for a certain time and then increasing; pattern 3 (H3), heat flow constant for a certain time and then decreasing; pattern 4 (H4), heat flow increasing to a certain time and then constant; and pattern 5 (H5), heat flow decreasing to a certain time and then constant. Figure 11.32 shows graphically the five heat flow patterns for well COST-1 with $\alpha = 0.01$ and $T_c = 295$ K. Figures 11.33 and 11.34 show the profiles of TTI and R_0 versus depth for different heat flow patterns. From Figs. 11.33 and 11.34 we see that heat flow patterns 1, 2, and 3 cause little change in the TTI, R_0, and generation calculations; heat flow pattern 4 gives the highest values in the maximum TTI, maximum R_0, kerogen transformation ratio, and total amount of hydrocarbon; and pattern 5 gives the lowest values in these output variables. These features can be explained as follows: even though heat flow patterns 1, 2, and 3 have different values before 23.7 MYBP, they keep the same value (1.7 HFU) from the present to 23.7 MYBP (Fig. 11.32). This period is the most effective time for the bottom layer to mature and generate hydrocarbons, because the bottom layer was in the effective maturation zone. Therefore these three heat flow patterns give the same effective temperature history, resulting in the same maturation and generation features for the bottom layer. However, heat flow pattern 4 provides the highest heat flow during the period from the present to 23.7 MYBP, and pattern 5 provides the lowest heat during this period; hence they give the highest values and the lowest values for the TTI, R_0, kerogen transformation ratio, and total amount of hydrocarbons, respectively.

Similar sensitivity test results for well NWTEST301 with similar heat flow patterns, as shown in Fig. 11.35, produce the profiles of TTI and R_0 versus depth for these different heat flow patterns given in Figs. 11.36 and 11.37. We see the same features as we have seen from well COST-1. Pattern 4 gives the highest values in the maturation and generation, and pattern 5 gives the lowest values. Patterns 1, 2, and 3 also give the same

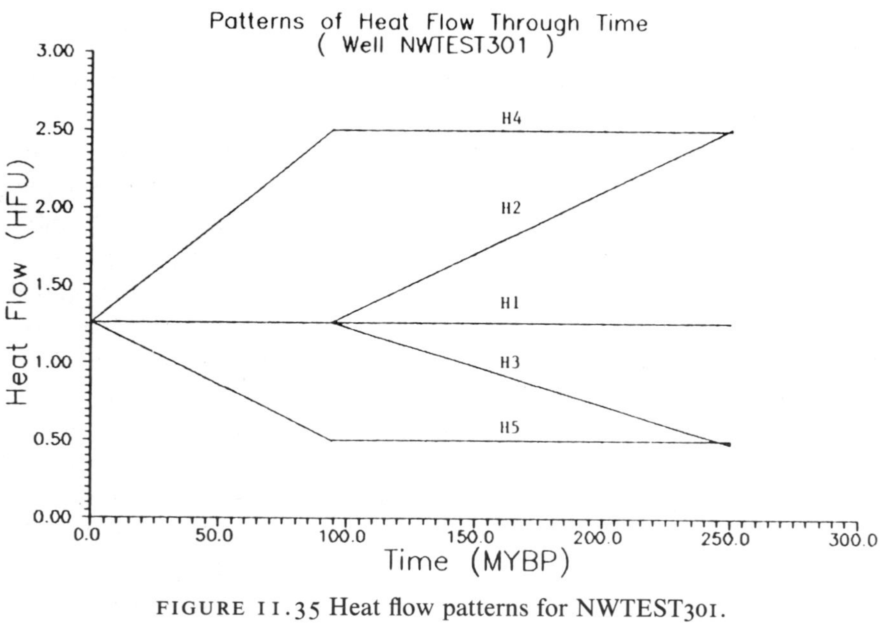

FIGURE 11.35 Heat flow patterns for NWTEST301.

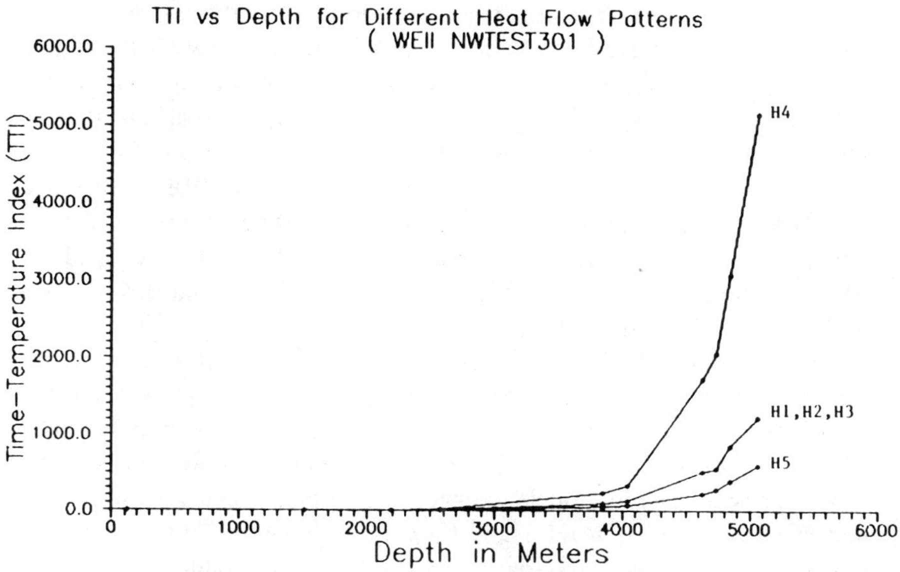

FIGURE 11.36 TTI versus depth for different heat flow patterns for NWTEST301.

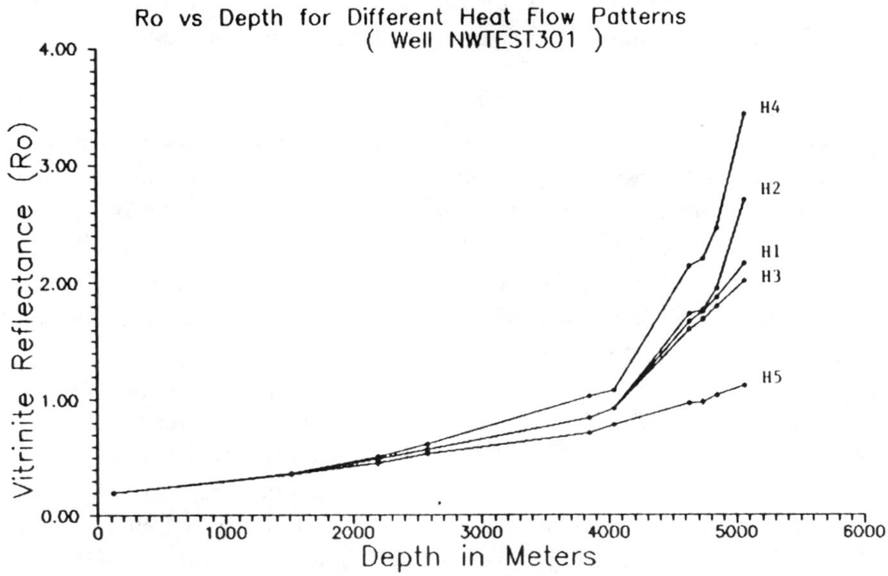

FIGURE 11.37 R_0 versus depth for different heat flow patterns for NWTEST301.

values in TTI, kerogen transformation ratio, and total amount of hydrocarbons but slightly different values in R_0 (Fig. 11.37) with the highest value (2.155) for pattern 2 and the lowest value 2.007 for pattern 3. This difference in the R_0 value for patterns 1, 2, and 3 suggests that the effective temperature range for R_0 is different from the one for the TTI and hydrocarbon generation. Since the effective temperature starts from the T_c value, the bottom layer enters the effective temperature range earlier for pattern 2 (which gives the highest heat flow before 94 MYBP among patterns 1, 2, and 3) than for patterns 1 and 3 and obtains more energy to mature the vitrinite.

From the above discussion, it is obvious that different heat flow patterns do influence significantly the temperature history, especially the heat flow variations during the time the sediment rocks are deeply buried. Therefore, it is crucial to develop nonlinear and nonexponential heat flow functions with time in models in order to give more accurate thermal and generation histories. And such heat flow functions must surely be constrained by present-day data in order to be acceptable.

E. Summary of Sensitivity Tests

Table 11.12 gives a summary of the sensitivity analysis from this study and the more extensive investigation of Cao (1987). The meanings of symbols in Table 11.12 are as follows: —, no sensitivity; *, slightly sensitive; **, sensitive; ***, very sensitive; blank, no sensitivity tests were done.

In Table 11.12 the input variables on which the sensitivity is tested are listed vertically and the output variables are listed horizontally. Sensitivity of an output variable to an input variable means that the input variable directly influences the output variable without any effects from other variables. For example, the sensitivity test of the maturation and generation calculations on the formation depth is performed with a fixed heat flow function to take out the effect of variable heat flow on the maturation and generation calculations.

From Table 11.12 it can be seen that among the input data variables, the formation depth, eroded thickness, lithology, and present-day heat flow have strong influences on the output variables which describe the thermal history, maturation, and generation history. The formation age and erosion time have relatively slight effects on these output variables. The sediment surface temperature, bottom-hole temperature, and thermal gradient influence the present-day heat flow. Since output results for maturation and generation are sensitive to the input variable of present-

Table 11.12 Sensitivity Analysis Summary[a]

	Output variable															
		Average values for the bottom layer						Maximum			TTI = 15		R_0 = 0.6			
Input variable	Total depth	Permeability	Porosity	Pressure	Heat Q_0	β	MSR	R_0	TTI	Temp.	Time	Depth	Time	Depth	KTR	Total HC
Formation depth	***	*	**	***	**	**	**	**	**	**	**	**	**	**	**	**
Formation age	—	—	—	—	—	*	**	*	—	—	—	—	*	*	—	—
Unconformity eroded thickness	**	*	*	**	**	**	**	**	**	**	*	*	*	*	**	**
Unconformity erosion time	*	*	*	*	*	*	*	*	*	*	*	*	—	—	—	—
Lithology	**	***	**	***	**	**	**	**	**	**	**	**	**	**	**	**
Present heat flow	—	—	—	—	***	***	***	***	***	***	***	***	***	***	***	***
Surface temperature	—	—	—	—	***											
BHT	—	—	—	—	***											
Thermal gradient	—	—	—	—	***											
Kerogen composition	—	—	—	—	—	—	—	—	—	—	—	—	—	—	**	**
Surface porosity	**	*	***	**												
Surface permeability	**	***	**	**												
Surface frame pressure	**	**	**	**												
Parameter A	***	**	***	**												
Parameter B	**	***	**	***												
Parameter T_c	—	—	—	—	—	**	**	***	—	—	—	—	***	***	—	—
Parameter T_D	—	—	—	—	—	*	**	**	—	—	—	—	*	*	—	—
Thermal conductivity	—	—	—	—	***	**	**	**	***	***	***	***	***	***	***	***
Patterns for $T_0(t)$	—	—	—	—	—	—	—	**	**	—	**	**	**	**	**	**
Patterns for $Q(t)$	—	—	—	—	—	—	—	**	**	—	**	**	**	**	**	**

[a] See text for meanings of symbols.

day heat flow, the sediment surface temperature, bottom-hole temperature, and thermal gradient must be determined as accurately as possible. For the equation parameter tests, the burial history (reflected by the total depth, formation permeability, porosity, and pressure) is sensitive to the five input variables of sediment surface porosity, permeability, frame pressure, parameter A, and parameter B. It is obvious that these five input variables have major influences on thermal history and hydrocarbon generation history because they are partly controlled by the burial history. The parameters T_c and T_D influence the R_0 calculation and also have some effect on β and MSR determinations. Likewise, the thermal conductivity is a very sensitive input variable for thermal and generation histories, which are sensitive to the paleotemperature and heat flow in the past.

A comprehensive sensitivity analysis is important in the examination of a simulation model for basin analysis in terms of the various disturbances in the input data, equation parameters, and assumptions of the model. Also, by a sensitivity analysis we can obtain more quantitative measures of the connections between the geological and geochemical variables which influence the geohistory, thermal history, and hydrocarbon generation and migration history in a sedimentary basin. For example, obvious concerns are the influence of depth and lithology of a formation on its thermal maturity and hydrocarbon generation potential, the effect of thermal conductivity on maturation and generation histories, and the relationship between bottom-hole temperature and present-day heat flow.

Any simulation model has its own limitations and weaknesses because of (1) the differences in understanding of the geological and geochemical processes related to hydrocarbon generation and migration, and (2) the choices of numerical methods for solving the equations describing these processes. Sensitivity analysis is a powerful tool for determining these limitations and weaknesses and hence makes the model more useful and effective in basin analysis.

For a complicated simulation model which involves different aspects of geology, geophysics, geochemistry, thermodynamics, and hydrodynamics, it is not easy to use direct methods (Tomovic, 1963; Yukler, 1979; Fiacco, 1984) to run a comprehensive sensitivity analysis. In this case a perturbation-type method should be undertaken even though more computer time has to be used.

Finally, in Cao's (1987) study, the sensitivity of hydrocarbon generation was run only for Tissot's model and showed that the hydrocarbon generation was sensitive to the burial history, thermal history, and kerogen composition (of kerogen type and content) of source rocks.

So far, we are unaware of any exhaustive sensitivity study of the many hydrocarbon generation models suggested or the equation parameters in the generation models, perhaps because too many parameters would need to be examined. For example, even in Tissot's model each type of kerogen has six components to degrade to oil and each component has three free parameters [genetic potential (X_{i0}), activation energy (E_i), and frequency factor (A_i) (Tissot and Welte, 1978)]. A pressing concern is to determine how strongly hydrocarbon generation and migration vary during basin evolution history as the models and parameters are changed.

Epilogue

WE WOULD BE REMISS if we did not at least attune the reader to some limitations of the procedures examined. The general logic of examining the convoluted evolutionary interplay between sediment burial, thermal history, and hydrocarbon generation, migration, and accumulation will, presumably, always be one of the more dominant foci of quantitative basin analysis. Some of the methods of examination have been the major thrust of this book, with case histories to exemplify particular points at each stage of development. In order to provide such a pattern of interlocked behaviors, we had to forgo a more detailed investigation of many major factors controlling, or modifying, particular behaviors and processes in specific basins. Thus, effects due to diagenesis, salt movement, shale diapirism, flexural basement response, carbonate development, igneous dikes and sills, overthrust domains, rheological creep, glaciation and postglacial rebound, and so on were treated either rather cursorily or not at all. These omissions should not be construed as a statement that such effects are not of import. Indeed, quite the contrary is the case, as is clear from many case histories around the world. In each individual situation the impact of physical, chemical, and geological processes must be examined for relevance, resolution, uniqueness, and precision, including processes not addressed here. The point is that the pattern of logic connecting sediments, thermal effects, and hydrocarbons remains invariant, while the techniques and details change, dependent on the quantification of processes being investigated.

Apart from the limitations in any particular application of basin analysis techniques due to the finite number of processes included, the reader should not form the impression that the quantitative techniques of basin analysis are a universal panacea. Just as in any other area of science, the results of quantitative analysis are only as good as (1) the quality, quantity, and sampling frequency of the input data; (2) the accuracy and precision of input parameters; and (3) the capability of underlying assumptions to provide appropriate underpinning keystones on which the veracity of quantitative analyses rest.

But even with the limitations, and resultant uncertainties, of basin analysis outputs, two very positive factors are apparent. First, as the capability to model more complex and realistic situations expands, including more data and different types of data, the basic understanding of basinal evolution increases. Thus we are in a mode of growth, converting qualitative perceptions of basins and their evolution into quantitative assessments of the relevant evolutionary behaviors. As with all branches of science, this objective approach sharpens and corrects our preconceived ideas and perceptions, hones the quantitative tools and devices at our disposal, and suggests further methods and data needed to improve on our capability for accurately assessing basinal evolution—a Baconian philosophy if you will.

Second, in many basins we already know where oil and gas accumulations occur. It then becomes a relatively simple matter to do "blind" tests of proposed models or processes, comparing predictions against observations until a clearer understanding is achieved of the limits of accuracy, resolution, uniqueness, and uncertainty of current methods of quantitative basin analysis. Such testing schemes have proved inordinately useful in sorting out viable methods of analysis from others that appear equally acceptable at first glance.

As we succeed at understanding correctly as many observations as possible in as many basins as possible, we have the satisfaction of knowing that the integrated pictures based on basin analysis methods are progressively better and more accurate. The application of such techniques in frontier basins then provides a better basis for risk assessment than would otherwise be possible. We must cut the cost, we must be more accurate in locating hydrocarbons, and we must suggest, in as expert a manner as possible, potential areas of hydrocarbon proneness for future generations. It is toward these ends that quantitative methods in basin analysis are geared.

References

Abbott, G. D., Lewis, C. A. and Maxwell, J. R., (1985). Laboratory models for aromatization and isomerization of hydrocarbons in sedimentary basins. *Nature* **318,** 651–653.

Agassiz, L., (1840). "Études sur les glaciers." Neuchatel, Switzerland (privately published).

Airy, G. B. (1855). On the computation of the effect of the attraction of mountain-masses as disturbing the apparent astronomical latitude of station of geodetic surveys. *Philos. Trans. R. Soc. London* **145,** 101–104.

Aki, K., and Lee, W. H. K. (1976), Determination of three-dimensional anomalies under a seismic array using first P arrival times from local earthquakes; a homogeneous initial model. *JGR, J. Geophys. Res.* **81,** 4381– 4399.

Allen, J. R. L. (L964), Studies in fluviatile sedimentation:Six cyclothems from the Lower Old Red Sandstone, Anglo–Welsh Basin. *Sedimentology,* **3,** 163–198.

Anders, D. E., and Gerrild, P. M. (1984). Hydrocarbon generation in lacustrine rocks of Tertiary age, Uinta Basin, Utah—organic carbon, pyrolysis yield, and light hydrocarbons. In "Hydrocarbon Source Rocks of the Greater Rocky Mountain Region" (J. Woodward, F. F. Meissner, and J. L. Clayton, eds.), pp. 513–529. Rocky Mt. Assoc. Geol., Denver, Colorado.

Anderson, D. W., and Picard, M. D. (1974). Evolution of synorogenic clastic deposits in the intermontane Uinta Basin of Utah. Spec. Publ.—*Soc. Econ. Palentol. Mineral.* **22,** 167–189.

Anderson, R. B. (1965). The sensitivity problem in control system optimization. Ph.D. Dissertation, Georgia Institute of Technology, Atlanta.

Andrew-Speed, C. P., Oxburgh, E. R., and Cooper, B. A. (1984). Temperature and depth-dependent heat flow in west North Sea. *Am. Assoc. Pet. Geol. Bull.* **68,** 1764–1781.

Anthony, D. B., and Howard, J.B. (1976). Coal devolatilization and hydrogasification. *AIChE J.* **22,** 625.

Arana, V., and Vegas, R. (1974). Plate tectonics and volcanism in the Gibralter arc. *Tectonophysics* **54,** 197–212.

Armagnac, C. (1985). Thermal history of the National Petroleum Reserve, Alaska. M.S. Thesis, University of South Carolina, Columbia.

Armagnac, C., Bucci, J., Kendall, C. G. St. C., and Lerche, I. (1988). Estimating the thickness of sediment removed at an unconformity using vitrinite reflectance data. In "Thermal History of Sedimentary Basins" (N. Nayser and T. H. McCulloh, eds.). Springer-Verlag, Berlin and New York.

Armagnac, C., Kendall, C. G. St. C., Kuo, C., Pantano, J., and Lerche, I. (1988). Determination of paleoheatflux from vitrinite reflectance data, and sterane and hopane isomer data. *J. Geochem. Exp.* **30,** 1–28.

Athy, L. F. (1930a). Density, porosity, and compaction of sedimentary rocks: *Am. Assoc. Pet Geol. Bull.* **14,** 1–24.

Athy, L. F. (1930b). Compaction and oil migration. *Am. Assoc. Pet. Geol. Bull.* **14,** 25–35.

Atwater, G. I. (1956). Future of Louisiana offshore oil province. *Am. Assoc. Pet Geol. Bull.* **40,** 2624–2634.

Atwater, G. I., and Miller, E. E. (1965). The effect of decrease in porosity with depth on future development of oil and gas reserves in south Louisiana. *Am. Assoc. Pet. Geol. Bull.* **49,** 334 (abstr.).

Bachmann, G. H., and Koch, K., (L983). Alpine front and Molasses Basin, Bavaria. *Stud. Geol. (Tulsa, Okla.)* **15(3),** 3.4–1–27.

Bally, A. W. (1975). A geodynamic scenario for hydrocarbon occurrences. *Proc. World Pet. Congr.* **9,** 33–44.

Bally, A. W. (1981). Basins and subsidence. *Am. Geophys. Union, Geodyn. Ser.* **1,** 5–20.

Barker, C. E. (1983). Influence of time on metamorphism of sedimentary organic matter in liquid-dominated geothermal systems, western North America. *Geology* **11,** 384–388.

Barnes, H. L., ed. (1979). "Geochemistry of Hydrothermal Ore Deposits." Wiley, New York.

Battersby D. G. (1976). Cooper Basin gas and oil fields. *Monogr. Ser—Aust. Inst. Min. Metal.* **7,** 321–368.

Beaumont, C. (1981). Foreland basins. *Geophys. J. R. Astron. Soc.* **65,** 291–329.

Berger, A. L. (1976). Obliquity and precession for the last 5,000,000 years. *Astron. Astrophys.* **5,** 127–135.

Berger, A. L. (1978). Long-term variations of caloric insolation resulting from the earth's orbital elements. *Quat. Res.* **9,** 139–167.

Bethke, C. M. (1985). A numerical model of compaction-driven groundwater flow and heat transfer and its application to the paleohydrology of intracratonic sedimentary basins. *JGR, J. Geophys. Res.* **90,** 6817–6829.

Biju-Duval, B., and Montadert, L. (1977). Introduction to the structural history of the Mediterranean basins. In "The Structural History of the Mediterranean Basins" (B. Biju-Duval, and L. Montadert, eds.), pp. 1–12 Editions Technip, Paris.

Bird, K. J. (1981). Petroleum exploration of the North Slope, Alaska. In "Petroleum Geology in China, Principal Lectures Presented to the United Nations International Meeting on Petroleum Geology, Beijing, China 18–25

March 1980" (J. F. Mason, ed.), pp. 233–248. Penn Well Books, Tulsa, Oklahoma.

Bissell, E. R., Burnham, A. K., and Braun, R. L. (1985). Shale oil cracking kinetics and diagnostics *Ind. Eng. Chem. Process Des. Dev.* **24,** 381.

Blanchard, D. C., and Tailleur, I. L. (1962). Temperatures and interval geothermal-gradient determinations from wells in National Petroleum Reserve in Alaska. *Geol. Surv. Open-File Rep. (U.S.)* **82–391,** 1–79.

Bomford, G. (1971). "Geodesy," 2nd ed. pp. 441–443. Oxford Univ. Press, London and New York.

Bond, G. (1978). Speculations on real sea level changes and vertical motions of continents at selected times in the Cretaceous and Tertiary periods. *Geology* **6,** 247–250.

Bond, G. C., and Kominz, M. A. (1984). Construction of tectonic subsidence curves for the early Proterozoic Miogeocline, Southern Canadian Rocky Mountains: Implications for subsidence mechanisms, age of breakup, and crustal thinning. *Geol. Soc. Am. Bull.* **95,** 155–173.

Bonham, L. C. (1980). Migration of hydrocarbons in compacting basins. *Am. Assoc. Pet. Geol. Bull.* **64,** 549–567.

Born, M., and Wolf, E. (1964). "Principles of Optics." Pergamon, New York.

Bradley, W. H. (1925). A contribution to the origin of the Green River Formation and its oil shale. *Am. Assoc. Pet Geol. Bull.* **9,** 247–262.

Bradley, W. H. (1929). The varves and climate of the Green River epoch. *Geol. Surv. Prof. Pap. (U.S.)* **158–E,** 87–110.

Bradley, W. H. (1931). Origin and microfossils of the oil shales of the Green River Formation of Colorado and Utah. *Geol. Surv. Prof. Pap. (U.S.)* **168,** 58.

Bradley, W. H. (1970), Green River oil shale—concept of origin extended. *Geol. Soc. Am. Bull.* **81,** 985–1000.

Braun, R. L., and Burnham, A. K. (1987). General kinetic model of oil shale pyrolysis. *Energy Fuels* **1,** 153–161.

Brederhoeft, J. D., and Hanshaw, B. B. (1968). On the maintenance of anomalous fluid pressures. I. Thick sedimentary sequences. *Geol. Soc. Am. Bull.* **79,** 1097–1106.

Brent, R. P. (1973). "Algorithms for minimization without Derivatives." Prentice-Hall, Englewood Cliffs, New Jersey.

Brewer, J. (1981). Thermal effects of thrust faulting. *Earth Planet Sci. Lett.* **56,** 233–244.

Broecker, W. S. (1966). Absolute dating and the astronomical theory of glaciation. *Science* **151,** 299–304.

Broecker, W. S., and Van Donk, J. (1970). Insolation changes, ice volumes, and the O^{18} in deep-sea cores. *Rev. Geophys. Space Phys.* **8,** 169–198.

Brooks, J., and Welte, D. (1984). "Advances in Petroleum Geochemistry." Academic Press, London.

Brownlee, K. A. (1960). "Statistical Theory and Methodology in Science and Engineering." Wiley, New York.

Bucci, J. (1985). Basin analysis, thermal history and hydrocarbon generation of the Rharb Basin, Morocco. M.S. Thesis, University of South Carolina, Columbia.

Buffler, R. T., Shipley, T. H., and Watkins, J. S. (1978). Blake continental margin. *Am. Assoc. Pet. Geol., Seismic Line* **2**.

Burnham, A. K. (1981). Chemistry of shale oil cracking. *ACS Symp. Ser.* **163,** 39.

Burnham, A. K., and Singleton, M. F. (1983). High-pressure pyrolysis of Green River oil shale. *ACS Symp. Ser.* **230,** 335.

Burnham, A. K., Clarkson, J. E., Singleton, M. F., Wong, C. M., and Crawford, R.W. (1982). Biological markers from Green River kerogen decomposition. *Geochim. Cosmochim. Acta* **46,** 1243–1251.

Burnham, A. K, Braun, R. L., Gregg, H. R., and Samoun, A. M. (1987). Comparison of methods for measuring kerogen pyrolysis rates and fitting kinetic parameters. *Energy Fuels* **1,** 452–458.

Burruss, R. C., Cercone, K. R., and Harris, P. M. (1983). Fluid inclusion petrography and tectonic-burial history of the Al Ali no. 2 well: Evidence for the timing of diagenesis and oil migration, northern Oman foredeep. *Geology* **7,** 567–570.

Bushnell, H. (1981). Unconformities—key to North Slope oil. *Oil Gas J.* **79,** 114–116.

Campbell, J.H., Koskinas, G. J., and Stout, N. D. (1978a). Kinetics of oil generation from Colorado oil shale. *Fuel* **57,** 372–376.

Campbell, J. H., Koskinas, G. J., Stout, N. D., and Coburn, T. T. (1978b). Oil shale retorting: Effects of particle size and heating rate on oil evolution and intra-particle oil degradation. *In Situ* **2,** 1–47.

Campbell, J. H., Gallegos, G., and Gregg, M. (1980). Gas evolution during oil shale pyrolysis. 2. Kinetic and stoichiometric analysis. *Fuel* **59,** 727–732.

Cao, S. (1985). A quantitative dynamic model for basin analysis and its application to the Northern North Sea basin. M.S. Thesis, University of South Carolina, Columbia.

Cao, S. (1987). Sensitivity analysis of 1-D dynamical model for basin analysis. Ph.D. Dissertation, University of South Carolina, Columbia.

Cao, S., and Lerche, I. (1987). Geohistory, thermal history and hydrocarbon generation of the northern North Sea basin. *Energy Explor. Exploit.* **5,** 315–355.

Cao, S., and Lerche, I. (1988). Geohistory, thermal history and hydrocarbon generation history of Navarin Basin COST No. 1 well, Bering Sea, Alaska. *J. Pet. Geol.* (in press).

Cao, S., Glezen, W. H., and Lerche, I. (1986). Fluid flow, hydrocarbon generation and migration: A quantitative model of dynamical evolution in sedimentary basins: *Proc. Offshore Technol. Conf.* **2,** Pap. 5182, 267–276.

Cao, S., Lerche, I. and Hermanrud, C. (1988). Formation temperature estimation by inversion of borehole measurements. *Geophysics* **53,** 979–988.

Carmichael, R. S., ed. (1982). "Handbook of Physical Properties of Rocks," Vol. 2. CRC Press, Boca Raton, Florida.

Carslaw, H. S., and Jaeger, J. C. (1947). "Conduction of Heat in Solids." Oxford Univ. Press, London and New York.

Carter, R. D., Mull, C. G., Bird, K.J., and Powers, R.B. (1977). The petroleum geology and hydrocarbon potential of Naval Petroleum Reserve No. 4, North Slope, Alaska. *Geol. Surv. Open-File Report (U.S.)* **77–475,** 1–61.

Castano, J. R. (1985). Uses of vitrinite reflectance in determining thermal history in sedimentary basins. *Am. Assoc. Pet. Geol. Bull.* **69,** 243 (abstr).

Cathles, L. M., and Smith, A. T. (1983). Thermal constraints on the formation of Mississippi Valley-type lead–zinc deposits and their implications for episodic basin dewatering and deposit genesis. *Econ. Geol.* **78,** 983–1002.

Cercone, K. R. (1984). Thermal history of Michigan Basin. *Am. Assoc. Pet. Geol. Bull.* **68,** 130–136.

Chapman, D. S., Keho, T. H., Bauer, M. S., and Picard, M. D. (1984). Heat flow in the Uinta Basin determined from bottom hole temperature (BHT) data. *Geophysics* **49,** 455–466.

Chappell, J. (1974). Geology of coral terraces, Huon Peninsula, New Guinea: A study of Quaternary tectonic movements and sea level changes. *Geol. Soc. Am. Bull.* **85,** 553–570.

Chappell, J., and Veeh, H. H. (1978). Late Quaternary tectonic movements and sea level changes at Timor and at Atauro Island. *Geol. Soc. Am. Bull.* **89,** 356–368.

Chatfield, J. (1972). Case history of Redwash Field, Uinta County, Utah, *Mem. Am. Assoc. Pet. Geol.* **16,** 342–353.

Chorafas, D. N. (1965). "System and Simulation." Academic Press, New York.

Churkin, J., Jr., Carter, C., and Trexler, J. H., Jr. (1980). Collision-deformed Paleozoic continental margin of Alaska—foundation for microplate accretion. *Geol. Soc. Am. Bull.* **91,** 648–654.

Coburn, T. T., and Campbell, J. H. (1977). Oil shale retorting. Part II. Variation in product oil chemistry during retorting of an oil shale block. *Lawrence Livermore Lab. Rep. UCRL* **UCRL-52256**, Pt. 2.

Connan, J. (1974). Time–temperature relation in oil genesis. *Am. Assoc. Pet. Geol. Bull.* **58,** 2516–2521.

Conybeare, C. E. B. (1965). Hydrocarbon generation potential and hydrocarbon yield capacity of sedimentary basin. *Bull. Can. Pet. Geol.* **13,** 509–528.

Cook, F. A., Brown, L. D., Kaufman, S., and Oliver, J. E. (1983). The COCORP seismic reflection traverse across the southern Appalachians. *Stud. Geol. (Tulsa, Okla.)* **14,** 61.

Cooper, B. S., and Barnard, P. C. (1984). Source rocks and oils of the central and northern North Sea. *Am. Assoc. Pet. Geol. Mem.* **35,** 303–314.

Crain, W. E., Mero, W. E., and Patterson, D. (1985). Geology of the Point Arguello discovery. *Am. Assoc. Pet. Geol. Bull.* **69,** 537–545.

Curnelle, R., and Marco, R. (1983). Reflection profiles across the Aquitaine Basin (North Pyrenean Basin). *Stud. Geol. (Tulsa, Okla.)* **15(1),** 1.2–3–1.
Curnelle, R., and Marco, R. (1983). Reflection profiles across the Aquitaine Basin (sedimentary sequences). *Stud. Geol.* **15**(1), 1.2–3–4.
Cutler, R. T., Langan, R. T., Love, P. L., and Bishop, T. N. (1986). A tomographic solution to the travel time problem in general inverse seismology. In "Advances in Geophysical Data Processing." JAI Press (in press).
Dahlberg, E. C. (1982). "Applied Hydrodynamics in Petroleum Exploration." Springer-Verlag, New York.
Darracott, B. (1974). On the crustal structure and evolution of southeastern Africa and the adjacent Indian Ocean. *Earth Planet. Sci. Lett.* **24,** 282–290.
Dastillung, M., and Albrecht, P. (1977). Δ^2 Steranes as diagenetic intermediates in sediments. *Nature* **269,** 678–679.
DeBuyl, M. and, Austin, J. (1983). "Non-Exclusive Geophysical Survey Offshore Southern Mozambique, Review of Exploration Opportunities." Western Geophysical Company of America, London.
Demaison, G. (1984). The generative basin concept. *Am. Assoc. Pet. Geol. Mem.* **35,** 1–14.
Dickey, P. A. (1975). Aqueous solubility of petroleum as applied to its origin and primary migration. *Am. Assoc. Pet. Geol. Bull.* **59,** 337–345.
Dickinson, G. (1953). Geological aspects of abnormal reservoir pressures in Gulf Coast Louisiana. *Am. Assoc. Pet. Geol. Bull.* **37,** 410–4321.
Dimian, M. V., Gray, R., Stout, J., and Wood, B. (1983). Hudson Bay Basin. *Stud. Geol.* **15(2),** 2.2–4–1.
Dow, W. G. (1977). Kerogen studies and geological interpretations. *J. Geochem. Explor.* **7,** 77–79.
Dozy, J. J. (1970). A geologic model for the genesis of the lead–zinc ores of the Mississippi Valley, U.S.A. *Trans.—Inst. Min. Metal.* **79,** 163–170.
Du Rouche, T.(1981). Stress field, a key to oil migration. *Am. Assoc. Pet. Geol. Bull.* **65,** 79–99.
Eaton, B. A. (1969). Fracture gradient prediction and its application in oil field operation. *J. Pet. Technol.* **21,** 1353–1360.
Ensminger, A., Van Dorsselaer, A., Spyckerelle, C., Albrecht, P. and Ourrison, G. (1974). Pentacyclic triterpanes of the hopane type as ubiquitous geochemical markers: Origin and significance. *Adv. Org. Geochem. Proc. Int. Meet., 6th, 1973* pp. 245–260.
Ensminger, A., Albrecht, P., Ourisson, G. and Tissot, B. (1977). Evolution of polycyclic alkanes under the effects of burial (early Toarcian shales, Paris Basin). *Adv. Org. Geochem. Proc. Int. Meet., 7th, 1975* pp. 45–52.
Ensminger, A., Joly, G., and Albrecht, P. (1978). Re-arranged steranes in sediments and crude oils. *Tetrahedron Lett.* **18,** 1575–1578.
Erxleben, A. W., and Carnahan, G. (1983). Slick Ranch area, Starr County, Texas. *Stud. Geol. (Tulsa, Okla.)* **15(2),** 2.3–1–22.

Evans, T. R. (1977). Thermal properties of North Sea rocks. *Log Anal.* March-April, 3–12.

Evernden, J. F., Curtis, G. H., Kistler, R. W., and Obradovich, J. (1960), Argon diffusion in glauconite, microcline sanidine, leucite and phlogipite. *Am. J. Sci.* **256,** 583–604.

Fabuss, B. M., Smith, J. O., and Satterfield, C. N. (1964). Thermal cracking of pure saturated hydrocarbons. *Adv. Pet. Geochem. Refin.* **9,** 157.

Falvey, D. A. (1974). The development of continental margins in plate tectonic theory. *APEA J.* **14,** 95–106.

Falvey, D. A., and Middleton, M. F. (1981). Passive continental margins: Evidence for a prebreakup deep crustal metamorphic subsidence mechanism. *Int. Geol. Congr., Colloque C3.3, Geol. Cont. Margins, 26th, 1980* to Vol. 4, Suppl., pp. 103–114.

Falvey, D. A., and Mutter, J. C. (1981). Regional plate tectonics and the evolution of Australia's passive continental margins. *BMR J. Aust. Geol. Geophys.* **6,** 1–29.

Fenske, M. R., Carnahan, F. L., Breston, J. N., Caser, A. H., and Rescorla, A. R. (1942). Optical rotation of petroleum fractions. *Ind. Eng. Chem.* **34,** 638–646.

Fertl W. H. (1976). Abnormal formation pressures: Implications to exploration, drilling, and production of oil and gas reservoirs. *Dev. Pet. Sci.* **2.**

Fiacco, A. V. (1984). "Sensitivity, Stability and Parametric Analysis," Math. Program. Study 21. North-Holland Publ., Amsterdam.

Fillon, R. H. (1984). Continental glacial stratigraphy, marine evidence of glaciation and insights into continental–marine correlations. In "Principles of Pleistocene Stratigraphy Applied to the Gulf of Mexico" (N. Healy-Williams, ed.), pp. 149–206. IHRDC, Boston, Massachusettes.

Fillon, R. H., and Williams, D. F. (1983). Glacial evolution of the Plio-Pleistocene: Role of continental and arctic ice sheets. *Paleogeogr., Paleoclimatol., Paleoecol.* **43,** 7–33.

Fillon, R. H., and Williams, D. F. (1984). Dynamics of meltwater discharge from Northern Hemisphere ice sheets during the last deglaciation. *Nature* **310,** 674–677.

Fischer, A. G. (1964). The Lofer cyclothem of the Alpine Triassic. *Bull.—Kans., State Geol. Surv.* **169,** 107–149.

Fleischer, R. L., Price, P. B., and Walker, R. M. (1965). Effects of temperature, pressure and ionisation on the formation and stability of fission tracks in minerals and glasses. *J. Geophys. Res.* **70,** 1497–1582.

Fleischer, R. L., Price, P. B., and Walker, R. M. (1975). "Nuclear Tracks in Solids." Univ. of California Press, Berkeley.

Flores, G. (1973). The Cretaceous and Tertiary sedimentary basins of Mozambique and Zululand. In "Sedimentary Basins of the African Coasts" (G. Glant, ed.), Pt. 2, pp. 81–111. South and East Coasts, Assoc. Afr. Geol. Surv., Paris.

Forster, R. (1975). The geological history of the sedimentary basin of southern Mozambique, and some aspects of the origin of the Mozambique channel. *Paleogeogr., Paleoclimatol., Paleoecol.* **17**, 267–287.

Fouch, T. D. (1975). Lithofacies and related hydrocarbon accumulations in Tertiary strata of the western and central Uinta Basin. In "Deep Drilling Frontiers in the Central Rocky Mountains," (D. W. Boyard, ed.), Spec. Publ., pp. 163–173. Rocky Mt. Assoc. Geol., Denver, Colorado.

Fouch, R. D. (1981). Distribution of rock types, lithologic groups, and interpreted depositional environments for some lower Tertiary and upper Cretaceous rocks from outcrops at Willon Creek–Indian Canyon through the subsurface of Duchesne and Altamont oil fields, southwest to north central parts of the Uinta Basin, Utah. *U.S. Geol. Surv. Oil Gas Invest. Map* Chart OC-81.

Fouch, T. D., and Cashion, W. B. (1979). Preliminary chart showing distribution of rock types, lithologic groups and depositional environments for some lower Tertiary, upper and lower Cretaceous and upper and middle Jurassic rocks in the subsurface between Altamont oil field and San Arroyo gas field, north central to southeastern Uinta Basin, Utah. *Geol. Surv. Open-File Report (U.S.)* **79-365,** 2 sheets.

Fox, F. G. (1983). Structure sections across the Parry Islands fold belt and the Vesey Hamilton Salt Wall, Arctic Archipelago, Canada. *Stud. Geol.* (*Tulsa, Okla.*) **15(3),** 3.4 1–54.

Furlong, K. P., and Edman, J. D. (1984). Graphic approach to determination of hydrocarbon maturation in overthrust terrains. *Am. Assoc. Pet. Geol. Bull.* **68,** 1818–1824.

Gairud, H. *et al* (1978). The Jan Mayen Ridge, synthesis of geological knowledge and new data. *Oceanol. Acta* **1,** 335–358.

Gamboa, L. A., Truchan, M., and Stoffa, P. L. (1985). Middle and upper Jurassic depositional environments at outer shelf and slope of Baltimore Canyon trough. *Am. Assoc. Pet. Geol. Bull.* **69,** 610–621.

Gardner, G. H. F., Gardner, L. W., and Gregory, A. R. (L974). Formation velocity and density; the diagnostic basis for stratigraphic traps. *Geophysics* **39,** 770–780.

Garrels, R. M., and Christ, C. L. (1965). "Solutions, Minerals, and Equilibria." Freeman, Cooper & Co., San Francisco, California.

Gaskell, S. J., and Eglinton, G. (1976). Sterols of a contemporary lacustrine sediment. *Geochim. Cosmochim. Acta* **40,** 1221–1228.

Gerling, E. K., Levskiy, L. K., and Morozova, J. M. (1963). On the diffusion of radiogenic argon from minerals. *Geochemistry* **6,** 551–554.

Gibbs, A. D. (1983). Balanced cross-section construction from the seismic sections in areas of extensional tectonics. *J. Struct. Geol.* **5,** 153–160.

Gleadow, A. J. W., and Brooks, C. K. (1979). Fission track dating, thermal histories and tectonics of igneous intrusions in east Greenland. *Contrib. Mineral. Petrol.* **71,** 45–60.

Gleadow, A. J. W., and Duddy, I. R. (1981). A natural long-term track annealing experiment for apatite. *Nucl. Tracks* **5**, 169–174.

Gleadow, A. J. W., and Duddy, I. R. (1984). Fission track dating and thermal history analysis of apatites from wells in the north-west Canning Basin. In *Pet. Explor. Soc. Aust., Symp. Cann. Basin, 1984* pp. 377–387.

Gleadow, A. J. W., and Lovering, J. F. (1974). The effect of weathering on fission track dating. *Earth Planet. Sci. Lett.* **22**, 163–168.

Gleadow, A. J. W., Duddy, I. R., and Lovering, J. F. (1983). Fission track analysis: A new tool for the evaluation of thermal histories and hydrocarbon potential. *APEA J.* **23**, 93–102.

Glezen, W. H. (1983). "Internal Technical Memo." Gulf Research & Development Co., Pittsburgh, Pennsylvania.

Glezen, W. H., and Lerche, I. (1985). A model of regional fluid flow: Sand concentration factors and effective lateral and vertical permeabilities. *Math. Geol.* **17**, 297–315.

Goff, J. C. (1983). Hydrocarbon generation and migration from Jurassic source rocks in the east Shetland basin and Viking graben of the northern North Sea. *J. Geol. Soc. London* **140**, 445–474.

Goodwin, P. W., and Anderson, E. J. (1985). Punctuated aggradational cycles: A general hypothesis of episodic stratigraphic accumulation. *J. Geol.* **93**, 515–534.

Grabau, A. W. (1936). "Paleozoic Formations in the Light of the Pulsation Theory," Vol. 1. Heath, New York.

Graf, D.L. (1982). Chemical osmosis, reverse chemical osmosis, and the origin of subsurface brines. *Geochim. Cosmochim. Acta* **46**, 1431–1448.

Grantz, A., and May, S. D. (1983). Rifting history and structural development of the continental margin north of Alaska. *Mem—Am. Assoc. Pet. Geol.* **34**, 923–105.

Grantz, A., Eittreim, S., and Dintar, D. A. (1979). Geology and tectonic development of the continental margin north of Alaska. *Tectonophysics* **59**, 263–291.

Grayson, J. F. (1975). Relationship of palynomorph translucency to carbon and hydrocarbons in clastic sediments. *Petrogr. Matiere Org. Sediments, Relat. Paleotemp. Potentiel Pet., Colloq. Int., 1973* pp. 261–273.

Green, P. F. (1981). "Track-in-track" length measurements in annealed apatites. *Nucl. Tracks* **5**, 121–128.

Green, P. F., and Duddy, I. R. (1988). Some comments on paleotemperature estimation from apatite fission track analysis. *J. Pet. Geol.* (in press).

Green, P. F., Duddy, I. R., Gleadow, A. J. W., Tingate, P. R., and Laslett, G. M. (1985). Fission-track annealing in apatite: Track length measurements and the form of the Arrhenius plot. *Nucl. Tracks* **10**, 323–328.

Green, P. F., Duddy, I. R., Gleadow, A. J. W., Tingate, P. R., and Laslett, G. M. (1986). Thermal annealing of fission tracks in apatite. 1. A qualitative description. *Chem. Geol.* **59**, 237–253.

Griffiths, W. C. (1967). "Scientific Method in the Analysis of Sediments." McGraw-Hill, New York.

Guidish, T. M., Kendall, C. G. St. C., Lerche, I., Toth, D. J., and Yarzab, R. F. (1984). An overview of basin evaluation from burial history. *Am. Assoc. Pet. Geol. Bull.* **69,** 92–105.

Guidish, T. M., Lerche, I., Kendall, C. G. St. C., and O'Brien, J. J. (1985). Relationship between eustatic sea level changes and basement subsidence. *Am. Assoc. Pet. Geol. Bull.* **68,** 164–177.

Gutjahr, C. C. M. (1966). Carbonization measurements of pollen-grains and spores and their application. *Leidse Geol. Meded.* **38,** 1–29.

Haas, J. L., Jr. (1978). An empirical equation with tables of smoothed solubilities of methane and aqueous sodium chloride solutions up to 25 weight percent, 360°C and 138 MPa. *Geol. Surv. Open-File Rep. (U.S.)* **78-1004.**

Halbouty, M. T. *et al.* (1970). Factors affecting formation of giant oil and gas fields, and basin classification. Part II. *Mem. Am. Assoc. Pet. Geol.* **14,** 528–555.

Hallam, A. (1977). Secular changes in marine inundation of USSR and North America through the Phanerozoic. *Nature* (*London*) **269,** 769–772.

Hallam, A. (1984). Pre-Quaternary sea-level changes. *Annu. Rev. Earth Planet. Sci.* **12,** 205–243.

Hamberg, L. R. (1983). Seismic profiles and a stratigraphic trap East Texas field. *Stud.Geol.* (*Tulsa, Okla.*) **15(1),** 1.2–2–15.

Hansen, W. R. (1984). Post-Laramide tectonic history of the eastern Uinta Mountains, Utah, Colorado, and Wyoming. *Mt. Geol.* **21,** 5–29.

Harbaugh, J. W., and Bonham-Carter, G. (1970). "Computer Simulation in Geology." Wiley, New York.

Hardenbol, J., Vail, P. R., and Ferrer, J. (1981). Interpreting paleoenvironments, subsidence history, and sea level changes of passive margins from seismic and biostratigraphy. *Oceanol. Acta Proc. Int. Geol. Congr. Geol. Cont. Margins Symp., 26th, 1980,* pp. 33–44.

Harding, T. P. (1984). Graben hydrocarbon occurrences and structural style. *Am. Assoc. Pet. Geol. Bull.* **68,** 333–362.

Harding, T. P. (1985). Seismic characteristics and identification of negative flower structures, positive flower structures and positive structural inversion. *Am. Assoc. Pet. Geol. Bull.* **69,** 582–600.

Harding, T. P., and Lowell, J. D. (1983). Structural styles, their plate tectonic habitats and hydrocarbon traps in petroleum provinces. *Stud. Geol.* (*Tulsa, Okla.*) **15(1),** 1–24.

Harding, T. P., Gregory, R. F., and Stephens, L. H. (1983). Convergent wrench fault and positive flower structure, Ardmore Basin, Oklahoma. *Stud. Geol.* (*Tulsa, Okla.*) **15(3),** 4.2–13.

Harper, M. L. (1971). Approximate geothermal gradients in the North Sea basin. *Nature* (*London*) **230,** 235–236.

Harris, P. M., Kendall, C. G. St. C., and Lerche, I. (1985). Carbonate cementation—A review for the hydrocarbon explorationist. *Spec. Publ.—Soc. Econ. Palentol. Mineral.* **36,** 79–95.

Harrison, M. T. (1981). Diffusion of ^{40}Ar in hornblende. *Contrib. Mineral. Petrol.* **78,** 324–331.

Harrison, M. T. (1982). The thermal significance of potassium feldspar K-Ar ages inferred from ^{40}Ar/^{39}Ar age spectrum results. *Geochim. Cosmochim. Acta* **46,** 1811–1820.

Harrison, M. T., and Be, K. (1983). ^{40}Ar/^{39}Ar age spectrum analysis of detrital microclines from the southern San Joaquin Basin, California: An approach to determining the thermal evolution of sedimentary basins. *Earth Planet. Sci. Lett.* **64,** 242–256.

Harrison, M. T., and McDougall, I. (1980). Investigations of an intrusive contact, northwest Nelson, N.Z. II. Diffusion of radiogenic and excess ^{40}Ar in hornblende revealed by ^{40}Ar/^{39}Ar age spectrum analysis. *Geochim. Cosmochim. Acta* **44,** 2005–2020.

Harrison, M. T., Armstrong, R. L., Naeser, C. W., and Harakal, J. E. (1979). Geochronology and thermal history of the coast plutonic complex near Prince Rupert, British Columbia. *Can. J. Earth Sci.* **16,** 400–410.

Hawley, B. W., Zandt, G., and Smith, R. B. (1981). Simultaneous inversion for hypocenters and lateral velocity variations: An iterative solution with a layered model. *JGR, J. Geophys. Res.* **86,** 7073–3086.

Hayes, J. B. (1979). Sandstone diagenesis—the hole truth. *Spec. Publ.—Soc. Econ. Paleontol. Mineral.* **26,** 127–139.

He, Z., and Lerche I. (1988). Inversion of multiple thermal indicators: Quantitative methods of determining paleoheat flux and geological parameters. IV. Case histories using thermal indicator tomography. *Math. Geol.* (in press).

Hedberg, H. D. (1936). Gravitational compaction of clays and shales. *Am. J. Sci.* **31,** 241–287.

Hellinger, S. J., and Sclater, J. G. (1983). Some comments on two-layer extentional models for the evolution of sedimentary basins. *JGR, J. Geophys. Res.* **88,** 8251–8269.

Horne, J. C., and Ferm, J. C. (1976). Carboniferous depositional environments: Eastern Kentucky and southern West Virginia, a field guide. Dept. of Geol., University of South Carolina, Columbia.

Horsfield, B. (1984). Pyrolysis studies and petroleum exploration. *Adv. Pet. Geochem.* **1,** 247–298.

Hubbert, M. K. (1953). Entrapment of petroleum under hydrodynamic conditions. *Am. Assoc. Pet. Geol. Bull.* **37,** 1954–2026.

Hunt, J. M. (1979). "Petroleum Geochemistry and Geology." Freeman, San Francisco, California.

Hunt, J. M., Huc, A. Y., and Whelan, J. K. (1980). Generation of light hydrocarbons in sedimentary rocks. *Nature (London)* **288,** 688–690.

Huntsberger, T. L., and Lerche, I. (1987). Determination of paleoheat flux from fission scar tracks in apatite. *J. Pet. Geol.* **10,** 365–394.

Hurford, A. J., Grunau, H. R., and Stocklin, J. (1984). Fission track dating of an apatite crystal from Hormuz Island, Iran. *J. Pet. Geol.* **7**, 365–380.

Ishiwatari, R., Ishiwatari, M., Kaplan, I. R., and Rohrback, B. G. (1976). Thermal alteration of young kerogen in relation to petroleum genesis. *Nature (London)* **264**, 347–349.

Jackson, J. D. (1975). "Classical Electromagnetism." Wiley, New York.

Jackson, S. A., and Beales, F. W. (1967). An aspect of sedimentary basin evolution: The concentration of Mississippi Valley-type ores during the late stages of diagenesis. *Bull. Can. Pet. Geol.* **15**, 393–433.

James, N. P. (1979). Shallowing-upward sequences in carbonates. In "Facies Models" (R. G. Walker, ed.), Geosci. Can. Reprint Ser., Vol. 1, pp. 109–119.

Jones, P. H. (1975). Geothermal and hydrocarbon regimes, northern Gulf of Mexico basin. *Proc.—Geopressured Geotherm. Energy Conf.* **1**, 15–89.

Jones, R. W. (1975). A quantitative geologic approach to prediction of petroleum resources. *Stud. Geol. (Tulsa, Okla.)* **1**, 186–195.

Jordan, T. E. (1982). Thrust loads and foreland basin evolution, Cretaceous, western United States. *Am. Assoc. Pet. Geol. Bull.* **65**, 2506–2520.

Jost, W. (1960). "Diffusion in Solids, Liquids, and Gases." Academic Press, New York.

Kantsler, A. J., Prudence, T. J. C., Cook, A. C., and Zwigulis, M. (1983) Hydrocarbon habitat of the Cooper/Eromanga Basin, Australia. *APEA J.* **23**, 75–90.

Kappelmeyer, O., and Haenel, R. (1974). "Geothermics: With Special Reference to Applications," Geoexplor. Monogr. No. 4. Borntraeger, Berlin.

Katahira, T. (1974). Stratigraphy of the Neogene Tertiary in the central and northern parts of the Niigata prefecture. *J. Jpn. Assoc. Pet. Technol.* **39**, 167–178.

Keen, C. E. (1978). Thermal history and subsidence of rifted continental margins—evidence from wells on the Nova Scotian and Labrador shelves. *Can. J. Earth Sci.* **16**, 505–522.

Keen, C. E., Beaumont, C., and Boutilier, R. (1981). Preliminary results from a thermo-mechanical model for the evolution of Atlantic-type continental margins. *Int. Geol. Congr., Colloq. C3.3, Geol. Cont. Margins, 26th, 1980* Vol. 4, Suppl., pp. 123–128.

Keenen, J. H., and Keyes, F. G. (1957). "Thermodynamic Properties of Steam." Wiley, New York.

Kendall, C. G. St. C., and Lerche, I. (1985). Determination of paleoheat flux from vitrinite reflectance data: A reply. *Am. Assoc. Pet. Geol. Bull.* **69**, 2149.

Kennedy, W. J., and Gentle, J. E. (1980). "Statistical computing." Dekker, New York.

King, P. B. (1948). Geology of southern Guadalupe Mountains, Texas. *Geol. Surv. Prof. Pap. (U.S.)* **215**, 1–183.

Kirschner, C. E., Gryc, G., and Molenaar, C. (1983). Regional seismic lines in the

National Petroleum Reserve in Alaska. *Stud. Geol. (Tulsa, Okla.)* **15,** 1.2.5–14.

Klein, G. DeV. (1970). Tidal origin of a Precambrian quartzite—the lower fine-grained quartzite (middle Dalradian) of Islay, Scotland. *J. Sediment. Petrol.* **40,** 973–985.

Knuth, D. E. (1981). "Seminumerical Algorithms," 2nd ed., Vol. 2, pp. 116ff. Addison-Wesley, Reading, Massachusettes.

Krumbein, W. C., and Grayhill, F.A. (1965). "An Introduction to Statistical Models in Geology." McGraw-Hill, New York.

Kuenen, P. H. (1939). Quantitative estimations relating to eustatic movements. *Geol. Mijnbouw* **8,** 194–201.

Lafourcade, P. (1984). Étude géologique et géophysique de la marge continentale du sud Mozambique (17°S à 28°S). Ph.D. Dissertation, Laboratoire de Géodynamique Sous-Marine, Villefranche-sur-Mer, Université Pierre et Marie Curie, Paris.

Laslett, G. M., Green, P. F., Duddy, I. R., and Gleadow, A. J. W. (1987). Thermal annealing of fission tracks in apatite. 2. A quantitative analysis. *Chem. Geol.* **65,** 1–13.

Laughlin, A. W. (1969). Excess radiogenic argon in pegmatite minerals. Ph.D. Thesis, University of Arizona, Tucson.

Lawler, D. L., and Robinson, W. E. (1967). Fatty acids in *n*-alkanes in Green River oil shale: Changes in depth. *Prepr. Pap—Am. Chem. Soc., Div. Fuel Chem.* **2,** 480–486.

Lawson, C. L., and Hanson, R. J. (1974). "Solving Least Squares Problems." Prentice-Hall, Englewood Cliffs, New Jersey.

Lerche, I. (1988a). Inversion of multiple thermal indicators: Quantitative methods of determining paleoheat flux and geological parameters. I. The theoretical development for paleoheat flux. *Math. Geol.* **20,** 1–36.

Lerche, I. (1988b). Inversion of multiple thermal indicators: Quantitative methods of determining paloeheat flux and geological parameters. II. The theoretical development for chemical, physical and geological parameters. *Math. Geol.* **20,** 73–96.

Lerche, I. (1988c). Paleotemperature estimation from fission track analysis—a reply. *J. Pet. Geol.* (in press).

Lerche, I., and Glezen, W. H. (1984). "Deposition, Compaction and Fluid Migration: Time-Dependent Models in One and Two Dimensions." Gulf Research & Development Co., Pittsburgh, Pennsylvania.

Lerche, I., and O'Brien, J. J., eds. (1987). "Dynamical Geology of Salt and Related Structures." Academic Press, Orlando, Florida.

Lerche, I., Yarzab, R. F., and Kendall, C. G. St. C. (1984). The determination of paleoheat flux from vitrinite reflectance data. *Am. Assoc. Pet. Geol. Bull.* **68,** 1704–1717.

Lewan, M. D., Winters, J. C., and McDonald, J. H. (1979). Generation of oil-like pyrolysates from organic-rich shales. *Science* **103,** 897–899.

Lewis, C. R., and Rose, S. C. (1970). A theory relating high temperatures and overpressures. *J. Pet. Technol.* **22,** 11–16.

Lopatin, N. V. (1971). Temperature and geologic time as factors in coalification. *Akade. Nauk SSSR, Izve. Ser. Geol.* **3,** 95–106.

Lucas, P. T., and Drexler, J. M. (1975). Altamont–Bluebell: A major fractured and overpressured stratigraphic trap, Uinta Basin, Utah. In "Deep Drilling Frontiers of the Central Rocky Mountains" (D. W. Boyard, ed.), pp. 265–273. Rocky Mt. Assoc. Geol., Denver, Colorado.

Ludwig, B., Hussler, G., Wehrung, P., and Albrecht, P. (1981). C_{26}–C_{28} triaromatic steroid derivatives in sediments and petroleums. *Tetrahedron Lett.* **22,** 3313–3316.

Lumley, J. L. (1970). "Stochastic Tools in Turbulence." Academic Press, New York.

MacKenzie, A. S., and McKenzie, D. (1983). Isomerization and aromatization of hydrocarbons in sedimentary basins formed by extension. *Geol. Mag.* **120,** 417–528.

MacKenzie, A. S., Patience, R. L., Maxwell, J. R., Vandenbroucke, M., and Durand, B. (1980). Molecular parameters of maturation in the Toarcian shales, Paris Basin, France. I. Changes in the configurations of acyclic isoprenoid alkanes, steranes and triterpanes. *Geochim. Cosmochim. Acta* **44,** 1709–1721.

MacKenzie, A. S., Hoffman, C. F., and Maxwell, J. R. (1981). Molecular patterns of maturation in the Toarcian shales, Paris Basin, France. III. Changes in aromatic steroid hydrocarbons. *Geochim. Cosmochim. Acta* **45,** 1345–1355.

MacKenzie, A. S., Lamb, N. A., and Maxwell, J. R. (1982). Steroid hydrocarbons and the thermal history of sediments. *Nature (London)* **295,** 223–226.

Magara, K. (1976a). Water expulsion from clastic sediments during compaction, directions and volumes. *Am. Assoc. Pet. Geol. Bull.* **60,** 543–553.

Magara, K. (1976b). Thickness of removed sedimentary rocks, paleopore-pressure and paleotemperature, southwestern part of western Canada Basin. *Am. Assoc. Pet. Geol. Bull.* **60,** 554–565.

Magara, K. (1978a). "Compaction and Fluid Migration—Practical Petroleum Geology," pp. 11–25. Am. Elsevier, New York.

Magara, K. (1978b). Timing of oil generation and migration and oil window. *Bull. Can. Pet. Geol.* **26,** 152–155.

Marlow, M. S. (1979). Hydrocarbon prospects in Navarin Basin province, northwest Bering Sea shelf. *Oil Gas J.* 190–196.

Marsaglia, G., MacLaren, M. D., and Bray, T. A. (1964). A fast procedure for generating normal random variables. *Commun. ACM* **7,** 4–10.

Mauger, R. L. (1977). K-Ar ages of biotites from tuffs in Eocene rocks of the Green River, Washakie, and Uinta Basins, Utah, Wyoming, and Colorado. *Contrib. Geol.* **15,** 17–41.

Maxwell, J. R., MacKenzie, A. S., and Volkman, J. R. (1980). Configuration at C-24 in steranes and sterols. *Nature* **286,** 694–697.

McClelland Engineers Inc. (1982). Alaskan North Slope Generalized Subsurface Biostratigraphic Chart. McClelland Engineers Inc., Houston, Texas.

McCulloh, T. H., and Fan, J. J. (1984). Burial thermal histories, vitrinite reflectance, and laumonite isograd. *Am. Assoc. Pet. Geol. Bull.* **69,** 285. (abstr.).

McDowell, A. N. (1975). What are the problems in estimating the oil potential of a basin? *Oil Gas J.* 85–90.

McKenzie, D. (1978). Some remarks on the development of sedimentary basins. *Earth Planet. Sci. Lett.* **40,** 25–32.

Meissner, F. (1982). "Basin Geochemistry—Case Histories," Notes from AAPG Geochemistry for Geologists School, Houston, Texas.

Menke, W. (1984). "Geophysical Data Analysis: Discrete Inverse Theory." Academic Press, Orlando, Florida.

Merrihue, G., and Turner, G. (1966). Potassium–argon dating by activation with fast neutrons. *J. Geophys. Res.* **71,** 2853–2857.

Mesolella, K. J., Matthews, R. K., Broecker, W. S., and Thurber, D. L. (1969). The astronomical theory of climatic change: Barbados data. *J. Geol.* **77,** 250–274.

Middleton, M. F. (1980). A model of intracratonic basin formation, entailing deep crustal metamorphism. *Geophys. J. R. Astron. Soc.* **62,** 1–14.

Middleton, M. F. (1984). Seismic geohistory analysis—a case history from the Canning Basin, Western Australia. *Geophysics* **49,** 333–343.

Middleton, M. F., and Schmidt, P. W. (1982). Paleothermometry of the Sydney Basin. *JGR, J. Geophys. Res.* **87,** 5351–5359.

Milankovich, M. (1941). Canon of insolation and the ice-age problems. *R. Serb. Acad. Sect. Math. Nat. Sci., Spec. Publ.* **132,** 33 (transl. by U.S. Dept. of Commerce).

Miller, B. M. *et al.* (1975). Geological estimation of undiscovered recoverable oil gas resources in the United States. *Geol. Surv. Circ. (U.S.)* **725,** 78.

Milliman, J. D. (1974). "Marine Carbonates," Part 1. Springer Verlag, New York.

Molenaar, C. M. (1981a). Depositional history of the Narushuk Group and related strata. *Geol. Surv. Circ. (U.S.)* **823-B,** B4-B5.

Molenaar, C. M. (1981b). Depositional history and seismic stratigraphy of Lower Cretaceous rocks, National Petroleum Reserve in Alaska and adjacent areas. *Geol. Surv. Open-File Rep. (U.S.)* **81-1084,** 1–42.

Momper, J. A. (1978). Oil migration limitations suggested by geological and geochemical considerations. *Am. Assoc. Pet. Geol. Cont. Educ. Course, Note Ser.* **8,** B1-B60.

Montgomery, J. A., and Chandler, D. (1979). Trajectory analysis of a kinetic theory for isomerization dynamics in condensed phases. *J. Chem. Phys.* **70,** 4056.

Moore, W. S. (1982). Late Pleistocene sea level history. *In* "Uranium-Series Disequilibrium: Applications to Environmental Problems in the Earth

Sciences" (M. Ivanovich and R. Harmon, eds.), pp. 481–496. Oxford Univ. Press, London and New York.

Moshier, S. O., and Waples, D. W. (1985). Quantitative elevation of Lower Cretaceous Mannville Group as source rock for Alberta's oil sands. *Am. Assoc. Pet. Geol. Bull.* **64,** 754.

Mougenot, D., Virlogeux, P., Vanney, J.-R., and Malod, J. (1986). La marge continentale au nord du Mozambique, résultats preliminaires de la campagne md40/macamo. *Bull. Soc. Geol. Fr.* **8,** 419–422.

Mull, C. C. (1980). Evolution of Brooks Range thrust belt and arctic slope, Alaska. *Am. Assoc. Pet. Geol. Bull.* **64,** 754.

Naeser, C. W. (1979). Fission track dating and geologic annealing of fission tracks. *In* "Lectures in Isotope Geology" (E. Jager and J. C. Hunziker, eds.), pp. 154–169. Springer-Verlag, Berlin and New York.

Naeser, C. W. (1981). The fading of fission tracks in the geologic environment—data from deep drill holes. *Nucl. Tracks* **5,** 248–250.

Naeser, C. W., and Faul, H. (1969). Fission track annealing in apatite and sphene. *J. Geophys. Res.* **74,** 705–710.

Nagpaul, K.K., Mehta, P. P., and Gupta, M. L. (1974). Annealing studies on radiation damage in biotite, apatite and sphene and corrections to fission track ages. *Pure Appl. Geophys.* **112,** 131–139.

Nakayama, K. (1986). A dynamic, two dimensional, fluid flow model for basin analysis. Ph.D. Dissertation, University of South Carolina, Columbia.

Nakayama, K., and Van Siclen, D. C. (1981). Simulation model for petroleum exploration. *Am. Assoc. Pet. Geol. Bull.* **65,** 1230–1255.

Narr, W., and Currie, J. B. (1982). Origin of fracture porosity—example from Altamont Field, Utah. *Am. Assoc. Pet. Geol. Bull.* **66,** 1231–1247.

Nelson, R. E. (1981). Paleoenvironments, during deposition at a section of the Gubik Formation exposed along the lower Colville River, North Slope. *Geol. Surv. Circ. (U.S.)* **823-B,** 9–11.

Newendorp, P. D. (1975). "Decision Analysis for Petroleum Exploration." Petroleum Publ. Co., Tulsa, Oklahoma.

Nind, T. E. W. (1981). "Principles of Oil Well Production." McGraw-Hill, New York.

Noble, E. A. (1963). Formation of ore deposits by water of compaction. *Econ. Geol.* **58,** 1145–1156.

Norton, M. A. (1983). Kemmerer Area, Lincoln County, Wyoming, 1983. *Stud. Geol.* (*Tulsa, Okla.*) **15,(3),** 3.4-1–45.

Nunn, J. A. (1981). Thermal contraction and flexure of intracratonic basins: A three dimensional study of the Michigan Basin. Ph.D. Thesis, Northwestern University, Evanston, Illinois.

Nunn, J. A., Sleep, N. H., and Moore, W. E. (1984). Thermal subsidence and generation of hydrocarbons in Michigan Basin. *Am. Assoc. Pet. Geol. Bull.* **68,** 296–315.

O'Brien, J. J., and Lerche, I. (1984). The influence of salt domes on paleotemperature distributions. *Geophysics* **49,** 2032–2043.

Ogura, N. (1981). Estimation of maturation history of petroleum source rocks in Niigata sedimentary basin. *Res. Org. Geochem.* **3,** 29–35.

Ourisson, G., Albrecht, P., and Rohmer, M. (1979). The hopanoids: Paleochemistry and biochemistry of a group of natural products. *Pure Appl. Chem.* **51,** 709–729.

Palcianskas, V. V., and Domenico, P. A. (1980). Microfracture development in compacting sediments: A relation to hydrocarbon maturation kinetics: *Am. Assoc. Pet. Geol. Bull.* **64,** 927–937.

Palmer, A. R. (1983). The decade of North American geology: 1983 geologic time scale. *Geology* **11,** 503–504.

Pantano, J. (1985). Stratigraphic age determination from inversion of sterane isomer data and vitrinite reflectance data. M. S. Thesis, University of South Carolina, Columbia.

Pantano, J., and Lerche, I. (1988). Inversion of multiple thermal indicators: Quantitative methods of determining paleoheat flux and geological parameters. III. Age determination from inversion of sterane isomer data and vitrinite reflectance data. *Math. Geol.* (submitted for publication).

Parsons, B., and Sclater, J. G. (1977). An analysis of the variations of ocean floor bathymetry and heat flow with age. *JGR, J. Geophys. Res.* **82,** 803–827.

Paten R. J. (1969). Palynologic contributions to petroleum exploration in Permian formations of the Cooper Basin, Australia. *APEA J.* **9,** 79–87.

Peng, D.-Y., and Robinson, D. B. (1976). A new two-constant equation of state. *Ind. Eng. Chem. Fundam.* **15,** 59–64.

Petroleum Information (1981). "Geochemical Data File Developed by Petroleum Information for N.P.R.A., North Slope," 2 vol. Petroleum Information, Tulsa.

Petrov, A. A., Pustil'nikova, S. D., Abriutina, N. N., and Kagramonova, G. R. (1976). Petroleum steranes and triterpanes. *Neftekhimia* **16,** 411–427.

Phelps, E. H., and Roripaugh, C. C., Jr. (1983). Carbonate shelf, central Louisiana. *Stud. Geol.* (*Tulsa, Okla.*) **L5,(1),** 1.2.4–1.

Philippi, G. T. (1965). On the depth, time and mechanisms of petroleum generation. *Geochim. Cosmochim. Acta* **25,** 1021– 1049.

Pieri, M. (1983). Three seismic profiles through the Po plain. *Stud. Geol.*(*Tulsa, Okla.*) **15(3),** 3.4–1–27.

Pitcher, M. G. (1976). U.S. discovery rate tied to technology. *Oil Gas* J. 34–35.

Pitman, J. K., Fouch, T. D., and Goldhaber, M. B. (1982). Depositional setting and diagenetic evolution of some Tertiary unconventional reservoir rocks, Uinta Basin, Utah. *Am. Assoc. Pet. Geol. Bull.* **66,** 1581–1596.

Pitman, W. C. (1978). Relationship between sea-level change and stratigraphic sequences of passive margins. *Geol. Soc. Am. Bull.* **89,** 1389–1403.

Press, W. H., Flannery, B. P., Tenkosky S. A., and Vetterling, W. T. (1986).

"Numerical Recipes—The Art of Scientific Computing." Cambridge Univ. Press, London and New York.

Price, L. C. (1982). Time as a factor in organic metamorphism and use of vitrinite reflectance as an absolute paleogeothermometer. *Am. Assoc. Pet. Geol. Bull.* **66,** 619–620.

Price, L. C. (1985). Examples, causes, and consequences of vitrinite reflectance, suppression in hydrogen-rich organic matter—a major unrecognized problem. *Am. Assoc. Pet. Geol. Bull.* **69,** 298 (abstr.).

Rabinowitz, O., Coffin, F., and Falvey, D. (1983). The separation of Madagascar and Africa. *Science* **220,** 67–69.

Rafavich, F., Kendall, C. G. St. C., and Todd, T.P. (1984). The relationship between acoustic properties and the petrographic character of carbonate rocks. *Geophysics* **49,** L622–L636.

Ratcliffe, E. H. (1960). The thermal conductivities of ocean sediments. *J. Geophys. Res.* **65,** 1535–1541.

Reaves, C. M., and von Rosenberg, D. U. (1985). Determination of paleoheat flux from vitrinite reflectance data: Discussion. *Am. Assoc. Pet. Geol. Bull.* **69,** 2148.

Reeckman, S. A., and Mebberson, A. J. (1984). Igneous intrusions in the north-west canning basin and their impact on oil exploration. *In* "Proceedings of the Canning Basin Symposium, Perth, 1984," pp. 45–52. Pet. Explor. Soc. Aust. Ltd.

Reed, W. E., and Henderson, W. (1971). Proposed stratigraphic controls on the composition of crude oils reservoired in the Green River Formation, Uinta Basin, Utah. *In* "Advances in Geochemistry," pp. 499–515. Pergamon, Oxford.

Rhead, M. M., Eglinton, G., and Draffan, G. H. (1971). Hydrocarbons produced by the thermal alteration of cholesterol under conditions simulating the maturation of sediments. *Chem. Geol.* **8,** 277–297.

Rich, J. L. (1921). Moving underground water as a primary cause of the migration and accumulation of oil and gas. *Econ. Geol.* **16,** 347–371.

Roadifor, R. (1975). A probability approach to estimate volumes of undiscovered oil and gas. *In* "Probability Methods in Oil Exploration" (J. C. Davis *et al.*, conveners), AAPG Res. Symp., p. 18. Stanford University, Stanford, California.

Robin, P. L., and Rouxhel, P. G. (1978). Characterization of kerogens and study of their evolution by infrared spectroscopy: Carbonyl and carboxyl groups. *Geochim. Cosmochim. Acta* **42,** 1341–1349.

Robinson, W. E. (1969). Kerogen of the Green River Formation, *In* "Organic Geochemistry" (G. Eglington and M. T. J. Murphy, eds.), pp. 619–637. Springer-Verlag, New York.

Roy, K. J., Procter, R. M., and McCrossen, R. C. (1975). Hydrocarbon assesment using subjective probability. *In* "Probability Methods in Oil Exploration"

(J. C. Davis *et al.,* conveners) AAPG Res. Symp, pp. 56–60. Stanford University, Stanford, California.

Royden, L., and Keen, C. E. (1980). Rifting process and thermal evolution of the continental margin of eastern Canada determined from subsidence curves. *Earth Planet. Sci. Lett.* **51,** 343–361.

Royden, L., Sclater, J. G., and Von Herzen, R. P. (1980). Continental margin subsidence and heat flow: Important parameters in formation of petroleum hydrocarbons. *Am. Assoc. Pet. Geol. Bull.* **64,** 173–187.

Ru, K., and Pigott, J. D. (1986). Episodic rifting and subsidence in the South China Sea. *Am. Assoc. Pet. Geol. Bull.* **70,** 1132–1151.

Rubey, W. W., and Hubbert, M. K. (1959). Role of fluid pressure in mechanics of overthrust faulting. *Geol. Soc. Am. Bull.* **70,** 167–206.

Rubinstein, R. Y. (1981). "Simulation and the Monte Carlo Method." Wiley, New York.

Ruddiman, W. F., and McIntyre, A. (1981). Oceanic mechanism for amplification of the 23,000-year ice-volume cycle. *Science* **212,** 617–627.

Ryder, R. T., Fouch, and T. D., Elison, J. H. (1976). Early Tertiary sedimentation in the western Uinta Basin, Utah. *Geol. Soc. Am. Bull.* **87,** 496–512.

Saxby, J. D., and Riley, K. W. (1984). Petroleum generation by laboratory-scale pyrolysis over six years simulating conditions in a subsiding basin. *Nature (London)* **308,** 177–179.

Schlee, J. S., and Fritsch, J. (1983). Seismic stratigraphy of the Georges Bank basin complex, offshore New England. In "Studies in Continental Marine Geology" (J. S. Watkins and C. L. Drake, eds.). *Mem.—Am. Assoc. Pet. Geol.* **34,** 223–251.

Sclater, J. G., and Christie, P. A. F. (1980). Continental stretching: An explanation of the post-mid-Cretaceous subsidence of the central North Sea basin. *JGR, J. Geophys. Res.* **85,** 3711–3739.

Segoufin, J. (1978). Anomalies mèsozoiques dans le bassin de Mozambique. *C. R. Hebd. Seances Acad. Sci., Ser. D* **287,** 109–112.

Seifert, W. K., Carlson, R. M., and Moldowan, J. M. (1983). Geomimetic synthesis, structure assignment and geochemical correlations. Applications of monoaromatized petroleum steranes. Adv. Org. Geochem. Proc. Int. Meet., 10th, 1981 pp. 33–54.

Seitz, F. (1946). "The Modern Theory of Solids" McGraw-Hill, Reading, Massachusetts.

Sengbush, R. L., Lawrence, P. L., and McDonal, F. J. (1961). Interpretation of synthetic seismograms. *Geophysics* **26,** 138–157.

Shackleton, N. J., and Opdyke, N. D. (1973). Oxygen isotope and paleomagnetic stratigraphy of equatorial Pacific core V28-238: Oxygen isotope temperature and ice volumes on a 10^5 year and 10^6 year scale. *Quat. Res.* **3,** 39–55.

Sharp, J. M., Jr. (1976). Momentum and energy balance equations for compacting sediments. *Math. Geol.* **8,** 305–322.

Sharp, J. M., Jr. (1978). Energy and momentum transport model of the Ouachita

Basin and its possible impact on the formation of economic mineral deposits. *Econ. Geol.* **73,** 1057–1068.

Sharp, J. M., Jr., and Domenico, P. A. (1976). Energy transport in thick sequences of compacting sediments. *Geol. Soc. Am. Bull.* **87,** 390–400.

Shchigolov, B. M. (1965). "Mathematical Analysis of Observations" (translated by Scripta Technical Inc.). Am. Elsevier, New York.

Shih, S. M., and Sohn, H. Y. (1980). Nonisothermal determination of the intrinsic kinetics of oil generation from oil shale. *Ind. Eng. Chem. Process. Des. Dev.* **19,** 420.

Sibley, D. F., and Blatt, H. (1976). Intergranular pressure solution and cementation of the Tuscaloosa orthoquartzite. *J. Sediment. Petrol.* **46,** 881–896.

Skinner, J. K., and Wolynes, P. G. (1980). General kinetic models of activated processes in condensed phases. *J. Chem. Phys.* **22,** 4913.

Sleep, N. H. (1971). Thermal effects of the formation of Atlantic continental margins by continental breakup. *Geophys. J. R. Astron. Soc.* **24,** 325–350.

Smith, J. W. (1974). "Geochemistry of Oil-Shale Genesis in Colorado's Piceance Creek Basin," Guidebook, pp. 71–79. Rocky Mt. Assoc. Geol., Denver, Colorado.

Smith, L. F. (1961). "Wave Mechanics of Crystalline Solids." Pergamon, New York.

Smith, M. B. (1968). Estimate resources by using computer simulation method. *Oil Gas J.* 81–84.

Snowdon L. R. (1979). Errors in extrapolation of experimental kinetic parameters to organic geochemical systems. *Am. Assoc. Pet. Geol. Bull.* **63,** 1128–1138.

Spencer, C. W. (1987). Hydrocarbon generation as a mechanism for overpressuring in the Rocky Mountain region, *Am. Assoc. Pet. Geol. Bull.* **71,** 368–388.

Steckler, M. S., and Watts, A. B. (1978). Subsidence of the Atlantic-type continental margin off New York. *Earth Planet. Sci. Lett.* **41,** 1–13.

Steffy, D. A., Turner, R. F., Martin, G. C., and Flett, T. O. (1985). Evolution and petroleum geology of the Navarin Basin, Bering Sea, Alaska. *Oil Gas J.* 116–124.

Steinen, R. P., Harris, R. S., and Matthews, R. H. (1973). Eustatic low stand of sea level between 125,000 and 105,000 B. P. Evidence from the subsurface of Barbados, West Indies. *Geol. Soc. Am. Bull.* **84,** 63–70.

Stille, H. (1924). "Grundfragen der vergleichenden Tektonik." Borntraeger, Berlin.

Stoian, E. (1965). Fundamentals and applications of the Monte Carlo method. *J. Can. Pet. Technol.* **4,** 120–129.

Stokes, W. L. (1973). "Geologic Map of Utah," 2 sheets. College of Mines and Mineral Industries, University of Utah, Salt Lake City.

Stone, D. S. (1983a). North Fork area, Powder River basin, Wyoming. *Stud. Geol.* (*Tulsa, Okla.*) **15,(3),** 3.2–2–15.

Stone, D. S. (1983b). Seismic profile: South Elk Basin. *Stud. Geol.* (*Tulsa, Okla.*) **15(3),** 3.2.2–20.

Stout, N. D., Koskinas, G. J., Raley, J., Santor, S. D., Opila, R. J., and Rothman, A. J. (1976). Pyrolysis of oil shale: The effects of thermal history on oil yield. *Oil Shale Symp. Proc.* **9,** 17–26.

Suess, E. (1906). "The Face of the Earth." Oxford Univ. Press (Clarendon), London and New York.

Suzuki, U. (1983). The volcanic mound. *Stud. Geol.* (*Tulsa, Okla.*) **15(1),** 1.3–15.

Sweeney, J. J., Burnham, A. K., and Braun, R. L. (1987). A model of hydrocarbon generation from type I kerogen: Application of Uinta Basin, Utah. *Am. Assoc. Pet. Geol. Bull.* **71,** 967–985.

Swift, A., Sawyer, D. S., Grow, J. A., and Klitgord, K. D. (1987). Subsidence, crustal structure and thermal evolution of Georges Bank Basin. *Am. Assoc. Pet. Geol. Bull.* **71,** 702–718.

Thomson, A. (1983). Salt rollers—East Texas salt basin. *Stud. Geol.* (*Tulsa, Okla.*) **15(2),** 2.3–2–18.

Thorne, J., and Watts, A. B. (1984). Seismic reflectors and unconformities at passive continental margins. *Nature* (*London*) **311,** 365–367.

Tissot, B. (1969). Premières données sur les mécanismes et la cinètique de la formation du petrole dans les sediments: Simulation d'un schema réactionnel sur ordinateur. *Rev. Inst. Fr. Pet.* **24,** 470–501.

Tissot, B., and Espitalie, J. (1975). L'évolution thermique de la matière organique des sediments: Applications d'une simulation mathématique. *Rev. Inst. Fr. Pet.* **30,** 743–777.

Tissot, B., and Welte, D. H. (1978). "Petroleum Formation and Occurrence" pp. 500–521. Springer Verlag, Berlin and New York.

Tissot, B., Deroo, G., and Hood, A. (1978). Geochemical study of the Uinta Basin: Formation of petroleum from the Green River Formation. *Geochim. Cosmochim. Acta* **42,** 1469–1485.

Tomovic, R. (1963). "Sensitivity Analysis of Dynamic Systems." McGraw-Hill, New York.

Toth, D. J. (1980). Cross-formational gravity flow of groundwater: A mechanism of the transport and accumulation of petroleum. *In* "Problems of Petroleum Migration" (W. H. Roberts and R. J. Cordell, eds.), pp. 121–167. Am. Assoc. Pet. Geol., Tulsa, Oklahoma.

Toth, D. J., Lerche, I., Petroy, D. E., Meyer, R. J., and Kendall, C. G. St. C. (1983). Vitrinite reflectance and the derivation of heat flow changes with time. *Adv. Org. Geochem., Proc. Int. Meet., 10th, 1981* pp. 588–596.

Turcotte, D. L., and Ahern, J. L. (1977). On the thermal and subsidence history of sedimentary basins. *JGR, J. Geophys. Res.* **82,** 3762–3766.

Turner, G. (1970). Argon-40/Argon-39 dating of lunar rock samples. *Geochim. Cosmochim. Acta, Suppl.* **1,** 1665–1684.

Turner, G., Miller, J. A., and Grasty, R. L. (1966). The thermal history of the Bruderheim meteorite. *Earth Planet. Sci. Lett.* **1,** 155–157.

Turner, R. F., McCarthy, C. M., Steffy, D. A., Lynch, M. B., Martin, G. C.,

Sherwood, K. W., Flett, T. O., and Adams, A. J. (1984). "Geological and Operational Summary, Navarin Basin COST No. 1 Well, Bering Sea, Alaska," OCS Rep. MMS 84-0031. U.S. Department of the Interior, Mineral Management Service, Anchorage.

Umbgrove, J. H. F. (1939). On rhythms in the history of the earth. *Geol. Mag.* **76,** 116–129.

Ungerer, P. *et al.* (1984). Geological and geochemical methods in oil exploration, principles and practical examples. *Mem.—Am. Assoc. Pet. Geol.* **35,** 53–77.

Vail, P. R., and Hardenbol, J. (1979). Sea level changes during Tertiary. *Oceans* **22,** 71–79.

Vail, P. R., and Todd, R. G. (1981). Northern North Sea Jurassic unconformities, chronostratigraphy and sea level changes from seismic stratigraphy. *In "Proceedings of the Petroleum Geology Conference of the Continental Shelf of Northwest Europe" (L. V. Illing and G. D. Hobson, eds.),* pp. 216–235. Heyden, London.

Vail, P. R., Mitchum, R. M., Jr., and Thompson, S., III (1977). Seismic stratigraphy and global changes of sea level. Part 4. Global cycles of relative changes of sea level. *Mem.—Am. Assoc. Pet. Geol.* **26,** 83–97.

Vail, P. R., Hardenbol, J., and Todd, R. G. (1984). Jurassic unconformities, chronostratigraphy and sea level changes from seismic and biostratigraphy. *Mem.—Am. Assoc. Pet. Geol.* **36,** 347–363.

van Graas, G., Baas, J. M. A., Van der Graaf, B., and De Leeuw, J. W. (1982). Theoretical organic geochemistry. I. The thermodynamic stability of several cholestane isomers calculated from molecular dynamics. *Geochim. Cosmochim. Acta* **46,** 2399–2402.

van Heek, K. H., and Juntgen, H. (1968). Determination of reaction-kinetic parameters from nonisothermal measurements. *Ber. Bunsenges. Phys. Chem.* **72,** 1223; available as English Translation in Lawrence Livermore National Laboratory Report UCRL-TRANS-10974 (1975).

van Hinte, J. E. (1976a). A Jurassic time scale. *Am. Assoc. Pet. Geol. Bull.* **60,** 489–497.

van Hinte, J. E. (1976b). A Cretaceous time scale. *Am. Assoc. Pet. Geol. Bull.* **60,** 498–516.

van Hinte, J. E. (1978). Geohistory analysis—application of micropaleontology in exploration geology. *Am. Assoc. Pet. Geol. Bull.* **62,** 201–222.

van Hinte, J. E. (1983). Synthetic seismic sections from biostratigraphy. *Mem.—Am. Assoc. Pet. Geol.* **34,** 675–685.

Van Krevelen, D. H. (1961). "Coal," pp. 342–372 Am. Elsevier, New York.

Vizgirda, J., O'Brien, J. J., and Lerche, I. (1985). Thermal anomalies on the flanks of a salt dome. *Geothermics* **14,** 553–565.

Voge, H. H., and Good, G. M. (1949). Thermal cracking and high paraffins. *J. Am. Chem. Soc.* **71,** 593.

von Rad, U., and Exon, N. F. (1983). Mesozoic–Cenozoic sedimentary and volcanic evolution of the starved passive margin off Northwest Australia. *Mem.—Am. Assoc. Pet. Geol.* **34,** 253–281.

Wagner, G. A., and Reimer, G. M. (1972). Fission track tectonics: the tectonic interpretation of fission track apatite ages. *Earth Planet. Sci. Lett.* **14,** 263–268.

Walker, R. G., and Harms, J. C. (1971). The "Catskill Delta": A prograding muddy shoreline in central Pennsylvania. *J. Geol.* **79,** 381–399.

Walstrom, J. E., Mueller T. D., and McFarlane, R. C. (1967). Evaluating uncertainty in engineering calculation. *J. Pet. Technol.* **19,** 1595–1603.

Wang, X., Lerche, I., and Walter, C. (1989). The effect of igneous intrusions on sedimentary thermal maturity. *Org. Geochem.* (in press).

Waples, D. W. (1980). Time and temperature in petroleum formation: Application of Lopatin's method to petroleum exploration: *Am. Assoc. Pet. Geol. Bull.* **64,** 916–926.

Waples, D. W. (1983). private communication.

Waples, D. W. (1984). Thermal models for oil generation. *Adv. Pet. Geochem.* **1,** 7–68.

Waples, D. W. (1985). Predicting thermal maturity. *In* "Treatise of Petroleum Geology Reprint Series" (N. N. Foster and E. A. Beaumont, compilers) No. 1, pp. 249–282. Am. Assoc. Pet. Geol., Tulsa.

Waples, D. W. (1988). Maturity modeling of sedimentary basins: Approaches, limitations, and assessment of future developments. *Workshop Quant. Dyn. Stratigr., 1988 p.* 19.

Ward, W. T., and Chappell, J. (1975). Geology of coral terraces, Huon Peninsula, New Guinea: A study of Quaternary tectonic movements and sea level changes: Discussion and reply. *Geol. Soc. Am. Bull.* **86,** L482–L486.

Warren, J. E. (1979). Basin evaluation. *SPE Econ. Eval. Symp.,* 1979.

Watkins, J. S. *et al.* (1976). Southern Gulf of Mexico. *Am. Assoc. Pet. Geol., Seismic Line* 1.

Watts, A. B. (1982). Tectonic subsidence, flexure and global changes of sea level. *Nature (London)* **297,** 469–474.

Watts, A. B., and Steckler, M. S. (1979). Subsidence and eustacy at the continental margin of eastern North America, in Deep drilling results in the Atlantic Ocean; continental margins and paleoenvironment. *Maurice Ewing Ser., Proc. Symp.* No. 3, p. 218.

Weeks, L. G. (1952). Factors of sedimentary basin development that control oil occurrence. *Am. Assoc. Pet. Geol. Bull.* **36,** 2071–2124.

Wei, Z. P., and Lerche, I. (1988). Quantitative dynamical geology of the Pinedale Anticline, Wyoming, U.S.A.: An application of a two-dimensional simulation model. *Appl. Geochem.* **3,** 423–440.

Welte, D. H., and Yukler, M. A. (1981). Petroleum origin and accumulation in basin evolution—a quantitative model. *Am. Assoc. Pet. Geol. Bull.* **65,** 1387–1396.

Williams, B. P. J., and Wild, E. R. (1985). The Tirrawarra sandstone and Merrimelia Formation of the southern Cooper Basin, South Australia—the sedimentation and evolution of a glaciofluvial system. *APEA J.* **25,** 50–66.

Williams, D. F., Moore, W. S., and Fillon, R. H. (1981). Role of glacial Arctic Ocean ice sheets in Pleistocene oxygen isotope and sea level records. *Earth Planet. Sci. Lett.* **56,** 157–166.

Willumsen, P. S., and Cote, R. P. (1983). Tertiary sedimentation in the southern Beaufort Sea, Canada. *Mem—Am. Assoc. Pet. Geol.* **34,** 283–293.

Witmer, R. J. (1981). Biostratigraphic correlations of selected test wells of National Petroleum Reserve in Alaska. *Geol. Surv. Open-File Rep. (U.S.)* **81-1165**, 1–89.

Wood, J. R., and Surdam, R. C. (1979). Application of convection–diffusion models to diagenetic processes. *Spec. Publ.—Soc. Econ. Paleontol. Mineral.* **26,** 243–250.

Yarzab, R. F., Lerche, I., and Kendall, C. G. St. C. (1983). Relationship of vitrinite reflectance to heat-flow history of North Sea. *Am. Assoc. Pet. Geol. Bull.* **67,** 574 (abstr.).

Yen, T. G., and Chilingarian, G. V., eds. (1976). "Oil Shale." Elsevier, Amsterdam.

Yukler, M. A. (1979). Sensitivity analysis of groundwater flow system and an application to a real case. *In* "Geomathematical and Petrophysical Studies in Sedimentology (Computer Geology)" (D. F. Merriam, ed.). Pergamon, New York.

Yukler, M. A., and Kokesh, F. (1984). A review of models used in petroleum resource estimation and organic geochemistry. *Adv. Pet. Geochem.* **1,** 69–114.

Yukler, M. A., Cornford, C., and Welte, D. H. (1978). One dimensional model to simulate geologic, hydrodynamic and thermodynamic development of a sedimentary basin. *Geol. Rundsch.* **67,** 960–979.

Zeigler, P. A. (1975). Geologic evolution of North Sea and tectonic framework. *Am. Assoc. Pet. Geol. Bull.* **59,** 1073–1097.

Zeigler, W. H. (1975). Outline of the geological history of the North Sea. *In* "Petroleum and the Continental Shelf of Northwest Europe" (A. W. Woodland, ed.), Vol. 1, pp. 165–190. Applied Science Publishers, London.

Zierfuss, H., and van der Vliet, G. (1956). Laboratory measurements of heat conductivity of sedimentary rocks. *Am. Assoc. Pet. Geol. Bull.* **40,** 2475–2488.

Index